Handbook of Package Engineering

Handbook of Package Engineering

JOSEPH F. HANLON

McGRAW-HILL BOOK COMPANY

New York St. Louis San Francisco Düsseldorf Johannesburg
Kuala Lumpur London Mexico Montreal New Delhi
Panama Rio de Janeiro Singapore Sydney Toronto

SPONSORING EDITORS Harold B. Crawford/Daniel N. Fischel
DIRECTOR OF PRODUCTION Stephen J. Boldish
EDITING SUPERVISOR Linda B. Hander
EDITING AND PRODUCTION STAFF Gretlyn Blau,
Teresa F. Leaden, George E. Oechsner

HANDBOOK OF PACKAGE ENGINEERING

 Library of Congress Catalog Card Number 70-124138

07-025993-3

4 5 6 7 8 9 KPKP 7 9 8 7

To my devoted wife, Clara, who learned to type just so that she could help me with my manuscript

Preface

This Handbook is intended as a useful source of information on the materials and structures of packaging, for the engineer, designer, purchasing agent, production manager, quality control inspector, and others who, in one way or another, are involved in packaging. There are very few books on the subject of packaging, and although many articles on specific aspects of packaging have appeared, very little of this information has been summarized so far. It is the purpose of this volume to bring together as much pertinent information as possible, without overburdening the reader with extraneous text.

It has been a difficult task to sort out the significant data from the voluminous flow of material that has been issued by suppliers of packaging materials and is being reported in the trade journals. Much information is understandably intended to promote a particular manufacturer's product, but in this book every effort has been made to give a true evaluation of the properties and applications of each of the materials. The author has tried also to point out some of the risks and pitfalls in working with these materials—the kind of information that too often comes only with bitter experience.

This work has been divided into sections according to type of material. A preliminary discussion covers some of the broad aspects of education, training, and management of the packaging function. Most of the following discussions are devoted to specific materials, such as glass, metal, paperboard, plastics, and wood, or, if the structure is the important factor, to forms of packaging, such as bags and envelopes, corrugated containers, and folding cartons. There are also brief discussions of some related subjects that were considered pertinent to the design and development of packages, such as quality control, laws and regulations, and the equipment of packaging.

It is hoped that workers in the field of packaging will find this handbook a valuable addition to their technical library, and that students will use it as a foundation for further studies. Comments and suggestions are always welcome, and readers who have data they would like to contribute for inclusion in subsequent editions are invited to submit such information to the author, in care of the publisher.

Joseph F. Hanlon

Contents

Preface vii

Section 1. ELEMENTS OF PACKAGING **1-1**

Position of packaging in our economy—Present state of packaging technology—Challenge of packaging—Training required—Organization for packaging development—Use of consultants, suppliers, and contract packagers—Professional approach to package development—Packaging values—Scope of packaging

Section 2. FOLDING CARTONS AND SETUP BOXES **2-1**

History and statistics of the industry—Materials—Constructions—Design hints—Manufacturing processes—Decorating methods—Specifications and quality control—Cost calculations

Section 3. FILMS AND FOILS **3-1**

Properties of flexible materials—Production methods—Specifications—Identification tests—Heat sealing—Shrink films—Cellophanes—Cellulosics—Polyolefins—Specialty films—Metal foils

Section 4. PAPER AND PAPERBOARD **4-1**

History of papermaking—Fundamentals of paper structure—Methods of manufacture—Properties of paper and paperboard—Nomenclature of paper products—Relative costs—Industry practices

Section 5. CORRUGATED FIBREBOARD 5-1

Types of corrugations—Combining methods—Box constructions—Selection and design—Cost calculations—Manufacturer's joint—Closing and sealing—Interior packing—Purchase specifications

Section 6. GLASSWARE 6-1

Manufacturing processes—Chemistry of glass—Strength and other properties—Design hints—Coatings—Decoration—Quality control—Glass defects—Tubing products—Manufacturers' identification marks—Types of finishes—Dimensions and tolerances

Section 7. METAL CONTAINERS 7-1

Tinplate and tin-free steel—Alloys and tempers—Standard dimensions—Coatings and linings—Sanitary cans—Aluminum cans—Aerosols—Collapsible tubes—Trays—Pails—Drums—Costs

Section 8. PLASTICS 8-1

Chemistry—Processes—Additives—Copolymers—Properties—Costs—Blow molding—Injection molding—Compression molding—Thermoforming—Molds—Decoration

Section 9. CLOSURES, APPLICATORS, FASTENERS, AND ADHESIVES 9-1

Cap threads—Torque standards—Cap liners—Shrink bands—Stoppers and plugs—Rubber bulbs—Tapes—Adhesives

Section 10. BAGS AND ENVELOPES 10-1

Multiwall sacks—Baler bags—Sewn sacks—Valve sacks—Pasted bags—Plastic bags—Costs

Section 11. AEROSOLS 11-1

Principles of operation—Propellants—Valves—Containers—Testing—Laws and regulations—Suppliers—Spray patterns—Powder aerosols—Metering valves

Section 12. FIBRE TUBES, CANS, AND DRUMS 12-1

Paper tubes—Composite containers—Materials—Costs—Fibre drums—Standard sizes—Laws and regulations

Section 13. COATINGS AND LAMINATIONS 13-1

Waxed papers—Solvent and emulsion coatings—Heat-seal coatings—Papers, films, and foils for laminating—Barrier calculations—Costs—Suggested combinations

Section 14. LABELS AND LABELING 14-1

History—Merchandising value—Types—Materials—Processes—Direct printing—Heat transfer labels—Graphics—Adhesives—Production problems

Section 15. WOOD CONTAINERS 15-1

Wood types—Nailed boxes—Crate design—Wirebound boxes—Baskets—Barrels

Section 16. LAWS AND REGULATIONS 16-1

Carrier requirements—State laws—Hazardous materials—Post Office rules—Patents—Trademarks—Food and Drug Administration—Federal Trade Commission—Department of Commerce—Interstate Commerce Commission—Department of Transportation

Section 17. CUSHIONING 17-1

Vibration—Impact—Fatigue—Temperature—Humidity—Design hints—Calculations

Section 18. TEST METHODS 18-1

Fundamentals—Design of experiments—Process control—Sampling systems—Compendium of tests

Section 19. QUALITY CONTROL 19-1

Probabilities—Variables—Attributes—AQL—Significance—Sample size—Random numbers

Section 20. MACHINERY AND EQUIPMENT 20-1

Selection—Contracts—Justification—Costs—Savings—Flexibility—Changeover—Specifications

Index follows Section 20.

Section 1

Elements of Packaging

Introduction 1-1
State of the Art 1-3
The Challenge 1-6
Scope of Packaging 1-6
Preparation 1-8
Organization 1-8
Consultants 1-9
Suppliers 1-10
Contract Packagers 1-10
Approach to Problems 1-11
  Ethics 1-11
  Know Your Product 1-13
  Analyze the Market 1-13
  Competitive Packages 1-13
  Recognize Needs 1-13
  Innovations 1-15
  Integrated Systems 1-15
  Small or Large Volume 1-16
  Export Problems 1-16
  Vending Machines 1-17
Functions of a Package 1-17
Elements of a Package 1-20
  Structure 1-22
  Copy 1-22
Technology 1-23

INTRODUCTION

Our economy is a complex structure with many facets, and the importance of packaging within this system is becoming increasingly significant. Current methods of preserving and distributing goods are such an essential part of our way of life that we take them for granted and hardly realize how much modern techniques contribute to our standard of living. Some idea of the magnitude of packaging's share in our economy is revealed by the fact that expenditures for packaging are approaching $25 billion per year in this country, of which 60 percent is for materials and the remainder divided among labor, utilities, and other support costs. Packaging, along with better transportation, has made it possible to centralize production facilities and take advantage of the economies of large-scale operations. The product and the package are becoming so interdependent that we cannot consider one without

TABLE 1 Packaging Expenditures in the United States in Percentages*

Type of package	Apparel	Petroleum products	Hardware	Toys	Food	Beverages	Confections	Tobacco	Drugs and cosmetics	Miscellaneous	Total
Metal	. . .	0.60	. . .	. . .	8.00	3.30	1.60	. . .	0.30	10.60	24.40
Glass	. . .	. . .	. . .	. . .	. . .	3.00	0.44	. . .	0.27	3.69	7.40
Plastic . . .	. . .	0.10	. . .	. . .	0.39	. . .	. . .	. . .	1.05	4.26	5.80
Paperboard.	0.50	. . .	0.50	0.25	7.00	0.80	0.60	0.15	1.70	40.80	52.30
Flexible . . .	0.40	. . .	. . .	. . .	5.20	. . .	0.55	0.40	. . .	3.55	10.10
Total . . .	0.90	0.70	0.50	0.25	20.59	7.10	3.19	0.55	3.32	62.90	100.00

* To convert the figures to dollars, multiply by the total current expenditures, which at this writing are around $25 billion.

the other. The amount of money being spent in this country for packaging is indicated in Table 1 as percentages of the total national figure, showing which industries are the largest users of packaging materials.

The growth of packaging has outrun the technology needed for orderly progress and development. Very little literature is available to the serious student. The workers in this field must lean heavily on the periodicals to keep abreast of the rapid changes in this field, and they often have to develop their own disciplines to suit the situation at hand. It also becomes necessary to borrow from the other sciences to supplement the meager knowledge that has been accumulated in this area, and the skill of the packaging operator in making intelligent decisions depends on his breadth of knowledge in many fields and his depth of experience within the packaging industry.

Since packaging accounts for an increasing share of the cost of goods, it follows that the worker with the responsibility for making recommendations and influencing decisions needs all the knowledge and skill he can muster. Further, he should approach his task with an open mind and a broad perspective if he is to take full advantage of the great variety of materials and techniques that are available. To become preoccupied with the routine solutions to problems is to miss the opportunities that exist for truly great accomplishments in a dynamic and exciting field.

STATE OF THE ART

There is beginning to come into existence a body of knowledge related to packaging that might be considered the foundation for a technology which will someday take its place with the older established sciences. Food technology, for example, is getting increasing attention in colleges; it is largely concerned with the preservation of foods by packaging and processing, with emphasis on the maintenance of quality in addition to simply preserving the product. Nuclear engineering is directing more of its attention to the treatment of the nondurable goods that are necessary to the health and welfare of our society. Chemical engineering, industrial engineering, and mechanical engineering also are devoting a larger share of their efforts in this direction. Out of this will eventually come a unifying system for applying the fragmentary operations of each of these disciplines to the total problem of delivering goods to the point of use with the maximum efficiency. (See Fig. 1.)

Most of the progress in packaging in the past has been by trial-and-error methods. There is a great need for a more reliable basis for decisions, and this can come only from research in sufficient depth to give a solid foundation to the techniques required.

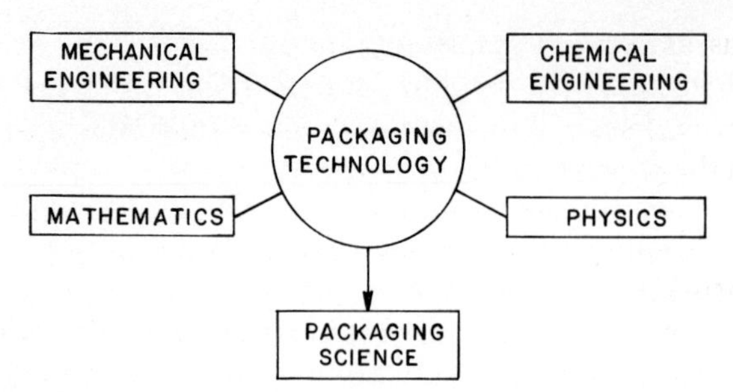

Fig. 1. State of the Art. Older disciplines are being applied to nondurable goods, out of which packaging technology is emerging. This will ultimately lead to a packaging science, directed toward the processing and preservation of food, clothing, and other maintenance items.

Some progress has been made in mass production and mass distribution of packaged goods up to the wholesale level. Further work being done with *containerization* will provide still greater efficiency in the primary channels of distribution. (See Fig. 2.) Much less advanced techniques are being utilized in the secondary movement of manufac-

Fig. 2. Modern transportation. A containership can move goods at a lower cost and with less damage than conventional methods of handling. The containers are made in various standard sizes; those on the deck of the ship shown are mostly 40- by 8- by 8-ft units, which are the largest in general use. Loaded in the shipper's warehouse, they are transported by rail or truck to the dock, put aboard the ship, and taken ultimately to the consignee's warehouse, without ever being unpacked. Damage and pilferage, which used to run up to 10 percent, are reduced to less than 1 percent, insurance rates are usually cut almost in half, and the delivery time may be reduced as much as one-third. (*Flying Camera, Inc.*)

tured goods, as between wholesaler and retailer, where regrouping of diverse products takes place. The final selection and transfer of goods to the point of end use are the area of greatest inefficiency and waste, compared with the sophistication at the producing end. (See Fig. 3.) There is a great opportunity for developing new and better methods for determining needs, and fulfilling those needs. For example, the operations that a consumer goes through in choosing goods in a supermarket, transporting them to a check-out station, waiting in line, tallying the cost, making change, jumble packing into single-trip bags for the short journey to the waiting automobile, and an expensive ride to the point of final storage and consumption are all highly inefficient. A similar situation exists with institutional items and some industrial products. An indication of the possibilities in these areas can be seen in the progress that has been made with vending machines. Here the goods are brought to the point of use in quantity; mechanical means are used to assist in the selection and make the transaction, at the time and place of consumption, with a minimum waste of labor and materials, and with efficient use of modular units and standardized components.

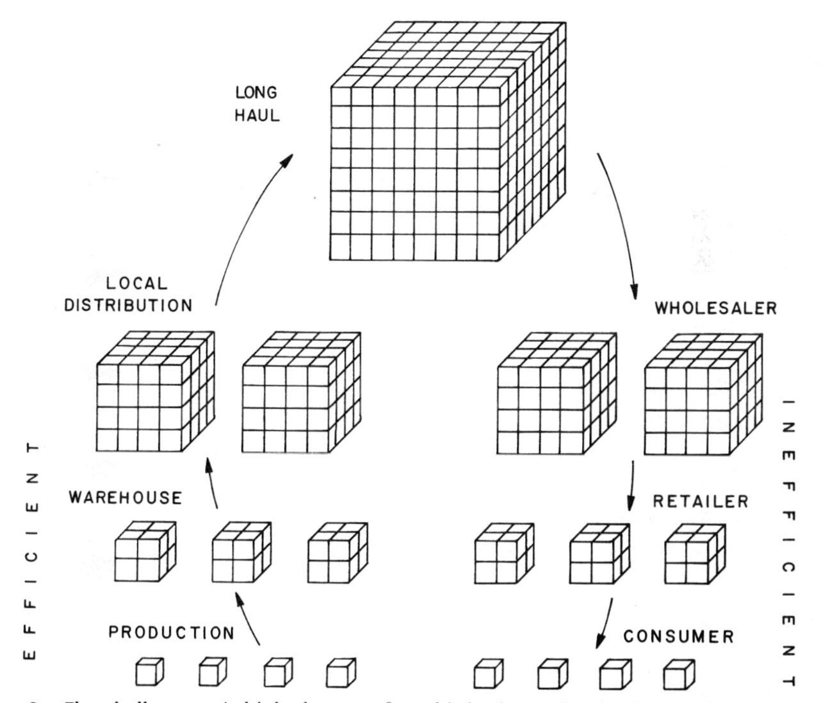

Fig. 3. The challenge. A high degree of sophistication exists in the production and warehousing of packaged goods. Efficient methods of assembling and moving these items have been developed; but at the consumer end, the techniques for subdividing into smaller units of sale are crude and wasteful.

The point here is to look upon packaging not as a means of merely containing products, but rather in its broader sense as a process of getting goods from the source to the point of use in the most beneficial manner. We must try to grasp the significance of the package and its position in the whole scheme of things. This entails all the aspects of handling, storage, preservation, distribution, advertising, sales, promotion, display, dispensing, preparation, and the various other facets of our industrial system.

THE CHALLENGE

The boon of packaging carries with it the bane of waste disposal. While sources of raw materials appear to be inexhaustible, the accumulation of by-products is beginning to draw attention to the more sordid aspects of packaging. Our highways and public places are becoming littered with debris, mostly packaging, and there has been a flurry of legislation to combat this menace. Air pollution and the fouling of our waterways with trash also are of increasing concern to government and industry alike. In testimony before a committee of the U.S. House of Representatives studying waste disposal problems, it was revealed that over 50 million tons of packaging materials are sold each year in this country. The worker in packaging must not ignore this situation, for it is an important part of the complete cycle that makes up the total packaging system. If a method of disposal can be built into a container when it is designed, it will be another step toward a better way of life through packaging.

One other item to be considered from this viewpoint is the matter of packages that damage themselves and each other. A package that is designed to protect its contents from damage by other packages, should not, in the broadest sense, inflict damage on the other packages if it is properly designed.

SCOPE OF PACKAGING

Packaging has many faces. In its more familiar forms it is the box on the grocer's shelf and the wrapper on a candy bar. It can also be the crate around a machine or a bulk container for chemicals. It is art and science; it is materials and equipment; it is protection, promotion, law, logistics, manufacturing, and materials handling all rolled into one. It is many things to many people, and a very difficult concept to describe and define.

Nevertheless we shall proceed to dissect packaging along certain arbitrary lines in order to study its composition. There are three broad categories that require very different technologies and talents for their accomplishment: (1) consumer packaging, (2) industrial packaging, and (3) military packaging. The first is concerned generally with small units in large numbers, often decorated in an attractive manner. Industrial packaging, in contrast, is usually made up of larger and heavier units, with no attempt to make them appealing to the eye. The third category, military packaging, is a highly specialized type of protective packaging in which all the elements have been worked out by the government and documented in the most intricate and sometimes vexing detail.

Within these major groupings there are various subdivisions, and in the case of the small retail packages there are two main sections to be considered: retail items and institutional items. In the first, the emphasis is on the visual appearance and value of the package as a sales tool. In institutional packaging, on the other hand, the importance of the various aspects would more likely be protection first, then cost, convenience, and appearance in about that order. Items that are intended for select classes of trade, such as veterinarians, beauty operators or mechanics, would also be classed as institutional packages under this description.

A further breakdown in the consumer field leads to subsections on food, pharmaceuticals, soft goods, hardware, cosmetics, and toiletries. Each of these has its own special problems, and the knowledge and skills required in each of these industries will vary considerably. The problems within a company also may be quite varied, so that the greatest need might be in automatic equipment rather than in the packages themselves, or it could be a matter of logistics requiring an industrial engineering approach, or design and decoration that could be handled only by an able stylist.

It is not possible for any individual to be skilled in all these areas, and most situations require the combined talents of a team of experts. These people may come from within the company, or they may be consultants commissioned to work on a particular assignment, or a combination of staff and consultants working together. The suppliers of packaging materials sometimes offer assistance on special problems, and they can often provide a valuable supplement to the efforts of the staff specialists.

There are great opportunities in research, apart from the day-to-day solving of practical problems, which should be of interest to the serious student of packaging. This area has not been adequately covered, partly because the rapid growth of packaging has not allowed sufficient time for any study in depth, and also because the few operators that are available for this work get siphoned off by the more practical needs of applied packaging. There has been too little effort spent in pulling together the disconnected data that are pouring out of the development laboratories of industry, and no solid foundation of timeless knowledge is being built up to undergird the bits and pieces of technical information that are accumulating. The subject deserves more attention from our academic institutions, and it warrants the serious support of the industries that depend on packaging for their existence.

PREPARATION

The education and experience required of a worker in packaging will depend on the particular situation, and the range of possible skills and talents that might be required will extend into every branch of science and technology. A young person starting a career in packaging would do well to get the broadest possible foundation at the outset, and then specialize as the job requires it, while maintaining and expanding a wide perspective.

Certain disciplines are basic to almost any branch of packaging; the education of a technologist should include these fundamentals: mechanical engineering, chemistry, physics, mathematics, and economics. All these are applicable in some degree to every packaging problem. Optional subjects that might be included are marketing, art, graphics, industrial engineering, psychology, law, and bacteriology.

An educational program in packaging must be a good balance among these divergent subjects in order to anticipate the demands of different manufacturing, marketing, and research conditions. Several colleges have established curricula in packaging, notably Michigan State University and Stout State University at the undergraduate level, and Rutgers University in its graduate school. Other colleges with more limited programs in this field are the University of Wisconsin, Purdue University, University of Georgia, Oregon State University, and the University of Connecticut.

Numerous evening courses also are offered by schools and sponsored by trade associations. In most cases these are intended as continuing education for the worker who is already engaged in packaging and wants to expand his knowledge in certain directions. They are useful when the basic education has been in another field and the exigencies of the job require some familiarity with packaging technology.

ORGANIZATION

Because packaging has roots in nearly all the different departments of an organization, the proper placement of this function on the management chart is not so clear-cut as it might be, and the choice is often a personal matter instead of the logical result of the needs of the situation. The degree of success or failure of the packaging function will certainly be influenced by the line of authority that its location brings with it, and even more by the personality of the people involved and the support it receives from upper management.

In the consumer goods field, packaging design and development are related to manufacturing as well as to marketing. The procurement, processing, and shipping operations within manufacturing are all tied in with packaging to some degree. Since by its nature it is essentially a staff type of function, packaging development most often becomes associated with the engineering department in one or another of its various forms. Depending on the personality and orientation of the company, this could be the equipment, plant, process, design, or industrial engineering department. Very rarely does packaging have sufficient status to stand by itself alongside these other engineering functions.

A case can be made for placing packaging in other areas, such as purchasing, research and development, merchandising, or traffic, but the results are often disappointing. The ideal location would be completely apart from the influence of the other departments, with packaging directly responsible to a top officer in the company. In only two large corporations is this actually the case, but it is likely that packaging will become increasingly important to our economy and its prestige will grow accordingly.

CONSULTANTS

A large body of experts is growing up within the packaging field. Many of them are quite sophisticated and serve a very useful function as professional consultants. Their talents are quite varied, and they can give advice on any aspect of packaging. The fees range from the free services of material suppliers to the six-figure charges of the top designers for a complex job. There are also some advertising agencies, market research organizations, and industrial psychologists who work on packaging problems with varying degrees of proficiency. A firm that specializes in packaging as a coordinated product of art, science, and technology, with no special interests in any particular form or material of packaging, is most likely to provide a satisfactory service to its clients. The fees sometimes run high, but when such consultants are carefully chosen and used intelligently, the results are well worth the price. The proper use of the services of consultants requires a clear definition of the problem and as much supporting data as can be made available by the company that is employing them.

It would be wise at the outset to arrive at a clear understanding with a consultant as to the limitation of charges, schedule of delivery, rate of payment at various stages, completeness and form of the designs to be delivered, assignment of patents, secrecy, and conflict with competitive accounts.

SUPPLIERS

The service of suppliers in developing and producing packaging components is an important part of launching any new item or improving an existing one. While suppliers may not have the broad viewpoint of the consultants previously mentioned, they will be more knowledgeable within their own field and thus will be more practical in their recommendations. The quality of work will vary among different companies, and most vendors tend to specialize in a particular class of trade. Some have excellent art staffs and are capable of top-notch styling, but they need the guidance of and close liaison with the customer for best results.

The matter of *ethics* in dealing with suppliers should be taken very seriously for the benefit of all concerned. Any submission of ideas, designs, samples, or artwork that is offered on speculation should be treated as confidential. Unless full payment is made to the vendor on a mutually satisfactory basis, the customer does not have the moral right to disclose these designs to a rival supplier. Nor should the customer use these ideas unless the submitting company is given the opportunity to recover their development expenses through orders of sufficient size to include these costs.

A good working relationship with suppliers must include good communications. This means adequate specifications as well as agreeable acceptance and rejection standards. The importance of a clear understanding between vendor and customer, before any commitments are made, cannot be overemphasized. Such things as delivery dates and penalties, limits on overruns and underruns, cancellation charges, ownership of tools, secrecy from competition, and protection from patent suits should all be put in writing to avoid later disagreements.

CONTRACT PACKAGERS

Modern industry is an unstable mixture of markets, materials, and techniques that requires a certain amount of flexibility to be successful. One of the sources of the necessary operating know-how and production facilities is the contract packager. Nearly all companies, from the smallest to the largest, avail themselves of such a satellite manufacturing plant. A typical job packager will limit himself, say, to pharmaceuticals and will strictly avoid pesticides for fear of contamination. Or he may handle aerosols only, and not go in for dry products. Such specialization is fairly general and makes it necessary to choose carefully among the different co-packers.

As a supplement to a going operation, or even as the sole source of packaged products, the contract packager has earned a permanent place on the industrial scene. He provides equipment and labor at a competitive price and is the means of starting many new products with a minimum amount of risk. The contract packager manipulates a labor force which sometimes, but not usually, consists of part-time housewives, handicapped or deprived workers, or other special groups in order to keep his costs below those of his customers, who often have much higher overhead rates to support their more sophisticated administration and benefit programs.

When a product has become established or has at least shown some potential for growth, the manufacturer will want to consider taking it away from the contract packager and bringing it into his own plant. This would provide for better controls and opportunities for cost improvements through mechanization and purchasing power. In some instances the transition may not be justified, and the operation will be continued indefinitely on a contract basis. Aerosols, for example, require a large capital investment in equipment and very specialized knowledge for its operation. Products such as lipsticks and nail enamel, or bath soaps that require special skills in color matching, fragrance blending, or difficult formulating, are usually best left in the hands of experts in their field.

APPROACH TO PROBLEMS

A professional worker in packaging should approach his problems with a good knowledge of the materials and processes used in all branches of the industry, and with a high degree of skill built up by experience within the different areas of glass, metal, paper, and plastics. In addition he must exercise some ingenuity in adapting these materials to a solution of the problem at hand. He should also practice a code of ethics and use good judgment in working with suppliers if he is to be a credit to his profession.

Ethics. For his own good and the future of his company, the packaging specialist should know and practice a standard of behavior that fulfills his obligations to his profession and the business society in general. He will not design or endorse packages that are deceptive or misleading, even though they may be within the legal limits. A package should be honest and forthright, and must not deliberately give an impression of containing or delivering something that is not there. This is not to say that it should not enhance the product or make it more attractive. Certainly this is an important function of packaging. But fake bottoms, excess packing, and similar tricks to fool the consumer should be scrupulously avoided.

The ideal packaging specialist will serve his employer or his clients with devotion and loyalty. He will be honest in his estimates and sincere in his advice and counsel. There will not be any padding or overdesign with intentions of making dramatic savings at a later date. Any assignments which are beyond his abilities, in whole or in part, will be frankly admitted at the earliest moment, and a questionable suggestion or recommendation will be offered with all the pertinent data, without bias or prejudice. He will cooperate with any and all experts or consultants hired by his employer or client.

He will hold confidential information in secret and will not disclose designs, processes, plans, or intentions except with the express permission of employer or client. He will not accept assignments from different parties at the same time whose interests may be competitive or in conflict, so as to avoid any chance of cross-fertilization. Any knowledge of methods and techniques obtained in the course of a project will be used only for the benefit of that project and not to help a direct competitor.

A packaging professional will not accept gifts or gratuities intended to influence his recommendations or specifications. He will avoid having vendors spend large amounts of time or money on speculative presentations. He will not ask for samples or quotations from a supplier if he knows that no order is likely to result, unless the vendor is made aware of the circumstances.

Every effort will be made to pay for design and engineering samples, and those companies that charge for these services will be encouraged and no discrimination will be made against them. Credit for ideas and designs will be given where due, and the packaging specialist will not take credit for work that is not his own.

He will lend encouragement and will help to train novices in the field of packaging. In his relations with fellow workers he will cooperate freely in the exchange of knowledge and will give advice and counsel generously when requested. He will not injure the reputation or prospects of another packaging specialist, but if he has proof of unethical practices, he will bring this information to the proper authorities.

He will contribute time, effort, and funds to further the progress of packaging technology, through publications and membership in professional societies. He will participate actively in seminars and other educational programs to share his knowledge and experience with other members of his profession.

Know Your Product. It may seem elementary to suggest that a thorough knowledge of the product is essential in designing containers for the product, but too often a packaging specialist does not learn all he should before starting his design work. There is a temptation to jump ahead on the assumption that incomplete information is adequate for the purpose. Whether it be market data or physical properties of the formula, it pays to take the time to become familiar with all the available information.

Analyze the Market. A clear understanding of the segment of the market which the package is expected to serve is necessary before a good design can be created. Whether the potential customer is male or female, young or old, rich or poor, urban or rural will often affect the type of design to be recommended. Unless it is correctly placed, the package cannot do the best possible job at the point of sale or at the point of use. Merchandising methods, display techniques, and advertising support will all have an important influence on the final design.

Competitive Packages. The appearance of a package on a shelf may be quite different from the way it looks on the drawing board or on the client's desk. It would be wise to view a new package in a trade situation, surrounded by competing products, before finalizing the design. If the surrounding packages are all dark, or all tall, or all decorated in a particular way, it would be well to choose a contrasting format to distinguish it from the competing brands. A word of caution, however, before adopting a design just to be different: If the customer is accustomed to getting a product in a certain type of container, he may not accept a design too far removed from the expected size, shape, or color.

Recognize Needs. A well-designed package will cater to the needs of the purchaser. These may be either psychological desires or practical applications; but whatever they are, they must be recognized and satisfied by the package designer. To be more specific, the design of the package must show at once the intended use, method of application, and promised results. A container of talcum powder should not look as if it contains scouring powder, nor should face cream jars resemble shoe polish containers. It is sometimes possible, however, to upgrade a household product by borrowing some of the design elements from the cosmetic and toiletry field, without losing the identity of the contents.

Besides projecting its contents visually, a package can convey the desired mood of the item. If dependability is one of the selling points, a design that uses old-fashioned elements may create an impression of proven efficacy. Toiletries can take on an aura of glamour and luxury to their advantage, and cosmetics will often stress romantic themes. Pharmaceuticals can play up the research angle, soft goods can go high style, and food can

MEDIEVAL
feminine
EMPHATIC
DIGNIFIED

Fig. 4. Power of persuasion. Choice of typeface can help to set the mood of a package. Different styles connote various impressions in the mind of the viewer, and a type can usually be found to suggest the precise shade of meaning that is desired.

emphasize the candlelight and silver mood of a dinner party. (See Fig. 4.)

A promise should be implied in the text on the main panel. The trade name and the subtitle or supporting phrases must appeal to the natural desires of the purchaser, either the need for sustenance and comfort, or, better still, the desire for self-improvement, status, or indulgence for self or family. The body of the text can expand on this theme and provide more specific information about the product and its usefulness. The sales message should not obscure the instructions for use and other pertinent data, however, and the questions in the mind of the purchaser must be anticipated and answered clearly and succinctly. Otherwise the package may be put back on the shelf and the sale missed.

Consider all five senses when designing a package, and not merely the visual impact. It is often possible to appeal to the sense of touch by choosing a material with a special texture, or by embossing or forming it to make the package feel attractive. The rustle of silk, the pop of a cork, the clink of glassware all add to the enjoyment of certain products. It is possible to put perfume in the ink that is used for printing a package, and in this way to add subliminal appeal to the potential purchaser. We have only begun to explore the possibilities for putting more sales appeal in packaging, and we may be missing opportunities for creativity by slavishly copying packages that are already on the market.

The size and shape of a package can be important elements of design for several reasons. The value of a product may be measured subconsciously by the impression of its size, and in a competitive situation the consumer will often reach for the larger of two rival brands. The space limitations in the store or in the home also must be considered, and a drug product that does not fit in a medicine cabinet or a box of cereal that is too large for a kitchen cabinet can have an adverse effect on repeat sales.

It may be useful to consider the *needs* and *wants* of the consumer separately. First, it is most important to satisfy the needs which are fundamental and absolutely essential to the success of the package. When these have been clearly and accurately delineated, then it may be possible to add the desirable but less critical factors that are contributed by creative design and copy. In this way there is less risk of ending up with a package that wins awards among the practitioners of the packaging arts, but is a miserable failure in the marketplace. Too often the arty and cute designs are so subtle that they completely miss the mark; it takes a real professional to blend successfully the psychological elements with the practical aspects of packaging.

Innovations. In some rare cases the package actually becomes a part of the product, and in this way serves to augment and enhance its usefulness. Containers and closures that measure, or mix, or dispense the product can, in a sense, be considered an integral part of the product concept. As an example, hair spray without the valve-container-propellant system would not be nearly so popular as it has become; it was in fact available, but little used, long before the advent of the pressurized package. Popcorn in a disposable cooking utensil, charcoal for a barbecue that uses the container as a starter, and TV dinners in a serving tray are all innovations that won enthusiastic acceptance by the consumers.

A radical design concept is more likely to bring outstanding success, by its very nature, than a conventional solution to a problem will. Therefore the extra effort that is required to brush aside the obvious answers and to probe deeper into the requirements of the package will usually yield great rewards. The successful designer must be constantly asking what, how, when, and where for every aspect of the problem, and he must shut his mind to the obvious answers to problems long enough to explore new avenues and assure himself that there is not a great idea lurking close at hand, waiting to be uncovered.

Integrated Systems. The design of a package is not an isolated thing, but a part of a larger effort to carry products to their end use in a successful manner. As such, the package must fit into the manufacturing-marketing system and contribute its share to the effective operation of all the segments of the whole process. In order to do this, the design of a package should be measured against the capabilities of the engineering, advertising, distribution, and final application conditions within which it must function. The professional package developer will think ahead to the various situations which will confront the product and its container through the many channels of procurement, manufacturing, storage, shipment, and use, in order to achieve the optimum results overall.

It is important to have good communications with all the areas previously mentioned as concerned with the final package, and to gain the cooperation of the people who can contribute information and assistance to each project. The degree to which this is carried out will depend somewhat on the magnitude and complexity of the individual problems, but the likelihood of doing a job *too* thoroughly is not nearly so great as the chance of making only a superficial analysis that yields mediocre results. The experienced designer takes a broad view of every problem and uses good judgment in choosing the areas to be explored in depth. The novice would do well to practice looking beyond the obvious and immediate answers to problems, trying to improve his perspective, and being meticulous in his analytical work.

Small or Large Volume. The requirements of a package design depend greatly on the quantities involved. The small-volume item, for example, will not justify special molds and sophisticated equipment, but must utilize stock containers and semiautomatic or manual assembly operations. This is not to say that this type of assignment is any easier; on the contrary, the limitations of the small volume may very well tax the knowledge and skill of the developer far more than the package that can be designed from the ground up. The high-volume item, on the other hand, will take a more specialized type of skill and permit a more thorough investigation of each aspect of the problem.

The design for a beverage container, for example, requires a very different approach from that for a similar container to hold a chemical intermediate of very limited application. The physical needs may be equally rigid, and yet the opportunities for design will be quite different. Costs will be more critical in the first case, and some risks of failure in shipment or storage can be taken if they can be justified on an economic basis. The small-volume item often demands more ingenuity in adapting existing materials to the purpose, and with less money involved in total; the best approach is to overdesign rather than to take a chance on failure of this type of package. Some idea of the portion of cost which is put into the package, for various kinds of products, is shown in Tables 2 and 3.

Export Problems. If a package is intended for use outside the country of origin, there is a whole host of special problems. As might be expected, the hazards of transportation are generally much greater when we get beyond local rail or truck movement. In some cases the handling of goods with slings on and off ships puts special strains on

TABLE 2 Packaging Costs as a Percentage of the Manufacturer's Selling Price in Various Industries*

Product	Percentage
Cosmetics	30
Motor oil	28
Drugs	27
Foods	22
Toys	9

* From an industry survey of 500 companies.

TABLE 3 Packaging Costs as a Percentage of Manufacturer's Total Costs for Various Products

Product	Percentage
Aerosols	80
Bar soap	50
Beer	27
Potato chips	17
Women's hosiery	10
Cigarettes	3

packages, and in some seaports cargoes must be transferred to lighters before they can be brought to dockside. Moreover, considerable hand labor is used to load and unload ships, and the stevedores that do this work bring about special problems of damage and pilferage. The growing use of large cargo containers is changing this to some degree, but unless it is known that containerization will be employed, it is the usual practice to make packages 1½ to 2 times as strong for this type of service. Single-wall domestic corrugated boxes, for example, have double or triple walls for export use. Wood containers may be substituted for corrugated, and textile bags will sometimes replace multiwall paper bags, or at least two extra plies of paper are often used.

Markings and decorations become more complex, and the laws, customs, and traditions in foreign markets are pitfalls for the uninitiated package designer. Certain colors have special significance in other lands, and good English words can become completely unacceptable in translation to another language. This is a highly specialized area, and the best advice is to move cautiously and to seek help from people with firsthand knowledge of the situation.

Vending Machines. If there is any possibility that an item might be used in vending machines, it would be well to consider this fact in the design stages. A rigid package is much more practical for this purpose than a flexible pouch or bag. Very thin items are difficult to dispense, and oversize packages may limit or completely eliminate this outlet for the product.

The value of the unit package should lend itself to the standard prices in the vending trade. Investigation will determine also whether there is any special modular system for dimensions, or whether trade practices or mechanical operation of the machines puts special limits on the materials that can be used. A little forethought in the developmental stages may avoid embarrassing compromises or costly changes at a later date. It is not always necessary to design specifically for the vending trade, but by recognizing the existence of this segment of the market, the designer will at least be aware of the implications and future possibilities for his packages.

FUNCTIONS OF A PACKAGE

Of the several ways that we could analyze the purpose of packaging, it is unimportant whether we take one or another; the answers turn out pretty much the same. The functions of a package are basically to *contain*, *carry*, and *dispense*. Shells and the skins of animals served as packages for these purposes for primitive man. As time went on, there were added other requirements such as to *preserve*, and to *measure*, and later to *communicate* and to *display*. We have now entered into an era when the package is called upon to *motivate, promote, glamorize*, and sometimes to *build up* or even *disguise* the contents.

To transfer a particular product, for example, a powder, from the place of manufacture to the point of use requires some kind of container. At the manufacturing stage the product is usually in a bulk tank of some type, and it becomes necessary to subdivide it into more convenient units for transportation and handling. Without the means to *contain* a powder product it would be impossible to transport it beyond the point of origin. At the wholesale level, and again at the retail level, it may be necessary to reduce the size of the units even more to accommodate the demands of the trade.

The sugar barrel and the butter tub were once the standard packages in the grocery trade. Much of the subdividing in those days was done at the retail level. This method has given way to a system that includes a reduction in the size of the units produced at the manufacturing plant. The product is put into small boxes or bottles, which are then gathered into groups by means of chipboard packers or by bundling. These are further overpacked into case lots for shipment out of the manufacturing plant. Thus it is made much easier to reduce the size of the unit to be handled at each stage along the way, and the function of *carrying* is accomplished more effectively toward the point of ultimate use.

In the hands of the consumer the product may not be used all at once, but only partially consumed. Therefore it becomes necessary to remove a portion of the product without destroying the container, so that the remainder of the product can be kept safely for future use. The *dispensing* of the contents is an important function of the package, along with containing and carrying. In most cases the commodity being packaged will not be immediately consumed, and it must therefore be protected and *preserved* for an extended period of time. As we learn better methods of protecting such products by means of more secure packages, aided by such techniques as refrigeration and sterilization processes, we are able to do a more effective job of preserving products for future use.

The advances that have been made in prepackaging goods have produced some additional benefits. One of these is the opportunity to standardize the quantity of material in a package to obviate dealing with arbitrary amounts of goods. A uniform system was established whereby an item such as butter could be bought in even pound lots, and milk in exact quart containers. The package therefore became a *measuring* device in addition to its other functions.

The opportunity to identify the contents by the shape, color, and decoration on the container was realized a long time ago, and this has been exploited to an ever-increasing degree until it has now become one of the prime functions of the package. Symbols, trademarks, slogans, and other devices are being used to help the manufacturer to *communicate* with the consumer. They may also take the form of instructions, warnings, guarantees, and specifications. This is especially effective because the message remains with the product through all stages of transportation, barter, and consumption, and is accessible whenever needed. Trademarks on packages have been very useful in promoting sales to consumers; and by guaranteeing consistent quality, they have been the means of developing a loyalty among purchasers for brands such as Campbell's soup, Dole's pineapple, Heinz pickles, and other well-known names. This could not have been accomplished when butter was sold from a tub and sugar from a barrel; it became possible only with the advent of unit packaging.

As the package and its contents move through commerce, the message on its surface will be seen by a great number of people. (See Table 4.) Exposure of the message on the package to the passing traffic is a very effective form of advertising, and we are constantly learning better ways to take advantage of the *display* value of the package to attract attention.

At the point of sale the package and its message play an important role in helping the purchaser to decide whether a need exists, whether the contents of the package will satisfy that need, and which of a variety of sizes or brands will best serve the purpose. Thus the package becomes a potent force in *motivating* the consumer to make a purchase and to promote the sale of one brand rather than another.

With changes in methods of merchandising and an increasing amount of self-selection, without the benefit of a sales clerk, the effect of the package on sales has assumed very significant proportions. This has given the manufacturer and the dealer an opportunity to enhance and *glamorize* the product through the medium of the package. Particularly in the cosmetic and toiletries field, foils, fabrics, decorating and embossing techniques, and exotic closures are used to create an illusion of something very precious.

TABLE 4 Cost of Packaging as an Advertising Medium*

Medium	Cost per thousand impressions
Container, printed on four sides	$0.05
Newspaper, full page	3.10
Magazine, full page	3.40
Radio, quarter hour	5.00
Television, quarter hour	15.00
Direct mail, 11″ × 17″ self-mailer	47.00

* Based on impressions sent out, not on messages seen or heard.

Using the package for purposes of *deception* is a practice to be avoided, but there are many examples in the marketplace that border on outright deception. The packages for toys and games, for example, frequently are expanded out of all proportion to the physical requirements of the contents. Although the container serves as a display piece, and is often treated like a billboard, this is scarcely an excuse for the overpackaging that is the rule and not the exception in the toy industry. It can be argued that custom makes such a practice acceptable and that the customer is not being deceived, but the industry might do well to set up some standards in this area before the legislators do it for them.

Variety store items and other self-service articles that are liable to be pilfered at the retail level are often put into oversize containers or mounted on a display card to combat this problem. Small articles that can be concealed easily by the shopper, by palming or by slipping into a purse or pocket, are a serious problem to the shopkeeper. Some dealers add 7 percent to their operating costs as an accounting procedure, to cover such losses. There is some justification in this case for an oversize package intended to frustrate the shoplifter, and the use of windows or transparent covers takes away any suggestion of deceptive tactics, in most instances.

ELEMENTS OF A PACKAGE

Several essential parts make up a package, and although not all of them appear in every type of package, most will be found in the majority of packages. In some cases where they are not, there is good reason to think that perhaps they should be. For example, there are opportunities for printed instructions and for decoration on industrial-type packages that are not being fully exploited, and manufacturers of these items would do well to give more thought to their packaging. Shipping containers become traveling billboards as they move through the channels of commerce, and with very little effort or expense they could become a means of building goodwill for the manufacturer or contributing to better trade relations. It might be worthwhile to evaluate every package in terms of certain basic elements to see whether any possibilities are being overlooked. Toward this end we will discuss separately the functions of *structure, aesthetics, style, communication,* and *legal* requirements.

In its broadest sense a package can be considered to be any *structure* that contains or limits its contents. This would include crates, nets, and cocoons, as well as displays, utensils, and conveyances. There is a special connotation of the term "packaging" in the electronics world that is outside our scope, but since it causes some confusion it will be mentioned here. This is the assembly of wires, controls, and other electronic gear onto a chassis or within a cabinet. Our treatment of the subject will include only the more familiar boxes, bottles, and cans of consumer and industrial packaging.

An open crate is a container of sorts; wooden structures are treated in detail in Sec. 15, "Wood Containers." Metal packages range in size from purse-size aerosols and lipstick cases to 55-gal steel drums. Flexible bags and pouches of paper and cellophane also are structures that fulfill all the purposes of packaging; the materials and processes used for this type of packaging are discussed in Sec. 10, "Bags and Envelopes."

The subject of packaging must inevitably bring us to a consideration of *aesthetics* and the part that appearance plays in the design of containers. Since the effect that the visual impact of a package has on the beholder is only just beginning to be understood, it is difficult to establish criteria for artistic design as it is applied to packaging. Much has been written on the psychology of color, copy, and composition, as well as the styling of containers; and creative design has accomplished some interesting results. While some conclusions can be drawn from these experiences, there is nevertheless a tendency among designers to copy one another's successful creations, and this usually leads to fashions and fads instead of any real effort toward satisfying the needs of the trade. Thus we have packages of cake mix that all try to incorporate "appetite appeal" with a color photograph of the finished cake from which a slice has been removed. This generally results in a sameness that cancels out any competitive advantage that one brand might have over another. All the liquid detergents, with few exceptions, have been put into narrow-waist plastic bottles with paper labels that differ very little from one another on the retail shelf. However, a few pioneers among the designers are making good use of the graphic processes and structural styling, and their approach to marketing problems is worthy of careful study.

The matter of *style* in packaging deserves some explanation. While it is closely associated with aesthetics, and is a term that is sometimes applied to periods of design such as baroque, Bauhaus, sculptured, streamlined, or rococo, our use of the term is somewhat different. What is meant by styling in this text is the total impression of the shape, color, texture, and typography of the package. Thus it is possible to have a design that is obviously feminine, delicate, and romantic for a cosmetic item, or in another case, a package that is bold, rugged, and strongly masculine for a product in the hardware field. In the first case the colors might be pastel shades, the typography might be a dainty script style, and the structure might be delicate and frilly. On the other hand, to appeal to men a product would more likely be sturdy-looking, with strong dark colors and bold printing. (See Fig. 4.)

The proper choice of design elements for a package can provoke a feeling of nostalgia, or whet the appetite, or create an atmosphere of opulence and luxury, according to the way these components are handled. It is this projection of a feeling for the product contained in the package that gives packaging some of its power for promoting sales and increasing profits. On the contrary it can be a deterrent to sales if it is not handled properly. In this image building through design it is necessary to stay within certain boundaries: not to limit ingenuity and creativeness, necessarily, but to keep within the recognizable area of design that is associated with a particular class of trade. The point here is that a package for a certain product should look as if it belongs to that kind of product. It would be completely out of character, for example, to put floor wax into a footed crystal bottle with a gold tassel and an exotic closure, just as it would be wrong to have an expensive perfume packaged in an amber prescription bottle.

The aesthetic design of a package should be used to dress up or enhance the product it contains, but it must not be used to mislead or deceive the purchaser. The power of a package design to attract attention and encourage sales should not be abused at the risk of alienating the buying public, or bringing the forces of the govenment against the packaging community.

Structure. The shape of the package may be dictated by the available containers and the nature of the product, or it may offer opportunities for variety of size and proportions, depending on quantities, physical requirements, and other factors. Designers are often carried away by their desire to give the maximum size impression, and they fail to recognize the need for stability on the shelf and the space limitations in the home. Another fairly common error is to overlook the practical aspects of manufacturing, and to create designs that cannot be run on the packaging line at a good speed because the containers topple over, or jam between the guide rails, or have an opening too small for filling.

It may be trite to say that good communications should be maintained with all departments during the development of a new package structure, but it is a very true statement nevertheless. The problem is not usually a lack of information; more often it is either an unwillingness to seek out the information or an ineptness on the part of the designer in not recognizing the practical requirements of mass production. A professional package developer will measure his design against the needs of procurement, manufacturing, storage, shipment, and sales to be sure he has not overlooked any potential trouble spots. Although a compromise may be necessary where the various requirements are in conflict, a capable designer will try to make his judgments in the light of the facts.

Copy. The handling of the text on a package should be given very careful study. The least amount of copy is nearly always the most effective, and particularly on the display panel of a package, the barest minimum of wording should be used. The more elements that are fighting for the attention of the viewer, such as the trademark, description, catchphrases, quantity designation, and manufacturer's name, the weaker will be the total effect.

It is important to have enough information on the package to answer the potential buyer's questions quickly and clearly. Too much "sell" often ruins otherwise informative copy, and this part of a package design requires careful handling. Usually the time and effort that are needed to edit and arrange text, and to test it adequately, are well spent. Store shelves are full of examples of wrong things being emphasized, surfaces cluttered with nonessential trivia, and important information lacking. A good testing program could have corrected most of these packaging errors, but a lack of humility among creative people often prevents them from seeking this type of assistance.

Although the preparation of the copy for a package is beyond the scope of this book, a few details should be mentioned at this point. Some of the legal and regulatory aspects of the printed package are covered in Sec. 16, "Laws and Regulations," and some requirements are well known and quite obvious. Special attention should be directed to the following essential information: inclusion and proper placement of a descriptive name as well as a trade name, so that misuse will be avoided and, in case a mistake is made, the misapplication can be corrected; suitable warnings against possible hazards, prominently displayed; a list of important ingredients if required by law or if identification might be needed in an emergency; name and address of the manufacturer; stock number and a clear designation of quantity for warehouse and dealer information; and directions for use or method of application. This last is one of the most frequent areas of errors and omissions.

TECHNOLOGY

There exists a body of knowledge that can be applied to the solution of problems in packaging, but it is somewhat limited in both quantity and quality. Partly because packaging is such a rapidly growing field, as well as a fast-changing one, it is very difficult to get this information documented before it becomes obsolete.

One of the most exciting branches of technology is the field of plastics, and an increasing percentage of all packaging is being made from these versatile materials. To the novice it may seem scarcely worthwhile to spend much time studying any other *materials*. However, the glamorous attraction of plastics should not obscure the usefulness of some of the other more mundane packaging materials. Paper and paperboard are still the most economical materials and will always make up the major share of containers. Metals have a high degree of strength and rigidity that are needed for certain applications; and for the ultimate in protection with flexible packaging, only foil will provide the absolute barrier. Glass bottles and jars are almost always chosen in preference to plastic for the most precious commodities, and it would be a mistake to assume that they will ever be replaced entirely by synthetic materials. The old reliables will always have an important place in packaging, and the student would do well to learn all he can about these basic packaging forms. See Table 1 for the relative amounts of various materials used in packaging.

Hand in hand with a broad knowledge of materials and forms of packaging must go an intimate knowledge of the *processes* used in the manufacturing and assembling of packages. A thorough understanding of the methods of making paper and paperboard will help in understanding their properties and in using them to the best advantage. The same can be said of metal, glass, and plastics, and the serious technologist will learn all he can about these industries.

There is also much to be learned about the fabrication and conversion of these materials into finished packages. Various levels of sophistication will be found in different fields, depending on the volume that is being handled and the demands for quality and uniformity. Methods will range from completely hand-assembled containers, through semiautomatic operations, to the high-speed, fully automatic production lines in the food and beverage industries.

Various techniques used in packaging depend on many areas of science and technology. To name a few of these techniques, there are filling, wrapping, sealing, and bundling of various types. Processing and sterilizing are used for some foods and for surgical items. Cleaning and corrosion prevention are important in military packaging, and the prevention of mold, mildew, and insect infestation also has its place in packaging. Sometimes the work of the packaging specialist involves material handling, warehousing, and carloading. The diverse nature of this technology makes it difficult to plan a curriculum of studies that will prepare the student for all situations, and it usually becomes necessary to choose an area of specialization before any significant depth of knowledge can be attained.

Some consideration of *equipment* is an important part of any packaging program. It may not be necessary to have a complete understanding of the mechanical operations involved, but some grasp of the possibilities and limitations of packaging machinery should be a part of every packaging engineer's background. In some jobs it is the most important part of the work, while in other cases it is necessary only to have the most superficial knowledge of this phase of packaging.

The best place to get firsthand information about equipment is at the trade shows and expositions. The next best source is the periodicals and directories that are published regularly in the packaging field. It may be useful for the engineer to accumulate a reference library on packaging equipment, and the manufacturers are usually quite generous in supplying descriptive literature covering their machinery and equipment.

Any study of packaging inevitably gets into a consideration of *costs.* It is essential to have a good understanding of the various elements that enter into the cost of a package—not only the material and labor costs, but also the fixed elements comprised in the overall cost of doing business. A few of the items of expense that may be overlooked are control costs for inspection and testing; operational costs involving sterilization, processing, or similar treatment of the finished package; maintenance and supervision; warehousing of raw materials as well as finished goods; and anticipated spoilage. Figure 5 shows some of the components that make up the prime cost of a packaged item.

Whether promotional material such as display containers should be charged to the cost of goods or to a special advertising account will depend on different companies' practices. It has been suggested that any excess packaging costs, beyond the barest minimum necessary for shipment, should be charged against advertising expenses. This may not be practical from an accounting standpoint, but the suggestion does point up the fact that the value of packaging as an advertising medium is not fully appreciated. The cost per exposure, as advertising is usually evaluated, is far less than in almost any other medium.

The costs of getting a product launched, or even into a test market, are sometimes mistakenly used for calculating profits. To decide whether it would be profitable to go to market with a new product, it is necessary to project the costs ahead several years so that capital investment in tools and equipment can be spread over a reasonable period of time. The advantage of large-volume purchases of materials should also be taken into account.

On this basis a *standard cost* can be estimated, which will be useful until an actual operating cost can be established. A standard cost provides a means of measuring the efficiency of the manufacturing operation in terms of plus or minus variances from the standard, which calls attention to such things as high scrap losses or excess downtime on machines.

The packaging developer or coordinator has an obligation to see that the *legal* requirements are met. There are many laws and regulations concerning packaging, and it is a constant problem to know the applicable laws and to keep abreast of the changes that are made so frequently. Some of the agencies that have jurisdiction over the different aspects of packaging, to name a few, are the Food and Drug Administration, Department of Commerce, Post Office Department, Interstate Commerce Commission, Federal Trade Commission, U.S. Coast Guard, various state departments of weights and measures, tunnel

COMPANY XP Div.	COST ESTIMATE		ESTIMATE NUMBER 247
PRODUCT NAME After Shave Lotion			SUGGESTED RETAIL PRICE $1.98
DESCRIPTION OF PRODUCT Scented alcohol solution			NET SELLING PRICE .99
DESCRIPTION OF BULK Free flowing liquid			FORMULA NO. S733RK
INITIAL PRODUCTION 200,000	ANNUAL SALES 500,000	REQUESTED BY Brand Manager	DATE 8/11/73

DETAIL OF PACKING MATERIAL

DESCRIPTION	QUANTITY	PRICE	UNIT COST
Bottle, 4 oz. square flint	1	6.64 gr	.0600
Cap, bla ck molded P/TF liner	1	17.83 M	.0178
Label, 2 color embossed	1	3.29 M	.0033
Carton, R.T. 2 color and varnish	1	23.75 M	.0238
Packer, .024 WPCN 1 color	1/12	46.94 M	.0039
Shipper, 200 test kraft corrug.	1/36	93.50 M	.0026
Spoilage 2%			.0018
AMORTIZATION (AMOUNT) $ (ABSORPTION BASIS) PCS.			
Total Packing Material Cost Per Piece			.1132

DETAIL OF LABOR AND OVERHEAD

MANUFACTURING

FORMULA NO. S733RK	BATCH SIZE 250.1.	HOURS PER BATCH 3
MANUFACTURING DEPARTMENT 26	AMORTIZATION %	ABSORPTION %
UNIT LABOR TIME .003	LABOR RATE 2.85	OVERHEAD RATE F 1.5500 3.7250

PACKING

MINIMUM RUN 20,000	HOURLY OUTPUT 5,000	NO. OF OPERATORS 8
PACKING DEPARTMENT 32	PRODUCT %	PRODUCTION %
UNIT LABOR TIME .0016	LABOR RATE 2.40	OVERHEAD RATE F. 1.7000 V. 4.4000

BULK COST

Raw Material	$ .2100 lb.
Mfg. Labor	.0190
Variable Mfg. Overhead	.0700
Direct Cost Per Pound	$ .2990
Fixed Mfg. Overhead	.0294
Total Cost Per Pound	$ .3284 lb.

PRODUCT COST

Packing Material	$ .1132 ea.
Packing Labor	.0038
Variable Pkg. Overhead	.0070
Contents - Direct (.2300 lbs. Pc.)	.0755
Direct Cost	$ .1995 ea.
Fixed Pkg. & Mfg. Overhead	.0342
Leveling Factor	-
Completed Cost	$ .2337 ea.
Tolerance 5%	.0117
Estimated Cost	$ .2454 ea.
Percent of Net Selling Price	25 %

ALTERNATE COST

Direct Cost	$ %
Fixed Overhead	
(% of Net Selling Price)	
Leveling Factor	
(% of Net Selling Price)	
Completed Cost	$ %
Tolerance 5%	
Estimated Alternate Cost	$ %

DETAIL OF ESTIMATED TRANSPORTATION COST TO CUSTOMER

Shipping Weight .4875 lbs. Pc. Rate $ .0295 Per lb.

Transportation Cost to Customer .0146 ea.

APPROVALS

PACKAGE ENGINEER (MAILS. & PRICES)	INDUSTRIAL ENGINEER (LABOR, TIME & SHIP. WGT.)	PACKAGING COMMITTEE	COST DEPT. (CALCULATIONS & RELEASE)
[illegible] DATE [illegible]	[illegible] DATE [illegible]	[illegible] DATE [illegible]	[illegible] DATE [illegible]

Fig. 5. Economic facts. Typical cost estimate showing the various elements in the cost of goods at the point of assembly. Administrative and selling costs are not included.

and bridge authorities, city fire departments, and carrier regulating bodies. For export items there are also countless rules and regulations of foreign countries to be considered. A brief review of the more important legal requirements for packaging is given in Sec. 16, "Laws and Regulations."

One of the important lessons to be learned in package development is the need for an adequate *testing* program. This is often discovered the hard way, and the measure of professionalism in this work might well be the thoroughness and adequacy of the test methods that are used. It cannot be overemphasized that a new or revised package should be checked under actual conditions of manufacture, storage, shipment, and use. Failure to do a careful job of testing can be very costly when its effects are multiplied by the large number of inadequate packages that might be produced and the extra cost of correcting the errors.

There is a feeling in some quarters that packaging is a stepchild of the engineering profession. This is far from the truth. Packaging is big enough and important enough to stand on its own two feet. It may be necessary to make a special effort to gain the recognition of other departments, and especially to get the backing of top management. This requires certain techniques which might be called *selling to management.* Whether it is an individual project, or a complete program, or the whole concept of package development, it must be done with finesse. It will take time and effort and money, but it is an investment that can bring big dividends.

Prepare in advance for your presentation to management; use lots of visual aids such as charts and illustrations; rehearse the presentation until it is letter-perfect; and put it across in a suitable setting. All this takes a certain amount of showmanship, but this should be a part of every designer's bag of tricks. It is not enough to dream up a great idea, or to design a sensational package, if you cannot sell it. The conception and execution of a new design are only half the job; selling it to management is just as important. A good idea that fails because it was not presented properly is a real tragedy, and it reflects on the competence of any man who thinks he is a professional.

The language of packaging reaches its purest form in the *specifications* that are used to document a completed design. This is the means of communicating, in precise terms, with purchasing, manufacturing, quality control, and all the other departments that are directly involved in the execution of the designer's intentions. It increases efficiency in all areas, helps to avoid errors, and should keep costs to a minimum if the specifications are prepared properly. It is the concluding phase of the development process and becomes a record of the results of the work that has gone on before.

When the components of a package are *purchased,* it is essential that as much information be given to the vendor as possible. A clear understanding at the very beginning may prevent costly errors and delays in delivery. Samples should be approved in writing before full-scale production is started. Tolerances for size, color, density, etc., should be agreeable to both sides, and the basis for acceptance or rejection should be spelled out. Limits should be placed on overruns or underruns, and penalties for late delivery, extra inspection, or rework ought to be established before the order is placed.

The best sources of *information* on packaging are the trade shows and the periodicals on the various aspects of packaging. There are annual expositions for materials of packaging as well as separate shows for machinery and equipment. Not only the ones that are devoted exclusively to packaging should be attended; there is much to learn at the pulp and paper exhibitions, as well as chemical engineering, design engineering, point of purchase, and other related shows.

Among the publications, the standard reference in the packaging field is the *Packaging Encyclopedia*,* which is updated each year. A similar reference work is the *Plastics Encyclopedia*,† which contains much information on plastic films, blow molding, plastic closures, foam plastics, and other packaging materials. There are numerous periodicals on packaging in general, as well as specialized magazines on glassware, paperboard, aerosols, and similar subjects. It is recommended that time be set aside for reading these periodicals on a regular basis.

* Encyclopedia issue of *Modern Packaging.*
† Encyclopedia issue of *Modern Plastics.*

Section 2

Folding Cartons and Setup Boxes

Folding Cartons 2-1
History 2-2
Advantages and Disadvantages 2-3
Materials 2-4
Styles of Boxes 2-5
Design 2-5
Processes 2-13
Decoration 2-19
Specifications 2-20
Costs 2-20

Setup Boxes 2-21
Introduction 2-21
Materials 2-21
Types of Construction 2-22
Decoration 2-22
Design Suggestions 2-22
Processes 2-22
Costs 2-25

FOLDING CARTONS

Probably the most popular among the various types of rigid packages is the folding carton. The reasons for such wide use are fairly obvious: it is very economical in cost of material as well as in fabrication and assembly. Being collapsible, it takes a minimum of space in shipment and in storage. The finest kind of printing and embossing can be used to make a very attractive package which will add value and sales appeal to the product. The versatility of these containers is evident in the great variety of sizes and styles, and the special features that are cut into the blank with little or no added cost.

The many functions that any package must serve—containing, protecting, selling, and carrying—are all performed very effectively by the paperboard carton. More than any other form of packaging, the folding box has made possible the modern self-service selling of branded merchandise which has become such a normal part of our way of life.

The starting point for the whole packaging industry is usually given as the introduction of the Uneeda Biscuit container—a folding box.

The characteristic which sets folding cartons apart from other paperboard boxes by definition is that they can be creased and folded to form containers. Generally they are made in such a way that they can be shipped and stored flat, to save space, and then erected at the point where they are to be filled. Since there are some other package styles about which these same statements can be made, such as corrugated and solid fibre shipping containers, we need to be more specific in our description. Folding cartons are usually much smaller than corrugated boxes—normally of a size that can easily be held in one hand. Incidentally the word *carton* is generally reserved for the folding variety that we are discussing, and the term *box* is applied to the larger shipping containers.

Perhaps the best way to describe folding cartons is to define the material from which they are made. The boxboard used for this purpose has a limited range of thickness, usually between 0.014 and 0.032 in., although occasionally it may be a bit heavier than this. Below this range it would be little more than heavy paper and would not have enough stiffness for the purpose. If it were any thicker than the maximum figure given, it could not be creased satisfactorily with standard folding box equipment.

The type of boxboards that are used will vary in composition, from filled sheets made from reprocessed or scrap materials to pure virgin pulp boards of high quality. All of them, however, must have the bending qualities that will allow them to be creased and folded without cracking. These are provided by means of materials added to the pulp when the paperboard is made. The kinds of softwood and hardwood pulps as well as the waxes and resins that are used for sizing and binder will affect this "bender," as it is called. When reprocessed materials are used, it is usually necessary to add some virgin pulp to give the necessary strength for bending. (See Sec. 4, "Paper and Paperboard," for more information on this subject.)

History. The earliest recorded use of paperboard packages goes back to the beginning of the nineteenth century. Carpet tacks were put up in boxes shaped over wooden forms, held together with tacks stuck through the folded ends, and then tied with string. Known familiarly as a "paper of tacks," this was the forerunner of the folding carton as we know it today. Paper boxes that were used in the 1840s required a great amount of hand cutting and gluing, and they were used only for luxury items. A mistake by a pressman in setting up to print some seed bags led to the technique that is now used for cutting and creasing. The faulty adjustment of the press caused the engraving to cut through the paper instead of printing it. To the credit of the Robert Gair Company, the significance of this was recognized, and after considerable experimentation a method was finally developed in 1879 for cutting and scoring boxes by machine instead of by hand. Previously the creasing had been done separately on a platen press, from which the ink rollers and fountains had been removed. The creasing die was made from brass printing rule locked into a form, and the cutting was done on a guillotine-type cutter. Robert Gair's contribution was combining the two operations on one machine. The first major user of this new-style box was the National Biscuit Company, which introduced an improved soda cracker called Uneeda Biscuit, in 1897. To protect the flavor and texture, these new crackers were put into a folding box with a waxed paper liner and overwrapped with printed paper. It was a resounding success, and other manufacturers quickly followed suit. In retrospect it is easy to see how this was so readily accepted: a good name, bold design, convenient packaging, and intelligent advertising—all recognized today as necessary ingredients of a successful marketing effort. But it took real genius in those days, without benefit of agencies and consultants.

By 1923 there were 200 manufacturers of folding cartons, and today the number has grown to about 600 plants, which use 3 million tons of paperboard, with sales of over a billion dollars. About 95 percent of the boxes produced are made to the customer's specifications, and only 5 percent are what might be called stock boxes, in standard sizes and styles for the small user. Table 1 gives a breakdown of the various industries and the relative quantities of folding cartons used.

Advantages and Disadvantages. The features that make folding cartons useful for the packaging engineer are the low costs, good strength properties, and excellent appearance on the retail shelf. In medium to large quantities the folding box has the lowest cost of all rigid packages. In small quantities the setup box can match it, because of the semiautomatic type of equipment that is used. In very large quantities the plastic package may approach it in cost, because the high tooling costs can be spread over a sufficient number of pieces to become

TABLE 1 Uses of Folding Cartons

Use	Percentage
Food products	38
Soap powders	11
Beverage carriers	9
Pharmaceuticals	7
Hardware	5
Textiles	5
Stock boxes	5
Miscellaneous	20

insignificant, but in a majority of cases the folding box offers the most package for the least money. Several boxes can be printed on a single sheet, and cut and creased at the same time. Mixed sizes or different printing can be combined in this way, in what is called a *combination run.* Quantities can be adjusted by the number "up" on the sheet, or by *drop-offs* in which the size of the sheet is reduced partway through the run, to skip the ones on the end. Since it is shipped and stored in collapsed form, the folding carton takes up a minimum of space, in contrast with other types of rigid containers that must be kept erect until used.

On the negative side, the folding carton is rather flimsy in comparison with a setup box or a plastic container. For luxury items it may not be considered elegant enough in appearance or feel, although there is no limit to the eye appeal that can be put into a folding box with good presswork. The strength of a chipboard carton is limited by the manufacturing process, which cannot fabricate stock that is much heavier than about 0.032 in. This allows only up to 2 or 3 lb of product to be packed in a box, with a maximum size of only a few inches on a side, whereas a corrugated or solid fibre box will hold 50 lb or more and can be made 2 or 3 ft on a side or larger.

Mechanical handling of folding boxes is very efficient, and equipment is available to open, fill, and seal at speeds well above 100 per minute. A great variety of sizes and styles can be made, with built-in platforms, windows, extended panels, curved scores, and many other novel treatments.

Materials. The paperboard used for folding cartons is made from furnish that is designed to give enough bender to the stock so that it can be folded without cracking along the score lines. There are different thicknesses and grades of board, depending on the requirements. A more complete description will be found in Sec. 4, "Paper and Paperboard." For food products or where a high grade of board is desired, it is best to use virgin material. This can be a *solid sulfate* or a *sulfite* board, which is a creamy white color all the way through. Although this is a high-priced stock, its strength and stiffness make possible the use of a lighter gauge, around 10 percent less in thickness than a filled board, for equivalent strength. See Table 2 for a comparison of the stiffness of different grades of boxboard.

The most widely used material for folding cartons is *cylinder* board, which is made up of several layers of pulp. The core is usually reprocessed scrap paper, with a layer of pure newspaper on one side and a skin of bleached virgin material on the other side. Better grades also have clay mixed with casein on the surface to give a smoother, whiter printing surface. As the quality goes up, so does the cost. It is false economy, however, to choose a stock strictly on the basis of price,

TABLE 2 Density as a Factor of Stiffness

Grade of boxboard	Lb per caliper point	Stiffness factor
Solid bleached sulfate	4.1–4.4	100
Low-density bleached sulfate	3.7–4.0	88
Ultra low density	3.3–3.6	80

because an inferior board will often increase production costs by causing more machine downtime and a higher scrap rate.

Sheet *plastic* has been used to a limited extent for folding cartons. Cellulose acetate or polyvinyl chloride that has been properly plasticized will make fairly good containers. There is a tendency to crack at the hinges, and the boxes will not stand much abuse in shipment. They have the one big advantage of transparency, and for luxury items they make a very glamorous presentation. A variation uses the plastic for a tube which is then combined with a paperboard tray to make a two-piece slide box. Laminations of plastic or foil with paperboard will provide an attractive printing surface on the outside; for greaseproofness or some other barrier function, the lamination can be turned to the inside.

Styles of Boxes. There are a great many variations in the construction of folding boxes. The most common style is the *reverse-tuck* carton. In this construction, the board is folded on four parallel lines with a small overlap, which is glued down to form a tube. The ends are cut and creased to provide flaps that cover the ends and lock in place to complete the package. A *seal-end* carton is similar except that the ends are glued down after the box is filled. There are many other variations in the end flaps, as shown in Figs. 1 and 2. Special locks are often put on the bottom flaps to keep heavy objects from sliding out of the bottom when the package is picked up. Extra tongues and slots are put on mailing boxes, to ensure that they will not open in transit.

The great number of modifications that can be made to the basic patterns is limited only by the designer's imagination. Cutouts and extensions are used for increasing visibility or for holding objects in position. Perforations and hinges, tabs and slots, easels and sleeves can all be incorporated in the design with little or no added cost. It is this versatility that makes the folding box such a useful packaging medium.

Design. The choice of material and the style of box will depend on the type of product and on the market requirements for that particular item. If the package structure is strictly utilitarian, such as a shelf packer or an industrial or institutional unit, cost is the primary consideration. The problem then becomes one of finding the minimum square area of the lightest gauge that will serve the purpose. The figures in Table 3 can be a starting point. Reference to Figs. 3, 4, and 5 will show

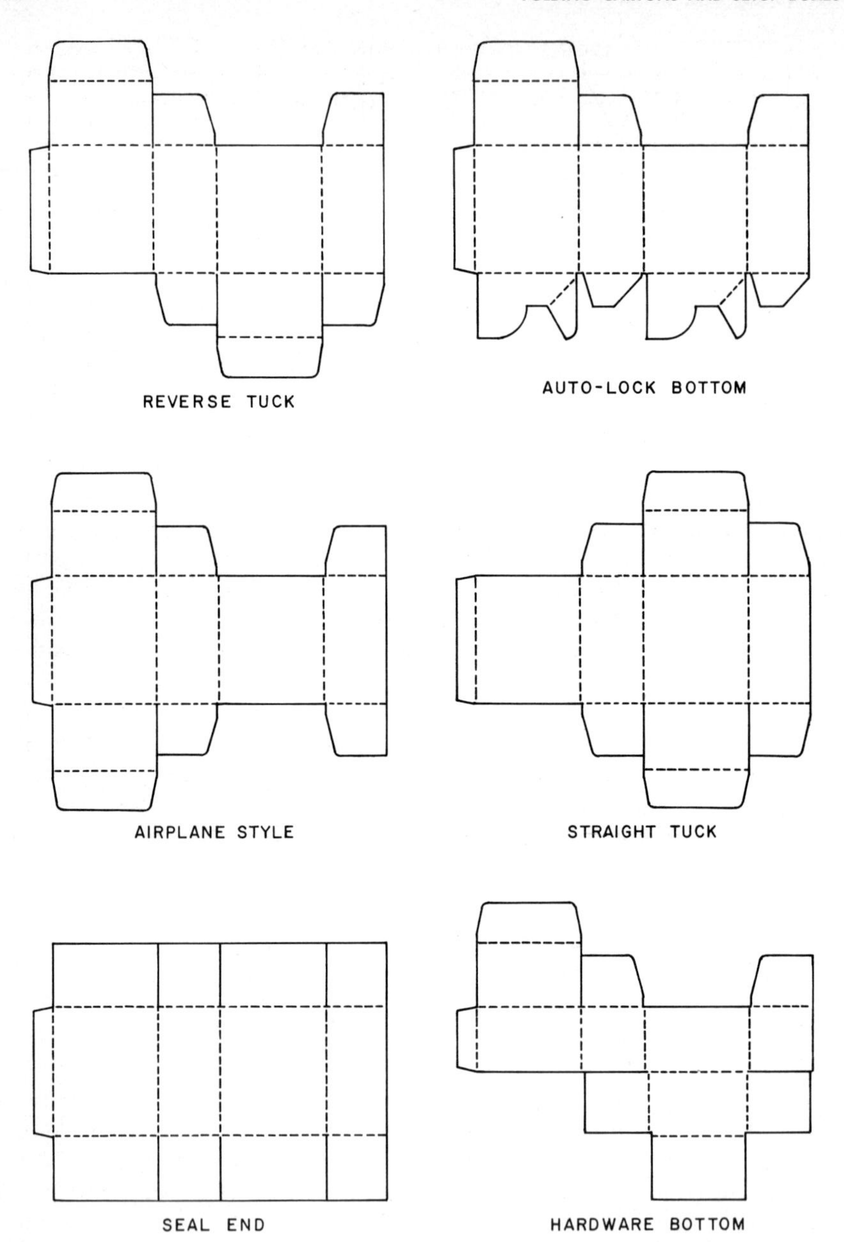

Fig. 1. Styles of folding cartons. These are the flat blanks that have been cut and creased, before they are folded and glued. The glue flap is usually attached to the back panel, so that the raw edge at the opposite end will face to the back.

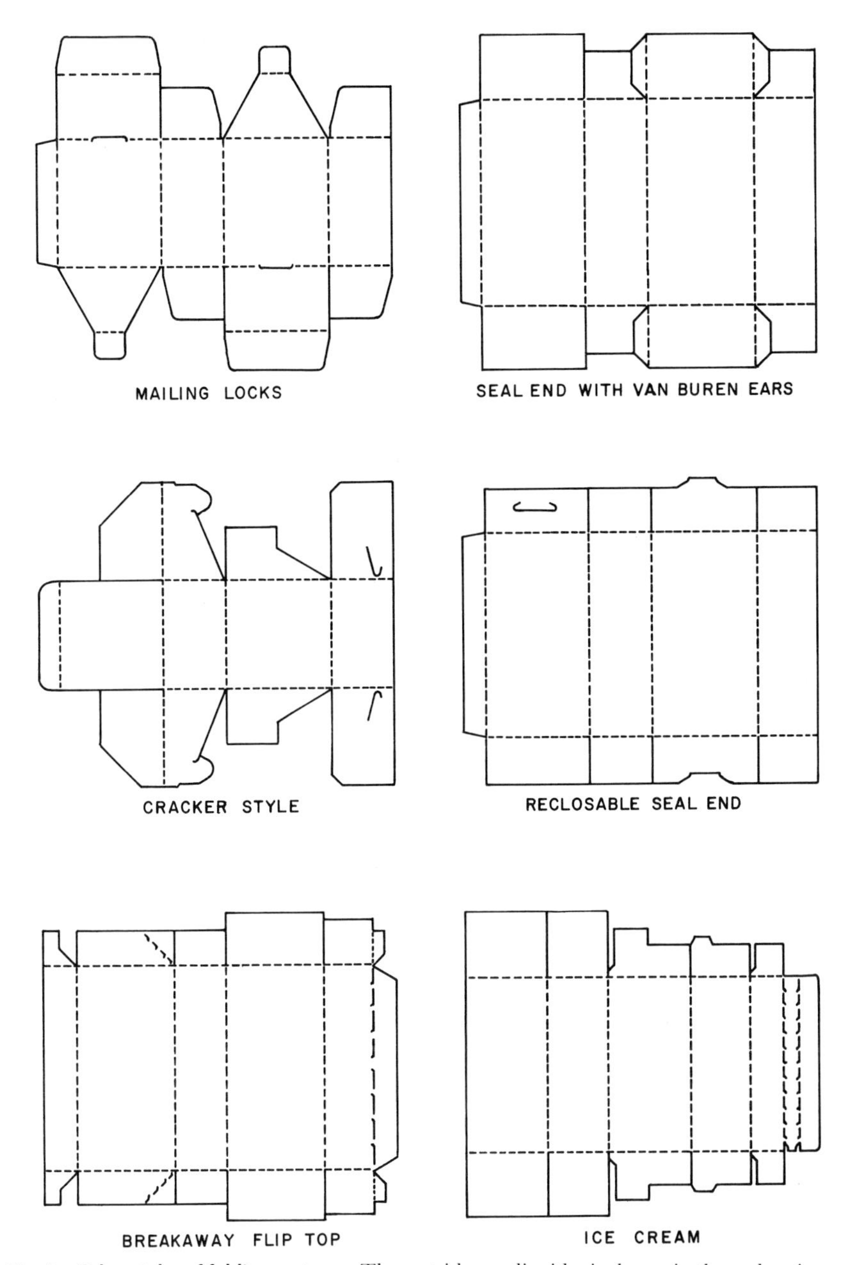

Fig. 2. Other styles of folding cartons. The outside, or die side, is shown in these drawings. There are a great many variations of these basic styles.

TABLE 3 Suggested Thickness of Boxboard for Nonsupporting Contents

Volume of carton, cu in.	Weight of contents, lb	Thickness of board, in.
Up to 20	Up to 1/4	0.018
20 to 40	1/4 to 1/2	0.020
40 to 60	1/2 to 3/4	0.022
60 to 80	3/4 to 1	0.024
80 to 110	1 to 1 1/4	0.026
110 to 150	1 1/4 to 1 1/2	0.028
150 to 200	1 1/2 to 2	0.030
200 to 250	2 to 2 1/2	0.032
250 to 300	2 1/2 to 3 3/4	0.036
300 to 375	3 3/4 to 5	0.040

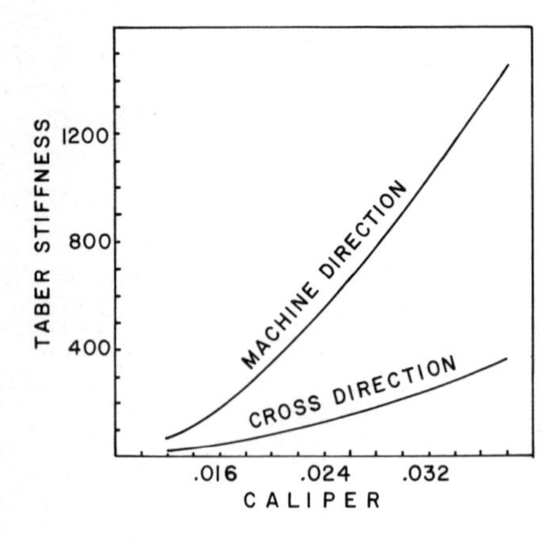

Fig. 3. Rigidity of paperboard. Stiffness is determined by the modulus of elasticity, grain, and cross section of the sheet. Wood pulp has a modulus of about 500,000 psi in the individual fibres. With most of the fibres pointed in the machine direction, the stiffness is greatest in that direction. As the caliper of the sheet increases, the moment of inertia increases in proportion to the square of the distance from the center of the sheet to the outside surface. The outer plies therefore have the greatest effect on the stiffness.

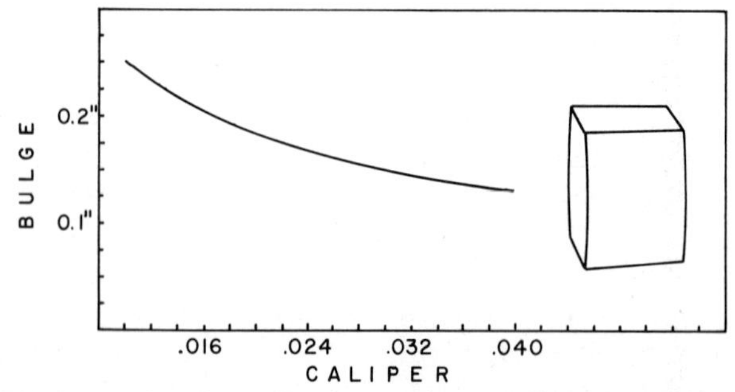

Fig. 4. Bulge factors. Large cartons of granular products will bulge from the internal pressure of the contents. The area of the panel is the most important factor. As the chart shows, increasing the thickness of the paperboard has little effect on reducing the bulge.

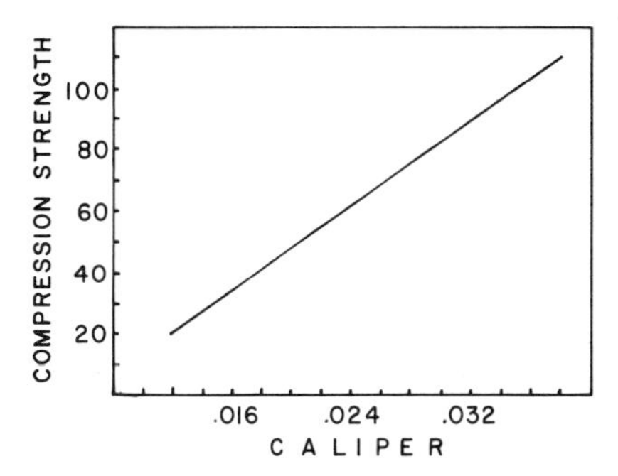

Fig. 5. Stacking strength. Most of the forces exerted on a carton in shipment are compressive forces. The chart shows how the strength of a carton increases with increasing caliper. This is based on the maximum top-to-bottom load in pounds that could be supported by a carton 8¼ by 3 by 11 in.

how the strength increases or decreases as the caliper changes. After the specifications have been worked out, tests should always be made to prove out the design. Drop tests and vibration tests are particularly important; they should be as realistic as possible, with the actual contents in place and full quantities packed in the intended shipping container.

If the folding carton will be required to do also a merchandising job, then other factors will have to be considered. The method of display is the first thing to take into account. On the dealer's shelf the package must be stable, so that it does not fall over easily. It should present a good face when it is stacked in mass display. It should be proportioned to give an impression of good value, but it should not be deceptive. Figure 6 shows how to find the maximum size of a carton for a collapsible tube. The size should fit between the standard store shelves. If it is likely to be located on a pegboard rack, then a hang tab or extended panel with a suitable hole will be needed.

For the convenience of the consumer it should be easy to pick up and hold. A heavy box can have a handle built in. A window to show the contents or a good illustration of the product may help to clinch a sale. To reduce pilferage, it may be necessary to enlarge the carton so that it cannot be palmed or hidden in another package. For the same reason it is best to glue the flaps of a box to deter the customer from returning the empty box to the shelf.

The strength of the carton, quality of board, and presswork all contribute to a good appearance on the shelf. It is poor economy to use inferior materials that quickly become shopworn. A comparison of the properties of the two main categories of boxboard is given in Table 4. Weather conditions can affect the strength of a carton; Fig. 7 shows the influence of humidity on the stiffness of boxboard.

Manufacturing processes are the next thing to consider when designing a folding carton. Even if the early production is expected to be a hand or semiautomatic operation, it is well to look ahead and plan for fully automated equipment. A snap-lock bottom, for instance, is

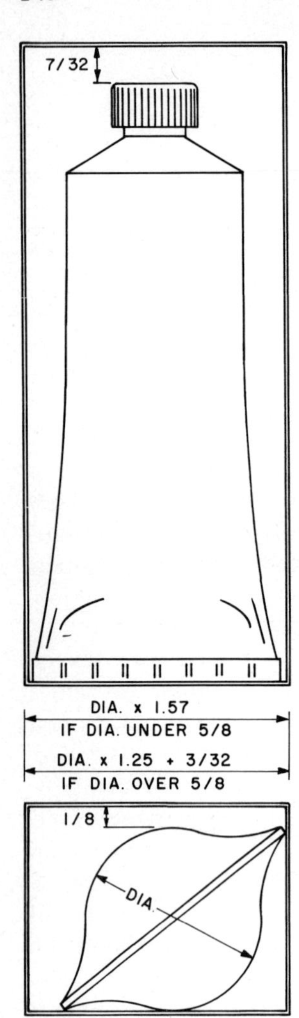

Fig. 6. Collapsible tube carton. Method of determining the maximum size of a carton for a tube, from the Bristol-Lund formula. This usually meets the requirements of the regulations pertaining to deceptive packaging.

TABLE 4 Comparison of Virgin Stock with Filled Board

Property	Solid Bleached Sulfate	Reprocessed Filler
Bulge	Bulges when open owing to less stiffness in the machine direction.	Stiffness is higher in the machine direction.
Scores	Scores hold well.	Scores reset with age.
Creep	Slow rate of creep. Less fatigue under compression.	High rate of creep. More fatigue under compression.
Gluing.	Expensive resin glue required.	Low-cost dextrin glue can be used.
Uniformity . .	Will machine more consistently because of the better-grade material.	Less uniform results due to waste material content

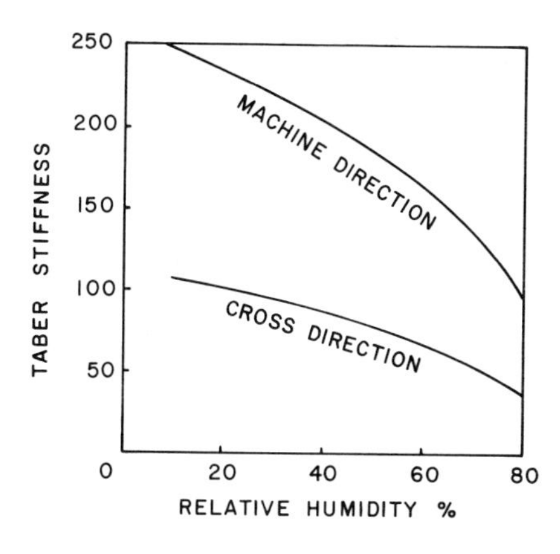

Fig. 7. Effect of humidity on stiffness of boxboard. Weather and storage conditions can reduce the stiffness as much as one-half.

ideal for hand loading but is impossible to close by machine. Instruction sheets, platforms, and filler pieces should be made so that conversion to a more sophisticated operation will not require new cutting dies and printing plates.

Cost is always an important consideration, and the choice of the design to be used and the type of material will largely control the final cost of the finished goods. In general any improvement in the properties of the board stock, or in the surface appearance, will increase the cost. It is in this area that the packaging engineer can perform a service to the marketing department by helping them to reach decisions. There are no hard-and-fast rules in this kind of work, and the selection of the final design will require a great many compromises. Reference to Table 4 will help in choosing between a filled board and one made from all virgin material.

At the outset of a specific project, the inside dimensions should be arbitrarily chosen; they may be determined by the size of the object to be contained. A small amount of clearance will be required, and usually $^1/_{32}$ to $^1/_{16}$ in. is added to each of the dimensions of the article that is cartoned. Thickness of the stock is the next choice to be made; 0.018 and 0.020 in. are the most commonly used gauges. A handmade sample is best for checking out a design at this early stage. To prepare such a sample, mark the score lines and cuts on the back of a sheet of board with pencil. The grain of the board is nearly always made horizontal; that is, it runs around the box perpendicular to the main score lines. This avoids bulging along the top edge and helps the main panels to lie flat. (See Fig. 8.)

To complete the sample, the outline is cut with scissors or a knife, and the scores are made by laying the blank facedown on a long piece of scoring rule, which can be obtained from your box supplier. This steel rule is mounted in a block of wood with the rounded edge up. A forked

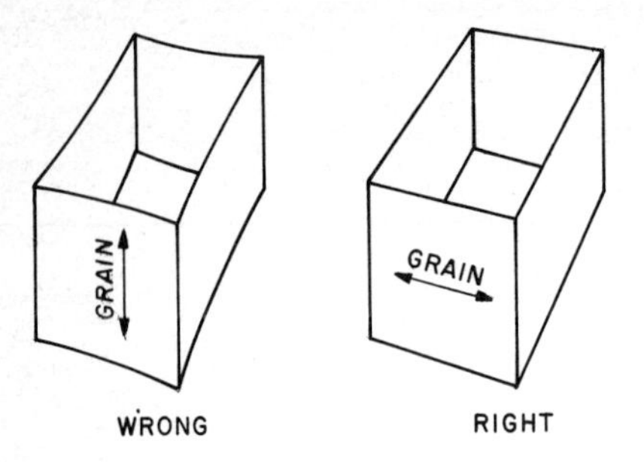

Fig. 8. Grain of board. Usually runs around the box, and perpendicular to the main scores. This provides more stiffness across the span from one score line to the other. It permits the folded edges to provide the stiffness in the opposite direction, to make up for the lack of rigidity with the grain.

stick is pressed down on the blank and pulled along the pencil lines, directly above the scoring rule. The stick is made from a piece of hardwood with a groove 1/16 in. wide and about 1/32 in. deep. This forms a bead along the score lines that breaks the rigidity of the boxboard at this point and permits it to be folded accurately. (See Fig. 9.)

If it is to be glued, the blank is folded and the adhesive applied, and a weight is placed on the carton in the collapsed position until the glue sets. If the cutting and scoring have been done carefully, the panels should line up square and true, and tucks and flaps should slide into place with just the right amount of friction. Very seldom is it necessary to trim or rescore for a good fit. This sample can then be used as a check against the dimensions that were originally chosen, to see whether it holds the contents as intended. Typical dimensions are given in Fig. 10.

For a carton that must work in automatic machinery, the next step would be to have the supplier make a cutting die. At least 500 samples should be made, and glued on a production gluer, for testing on the cartoning equipment. These do not need to be printed, but they should be made of the proper stock. If this test proves satisfactory, a die rub-off or a strike sheet should be made from the die and used to prepare the artwork. Using this kind of die sheet for layout purposes will ensure that the printing is properly spaced in each panel of the carton. The proper nomenclature in describing the various parts of a folding carton is shown in Fig. 11.

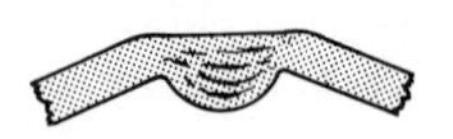

Fig. 9. Score lines. Paperboard is folded away from the score. The bead that is formed by the female groove in the cutting and creasing press breaks down the bond between the plies of paper. As the sheet is folded, the board delaminates at this point and bulges out into a bead that relieves the stress on the outside layer. Otherwise the outer surface would rupture.

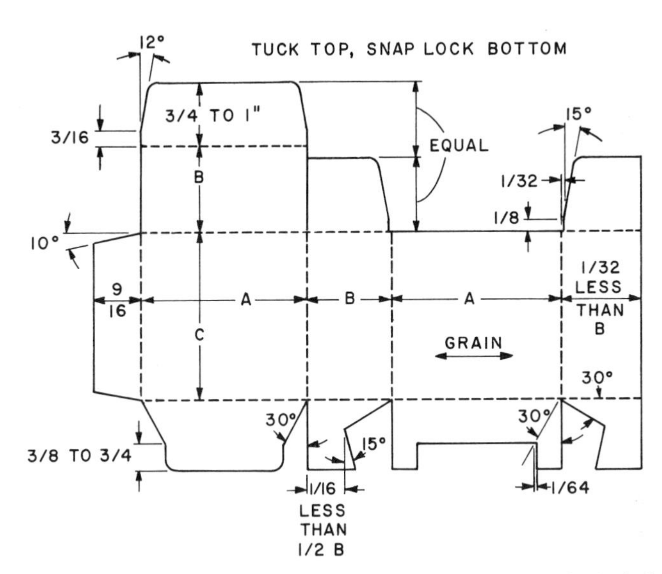

Fig. 10. Carton blank. Typical folding carton showing some of the principal dimensions. These are not standardized, and there are many variations within the industry.

Processes. In the carton plant a cutting and creasing die will be prepared for making several cartons at a time, depending on the size of the item and the quantity required. It may be a two-up die or twenty-up, as the case may be. For purposes of explanation, the cutting and creasing *die* can be likened to a large cookie cutter. It is made up of

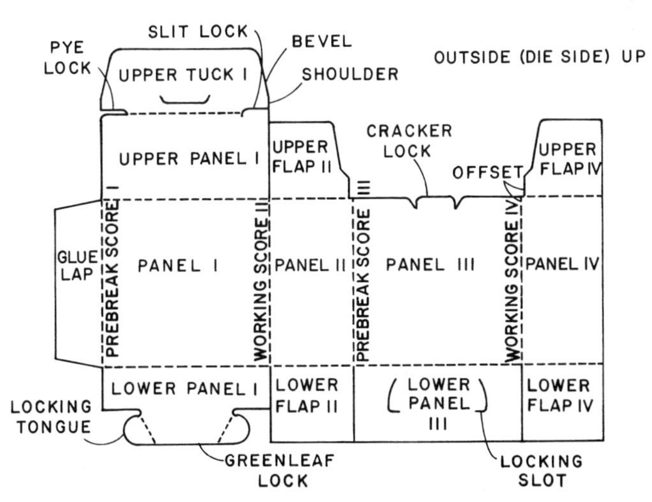

Fig. 11. Nomenclature. Names of the various parts of a folding carton. The correct use of the proper terminology will help to avoid errors in communication between purchaser and supplier.

wood blocks cut from ¾-in. plywood, birch, or cherry, with steel *rule* set on edge between and projecting $3/16$ in. above the blocks, held together inside a steel frame or *chase* by wooden wedges or *quoins*. (See Fig. 12.) Small blocks of sponge rubber or synthetic *cork* are placed on each side of the cutting rules, and especially where the rules join, to push the board free of these knives as soon as it is cut. None is required for the creasing rules. The cork is ¼ in. thick and is cut into ⅜-in.-wide strips. Pieces about ½ in. long are placed ½ in. apart and glued to the wood blocks.

Such a die can make up to 500,000 impressions before requiring reknifing. The layout of the die is made in such a way as to get the most economical use of board, and the different positions are butted and interlocked to keep waste to a minimum. If several items are run in *combination* on the same sheet, the number of positions for each one will be proportional to the order quantities. Sometimes the quantities will be adjusted by *dropping off* the end cartons; that is, a smaller sheet will be used partway through the run so that it does not include the end positions.

A number of printing plates, or *electros*, are made from the original engravings. These are mounted in the printing press to match the cutting die, which has been set up in the cutting and creasing press. Although the sheets will be printed before they are cut and creased, it is necessary to make a temporary setup of the cutting and creasing press

Fig. 12. Locking up the die. Pieces of cutting and creasing rule are placed between wood blocks inside a chase. Wedge-shaped quoins are driven between the furniture to lock up the die, prior to corking. (*Diemaking Diecutting and Converting*)

first. Cutting and creasing presses range in size from Thompson platen presses under 30 by 40 in. with an output of 1,000 sheets per hour, up to the 72-in. Miehle cylinder machines running at 2,400 sheets per hour. An *oil sheet* is used to get an impression of the cutting and creasing die by laying on carbon paper and rubbing against the die with a piece of wood. (See Fig. 13.) This sheet is taken to the printing press and used to guide the pressman in placing the printing plates in the proper location on the bed of the press. Allowance is made for *draw* in the direction around the cylinder; each row of scores pulls the sheet a small amount, which can accumulate to as much as four 2-point *leads* of space. A few blank *set* sheets are run through the cutting and creasing press, and put through the printing press to see whether the printing plates are in proper register with the cutting and creasing die, as well as with each other. Adjustments are made, if necessary, until everything is in alignment.

The printing press is then *made ready;* that is, the platen is built up wherever the impression is not heavy enough, by pasting bits of paper on the cylinder around which the paperboard is carried when it is being printed, until the impression of the ink on the paperboard is uniform, Solid colors and dark areas of halftones also are built up so that the light areas will not print too heavily. The ink fountains are then adjusted to get an even film of ink on the plates. All this may take from several hours to several days, depending on the complexity of the job. The stock is brought to the press on wood skids, about 1 ton of board on each. The steel straps and wrappings are removed, and the sheets are

Fig. 13. Making a die rub-off. Carbon paper is placed faceup on the die and covered with an oil sheet. A piece of wood is rubbed over the oil sheet to mark the impression of the die. This will be used at the printing press to locate the printing plates on the bed of the press. (*Diemaking Diecutting and Converting*)

then ready to be printed. All the sheets for the entire job are run on the printing press and then allowed to dry for a day or two before being cut and creased.

The cutting and creasing press is made ready in a similar manner, except that the objective is to get the knives to just cut through the paperboard without biting into the steel *jacket.* The pieces of gummed paper that are used to compensate for a light impression in spots are pasted on an *overlay* of 0.012-in. manila paper that is placed under the thin steel jacket. When this part of the job is completed, a piece of *counter* or *tympan* is glued on top of the jacket. This is a sheet of very dense paperboard a little less in thickness than the stock to be used for the cartons. After the glue has dried, the tympan is cut away to form a groove for the creasing rule in the die. The width of the groove should be equal to two thicknesses of carton stock plus the thickness of the rule. For example, if 0.022-in. paperboard is being creased with 2-point (0.028-in.) rule, the groove would be 0.072 in. in width. The tympan also is cut away completely in the area of the cutting rule. (See Fig. 14.)

The height of the *cutting rule* is typically 0.937 in., and the *creasing rule* 0.921 in. across the cylinder and a little less around the cylinder. This means that the creasing rule comes even with the top of the tympan, forcing the carton stock into the groove an amount equal to the stock thickness, while the cutting rule goes all the way to the bottom and just kisses the steel jacket. Cut scores are made with rule of a height that will cut about 75 percent through the board against the steel jacket, without any tympan in that section. The cutting knives are nicked with a very narrow grinding wheel in various places, so that the sheet will hold together after it leaves the cutting and creasing press, before it is stripped.

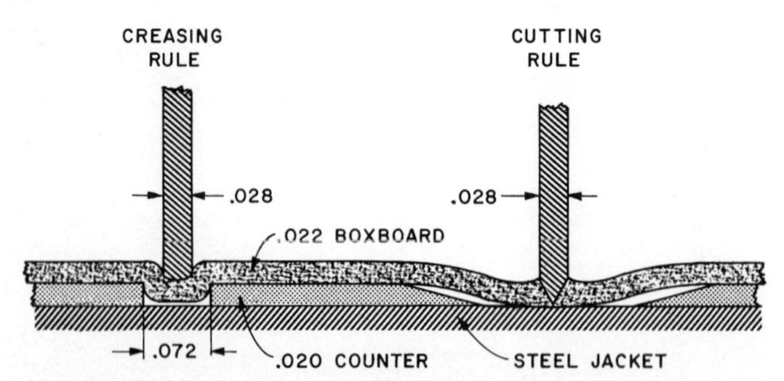

Fig. 14. Creasing and cutting boxboard with a steel rule die. The counter is cut away to form a groove for the creasing rule that is equal to the thickness of the rule, plus twice the thickness of the boxboard. The counter on each side of the cutting rule is skived to avoid marking the boxboard. Cork strips on each side of the cutting rule, to strip the stock away after cutting, are not shown.

Stacks of cut sheets are stripped by knocking off the trim around the edges and between the cartons with a mallet or, for large orders, an air hammer. The stacks are usually 1 to 3 ft high, although for intricate work or very small cartons they may be handled in lifts of only 100 sheets. The stripper first hits the stack in several places to lock the sheets together. Then he pulls on the outside trim while pounding it with his hammer to separate it from the carton blanks. Working all the way around the stack completely removes the trim, which is thrown into a scrap box. Next the individual lifts of blanks are separated from each other, and the scrap between them is knocked off. They are carefully stacked on skids ready for the gluer, with the different items of a combination run kept apart to avoid mix-up. If gluing is not required, they are packed directly into shipping cases.

Carton blanks are then put facedown into the magazine of the gluer. A feed wheel slides the blanks off the bottom, one at a time, where they are carried between belts moving at 1,200 ft per min, into the prebreak station. Here the first and third score lines are prebroken by stationary curved bars that plow up the end panels and then lay them back down again as the blank moves along on the belts. (See Table 5.) Next the glue flap passes over a glue wheel, which is about 3/16 in. wide, and receives a stripe of glue that is thick enough to spread out to the proper width when it is compressed further on.

The belts carry the blank through a folding section where guides or "swords" fold the end panels up on the second and fourth scores and bring them over on top of each other. They are then carried by the belts into the nip of a pair of wider belts that apply pressure as the cartons are carried along until the glue sets. These wide belts move much slower so that the cartons overlap in shingle fashion. When they leave the compression section, they are transferred to a delivery apron and counted. Every fiftieth carton is kicked out of line about 3/4 in. by a

TABLE 5 Effect of Different Degrees of Prebreaking of Carton Scores during Gluing

Prebreak angle	Bending resistance*
None	360
90°	280
135°	160
165°	130
180°	110

* Tests made with 0.016-in. patent-coated newsback stock on PCA Score Bend Tester.

timer mechanism so that the packer can put the exact quantities into each shipping case.

A *right-angle gluer* is more complex than the straight-line gluer just described. It also runs much slower. For some special types of cartons such as the diagonal-fold or the automatic lock-bottom styles, the glue must be applied in particular spots rather than as a stripe, and the folds must be made with moving arms and special tuckers. The right-angle machine uses timing chains instead of belts to convey the blanks through the various operations. Lugs on the chains push the blank so that it arrives at the right time and place for each operation. The glue wheel has a piece of cork cut to the right shape to print a spot of glue just where it is required.

After the blank is glued and folded, it is picked off the conveyor chain by another chain moving at right angles, and then transferred to belts. The second gluing and folding are similar to the first, modified as necessary to suit the style of the carton. Compression, delivery, and counting are much the same as with the straight-line gluer.

Cartons should be packed in sturdy shipping cases, with about 15 percent extra space to preserve the prebreak. If they are packed too tightly, the opposite panels will be pressed flat against each other; they should bow out slightly to make it easier to open for filling. Small cartons are best packed in trays, but large boxes may be simply separated with sheets of paperboard. The cartons must all face one way, especially if they are to be set up by automatic machinery. The shipping cases are usually made of corrugated boxboard and should be strong enough to withstand stacking 5 or 6 ft high without buckling, which might distort the cartons inside and cause malfunctioning of the cartoner.

A brief description of a typical erecting and filling operation on a *cartoner* will show the requirements of the user of folding boxes. There is a great variety of machinery: some very sophisticated, high-speed, fully automatic equipment, and some versatile, semiautomatic, moderate-capacity machines. The most common is probably the cartoner for tuck-end boxes, which we will use for our example. The collapsed cartons will be put into the magazine just as they come from the shipping case. They will be pulled off the bottom, one at a time, by vacuum cups attached to a swinging arm, and deposited in a pocket of the conveyor chain. At the same time a pusher bar will apply pressure to the trailing folded edge, forcing the leading edge against the bottom corner of the pocket of the conveyor chain. The combination of the suction cups pulling out at the center and the pressure on the folded edges causes the carton to open up ready for filling and closing.

As it moves along, the contents is pushed into the carton either by hand or from buckets in a chain traveling alongside the carton conveyor.

The dust flaps on the leading edge are wiped in by stationary bars as the carton moves along, but the trailing flaps must be folded forward by revolving fingers that move faster than the carton is traveling. Then the tucks are folded in by stationary curved bars that bend the score lines and guide the tucks into their slots. Since the cartons will be moving through the machine at more than 100 per minute, it is important that the design of tucks and flaps be correct and that the workmanship be uniformly good; otherwise the cartons will jam and cause a high percentage of scrap. An occasional jam is expected, but it should not occur more often than two or three times in an hour.

Decoration. Cartons can be printed, embossed, hot-stamped, silk-screened, or decorated with various coatings and attachments. The great majority are simply printed by letterpress, offset, or gravure processes. If a good grade of folding boxboard is used, a very high quality of reproduction is possible with halftones up to 120 screen and excellent *process* work. *Letterpress* gives sharpness and uniformity, *offset* is recommended for halftone work, and *gravure* is preferred for long runs that are not too complicated. Table 6 gives a comparison of these three methods.

Most carton printing is done on two-color letterpress equipment, and if more than two colors are required, the stock is put through twice. This gives the ink a chance to dry before the second pass, and it makes trapping of colors easier. It can cause problems in registration, however, if the sheets pick up moisture or dry out while the inks are drying; this would change the dimensions of the sheet. Additional colors can be done at the same time in limited areas across the sheet by splitting the ink fountains, that is, putting two or more colors in the same trough with temporary partitions to keep the inks separated. To provide an additional color without printing, clay coatings can be tinted for orders of 2 tons or more of board. Typical presses are Miller 28 by 41 in., running 3,000 impressions per hour, and Miehle 72 in., with an output of 2,000 sheets per hour. There are more sophisticated presses for large-volume or highly specialized work; for example, soap powder

TABLE 6 Printing Processes Compared

	Gravure	Offset	Letterpress
Plate costs	High	Low	Low
Makeready costs .	Medium	Low	High
Solid colors	Excellent	Fair	Excellent
Vignettes	Good	Excellent	Poor
Color control . . .	Excellent	Fair	Good
Varnish gloss . . .	Good	Fair	Excellent

cartons are sometimes run on a printing, cutting, creasing, and stripping machine at 800 a minute. This type of equipment is feasible only for orders of 100 million or more a year.

Artwork is prepared in the customer's art department or in an outside studio. Some carton companies will help with this part of the work. After *comprehensives* in full color are made by hand and approved, the final *mechanicals* are made in black and white. It is best if this is done on a *strike sheet* made from the cutting and creasing die, to ensure that the artwork is centered properly in each panel. The line work is made with black ink, and the proofs from the typesetters are cut and pasted in place. The paste-up is sent to the carton manufacturer with color swatches and a tissue overlay, marked to show where each color is to be used.

Specifications. The essential points to be covered in the specifications for folding cartons are (1) dimensions, (2) stock, and (3) presswork. The *dimensions* are always given in the order of length, width, and depth, measured from center of score to center of score. The first dimension is along the hinge of the cover, and is usually larger than the second figure, which is from the hinge to the tuck, for a tuck-end carton. The last dimension is parallel to the glue lap. If the style can be described in words, such as "tuck top, snap-lock bottom," no *drawing* is required. However, for intricate designs or for machine filling it is safer to make a detailed diagram.

The caliper of the *stock*, density or basis weight, coating, and brightness should all be spelled out. Both front and back should be designated; for example, white machine clay-coated, news back, bending boxboard.

If a particular method of *printing* is desired, it should be indicated; otherwise letterpress printing will probably be used. All colors and tints and shades should be identified by color swatches or a numbering system, along with the allowable deviation from the standard. If gloss ink, or spot or overall varnish is required, this too must be written into the specifications.

The method of sampling and the *acceptance quality levels* (AQL) should be shown; for example, critical defects 0.4 percent, major defect 1.6 percent, and minor defects 5.0 percent. See Sec. 19, "Quality Control," for details of statistical sampling and testing by attributes.

Costs. It is very difficult to estimate the cost of folding cartons, particularly if they are to be run in combination. The order quantity is a major factor in estimating prices, because there is such a high cost in setting up the printing press and the cutting and creasing press. In long runs these expenses wash out, but for small orders they are very significant.

TABLE 7 Comparative Costs of Different Grades of Boxboard*

Stock	Cost per thousand boxes
Machine clay-coated white back	$ 7.65
Solid bleached sulfate	8.10
Brush clay-coated	8.15
Brush clay-coated 75 brightness	8.70
Cast-coated 90 brightness	9.60
Gloss-laminated solid sulfate	10.10

* Based on a reverse-tuck carton 1 by 2 by 3 in. in 500,000 quantities.

In a typical carton plant, labor represents 10 percent of the selling price, while material is about 45 percent of the total. Table 7 shows the difference in cost for various types of paperboard. Many plants operate with a gross profit of 16 to 18 percent, of which about half is paid in taxes.

SETUP BOXES

Introduction. The setup box is similar in many ways to the folding carton described on the previous pages, but there are some major differences, both in the methods of manufacture and in the way it is used. By definition it is made in the shape in which it will be used, and it cannot be collapsed for shipment. This is in contrast to the folding carton, which is shipped flat and erected at the point of filling.

The manufacturing plants for setup boxes tend to be smaller and less sophisticated than the folding carton plants, but they have more versatility, which is one of the features of this type of packaging. The processes for setup boxes do not require expensive dies or complicated machinery. Thus they can easily accommodate small runs and unusual constructions. Costs are considerably higher than for folding cartons of equal size. However, there are some extra benefits in rigidity and appearance which cannot be matched with a folding carton. For these reasons the setup box is most often used for luxury goods.

Materials. Paperboard of the nonbending variety forms the base of the setup box. This runs from 0.016 to 0.062 in. in caliper, or as it is sometimes designated, from No. 35 to No. 120 board, according to the number of sheets in a 50-lb bundle. Most of it is in the 0.040-in. to 0.050-in. range. It is frequently lined or laminated with white paper, which is turned toward the inside of the box. The corners are usually reinforced with 30-lb gummed paper tape, known as stay paper.

The outside is typically covered with a 60-lb coated litho paper that has been printed or decorated beforehand. Other types of cover paper may be used, and there is an endless variety of foils, embossed or flocked papers, and fabrics that can be used.

Types of Construction. The simplest style of box is the telescope type, in which the lid is slightly larger than the base and fits down over it, either partly or full depth. Other variations include an extension edge on the bottom or a hinge on the cover, padding on the top surface, and many other adaptations. A few of these are illustrated in Fig. 15. There is almost no limit to the variety of drawers, trays, and platforms that can be included for special purposes.

Decoration. The cover paper is usually printed in the flat, and then wrapped around the box. Any graphic process may be used for the printing, such as letterpress, offset, hot stamping, embossing, or silk-screen decorating. Attachments can be added in the form of hinges, latches, medallions, ribbons, or cords to make the package more attractive as well as functional. The processes that are used in making setup boxes lend themselves to such innovations, and the designer has ample opportunity for developing unique and interesting creations.

Design Suggestions. In working out the design for a setup box, there are several things to keep in mind. A printed border close to the edge is not recommended, as it is difficult to keep parallel with the edge of the box and to maintain an equal distance all around. For the same reason a line or color separation should not occur right at the edge. Horizontal lines running around the four sides will seldom meet exactly at the corners. It is best to get a layout sheet from the box maker before making the final black-and-white paste-up, to be sure the spacing is right.

Bleed should be about 1/8 in. beyond the trim line, and at least 1/4 in. where the turn-in goes inside the cover or base. Offset printing should be specified for most purposes, but letterpress printing also can be used.

Processes. The first step in manufacturing setup boxes is to score the blanks for the lid and base. A sheet of boxboard about 24 × 36 in. is put through a scoring machine, then turned and put through another similar machine for scoring in the opposite direction. The scores are made halfway through the board, but the outside edges are cut completely through, and several lids are usually cut at the same time from one sheet.

A stack of lids or bases is then put into a corner cutter, where one corner of the stack is notched out at a time. The stack is turned and put back into the machine to notch each subsequent corner. Single blanks are then stayed in another machine, which bends up the flanges on the score line and puts a small strip of gummed stay paper around the

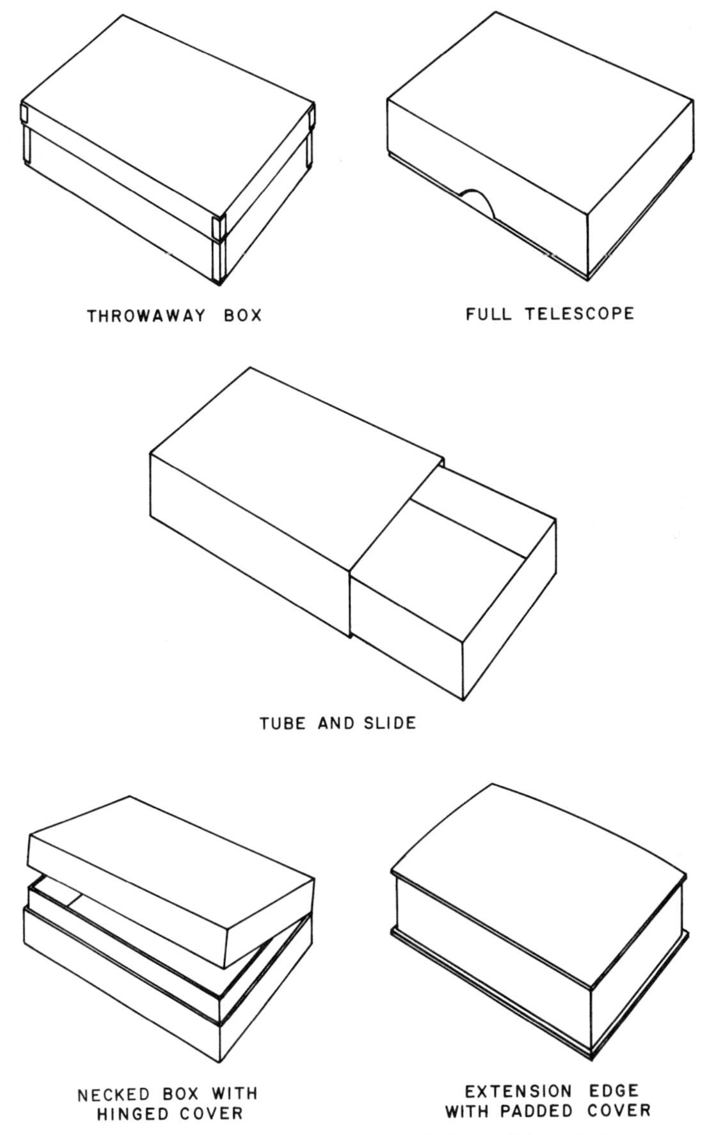

Fig. 15. Styles of setup boxes. There are many variations of these basic types.

corner to hold them in place. Reference to Fig. 16 will show how these parts go together.

The cover paper may be made in several pieces and applied to the box in separate operations; for example, the sides could be covered by a strip running all the way around, like the stripped neck in Fig. 16. This

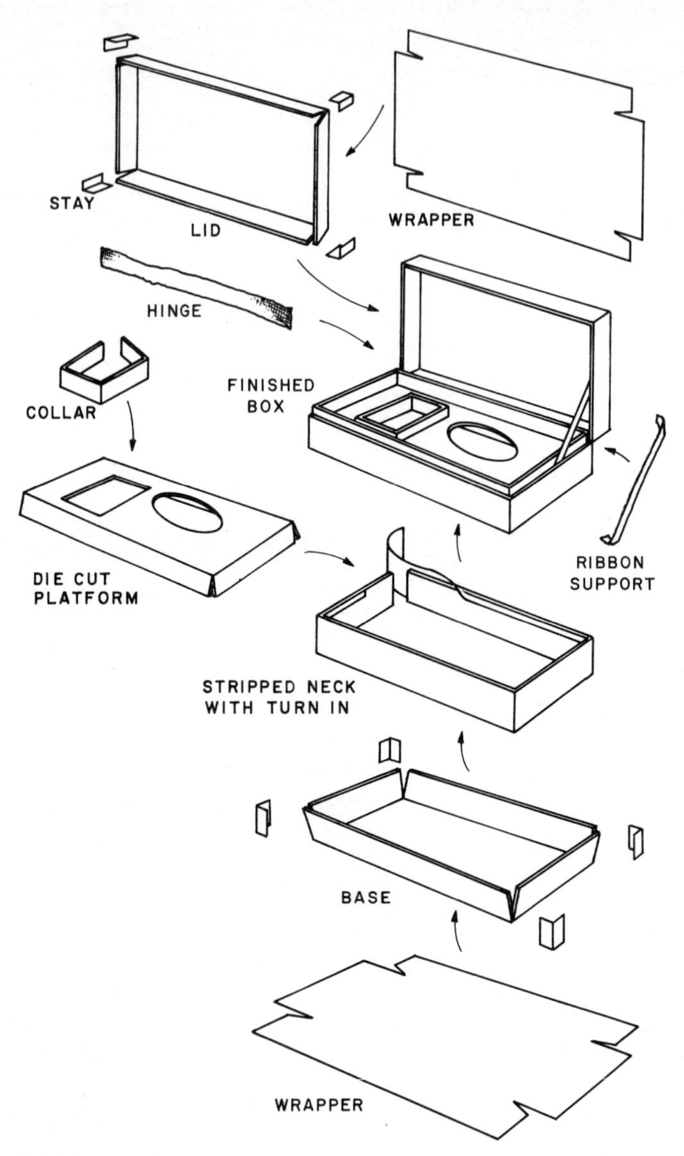

Fig. 16. Construction of a setup box. The various parts that make up the finished package are shown separately and again in the finished box.

would extend for a short distance onto the top, if this were a lid. A label would then be glued over the top, and run just to the corners of the lid, covering the edges of the side strip where they extend onto the top. A different method is to cover the top and sides with one piece, as shown for the lid and base in Fig. 16.

Much of this is handwork if the quantities are small, but it can be automated if there is sufficient volume to justify the cost of tooling up. The hinged cigarette boxes that are used for premium brands, for example, are made on high-speed machines that are completely automatic.

Costs. Setup box costs are not influenced as much by quantity as are some other package forms. In small quantities the costs are somewhat higher, but they compare very favorably with folding cartons in small quantities because of the high makeready costs. As the volume goes up, the cost of the setup box goes down only a small amount, whereas the folding carton becomes significantly cheaper.

For a rough comparison a hinged cigarette box would cost about 2 cents, a candy box about 10 cents, a cigar box about 20 cents, and a cosmetic gift box about 30 to 50 cents. Any deviation from a rectangular shape may double the cost, because of the increase in hand operations. A ribbon support costs about 7 cents extra, and a hinge adds around 5 cents. Platforms and fillers cost 2 to 10 cents, depending on the size and complexity. The box shown in Fig. 16 would probably cost about $1 in fairly large quantities.

Section **3**

Films and Foils

Common Packaging Films 3-1
Shrink Films 3-5
Cellophane 3-9
Cellulose Acetate 3-16
Ethyl Cellulose 3-17
Methyl Cellulose 3-19
Nylon 3-19
Polycarbonate 3-21
Polyester 3-23
Polyethylene 3-25
Polypropylene 3-34
Polyvinyl Chloride (PVC) 3-37
Rubber Hydrochloride 3-40
Saran (PVDC) 3-42
Styrene 3-44
Specialty Films 3-46
Cellulose Acetate-Butyrate 3-47
Cellulose Nitrate 3-47
Cellulose Triacetate 3-47
Chlorotrifluoroethylene 3-47
Edible Film 3-47
Ethylene Butene 3-47
EVA Copolymer 3-47
Fluorocarbon 3-47
Fluorohalocarbon 3-47
H-film 3-48
Ionomers 3-48
Methyl Methacrylate 3-48
Noryl 3-48
Parylene 3-48
Polyallomers 3-48
Polyethylene Oxide 3-48
Polyphenylene Oxide 3-49
Polysulfone 3-49
Polyurethane 3-49
Polyvinyl Alcohol 3-49
Polyvinyl Fluoride 3-50
Silicone 3-50
Identification Tests 3-50
Metal Foil 3-53
History 3-53
Alloys 3-54
Temper of Foil 3-55
Characteristics 3-55
Advantages and Disadvantages ... 3-56
Properties 3-56
Manufacturing 3-56
Adhesives 3-57
Coatings 3-58
Decoration 3-59
Primers 3-59
Laminations 3-59
Forming 3-59
FDA Approval 3-60
Costs 3-60
Suppliers 3-60

COMMON PACKAGING FILMS

About 20 percent of all the plastics manufactured in this country go into packages, mostly as films and coatings. We should start our discussion of these materials by differentiating between the terms "film" and "sheet." A plastic material up to 0.010 in. in thickness is considered a *film*, but above this thickness it is usually called *sheet*. When two or more discrete films are combined with an adhesive, they are called a *laminate*, but if they are extruded at the same time and combined, they are termed a *composite* film.

In this country we manufacture 2½ billion lb of film and sheeting per year, worth $1½ billion for materials and another $3 billion for processing and converting. About 85 percent of this is for film, and the other 15 percent is for sheet stock. There are about 300 converting plants which take the base films and decorate them. There are a dozen or so manufacturers of flexographic and gravure presses for printing on films, and nearly 200 companies are laminating or fabricating packaging materials from transparent film. The most rapid growth in the film industry is taking place in polyethylene, which is taking over some of the markets that were formerly dominated by cellophane; and yet the market for cellophane continues to grow.

The thickness of a film is usually specified in mils (a mil is 0.001 in.), although gauge numbers are sometimes used; thus 200-gauge film indicates a thickness of 2 mils. Cellophane weights are commonly designated by yield numbers; for example, a No. 195 film will provide 19,500 sq in. per lb. The various processes used to produce film are (1) knife coating, (2) reverse roll coating, (3) calender coating, (4) extrusion

TABLE 1 Film Yields

Type of film	Density, g/cc	Yield, sq in. per lb, 1 mil	Cost per lb	Cost, 1,000 sq in., 1 mil
Aluminum foil	2.70	10,250	$0.64	$0.055
Cellophane	1.50	19,500	0.64	0.033
Cellulose acetate	1.40	22,000	0.96	0.050
Methyl cellulose	1.23	22,500	2.10	0.090
Nylon	1.14	24,000	2.15	0.090
Paper	0.45	17,000	0.23	0.015
Polycarbonate	1.20	23,100	2.07	0.090
Polyester	1.38	20,000	1.80	0.112
Polyethylene, low-density	0.92	30,100	0.26	0.016
Polyethylene, high-density	0.96	29,200	0.36	0.021
Polypropylene	0.90	31,000	0.68	0.023
Polyvinyl chloride (rigid)	1.28	20,750	0.55	0.034
Rubber hydrochloride	1.10	25,000	0.95	0.033
Saran	1.70	16,300	1.08	0.066
Styrene	1.05	28,800	0.67	0.024
3/32-in. tear tape, 600L cello	....	94,000 lineal in.	1.36	

coating, (5) extrusion-blown tubing, (6) extrusion chill roll casting, (7) extrusion water bath, and (8) casting. (See Table 1.)

The production method to be used will depend upon the characteristics of the resin and the properties desired in the finished film; for example, polyethylene film is made by three methods: *extrusion-blown* tubing, extrusion onto a *chill roll,* and extrusion into a *water bath.* The latter two provide better clarity, since the rapid cooling inhibits the formation of crystals and spherulites which give the film a cloudy appearance. The same is true of polypropylene film. Vinyl film is generally *calendered,* because thermal degradation can be a serious problem with extrusion and with the other methods that might be used. Casting any film from solvent suspension on an endless polished steel belt will give the best surface finish, but it is the most costly process. It also permits the use of embossed patterns so that it is possible to simulate cloth, leather, wood, and other textures. Because of the growing interest in food preservation and the sterilization of medical supplies by means of ionizing radiation, the information in Tables 2 and 3 is included to show the effect of these processes on packaging films.

Certain properties of films can be improved by *orienting,* or stretching the film under carefully controlled temperatures. This causes a realignment of the molecules and yields a much tougher film. Styrene, for example, would be much too brittle to use as a film if it were not oriented. (See Fig. 1.) The film can be stretched in one direction only, or it can be stretched in both directions if this is desired. With the shrinkage equal in both directions, it is said to have a *balanced* shrink. An example of an unbalanced shrink is a polyethylene with 40 percent

TABLE 2 Degradation of Films by Irradiation*

Polystyrene	Unaffected
Rubber hydrochloride	1
Nylon	2
Polycarbonate	5
Saran	10
Polyethylene	20
Polypropylene	20
Polyacetal	50
Polyvinyl chloride	High
Paper	Very high†

* Figures are based on micromoles of gas produced from 1 g of film at 6 megarads of gamma ray radiation.

† Paper and similar cellulosic materials lose about 20 percent of their strength.

TABLE 3 Changes in Film Due to Radiation Treatment*

Type of film	Percentage change			
	Stiffness	Flexural strength	Tensile strength	Ultimate elongation
Polyethylene, density 0.920, melt index 0.2	−31	+12	−45	−99
Polyethylene, density 0.920, melt index 2.0	−16	−6	−29	−99
Polyethylene, density 0.947	−20	+13	+11	−97
Polyethylene, density 0.950	−63	−24	−43	−98
Polyethylene, density 0.960	−58	−40	+8	−99
Polypropylene, low-ash	....	....	−96	−87
Polypropylene, high-ash	....	....	−93	−96
Nylon, type 6	+181	+136	+107	−92
Nylon, type 6/6	+54	+111	+80	−95
Nylon, type 610	+52	+62	+49	−92
Polystyrene, general-purpose	−13	−24	−50	−45
Polystyrene-butadiene, high-impact	+99	+51	−35	−92
Polystyrene-acrylonitrile (SAN)	−5	−28	−34	−47
Acrylonitrile-butadiene-styrene (ABS)	+49	+5	−58	−93
Polyurethane	+176	+111	−59	−99

* All plastics received radiation of 5.8×10^{16} electrons/sq cm.

shrink in the machine direction and 15 percent in the cross direction. Some of the cross-linked polyethylene films have as much as 80 percent shrinkage. All extrusion-blown tubing has a small amount of shrinkage by virtue of the way it is produced, unless it is *heat-set* after it is blown. Oriented film is used for *shrink packaging*, which is a process in which the film is placed loosely around the articles to be packaged, and joined by heat-sealing the ends of the film together. Then it is passed through an oven which causes the film to tighten around the contents, making a snug package. This is described under "Shrink Films," page 3-5.

The forerunner of all packaging films was cellophane, and the techniques and equipment that were developed to handle this material

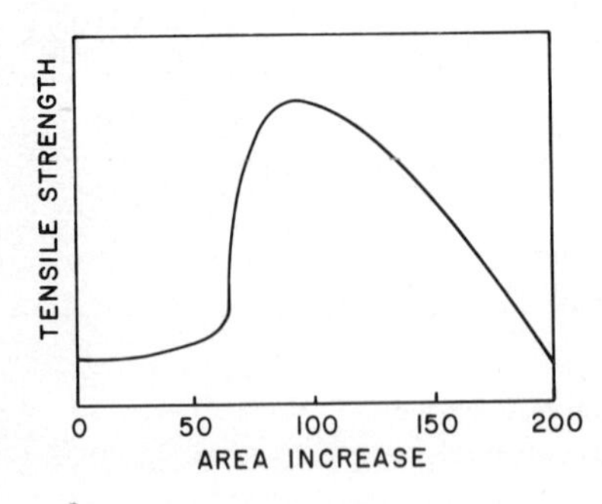

Fig. 1. Oriented film. Effect of orientation on strength of polystyrene film. There is almost no improvement up to a point, and then a rapid increase in strength takes place. This peaks at a ninetyfold increase in area and then falls off as the film is stretched further.

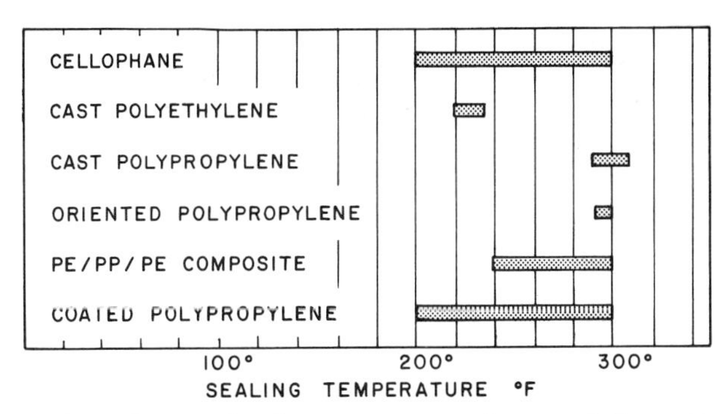

Fig. 2. Heat sealing. Range of sealing temperatures for several films. Free film has a narrow range and goes into holes if the temperature is exceeded. Coatings which have a lower melting point provide a broader heat-sealing range which is below the melting point of the base film.

became quite sophisticated. However, with the advent of the thermoplastic films such as styrene, rubber hydrochloride, polyamide, and the olefins, it was found that the same equipment and methods could not be used. Whereas cellophane could be pushed, the newer films had to be pulled. Whereas dwell times and temperatures had not been very critical, and the heat-seal jaws could be used for pulling the cellophane while they were sealing, the thermoplastic materials were much more demanding. At the same time, new structures became possible by means of these different properties. Side-seal bags, hot-wire cutoffs, stretch packaging and shrink packaging, blisters, and thermoformed containers all became possible as a result of the thermoplastic nature of these materials. (See Fig. 2.)

SHRINK FILMS

The ability of a film to draw down over the contents of a package when it is heated is a very useful attribute. Not only does it make an eye-appealing unit for display purposes, but it immobilizes the components, making them more secure for handling and shipping. Orientation of the film for this purpose is usually done in the manufacturing process, by extruding through a circular die and expanding the tube that is formed with compressed air while it is still molten. At the same time the tube is stretched in the long direction by pulling and winding up at a faster rate than the resin is extruding from the die. The amount of stretch can be nearly equal in both directions (balanced), or more in one direction, usually the machine direction, than in the cross direction (unbalanced). (See Fig. 3.)

For packaging purposes the film is wrapped around the contents and heat-sealed; then it is usually sent through a heated tunnel, where

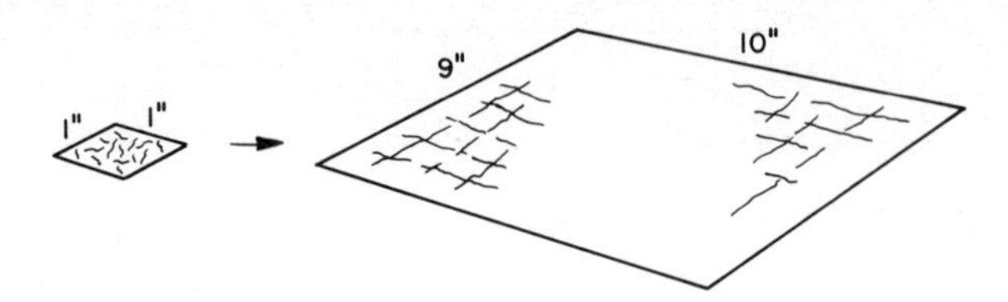

Fig. 3. Orientation. Polystyrene film changes from a stiff brittle material to a tough flexible film. It is stretched in both directions to 90 times the original area. The random arrangement of the molecules shown on the left becomes a more orderly arrangement as shown on the right. More important, they are parallel to the surface of the film after orientation.

radiant heat or heated air causes the film to draw up tight. Some equipment can be adjusted so that certain areas are given more heat than others, to avoid getting holes due to overheating. Immersion in hot water is another technique that is sometimes used, particularly when wrapping meat and poultry. The required thickness of the film will depend upon the size and weight of the item being packaged, but it is most often in the range of ½ to 1 mil. Film can be purchased already folded in half on the roll, so that it is necessary to seal only three sides when the product is placed between the two layers. Sealing with a hot wire or heated cutting bar makes the neatest wrap after shrinking, but unless a contoured seal is made, ears will stick out whereever there is a corner in the seal. Impulse-type sealers also can be used with shrink films, but they do not trim off the excess film the way the hot wire does. The least amount of heat that will do the job makes the best seals.

The properties shown in Table 4 are essentially the same for oriented film as for the unoriented variety. Impact strength may be improved by orientation, but clarity will probably not be as good as it was before orientation. Chemical resistance and permeability are not usually affected. *Sleeve packing* instead of sealing all around is an easier method to mechanize because only one edge needs to be sealed. It leaves openings at the ends which may allow dust to enter, but it provides for easier opening of the package. If the film extends beyond the ends, it will shrink down, reducing the size of the openings and making a reasonably neat package. Unbalanced shrinkage is good for sleeve packing, and also for flat packages like phonograph records. Bulky items which are almost as thick as they are long, however, require the balanced film, which shrinks in both directions almost equally.

When the film is wrapped around the package, it should be put on snugly so that the minimum of shrinkage is necessary to pull it tight. If the shrinkage can be kept down to 10 to 15 percent of the potential shrinkability, the loss of tensile strength will be less than 20 percent. Relaxing of the film with time will also be minimized, and it may be

TABLE 4 Properties of Films

Type of film	Cost, 1,000 sq in., 1 mil	Water-vapor trans-mission[a]	Gas permeability[b]			Water absorp-tion	Dust attrac-tion	Haze, percent	Gloss,[c] percent	Trans-parency,[d] percent	Print-ability	In-use temp. range, °F	Heat-seal temp., °F	Dielec-tric seal-ing[e]	Elon-gation, percent	Impact strength[f]	Tear strength notched[g]
			O_2	N_2	CO_2												
Cellophane, nitrocellulose-coated	$0.033	0.3	1	1	13	High	Low	1	95	. . .	E	24–300	225	. . .	20	5.0	8
Cellophane, polymer-coated	0.041	0.5	1/2	1/2	1/2	High	Low	1	95	. . .	E	24–300	250	. . .	20	6.0	8
Cellulose acetate, 10% pasticizer	0.050	150	35	40	1,000	High	Low	0	. . .	. . .	E	−15–140	350	230	40	2.5	15
Cellulose propionate, 10% plasticizer	0.050	150	60	75	High	High	Medium	0	. . .	. . .	. . .	−30–200	. . .	175	80	6.0	25
Ethyl cellulose, 10% plasticizer	0.040	10	2,000	600	5,000	High	Low	. . .	. . .	. . .	. . .	−70-250	. . .	115	30	4.0	20
Methyl cellulose	0.090	70	80	30	400	Sol.	Low	. . .	. . .	. . .	. . .	. . . . . .	325	650	. . .	5.0	20
Nylon	0.090	19	25	160	160	Med.	Medium	. . .	. . .	. . .	G	−100–200	. . .	250	100	2.5	75
Polycarbonate	0.090	11	300	50	1,000	Med.	Medium	0	. . .	86	. . .	−150–250	410	. . .	75	14.0	25
Polyester, oriented	0.112	1.7	4	1	16	Low	. . . . . .	. . .	. . .	. . .	G	−80–230	275	240	. . .	12.0	80
Polyethylene, low-density . .	0.016	1.3	550	180	2,900	Low	High	6	75	60	F	−70–180	250	2	400	5.0	350
Polyethylene, high-density .	0.021	0.3	600	70	4,500	Low	High	3	85	. . .	F	−20-250	275	1	100	2.0	300
Polypropylene	0.020	0.7	240	60	800	Low	High	1	90	80	G	0–275	350	5	300	1.0	330
Polyvinyl chloride, rigid . .	0.030	4	150	65	970	Low	High	0	. . .	88	E	−50–200	225	140	20	8.0	300
Rubber hydrochloride . . .	0.040	10	130	20	520	Med.	. . . . . .	3	. . .	. . .	. . .	to 200	250	. . .	. . .	10.0	1,600
Styrene, oriented	0.020	4	310	50	1,050	Low	Very high	0	. . .	. . .	E	−80–175	250	1	10	0.3	30
Saran	0.070	0.2	14	12	4	Low	Medium	. . .	. . .	90	P	0–200	280	570	60	6.0	20

[a] g loss/24 hr/100 sq in./mil at 95°F, 90% RH

[b] cc/24 hr/100 sq in./mil at 77°F, 50% RH; ASTM D1434-63.

[c] Gardner 60°.

[d] For comparison, glass is 92 percent. Thickness of 1/8 in. is used for tests.

[e] Loss current tangent × 10^4 at 10^6 cycles per second of high-frequency alternating current at 25°C (high numbers easiest to seal).

[f] Izod impact strength, ft-lb per in., notched; ASTM D256-541.

Table 5 Properties of Shrink Films

Type of film	Typical shrinkage, percent	Shrink temperatures, air °F	Sealing temperatures, °F
Polyester	35	350	275
Polyethylene	30	340	275
Polypropylene	60	425	350
Polyvinyl chloride . . .	60	325	225
Rubber hydrochloride .	45	300	250
Saran	45	350	280
Styrene	50	300	250

offset by further shrinkage during storage. This is more likely to be the case in summer weather, when the wrapper may become tighter over a period of time. (See Table 5 and Fig. 4.)

Selection of the right type of film for a particular application depends on the requirements of the product. For example, fresh produce which must breathe is usually wrapped in styrene film because of its high gas transmission rate. Red meats keep best in rubber hydrochloride film. Bakery items use cross-linked polyethylene, and hard goods are often put into polypropylene. For boil-in and bake-in products, polyester and nylon can be used. Some idea of the relative costs of the various kinds of shrink films can be seen in Table 4, although the processes of orientation and irradiation used to produce the shrink films will add something to the cost. *Polyethylene* film is available in various densities, irradiated and nonirradiated. The radiation treatment causes cross-linking of the molecules and improves some of the physical properties. It is low in cost, but is soft and not so easy to handle as some of the stiffer films. *Polypropylene* is one of the least expensive shrink films. Its impact

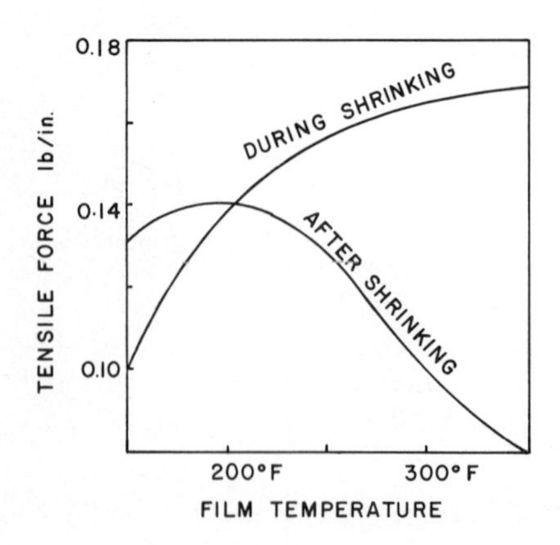

Fig. 4. Tightness of shrink film. The tension exerted by shrink film is dependent to some extent on the temperature used for shrinking. If the temperature is too high, the initial tension will be high, but the tension will not hold up in storage and handling.

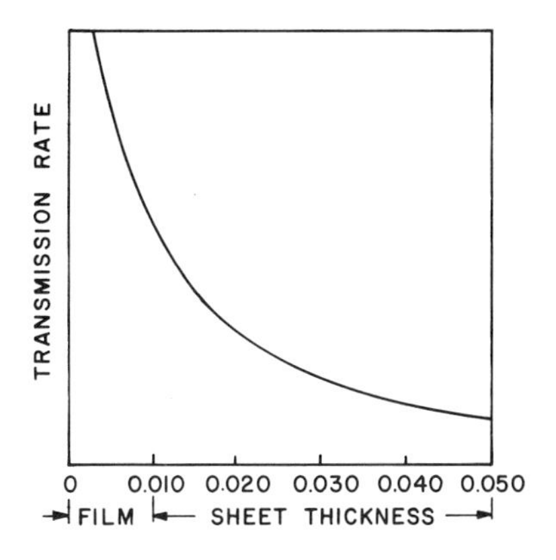

Fig. 5. Permeability. Gas transmission rate is inversely proportional to film thickness, for all practical purposes. Twice the thickness will give almost half the rate of transmission.

strength falls off at temperatures below 0°F, but it has good moisture barrier properties and grease resistance, and fair gas permeation rates. *Polyester* is a higher-priced film with exceptional tensile strength for heavy objects, and good chemical resistance. Permeation of moisture is only fair, and static buildup, with difficulties in heat sealing, may cause some problems. *Rubber hydrochloride* works well with red meats, but antioxidants in the plastic sometimes give the film an off odor. Clarity is excellent and impact strength is quite high except at freezing temperatures. Its grease resistance is good, but it is not a good gas barrier. *Styrene* is low in cost and has good optical properties. It has fairly good strength except at low temperatures, but chemical resistance is limited. Gas transmission rates are high, which may be an advantage for such things as fresh produce. *Saran* (polyvinylidene chloride) is one of the best barrier materials for moisture and gases. It is soft, clingy, and stretchy, and consequently difficult to handle, and the heat-sealing temperatures are rather critical. Cost is on the high side. *PVC* (polyvinyl chloride) is a tough, versatile, low-cost film that is widely used for meat, fresh produce, bundling of canned goods, and many other items. It has fairly good barrier properties and good grease resistance. (See Fig. 5.)

CELLOPHANE

Soluble cellulose was first produced in the laboratory in 1892, and a British patent which described a method of making thread and film from viscose was granted to C. H. Stearn in 1898. A Swiss chemist named Jacques Brandenberger designed a machine for making continuous film in 1911, and he called the product cellophane, from the first half of "cellulose" and the ending of "diaphane," the French word for transparent. The Du Pont Company secured the American rights to the Brandenberger process in 1923 and started making the film in Buffalo, New York, in 1924. The cellophane that was first produced was tacky, brittle, and lumpy, and users complained that cigars dried out and knives rusted when wrapped in this material.

A research chemist, Dr. W. Hale Church, was assigned by Du Pont in January, 1925, to the task of improving the properties of cellophane. He was joined 6 months later by Karl E. Prindle and others. By 1927 they had developed a coating made from two incompatible materials—nitrocellulose and paraffin wax—combined with a plasticizer and a blending agent. By means of volatile solvents they were able to deposit 1/20 mil on both sides of a 1-mil film, which made it moistureproof.

Up to this point, the most widely used wrapping materials had been wax paper and glassine. Today nearly 500 million lb of cellophane worth over $275 million is produced each year in about 130 different variations of coated and uncoated film. About 95 percent is the moistureproof type, and 94 percent is used for packaging, three-quarters of it for food products.

Chemistry. Cellophane is regenerated cellulose, and it is quite different from other transparent films. For one thing it is not thermoplastic, and it cannot be heat-sealed by itself. Only if a thermoplastic coating is used does it become heat-sealable. It is very sensitive to moisture, expanding and contracting as the humidity changes. Tensile strength and impact strength are excellent, but tear strength is poor and once a tear is started it propagates easily.

The basic molecule is $(C_6H_{10}O_5)_x$; this is made up of glucose molecules from which a molecule of water has been removed. The graphic formula incorporates the pyranose ring representation:

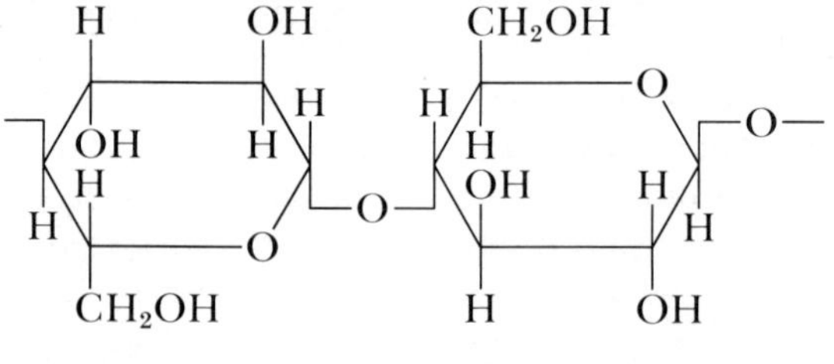

H OH CH_2OH
H H O
—O—
OH H H
H OH H
H O H
O
CH_2OH H OH

In the manufacturing process, cellulose is mixed with sodium hydroxide to form sodium cellulose. It is then reacted with carbon disulfide, which produces sodium cellulose xanthate. This would be colorless if it were not for the presence of impurities such as thiocarbonates, which give it a yellow or orange tint. This solution passes through a bath of dilute sulfuric acid and sodium sulfate which "salts out" the viscose and neutralizes the alkali, yielding a continuous web of hydrated cellulose.

Advantages and Disadvantages. Cellophane has excellent clarity and sparkle, it is easily machined, and it makes a strong package. Heat sealing is not critical, and the temperature can vary as much as 50°F and still give a good bond. When notched, it tears easily, which is both good and bad. For ease in opening by means of a tear tape, cellophane is unsurpassed by virtue of this easy tearing. It is a good substrate for coatings and laminations, since it is the only transparent film that does not soften with heat. The many varieties of cellophane attest to its versatility; it can be tailored to almost any degree of permeation or lack of permeation of moisture and gases, so that hard-crusted breads remain crisp, meats keep their color, and bags used for fresh produce do not fog up.

On the debit side, cellophane does not have the shelf life of the olefins; it shrinks and gets brittle under dry conditions in the winter; and the tearing characteristic which makes it easy to remove from a package can be a problem on the machine or in the marketplace, if the edge gets nicked or the wrapper starts to break. In comparison with polyethylene film, cellophane has better transparency and surface gloss or sparkle. The moisture transmission rate is about the same, but cellophane is a better barrier for oxygen, greases, and flavors. It is not as strong as polyethylene for heavy products, and the cost of cellophane is roughly one-third higher.

Manufacturing. Cellophane is made from wood pulp, mostly hemlock, although spruce and some hardwoods are sometimes used. The logs are chipped, boiled, and purified in pulp mills in the Northwest. Pulp sheets in bales are then shipped to the film plant, where they are steeped in a 20 percent sodium hydroxide solution at 68°F for about ½ hr. The excess liquid is pressed out, and the alkali cellulose is next sent to a shredder, where it is crumbled and put into cans for aging. After about 24 hr the aged crumbs are transferred to revolving drums, where carbon disulfide in an amount equivalent to about one-third the weight of the original pulp is added and the mixture is churned for about 2 hr until the vapor is all absorbed. This is transferred to a slurry tank, where the tail end of the caustic solution that was squeezed out by the steep press is added, with some additional water to produce a viscose solution. This is put through a mill to break up any remaining pieces, and is then filtered and sent to deaeration and ripening tanks, where it is stored for about 24 hr. During this ripening period the sodium cellulose xanthate decomposes, forming degradation products such as sulfides, sulfates, thiosulfates, thiocarbonates, and other sulfur compounds. The molecular weight of the cellulose which is in colloidal suspension at this point is around 95,000, with the molecule having about 600 units in its chain. The viscosity is around 50 poises, comparable to glycerin or castor oil.

The viscose solution is forced down through a 0.010-in. slit in the *casting* hopper, into a 10 percent solution of sulfuric acid buffered with

20 percent sodium sulfate at about 100°F. The extrusion die is about 50 in. long, and the opening is adjustable to deliver film of the desired thickness. The takeoff speed also helps to control the gauge of the film. The web passes in and out of tanks of decreasing concentration of acid, and then through a boiling hot water bath to drive off the remaining carbon disulfide. Next it is treated by a chlorine *bleach* bath and further washing, and then it goes into the *softener* solution. The softeners used are a 5 percent solution of glycerin, and for low temperatures, ethylene glycol, although propylene glycol and similar compounds may be included, along with hygroscopic salts as humectants. Cellophane with 15 percent glycerin is suitable for humidities above 30 percent RH and works well on packaging machinery. Under very dry conditions it is better to have about 20 percent glycerin content, and for twist wrapping of candies the higher amount is preferred, although there is a greater tendency for the film to block (stick together). If the film is to remain uncoated, a *sizing* of silica is dispersed in the softener solution so that when it is dry, there will be a thin dusting of powder on the surface of the film. This prevents blocking under humid conditions. Resins such as urea-formaldehyde may also be added to the softener solution to *anchor* the coating that is applied in a subsequent operation. These resins become polymerized by the heat during the drying of the film. Anchorage of the coating is necessary to avoid delamination when the film is used under wet conditions, as in the wrapping of fresh meat. About 30 percent of all cellophane produced is the anchored type. After passing over heated rolls in a *drying* oven, the cellophane is wound into mill rolls weighing almost 1,000 lb and containing nearly 5 miles of film.

For most purposes the cellophane is *coated* on one or both sides with nitrocellulose or polyvinylidene chloride, about 0.00005 in. thick, to give it moistureproofness and sealability. There is a great variety of coatings for different purposes, but a typical one might be pyroxylin, with dibutyl phthalate as a plasticizer, and a small amount of gum dammar and paraffin wax. These would be dissolved in a mixture of ethyl acetate and toluene, to make a solution with about 15 percent solids. The wax supplies the moistureproofing, and the nitrocellulose gives it the heat-sealing properties. The film passes into a dip tank and then between doctor rolls, which control the thickness of the coating. It is then carried up into a *drying* tower and goes through a *rehumidifying* chamber, and finally is rewound. This roll is later slit to the required widths or cut into sheets. About 95 percent of the cellophane is shipped as rolls and only 5 percent as cut sheets.

Properties. The polymer (saran)-coated cellophane has more luster than the nitrocellulose-coated film, but it is more expensive, being about

15 cents a pound higher. Cellophane has good tensile and impact strength, but it tears very easily once a tear starts. In dry weather it loses its strength and becomes quite brittle. It is not dimensionally stable, and any changes in humidity will cause a shrinkage of up to 5 percent in the machine direction and 7 percent in the cross direction. This is only partly reversible with a return to the higher humidity. Controlled permeability is one of the outstanding characteristics of cellophane. It is possible to apply different coatings to give just the right amount of barrier desired. For example, lettuce will fog up most kinds of film because of the condensation on the inside surfaces; but if some of the moisture is allowed to escape through the wrapper, the appearance is greatly improved. Fresh meat will turn dark rapidly unless there is sufficient oxygen present. This oxygen turns the myoglobin into oxymyoglobin, which is bright red, thus giving the meat the "bloom" that makes it attractive. Too much oxygen, however, changes the myoglobin into metmyoglobin, which is brown in color, and spoils the appearance. A type of cellophane has been developed with just the right amount of oxygen transmission for this purpose, and it will also retard the moisture loss that causes shrinkage. Hard-crusted bread requires a wrapper with a small amount of permeability to moisture. Otherwise the crust becomes soggy. In another case, antioxidants can be added to cellophane to retard rancidity in the film of oil that gets on the inside surface of snack packages. Fresh produce that is packed with a mixture of 95 percent nitrogen and 5 percent oxygen, to lengthen shelf life, is usually wrapped in cellophane with a low gas transmission rate. (See Table 6.)

Some static is generated by cellophane during dry weather, and machines should be equipped with static eliminators. Cellophane has a wide sealing range; since the basic sheet is not softened by heat, the temperature can be as much as 50°F above the melting point of the coating

TABLE 6 Cellophane Code Designations

A or B	Anchored (bonded)
C	Colored
D	Du Pont, or decreased moistureproofness
L	Intermediate moistureproofness
M	Moistureproof
O	One side coated
P	Plain uncoated
R	Vinyl-coated
S	Heat-sealable
T	Transparent
V, X, or K	Polymer (saran)-coated
WO	White opaque

with good results. Resistance to abrasion from such things as salty pretzels is good, especially with the polymer-coated cellophane.

Selection. The choice of the type of cellophane to be used will depend on its end use and the methods of fabrication. When there is doubt about the type of cellophane to use, MST-44 has most of the properties needed for the majority of cases. For better appearance, with a higher degree of durability and improved barrier properties, V-3 (K-207) is most likely to serve the purpose.

For a more careful choice, first consider the barrier properties. If moisture, grease, or gas resistance is important, the higher-priced polymer coatings may be justified. When cellophane-wrapped packages are bundled, the outer wrapper should have a lower sealing temperature so that the coating on the inside film will not be softened and stick to the outer wrapper; in this case a polymer-coated film should be used inside and a nitrocellulose-coated film for the bundling. (See Table 7.)

Processing. The most frequent problems in using cellophane have to do with *sealing.* Regular MST film should give seals that will withstand a stress of 60 g per in. of width when the film is properly conditioned, that is, above 35 percent RH. The sealing conditions should be around 3 sec

TABLE 7 Some Cellophane Types and Typical Uses

Type	Use	Product	Property
LST	Wrapping	Produce	Breathing
MSAT-10	Lamination	Lamination	Coated one side
MSAT-86	Wrapping	Wet products	Water-resistant
MSAT-87	Bags	Frozen food	Water-resistant
MST-44	General purpose	General purpose	General purpose
MST-51	Wrapping	Bread	Flexible
MST-52	Tear tapes	Oily or salty products	Rigid
MST-53	Bags	Sharp objects	Low temperature
MST-54	Wrapping	Baked goods	Grease-resistant
MST-58	Wrapping	Small items	Strong seal
MST-60	Bundling	Small, light items	Light duty
MST-66	Bundling	Small items	High slip
MT-31	Bundling	Cigarettes	Solvent seal
MT-33	Twist wrapping	Candy	Flexible
T-69	Bags	Sharp or heavy items	Tough
T-79	Bags	Cookies	Barrier
OF-16	Wrapping	Fresh meat, uncoated side in	Nonfogging
OX-511	Wrapping	Lightweight products	Moisture-resistant, grease-resistant
V-3	General purpose	General purpose	Barrier, appearance
V-4	Wrapping	Greasy products	Grease-resistant

dwell time at 275°F with a pressure of 20 psi. To penetrate more than two layers, the dwell time should be increased, not the temperature, in proportion to the square of the thickness. Bear in mind that temperatures above 300°F may damage the film. For problem cases, there are special films such as MSAT-86 and V-4 that are designed to provide exceptionally strong bonds. (See Fig. 6.)

For heat-sealing polymer-coated film, use heating surfaces of steel, and not aluminum or brass unless they are covered with Teflon. A voltage regulator of the Powerstat type will give more uniform heating if it is adjusted so that it is "on" about 90 percent of the time. Heating plates should be insulated with asbestos on the back and sides, to protect them from drafts. It is also important to have the thermostat as close to the sealing surface as possible, to be sure that you are getting a true reading.

The ideal *storage* conditions for cellophane are 35 to 50 percent RH and 70 to 75°F. You should avoid nicking the edges when handling it, as the tear strength is greatly reduced when it is notched, and this may cause problems on the machine. Three types of *feeding* methods are used on wrapping machines: pull-feed, belt-feed, and push-feed. The first two generally do not cause any difficulty if the unwind tension is right, but push feeding can be a problem if conditions are not ideal. There is usually a corrugating device to stiffen the film for push feeding, and if this is dirty or out of alignment, it will interfere with the proper operation of the machine. Cutting knives must be kept sharp, and the correct change parts should be used. When changing setups, there is a temptation to try to use some of the same parts to save time. This only leads to trouble that is sometimes hard to identify.

Some looseness in *wrapping* is necessary to allow for shrinkage of the cellophane under dry conditions. The amount of "pull-up" will vary

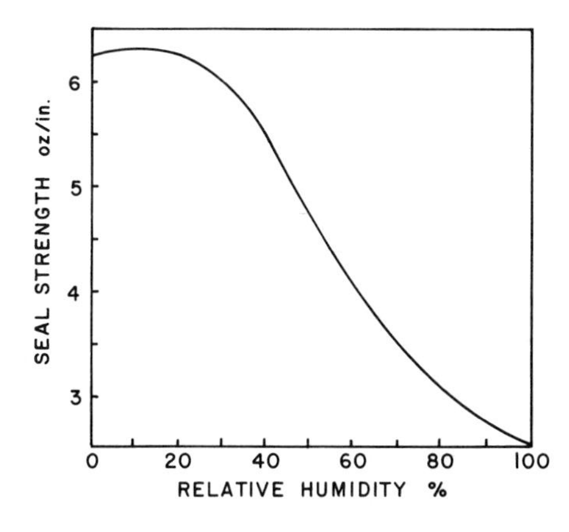

Fig. 6. Sealing conditions. Effect of moisture content on seal strength of cellophane. Seals made with cellophane stored in a dry atmosphere are stronger than with cellophane stored under humid conditions. Steam generated by the heat of sealing interferes with the adhesion of the heat-seal coating. Humidity also affects seals after they are made. They are strongest when the atmosphere is at 55 percent RH, falling off to about half at 80 percent RH.

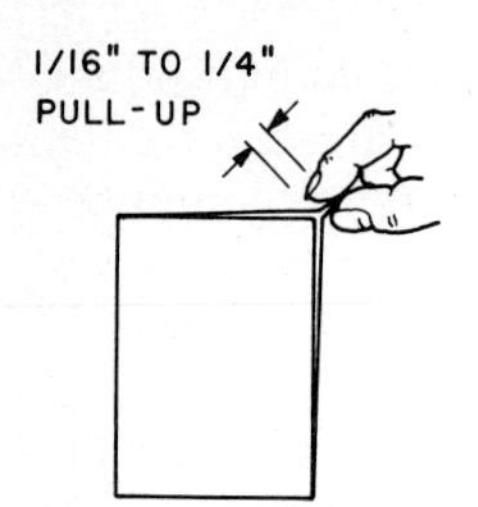

Fig. 7. Pull-up allowance. Looseness of cellophane wrapping is necessary to allow for shrinkage under dry storage conditions.

with the size of the package: from 1/16 in. on small packages to 1/4 in. on larger ones, such as cookie boxes. (See Fig. 7.) Problems with excess *drag* over formers can sometimes be helped by roughening the metal surface with a medium grade of emery cloth, rubbing in the direction that the film travels. If the surfaces become warm from being adjacent to heaters, they should be insulated to minimize the amount of drag. If static becomes a problem, conductive tinsel should be draped across the web of cellophane at the unwind end. In an emergency, a wet towel hung over the unwinding roll will usually correct the trouble.

If rolls of cellophane are found to be puckered from storage in a dry atmosphere, wipe the ends of the rolls with a damp cloth. On the other hand, if they are stored under humid conditions, the rolls may become "beaded" on the ends. In this case, snubber bars added to the machine to carry the film through an S curve just ahead of the feed rolls will "iron" out the film.

Costs. Cellophane has a density of 1.50 g/cc, and the lowest cost is 63 cents per pound. This makes the cost of 1-mil film, for comparison purposes, $0.033 per thousand square inches. Polymer-coated film is 81 cents per pound, or $0.041 per thousand square inches.

The yield per pound is about 19,500 sq in. for 0.001-in. film, and for other gauges it is usually indicated by the manufacturer's code number: thus 140 MST yields 14,000 sq in., and 195 MST yields 19,500 sq in.

The weight of a 9½-in.-diameter roll on a 3-in. core is about 3 lb per inch of width, and contains 4,850 lineal feet of 0.001-in.-thick film. The minimum order for a special size or type of cellophane is usually 500 lb.

Manufacturers. American Viscose Div., FMC Corp.; E. I. du Pont de Nemours & Co.; Olin Mathieson Chemical Corp.

CELLULOSE ACETATE

This material was first used for photographic film in 1912. Later it served as a coating for the fabric on airplanes. Although it is made from cellulose, the same as cellophane, cellulose acetate is quite a different material. For one thing, it is thermoplastic; that is, it will soften and melt when heated. It is nearly always used plain, without a heat-seal coating. Plasticizers such as diethyl phthalate are frequently added to improve its impact strength. It is crystal clear, tough, and easily processed. Barrier properties against moisture and gases are poor, but on the other hand, it is excellent for fresh produce and iced pastries because it breathes and does not fog up. For more details on composition and properties, see Sec. 8, "Plastics."

Manufacture. Cellulose acetate is made from cotton linters. First the bales of linters are opened, and the fibers are mixed with glacial acetic acid and acetic anhydride, with a small amount of sulfuric acid as a catalyst. The fibers dissolve slowly to form a solution of cellulose triacetate in acetic acid. Water is added to stop the reaction and to hydrolyze the triacetate. Boiling increases the hydrolysis until there is an acetyl content of about 51 percent. The curds or flakes of precipitated acetate are then washed and dried.

Acetate film is made by two processes: casting, which gives a smooth surface, and extrusion, which does not have as good optical qualities but is more economical. The acetate is dissolved in acetone with a suitable plasticizer as needed; for example, 15 percent dimethyl phthalate might be used. This is spread on a polished stainless steel band, and after the solvent has evaporated, the film is stripped off and sent through an oven for final drying and it is then wound on rolls. In the extrusion process, a small amount of solvent is used and the solution is heated to about 225°F. This is forced through an annular die, and when cool it is wound in rolls.

Wrapping machines for handling acetate may require static eliminators. Sealing is usually done with solvents, although heat sealing is possible.

ETHYL CELLULOSE

This material has not been widely used, for the reason that there are cheaper films that will do almost anything that ethyl cellulose will do. Its principal use has been for windows in envelopes and for folding cartons. Ethyl cellulose has been available for a long time, having been discovered in 1916 by an Austrian physician, Dr. L. Lilienfeld. The principal advantage of this material is its excellent impact strength at very low temperatures. Another outstanding characteristic is its deep-draw properties, and with the right plasticizers and proper temperatures the maximum ratio of depth to diameter is greater than for any other thermoplastic material. Since there is no degradation at elevated temperatures (up to 375°F), it is a useful constituent of hot-dip coatings, strippable coatings, and hot lacquers.

Chemistry. Ethyl cellulose is an ether in which ethyl groups $—OC_2H_5$ have replaced the hydroxyl groups $—OH$ of the cellulose molecule. While a complete etherification is possible, yielding a triethyl derivative, the commercial product averages 2¼ to 2½ ethoxyl groups per glucose unit, with the lower figure giving the greater toughness and lower solubility, but poorer compatibility with plasticizers and other additives.

Manufacture. Wood pulp or cotton linters are treated with a 17½ percent solution of sodium hydroxide, in a manner similar to the process for cellophane. The alkali cellulose is then treated with ethyl chloride to form ethyl cellulose. This is washed to remove the sodium chloride, which is one of the reaction products. It is dissolved in solvents to which plasticizers have been added, and is then cast into film by depositing on a polished steel belt.

Properties. Ethyl cellulose is colorless, odorless, and tasteless. Its strength decreases and the extensibility rises with increasing temperature. It has excellent strength at very low temperatures, and its flexural strength is outstanding. It ages well and is not seriously affected by sunlight, although some discoloration may result. Ethyl cellulose is a poor barrier for water vapor and gases, and it is soluble in most solvents except aliphatic hydrocarbons, glycols, and water. This suggests its use for packaging the ingredients used in chemical processing, where package-and-all is added to the batch.

Heat sealing is difficult unless the surface is activated with a solvent. Decoration is easily done with special types of inks. Good resistance to greases and oils has been reported for products such as butter and olive oil. Resistance to alkalies and weak acids is good, but strong acids may cause decomposition.

Processing. Static can be a problem on wrapping and laminating equipment. Electronic heat sealing is possible, but solvents or adhesives are more generally used for bonding.

FDA Approval. Some formulations of ethyl cellulose are permitted for certain food applications in accordance with section 121.1087 of the *Federal Register,* volume 27, 1962, page 4915. Specifications are given also in the *National Formulary XII,* 1965, page 164.

Costs. The specific gravity of ethyl cellulose is around 1.15, depending on the amount of etherification, and the yield of film is about 24,000 sq in. per lb. With a base price of 72 cents per pound, the cost is about 3 cents per cubic inch, and the film is around 4½ cents per thousand square inches per mil of thickness, which makes it a medium-price material.

Manufacturers. Dow Chemical Co.; Hercules, Inc.

METHYL CELLULOSE

This is a water-soluble film used for unit packages of products that can be added to mixes in the package; the package will dissolve. Toxic materials for agriculture, for example, can be added to spray tanks in premeasured amounts, with less hazard to the user. Other uses include rat poison, detergent powder, medicinal capsules, bubble bath powders, additives for bread dough, or a transfer medium for decorating molded desserts. Although it is soluble in water, the film retains its strength and does not become tacky under humid conditions. It is only when it comes in direct contact with "wet" water that it starts to disintegrate.

Manufacture. The process consists of treating cotton linters with sodium hydroxide solution as described under cellophane. The alkali cellulose that results is then reacted either with methyl sulfate to produce methyl cellulose, or with sodium chloroacetate to form carboxymethyl cellulose, both of which are water-soluble. After washing and drying, it is dissolved in organic solvents and cast on a stainless steel belt. It is then slit 28 in. wide and wound on large supply rolls.

Properties. The solubility rate varies with temperature, but methyl cellulose will dissolve even in the coldest water. At room temperature a 1-mil film dissolves in about 20 sec. The specific gravitity of methyl cellulose is 1.23, and the yield for a 1-mil film is about 22,500 sq in. per lb. It is affected by dilute alkalies, but not by strong alkali solutions. Resistance to animal and vegetable oils and greases and to petroleum hydrocarbons is good.

Trade Name. Methocel (Dow Chemical Co.).

NYLON

Introduced by Du Pont in 1938, nylon is better known as a fiber for woven or knitted fabrics than as a film. It is rather high-priced as films go, but its toughness and high-temperature characteristics make it useful for sterilizable packages. It is also well suited for oily and greasy products and for abrasive articles such as macaroni or salted pretzels. Boil-in-bag and oven-ready products can be put in this material. Nylon film is crystal clear and has good sparkle.

Chemistry. Nylons (polyamides) are made from dibasic acids combined with diamines. Since there are many dibasic acids and many different diamines, there can be various kinds of nylon. These are identified by the number of carbon atoms in the diamine molecule, followed by the number of carbon atoms in the acid. Thus we have nylon 6/6 with six carbons in each. However, when an amino acid is used, the numbers are added together to give a designation like nylon 12; for example, if a 10-carbon dibasic acid is used with a 6-carbon diamine, we have 610 nylon. Nylon 6 is the type most generally used for film, although in Europe nylon 11 is widely used because of its lower cost. For more information on the chemistry of nylon see Sec. 8, "Plastics." Nylon film can be made by any one of three methods: blown film, chill roll, and water quench.

Properties. The tensile strength of nylon film is about equal to that of cellophane or cellulose acetate. Yield strength is unusually high, and elongation is also very high, comparable to that of polyethylene. Bursting strength is so high that it is difficult to measure on a Mullen tester. Tear strength is good and notch sensitivity is low. Folding endurance is exceptionally good. For all practical purposes, nylon is considered to be three times as strong as polyethylene.

It is resistant to alkalies and dilute acids, but not to strong acids or oxidizing agents. Formic acid and phenol will dissolve nylon and can be used as cements. Gas transmission rates are quite low, but the rate for moisture transmission is fairly high. This is an advantage for sterilizable packages, since it allows the escape of moisture trapped inside. Normal changes in humidity will cause shrinkage or expansion of 1 to 2 percent, and under extreme conditions the film will absorb up to 8 percent moisture.

Nylon is tasteless, odorless, and nontoxic. Resistance to sunlight is only fair. It has a sharp melting point, and it heat-seals at about 470°F with seal strength about 85 percent of the ultimate film strength. Adhesives give joints with about 80 percent of the film strength.

Applications. Nylon film is used by the armed services for packaging spare parts because of its resistance to puncture, abrasion, grease, and solvents. Boil-in-bag foods and gas- or vacuum-packed products can be put into nylon film. Deep-drawn vacuum-type packages up to 2 in. deep are being used for meats. For heavy products, or where low moisture vapor is required, ½-mil nylon is coated with 2 mils of polyethylene. In this combination the nylon provides the strength to keep the polyethylene from stretching and breaking.

Sealing can be done with resistance heaters, but the temperature must be controlled closely because nylon has a sharp melting point. Impulse-type sealers are more dependable for this type of film, and high-frequency sealing equipment can also be used.

FDA Approval. Nylon film can be used for all food products with the exception of milk and dry milk products, for which approval has been temporarily withheld until it is determined whether it meets the cleanability standards of the Public Health Service. Nylon 6/6, nylon 6, nylon 6/10, nylon 11, and certain copolymers are cleared by FDA subject to limitations on extractables.

Costs. The price of nylon film is around $2.15 per pound. With a density of 1.14 the yield is 24,500 sq in. per lb for 1-mil film. This makes the cost per thousand square inches about 9 cents, a relatively high price for a packaging film.

Trade Names. Capran (Allied Chemical Corp.); Olin Mathieson Chemical Corp.; Rilsan (Organico-Rilsan); Zytel (E. I. du Pont de Nemours & Co.).

POLYCARBONATE

Polycarbonate was discovered in the early 1950s by Dr. Daniel W. Fox while working for General Electric Company. At about the same time the Farbenfabriken-Bayer AG of Germany also produced small quantities of this material. Mobay Chemical Company, partly owned by Farbenfabriken-Bayer, is now one of the manufacturers of polycarbonate in this country.

Its high price has deterred more widespread adoption for packaging purposes, but where toughness and a high softening temperature (270°F) are required, it serves very well. It can be classed with nylon film in cost and in many of its properties. The toughness of polycarbonate film suggests its use for hardware and similar sharp objects, and its heat resistance makes it useful for sterilization processes. It has all the qualifications for food packaging, being odorless and nonstaining. However, resistance to alkalies is poor, and permeability to moisture and gases is high. Polycarbonate is manufactured by reacting bisphenol A with phosgene.

FDA Approval. Approval of certain polycarbonate resins was published in the *Federal Register,* May 22, 1963, page 5083. Transparent films are generally made from approved resins, and if they meet the requirements set forth in the above regulation, they can be used in contact with meat, milk and milk products, and other foods without FDA review.

Properties. Polycarbonate has excellent clarity with a very slight yellowish tinge. It is exceptionally tough and has good flexibility and good dimensional stability. Its heat resistance is high, making it suitable for sterilization. For boil-in-bag foods it should be noted that the property of elongation drops almost to half in boiling water, although tensile strength is not affected, and it is satisfactory for the purpose.

Tensile strength is not affected significantly by changes in temperature. Mold shrinkage is very small. Impact strength is almost five times as much as for most other plastics, temperature range extends above that of most other polymers (270°F), and sub-zero performance also is superior to all others.

Moisture pickup will affect dimensions about 1/10 percent for a moisture increase of ½ percent. Polycarbonate has good resistance to water, organic and inorganic acids, solutions of neutral and acid salts, alcohols, ethers, aliphatic hydrocarbons, oxidizing agents such as bleach, and vegetable oils. It is attacked by solutions of alkalies and alkaline salts, and amines including ammonia. Resistance is poor to chlorinated hydrocarbons, ketones, esters, and aromatic hydrocarbons such as benzene, toluene, and xylene. Polycarbonate is a poor barrier for moisture and gases.

Methylene chloride, tetrachloroethane, and ethylene dichloride are used for solvent-cementing polycarbonate to itself; the first evaporates quickly and the others more slowly. Ketones and esters cause crystallization, and under stress will craze polycarbonate, as will carbon tetrachloride, hexane, heptane, and aromatic hydrocarbons.

Cost. Polycarbonate has a specific gravity of 1.2, and the yield for 1-mil film is 23,100 sq in. Cost per pound of film is $2.07.

Processing. For molding or extruding, the polycarbonate resin must be absolutely dry. Otherwise silvery streaks, splays, chicken tracks, or air bubbles may spoil the appearance of the finished package. Even when the material is dry at the start, if it is not protected from the atmosphere, within 10 min it can pick up enough moisture to affect production. Not only will appearance be affected, but the material will not have the toughness it should have, resulting in brittleness failure under impact.

Film is made by solvent casting or extrusion, the former giving better optical properties but at a higher cost. Polycarbonate can be heated rapidly with little risk of thermal degradation. For *thermoforming*, the sheet should be dried at 270°F for at least 2 hr. *Cold forming* of polycarbonate is possible, and bending, stamping, cold heading, drawing, coining, and rolling have been done successfully without heat. Cold rolling can be carried out with successive passes of 10 percent, up to a total of 50 percent reduction. Bending can be accomplished without heat by overbending; for a right-angle bend the amount of springback is about 15°.

Cementing of polycarbonate can be done with methylene chloride or with special adhesives made for the purpose. A solution of 5 percent polycarbonate in methylene chloride may be used, but a higher concentration than this may result in bubbles in the joint. For large joints and higher open time, add up to 40 percent ethylene dichloride to slow down the rate of evaporation. Other bonding methods include heat sealing, press welding, spin welding, hot-air welding, and ultrasonic welding.

Heat sealing of polycarbonate requires temperatures around 410°F; it can be done with settings of 450°F and about 2 sec dwell time. Seal strength is around 10 lb per in. Dielectric sealing is not feasible because of the low power factor.

Press welding of thick sections consists of heating the edges for 3 or 4 sec against a hot plate (660°F) and pressing them together. For *spin welding* the parts must have circular joints, and a shallow tongue and groove is preferred for easy alignment. When one piece is spun with a peripheral speed of about 50 ft per min and pressed against the mating part, which is held fixed, with a pressure of about 300 psi, friction produces a molten zone in a fraction of a second, which then becomes a weld when the rubbing stops.

POLYESTER

Polyester film was a British discovery, purchased by Du Pont and licensed to Imperial Chemical Industries, Ltd., for manufacture in Europe. The earliest work with this class of compounds is believed to have been done by W. H. Carothers around 1928. This set the stage for the discovery 12 years later of polyethylene terephthalate (polyester) by J. R. Whinfield and J. T. Dickson.

Chemistry. A condensation reaction between ethylene glycol and terephthalic acid is generally used to make polyester film, but other films can be produced with polymers from hydroxy compounds other than ethylene glycol, and with aromatic dibasic acids other than terephthalic acid.

$$-\overset{\overset{\displaystyle O}{\|}}{C}-R-\overset{\overset{\displaystyle O}{\|}}{C}-O-R-O$$

In its unoriented state, the film is somewhat brittle and is usually combined with other materials in laminations. When it is used alone, it should always be the oriented type of film.

Characteristics. Polyester has exceptional tensile strength over a wide temperature range, making it suitable for boil-in-bag and bake-in foods. It has good chemical resistance, usually contains no plasticizers, and has good transparency. Cost is rather high, being three times as much as some other transparent films, but its unusual strength makes it a very useful packaging material.

Advantages and Disadvantages. Its high tensile strength makes polyester film a good choice for packaging sharp objects, but its stiffness makes it a noisy film which might be objectionable for some applications.

Good aging characteristics, clarity, and chemical resistance also are among the merits of polyester film. It has a moderate moisture transmission rate, but this drops to near zero at freezing temperatures. Impact resistance is good, but tear and puncture resistance is not as good as that of the softer films. Polyester is a high-priced material, but the cost is offset somewhat by its strength, which allows its use in thinner gauges than other films. The uncoated film is not heat-sealable, but it can be solvent-sealed.

Forms and Modifications. Unoriented film is available for lamination, but when it is used alone, polyester film is always oriented. Coatings on one side or both sides are made with either vinylidene chloride or polyethylene. These coatings provide the sealability and improve the moisture barrier properties of the polyester film. A combination of ½-mil polyester film with 1½ mils of polyethylene is commonly used for heat-and-serve packaging of convenience foods.

Properties. The tensile strength of polyester film is over 20,000 psi and elongation is above 50 percent giving it good impact strength. The oriented film will shrink about 25 percent in both directions. Since there are no plasticizers, flexibility and dimensional stability are not significantly affected by changes in temperature and humidity. Polyester film retains its properties from −80 to 300°F for short periods; to 230°F for continuous exposure. Tear and puncture resistance is not as good as that of the softer films, but it is adequate for most purposes. Folding endurance is good up to 20,000 cycles.

Moisture permeability is fairly high, being comparable to that of the cellulosic materials, but since it drops at low temperatures, it is quite satisfactory for frozen foods. Polyester is a good barrier for odors and gases, and it is resistant to greases and oils. It has good resistance to weak acids and alkalies, alcohols, hydrocarbons, chlorinated hydrocarbons, ketones, and esters, but it is attacked by strong acids, strong alkalies, phenols, cresols, and benzyl alcohol. Irradiation for purposes of sterilization or pasteurization has no significant effect on its mechanical or physical properties.

Processes. Some types of polyester film can be used for shrink wrapping, and it is also possible to thermoform it, but a high temperature is needed to get it soft enough. Sealing polyester to itself with heat and pressure is possible but difficult. A very small amount of benzyl alcohol applied with a felt wick in combination with heat and pressure makes an excellent seal. Temperatures should be in the range of 335 to 385°F. Hot-wire seals also are possible, but they tend to be weak and are not liquidproof. A film that is specially coated for the purpose can also be used for heat sealing. Laminations make use of its great strength and good barrier properties, for gas packing of meats and cheeses and also for heat-and-serve goods.

FDA Approval. This material is widely used for packaging food products. FDA has stated that data indicate polyester film is acceptable for packaging foods and drugs at temperatures up to 250°F. Clearance for polyester film on meat products has been granted by the Meat Inspection Branch of the U.S. Department of Agriculture. Seventy-five percent of the polyester film currently used in packaging is for food items; the majority of this is used in laminations or coated films.

Costs. Relatively high in cost at $1.80 per pound, it is available in thin gauges; this helps to bring the cost down. With a yield of 20,000 sq in., a 1-mil film of polyester costs about 9 cents per thousand sq in.

Trade Names. American Viscose Div., FMC Corp.; Celanar (Celanese Plastics Co.); Kodar (Eastman Chemical Products, Inc.); Mylar (E. I. du Pont de Nemours & Co., Inc.); Scotchpak (3M); Terylene (Imperial Chemical Industries Ltd.); Videne (Goodyear Tire & Rubber Co.).

POLYETHYLENE

Discovered in England in the early 1930s, polyethylene reached commercial importance during World War II (see Sec. 8, "Plastics"). The earliest films were produced in 1945, were not very transparent, were subject to stress cracking, and were difficult to print. As improvements were made and the cost was reduced, the market expanded. By 1956 the price of 1½-mil polyethylene film had dropped below that of 195-gauge cellophane, and there was greatly increased activity in adapting wrapping machinery to handle this material. Cellophane, which had dominated the field for many years, was being threatened with obsolescence. Actually cellophane has suffered very little, and the growth of polyethylene has been in addition to, rather than at the expense of, cellophane.

Attempts to use cellophane wrapping machines for polyethylene film were not always successful. Temperatures had to be controlled more closely, the film had to be pulled instead of pushed because it lacks the stiffness of cellophane, hot plates had to be replaced with moving bands, and surfaces required coating with Teflon to prevent sticking. It was not until the wrapping machines were completely redesigned that good efficiency was achieved.

Polyethylene film offered a whole new set of properties that increased its usefulness far beyond that of cellophane. New techniques such as side-weld bags, hot-wire cutters and sealers, extruders, orienters, stretch, shrink, and blister packaging, and thermoforming, all became possible with the advent of thermoplastic films. Wicketed, rolled, and contoured bags were introduced. Soft packages with ponytails and chubs secured with metal clips became practical.

At the present time over 1 billion lb of polyethylene film is used each year. This does not include 5 million lb of heavy-gauge sheet material

TABLE 8 Uses of Polyethylene Packaging Film

Use	Million lb
Food:	
Fresh produce	160
Bread, cake	85
Meat, poultry	20
Candy	20
Frozen foods	19
Dairy products	12
Dried vegetables	10
Crackers, biscuits	8
Noodles, macaroni	7
Cereals	4
Snacks	2
Other foods	13
Nonfood:	
Shipping bags, liners	95
Textiles	90
Laundry, dry cleaning	75
Rack and counter	50
Paper	30
Miscellaneous	30

used for thermoforming. About 7 million lb of film is used for shrink packaging. The dollar value of polyethylene film manufactured in this country is around $400 million per year. About half the film is used for foods and half for nonfoods, such as shipping sacks, soft goods packaging, and laundry and dry-cleaning bags. In the food category, half is for fresh produce, an area that was not covered to any extent by cellophane, and one-fourth for baked goods, for which polyethylene has largely replaced cellophane. (See Table 8.)

Chemistry. Polyethylene is one of the simplest of polymers in composition, being essentially a straight-chain hydrocarbon:

$$
\begin{array}{cccccccccccc}
 & H & & H & & H & & H & & H & & H \\
 & | & & | & & | & & | & & | & & | \\
- & C & - & C & - & C & - & C & - & C & - & C & - \\
 & | & & | & & | & & | & & | & & | \\
 & H & & H & & H & & H & & H & & H
\end{array}
$$

Each molecule consists of hundreds, and sometimes thousands, of carbon atoms in each chain. A small amount of side branching and some cross-linking may be present. The length of the molecules and the amount of branching determine some of the characteristics of the film, and there are hundreds of formulations from which to choose. Additives in the form of UV stabilizers and slip agents are nearly always used, and thus further multiply the combinations that are available. For more information, see Sec. 8, "Plastics."

Characteristics. Polyethylene film is soft and flexible, making it ideal for packaging soft goods and bakery products; the pleasant feel of the package is a sales advantage. The stretchiness of the film is both good and bad. While it provides impact resistance to the package, it makes opening the package difficult for the consumer. Transparency varies, but good clarity can be obtained where necessary, at some sacrifice of other properties such as toughness. High-clarity films may be only half as durable as those which are formulated for impact strength.

As a moisture barrier it is excellent, but for gas transmission it is only fair. It is essentially odorless and tasteless, but for food applications it should be carefully checked. Polyethylene has good chemical resistance, except to some oils and greases. Stress cracking may be a problem with certain products, but there are special formulations that can be used under such circumstances.

Advantages and Disadvantages. Having the lowest cost of all the transparent films is the outstanding feature of polyethylene. Although not as sparkling as some, the slight haze is not a serious drawback for most purposes. Softness is characteristic of polyethylene, and for textile products and bakery goods this makes it an ideal packaging material. Low moisture transmission is an important property, and although the high gas rate may be a disadvantage for some applications, it is an advantage for packaging fresh produce, where escape of carbon dioxide and admission of oxygen prolong the shelf life of fruits and vegetables.

The gas transmission rate decreases as the temperature goes down, a fact to be considered in the case of frozen foods or refrigerated products. A high elongation factor gives good impact strength to flexible containers of polyethylene, especially for heavy items, but it is a source of annoyance to consumers who have to open these packages.

Selection Criteria. With so many varieties of polyethylene film to choose from, it is necessary to rely on the manufacturer for guidance in picking the material best suited for a particular application. A few points will be mentioned here that may help to narrow down the range, but the final selection should be made with the help of a supplier.

There are three important variables that affect the properties of polyethylene film: density, melt index, and molecular weight distribution, that is, the relative proportions of molecular chains of different lengths. Low-density films, for example, in the range of 0.915 to 0.925 have the highest *impact strength,* but are the most difficult to handle on sealing equipment because of their lack of *stiffness.* Films in the density

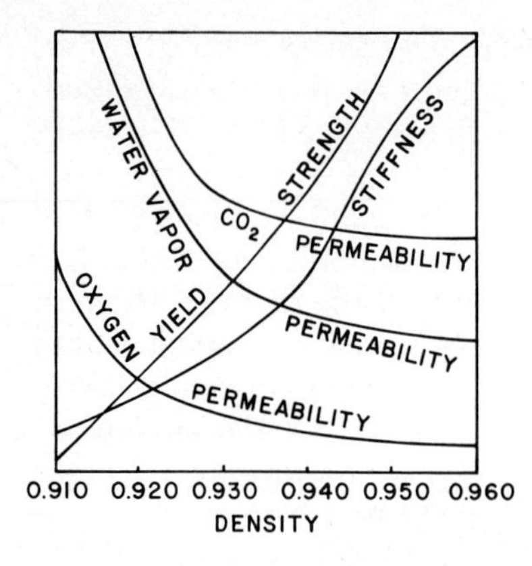

Fig. 8. Physical properties. Effect of density on polyethylene film. Strength and stiffness increase as density goes up, while permeability goes down. Curves cannot be compared, as each is drawn to a different scale; they were put on the same chart merely for convenience.

range of 0.925 to 0.935 are more *machinable,* and at 0.940 polyethylene film is three times as rigid as the lowest-density material. (See Table 9.) Impact strength is also dependent upon molecular weight distribution, and the narrowest distribution gives the *toughest* film, especially when coupled with a low melt index. (See Fig. 8.)

Higher-density materials are also less likely to *block,* so that *slip* agents are not usually required. High-density film makes a package easier to *open,* being less stretchy, and in fact often has a grain to it so that in the machine direction it tears in a perfectly straight line without the need of a tear tape. A high melt index will make for easier *machining,* but there will be some loss of physical properties.

Film that is cast on a chill roll has better *optical* properties but is not as tough as blown film. Extruded blown film becomes oriented in the blowing process so that the *tensile* strength is greatly improved. It also

TABLE 9 Stiffness of Typical Polethylene Films

Density	Modulus of elasticity, 10^5 psi
0.915	0.30
0.916	0.35
0.917	0.40
0.920	0.60
0.925	0.85
0.930	0.95
0.940	1.20
0.950	1.35
0.960	1.55

has a certain amount of shrink in it, unless it has been *heat-set,* and when heat is applied, it will pull down tight around the contents.

Forms and Modifications. The different types of polyethylene films run up into the hundreds. There are variations in density from 0.915, which is very soft and limp, suitable for twist wraps and similar applications, up to 0.960 with greater stiffness, better grease resistance, and a higher sealing temperature, but at the sacrifice of transparency and gloss. Additives to provide slip characteristics for better machining, or stabilizers to resist embrittlement from ultraviolet light rays, or anti-

Fig. 9. Blown tubular film. Polyethylene film is extruded upward from a circular die, and air pressure expands the tube while it is still hot. The polyethylene cools as it moves upward, and near the top the tube is flattened, guided over rollers, and brought down to the winding stand. For flat film instead of tubing, it is simply slit and separated before winding. The diameter of the die and the size of the opening can be varied; combined with take-away speed, such variations determine the final thickness of the film. The air pressure in the tube is carefully controlled to maintain accurate dimensions. The extrusion die oscillates or rotates to distribute variations in thickness and avoid high spots in the winding roll. (*Frank W. Egan & Co.*)

static compounds to minimize the attraction for dust are some of the possible ways to modify these films.

There are three methods of manufacturing film: blown tubing, flat die onto a chill roll, and flat die into a quench tank. Blown film is the cheapest and has the best strength characteristics. Cast film provides more accurate gauge control and better optical properties. (See Figs. 9 and 10.)

Properties. Polyethylene is resistant to most solvents, but at temperatures above 140°F it is attacked by some aromatic hydrocarbons. It is unaffected by acids and alkalies, with the possible exception of hot concentrated nitric acid. Because oils and greases sometimes cause the film to become sticky on the outside, the film should be carefully checked before use with this type of product. Mineral oil, for instance, will go through a 1½-mil low-density film in 4 or 5 days. Polyethylene is a good barrier for moisture, but it allows the passage of gases rather readily. It is a good dielectric, but this makes it difficult to seal by dielectric means because of the large amount of power required. The softening temperature is 210°F for low-density film and up to 260°F for film of 0.960 density. The surface of polyethylene is nonpolar, which means that it is difficult to get adhesives or inks to stick to it. This makes it necessary to treat the surface of the film with flame or corona discharge before printing. (See Figs. 11, 12 and 13.)

Odors and flavors are sometimes lost through polyethylene; this can be a partial loss in which some fractions move out more rapidly than others, causing peculiar results. The film itself is generally considered

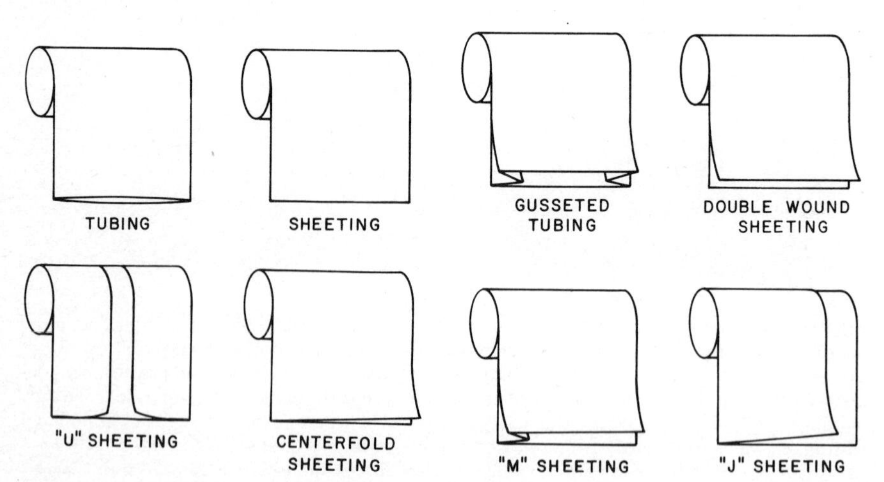

Fig. 10. Types of folds. Available forms of film in rolls. Treated surface for improved adhesion is normally toward the outside.

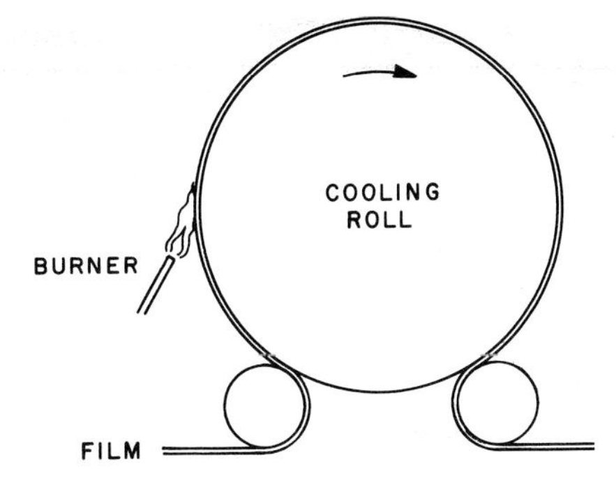

Fig. 11. Flame treatment of polyethylene film. The Kritchever process consists of direct contact of the flame with the polyethylene while the other side of the film is kept cool. An oxidizing flame with a temperature around 3600°F is used, and the cooling roll is maintained at about 75°F.

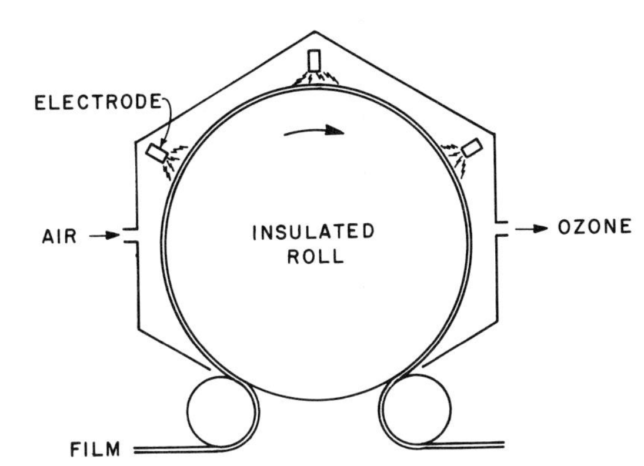

Fig. 12. Corona discharge treatment of polyethylene film. A high-voltage arc going from the metal bars to the drum impinges on the film and polarizes the surface.

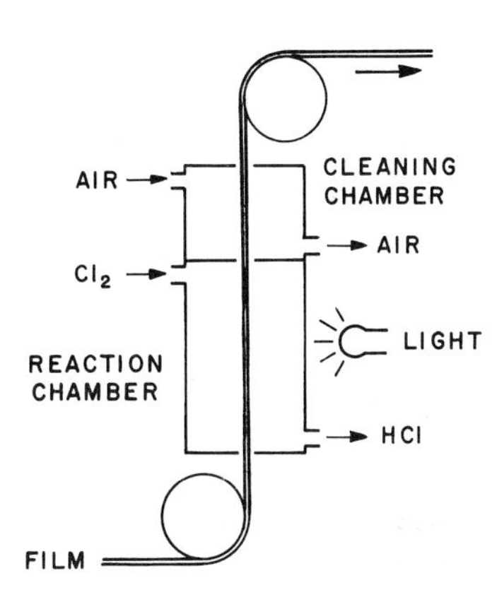

Fig. 13. Chlorination of polyethylene film. One of several chemical methods of making the film receptive to printing ink. Chlorine gas in the presence of light reacts with the molecules of the film to provide a polar surface.

TABLE 10 Effect of Density on Properties of Polyethylene Film

Property	Low density	High density
Water-vapor transmission rate, g/100 sq in./24 hr/mil	1.20	0.25
Oxygen transmission rate, cc/sq m/24 hr/mil	8,900	2,200
Carbon dioxide transmission rate, cc/sq m/24 hr/mil	27,000	5,400
Turpentine greaseproofness, hr for 2-mil film	2	168
Tensile strength, psi at 20 in./min	2,000	3,400
Impact strength, oz for 27-in. drop-dart for 2-mil film	4.5	1.5
Elmendorf tear strength, g/mil	150	75

to be odorless and tasteless, but if it is used with foods, it should be thoroughly tested. (See Tables 10 and 11.)

Processes. Oriented films can be used as shrink or stretch films, but the amount of stretch built into the film is generally not over 25 percent, which is less than in other oriented films such as polyvinyl chloride. Heat sealing can be accomplished with hot-wire, hot-bar, impulse, and band sealers, but not with dielectric sealers, at least not very efficiently. Hot-bar sealers require a Teflon covering to keep the film from sticking. (See Fig. 14.)

Printing by gravure or flexographic processes is easily done, provided the surface has been treated by flame or otherwise oxidized. Flexography is usually the method of choice, because of the lower plate costs. Tinted or colored film is readily available, and embossing can be done by means of engraved casting rolls at the time the film is made. (See Fig. 15.)

FDA Approval. Polyethylene itself is acceptable for packaging foods and drug products, provided no unacceptable additives or mold release agents are used in the manufacturing processes.

TABLE 11 Slip Characteristics of Polyethylene Film

Type	Kinetic coefficient of film-to-film slip
High slip	0.1–0.2
Medium slip	0.2–0.5
Low slip	0.5–1.0

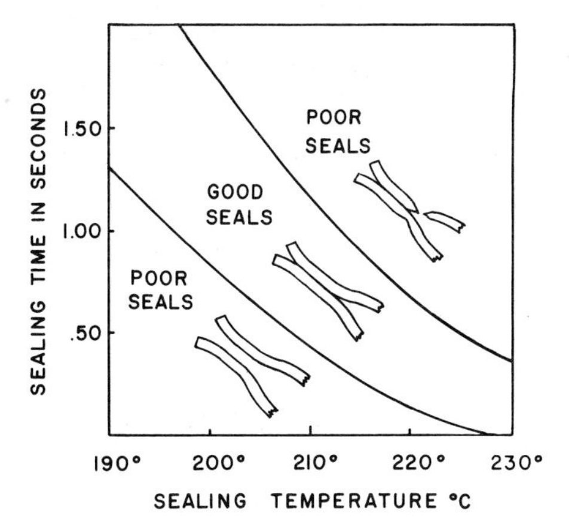

Fig. 14. Heat-seal conditions. Temperature and time relationship for sealing polyethylene. Insufficient time or temperature results in seals that separate at the interface. Too much will cause weakness adjacent to the seal area.

Fig. 15. Print designation. Variations in printing on roll. Direction must be specified, as well as distance between repeats and location of electric eye spot.

TABLE 12 Cost of Polyethylene Film

Type	Cost per lb
Frosty, low-clarity film	$0.34
High-impact, good-clarity film	0.37
Medium-density, stiff film	0.39
6-in. tubing, good clarity	0.42
4-in. tubing, good clarity	0.55
2-in. tubing, good clarity	0.75

Costs. Since polyethylene has the lowest specific gravity of all plastics except polypropylene, the yield per pound is quite good. The base cost will depend somewhat on the type of polyethylene used, but packaging film runs in the neighborhood of 30 cents per pound when the resin is selling below 15 cents per pound. Film with good clarity and toughness may be as high as 38 cents a pound. Generally the spread between resin and film prices is around 14 cents per pound. Because of the exceptionally good yield of 30,000 sq in. per lb in 1-mil thickness, the cost per thousand square inches is down around 1 cent. The minimum quantity of a special gauge or formulation is usually 500 lb. (See Table 12.)

Trade Names. Alamo Industries, Inc.; Alathon (E. I. du Pont de Nemours & Co.); Anar (Rohm & Haas Co.); Auburn Plastics, Inc.; Chicago Molded Products Corp.; Chippewa (Consolidated Thermoplastics, Inc.); Clopay Corp.; Conolene, Conolex (Continental Can Co., Inc.); Crown-Seal, Zee, Poly-Fresh (Crown Zellerbach Corp.); Deerfield Plastics Co.; Durethene (Sinclair-Koppers Co.); Ethylux (Westlake Plastics Co.); Extrudo Film Corp.; Firestone Plastics Co.; Fortiflex (Celanese Plastics Co.); Ger-Pak (Gering Plastics Co.); H & R Plastics Industries, Inc; Jodapak (Joseph Davis Plastics Co.); Katharon (Ross & Roberts, Inc.); Midwest Plastic Products Co.; Plicose (Harte & Co., Inc.); Polyfilm (Dow Chemical Co.); Poly-Plant, Roll-Tex (Texas Plastics, Inc.); Polython (Poly Plastic Products, Inc.); Relthene (Reliance Plastic & Chemical Corp.); Reynolds Metals Co.; Seilon, Inc.; Southern Plastics Co.

POLYPROPYLENE

One of the newer films which is rapidly gaining an important place in flexible packaging is made from polypropylene, a material discovered in 1954 by Professor Giulio Natta. For additional information on the basic resin, see Sec. 8, "Plastics." When film is cast onto a chill roll, a beautifully clear material is obtained that is widely used for textile items and paper goods in the luxury class. It is about twice as expensive as polyethylene, which is more easily made by extrusion blowing, but the superior optical properties have created a demand for polypropylene film in spite of its higher cost.

Chemistry. The molecular structure of polypropylene is a bit more complex than that of polyethylene, because of the CH_3 side groups that are spaced along the chain:

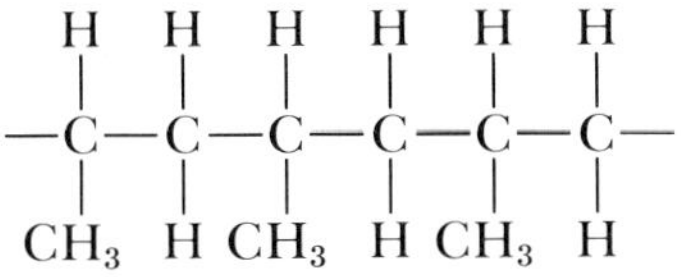

$$
\begin{array}{cccccccccccc}
 & H & & H & & H & & H & & H & & H \\
 & | & & | & & | & & | & & | & & | \\
- & C & - & C & - & C & - & C & - & C & - & C & - \\
 & | & & | & & | & & | & & | & & | \\
 & CH_3 & & H & & CH_3 & & H & & CH_3 & & H
\end{array}
$$

It is necessary to add antioxidants to prevent the quick deterioration that would result from the effects of oxygen and light on the sensitive hydrogen atoms in the side chains. When polypropylene, normally a translucent material, is extruded onto a chill roll, crystallization is prevented, so that an optically clear film is produced. A trace of metallic contamination from excess catalyst in the manufacturing process will have a catalytic effect on oxidation of the hydrogen atoms on the tertiary carbon molecules in the chain. Shelf life of the film is about 2 years.

Characteristics. Polypropylene film is similar to polyethylene in many respects. Stiffness and slip characteristics for machining are very much the same. Thin films are better than thick films at low temperatures, and orienting or adding 1 to 2 percent of polyethylene before extrusion greatly improves the low-temperature properties. Lightweight or soft products are not usually a problem; cold-weather difficulties occur more often with heavy, rigid products.

Heat-seal temperatures will be about 50° higher than for polyethylene, and seals are not as strong. Impact strength is not as good, and below 0°F it is very poor, although orientation of the film helps somewhat, and current practice is to heat-set the film after it is oriented. In appearance polypropylene is sparkling clear and has a soft, warm feel. Resistance to permeation by water vapor and gases is better than that of polyethylene, as is grease resistance.

Advantages and Disadvantages. Polypropylene has good clarity and a nice "hand" for packaging soft goods. Its high melting point makes it suitable for heat-processed products, and it is among the lowest-priced transparent films. It is difficult to heat-seal, especially through several thicknesses. Impact strength is good at normal temperatures, but drops off badly under freezing conditions. Polypropylene has good resistance to chemical attack and is an excellent moisture and gas barrier.

Properties. Polypropylene has greater tensile strength than polyethylene, but impact strength is less because it does not have as much stretchiness. It has good resistance to strong acids and alkalies, and is unaffected by most solvents at normal temperatures, except chlorinated hydrocarbons. At elevated temperatures benzene, xylene, toluene, and turpentine will attack polypropylene, as will strong nitric acid. Resistance to oils and greases is good, and there is none of the stress cracking that occurs with polyethylene. Permeability to moisture and gases is quite low. Flame treating is required prior to printing or using adhesives. As crystallinity decreases, clarity improves and the heat-seal range widens; but if extrusion temperatures are set too high to improve clarity, the strength of the film will be lowered.

Processes. Thermoforming and cold forming are possible with polypropylene. Heat sealing can be accomplished with hot-wire, impulse, or resistance heaters, but the softening range is narrow and the temperatures are critical. Sealing bar temperatures should not vary more than 20° for best results. Hot-bar cutoffs require a smaller radius than for polyethylene, as well as harder backup rolls, modified tensioning devices, and more takeoff belts. Treated film does not give as strong a seal as untreated film; it may differ by as much as 50 percent. The strength of the seal is helped by quenching to avoid crystal formation adjacent to the seal area. Sealing temperatures are 350°F and above, and Teflon coatings are usually needed on the heat-seal bars.

There is a tendency to get "angel hair" with hot-wire sealing, but this can be controlled with proper temperatures and speeds. Films heavier than 0.002 in. are difficult to seal through, especially with gusseted structures. A coating of polyethylene on one or both sides greatly improves the sealability of polypropylene. Dielectric heating is not practical, unless radiant heating also is used just prior to the final sealing. Special coatings, similar to those used on cellophane, are available to give hot-slip characteristics to the film for better machinability. Medium- or low-slip films are generally used for overwrapping; high-slip is better for bag machines.

Cut edges tend to weld together when several thicknesses are cut at the same time. This makes the mouth of a flush-cut bag difficult to open. This disadvantage cannot be entirely eliminated, but it can be lessened by using high-slip film, keeping knives sharp, and cutting quickly. (For more information, see Sec. 10, "Bags and Envelopes.")

Decoration. Embossed film can be made by engraving the chill rolls used in extruding the film. This gives a taffeta finish to the film and makes it appear thicker than it really is. Printing is usually done by flexography, although gravure is sometimes used for long runs. The solvent system for the ink can be equal parts of acetate, toluol, and alcohol. The surface of polypropylene is nonpolar and requires treatment with flame or corona discharge before it will accept ink successfully. This treatment will last from 4 to 6 months, and on medium- or low-slip films it may cause blocking or sticking of the film to itself in the roll. The test for good ink adhesion is to apply pressure-sensitive tape over the printed surface, and pull it off rapidly at right angles to the film. The ink should not be removed in the process; if it is, then the film was not treated properly.

FDA Approval. Some types of polypropylene will meet the requirements of Regulation 121.2501, provided that additives or mold release agents which do not comply are not used.

Costs. The base resin costs only a little more than polyethylene, but converting it into film is more difficult, since it cannot be extrusion-blown as tubing, but must be cast onto a roll or polished steel belt. The cost of the film becomes more than double that of polyethylene, but since polypropylene is the lightest of all plastics, the yield of 31,000 sq in. of film per pound of resin is exceptionally good, and the cost per thousand square inches is around 2 cents for 1-mil film, unoriented, and about 3½ cents for the oriented material. Minimum quantities are usually 500 lb of film for special formulations. (See Table 13.)

Trade Names. Allied Resinous Products, Inc.; Auburn Plastics, Inc.; Bexphane (BXL Industrial Products Group, Ltd.); Bicor (Mobil Chemical Co.); Chevron Chemical Co.; Clopay Corp.; E. I. du Pont de Nemours & Co.; Dynafilm (Alamo Industries, Inc.); Extrudo Film Corp.; FRF (Eastman Chemical Products, Inc.); General Tire & Rubber Co.; Gering Plastics Co.; Hercules, Inc.; H & R Plastics Industries, Inc.; Olefane (Avisun Corp.); Propafilm (Imperial Chemical Industries, Ltd.); Relpro (Reliance Plastic & Chemical Corp.); Southern Plastics Co.; Udel (Union Carbide Corp.); Vypro (W. R. Grace & Co.); Westlake Plastics Co.

POLYVINYL CHLORIDE (PVC)

Vinyl was produced in the laboratory in the early part of the last century by Henri Victor Regnault, but it was not available as a commer-

TABLE 13 Cost of Polypropylene Film

Size, in.	Type	Price per lb*
6 to 18	Unoriented	$0.59
18 and wider	Unoriented	0.64
3 to 72	Heat-set oriented	1.05
5 to 70	Polymer-coated oriented	1.42

* For less than 500 lb add $0.15 to the above prices.

cial product until 1927. Oriented shrink film was developed by Reynolds Metals Company in 1959. Polyvinyl chloride is produced as a rigid film or sheet, but more often it is plasticized to give a soft, pliable material. Present production in this country is about 100 million lb of film and an equal amount of sheeting. For more information see "Polyvinyl Chloride" and "Plasticizers" in Sec. 8, "Plastics," pages 8-33 and 8-41.

Chemistry. Vinyl is made by the chlorination of ethylene or acetylene to produce vinyl chloride $CH_2{=}CHCl$, which is then polymerized with a benzoyl peroxide catalyst at a temperature of about 130°F and some pressure to produce polyvinyl chloride:

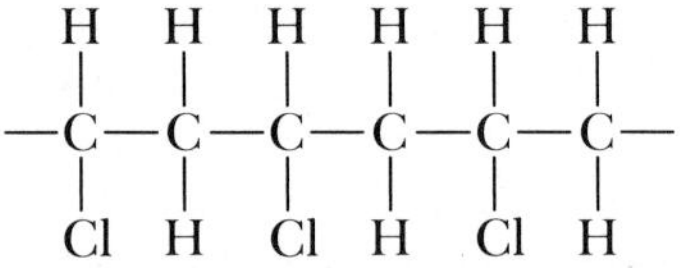

$$
\begin{array}{cccccccccccc}
 & H & & H & & H & & H & & H & & H \\
 & | & & | & & | & & | & & | & & | \\
- & C & - & C & - & C & - & C & - & C & - & C & - \\
 & | & & | & & | & & | & & | & & | \\
 & Cl & & H & & Cl & & H & & Cl & & H
\end{array}
$$

This is in the form of a latex, the polymer being insoluble in the monomer because of branching; it is coagulated with aluminum sulfate, followed by filtration and washing to yield a white powder of polyvinyl chloride.

Characteristics. Vinyl film without plasticizers is stiff and somewhat brittle. It can be given any degree of softness by adding oily materials such as dioctyl phthalate or tricresyl phosphate. PVC has excellent clarity and brilliance, and is among the lowest-price films. It has fairly good barrier properties, is resistant to oils and greases, and has good toughness. When it is oriented, it is one of the most popular shrink films.

Forms and Modifications. The earliest method of producing PVC film was by solvent spreading on a polished stainless steel belt. Because it is so heat-sensitive, the temperature must be closely controlled; consequently extrusion without degradation is almost impossible. Recent improvements in equipment and better techniques, however, have made flat die extrusion and even round die blown extrusion more feasible.

Orientation or stretching in one or both directions will improve its impact resistance. The moisture and gas barrier properties remain the same when the film is oriented, and also when it is put through a shrink tunnel to snug it around a package, these characteristics being a function of thickness only. The tensile strength of oriented film is about four times as much as for cast film of equal thickness, and impact strength goes up about ten times.

Shrink film should be stored in a cool place, as a hot summer day can cause as much as 5 percent shrinkage in 24 hr. This may result in rolls having floppy edges or sagging centers.

TABLE 14 Comparison of Rigid and Flexible Oriented PVC Films

Type	MVT in g per 100 sq in. per 24 hr	cc per 100 sq in. per 24 hr			Tensile strength, psi	Tear strength, g/mil	Elongation, percent	Slip coefficient
		O_2	N_2	CO_2				
Rigid PVC	4	20	3	140	16,000	5	50	½
Plasticized PVC .	14	183	30	785	8,000	60	124	5

Advantages and Disadvantages. Resistance to oils, greases, and waxes, as well as petroleum solvents, is good, provided that the plasticizers or other additives are not extracted by them. Heat in the vicinity of 280°F may degrade the film with the release of hydrochloric acid. Sunlight also can have a degrading effect, causing a yellowing of the plastic, unless stabilizers are added before the film is extruded. Transparency is exceptionally good, and as for cost, only the olefins and styrene are cheaper. Because unpleasant odors sometimes develop from oxidation of the stabilizers, the film should be carefully tested over a period of a week or more before it is used with food products.

Properties. The characteristics of polyvinyl chloride film are closely related to the additives that are used. These include plasticizers, antioxidants, UV stabilizers, and antistats. In its purest form, PVC turns yellow when exposed to heat or ultraviolet light. The surface is easily scratched or abraded, and high temperatures release corrosive degradation products. Rigid vinyl is resistant to acids and alkalies, except some oxidizing acids. It is not affected by oils, alcohols, and petroleum solvents, but it is attacked by aromatic hydrocarbons, halogenated hydrocarbons, ketones, aldehydes, esters, aromatic ethers, anhydrides, and molecules containing nitrogen, sulfur, or phosphorus. Oxidizing and reducing agents have no effect, nor does chlorine.

Unplasticized PVC has poor impact resistance, but it is a good barrier for moisture and gases. Flavors and odors are generally well preserved. To improve impact resistance and other properties, plasticizers are often added, and as much as 50 percent of a vinyl formulation can be made up of such additives. In selecting films for packaging, it is very important to consider the effects of the additives on the stability and compatibility of the contents. The best plasticizers are highly toxic and must not be used with food products. Some of these materials will migrate, either into other plastics with which they are in contact, or into the product, with disastrous results. Extreme care is necessary in selecting the right formulation, and an extensive test program under actual conditions of use is strongly recommended.

Processes. Vinyl film can be heat-sealed by hot-wire, impulse, resistance, or dielectric heaters, but heating bars should be protected from the corrosive effects of the film when it is heated. It can be adhesive-sealed or solvent-sealed quite readily. Shrink films are available with over 50 percent shrink, and they are used for both stretch packaging and shrink packaging. Machinability is good, and slip characteristics are generally satisfactory. Printing inks are usually formulated with vinyl resins, and flexography is the method generally employed, often with an overprint varnish of polyamide resin in an alcohol solvent to add gloss and scuff resistance.

FDA Approval. Some formulations of PVC are satisfactory for use with food products, but consideration must be given to the type of plasticizer, antioxidant, release agent, lubricant, stabilizer, or impact modifier that is used. There are materials that meet all requirements, but they may not have the best optical or processing characteristics.

Costs. The base resin costs 18 cents per pound for rigid PVC and 16 cents for the flexible type. When made into film, it costs around 70 cents per pound for rigid material and about 50 cents for the plasticized film, depending of course on the plasticizers used. Imported films have been reported as low as 36 cents a pound. Cast film with its better optical properties will run about 18 cents more per pound than the extruded film. With a yield of 21,500 sq in. per lb for 1-mil film, the cost on an area basis is around 3 cents per thousand square inches, putting it in a class with cellophane and about double the cost of polyethylene. Oriented film is about 60 percent higher in cost than the nonoriented film.

Trade Names. Allied Chemical Corp.; American Viscose Div., FMC Corp.; Cadillac Plastic & Chemical Co.; Cary Chemical Co.; Clopane (Clopay Corp.); Cobex (BXL Industrial Products Group, Ltd.); Joseph Davis Plastics Co.; Ethyl Corp.; Fandflex (Rand Rubber Co.); General Tire & Rubber Co.; W. R. Grace & Co.; H & R Plastics Industries, Inc.; Kaye-Tex Manufacturing Corp.; Koroseal (B. F. Goodrich Chemical Co.); Kypex (Rohm & Haas Co.); Oriex (Nixon-Baldwin Chemicals, Inc.); Panta-Pak (Pantasote Co.); Pervel Industries; Resinite (Borden Chemical Co.); Reynolon (Reynolds Metals Co.); Ross & Roberts, Inc.; Rucoam (Rubber Corp. of America); Seilon, Inc.; Shawinigan Chemicals, Ltd.; Southern Plastics Co.; Sumilite (Sumitomo Bakelite Co., Ltd.); Velon (Firestone Plastics Co.); Vitafilm (Goodyear Tire & Rubber Co.); Watahyde (Harte & Co., Inc.).

RUBBER HYDROCHLORIDE

This compound has been known since 1805, and laboratory synthesis was reported by Traun in 1859. A patent was issued in that same year to Engelhard and Havermann, for a method of chlorinating a benzene

Chemistry. Crepe rubber is milled and dissolved in an aromatic hydrocarbon solvent, and then chlorinated by bubbling hydrochloric acid gas through the solution:

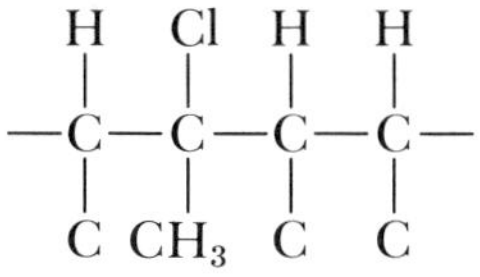

$$\begin{array}{ccccccccc} & \mathrm{H} & & \mathrm{Cl} & & \mathrm{H} & & \mathrm{H} & \\ & | & & | & & | & & | & \\ - & \mathrm{C} & - & \mathrm{C} & - & \mathrm{C} & - & \mathrm{C} & - \\ & | & & | & & | & & | & \\ & \mathrm{C} & & \mathrm{CH_3} & & \mathrm{C} & & \mathrm{C} & \end{array}$$

Rubber molecules tend to form coils, which give them the elastic property that is characteristic of rubber. When the HCl is added, however, the recurrent polar groups that are formed repel one another and discourage coiling of the molecule. In its crystalline form rubber hydrochloride is tough but somewhat brittle. Plasticizers and antioxidants are usually added, but because the hydrochloride is not completely saturated, it tends to harden, especially when exposed to sunlight.

Characteristics. Rubber hydrochloride film is glossy and transparent, but it has a brownish cast which increases with age. It usually has an odor from the antioxidant that is used. Plasticizers that may be used include dibutyl sebacate and dimethoxyethyl phthalate. When the film is stressed almost to the breaking point, it whitens or blushes. It will develop static when dry, but not so much as polyester or the olefins.

Advantages and Disadvantages. The clarity and brilliance of rubber hydrochloride film are excellent. Resistance to acids and alkalies is good, and the compatibility with greases and oils, except some essential oils, makes this a good choice for wrapping meats and meat products. Plasticizers may be extracted by certain types of oils, which leave the film quite brittle. Stretch wrapping and shrink wrapping characteristics are good, and impact strength is very high, so that large heavy objects can be packaged in this material.

The gas transmission rate for carbon dioxide is around 520 cc/24 hr/100 sq in./mil, which is not quite high enough for fresh vegetables. Odors and flavors can be lost through this material because of its low gas barrier properties. With age and exposure to light there is gradual degradation with release of hydrochloric acid, which may affect metals or other acid-sensitive materials. Moisture transmission rates vary from very low for unplasticized film to very high for the softest types. Extensibility and strong heat seals make this film suitable for shrink and skin packaging. Rubber hydrochloride softens at 200°F and cannot be used for boil-in-bag packaging, and at freezing temperatures the impact strength is poor. An objectionable odor is sometimes noticeable from the antioxidant that is used.

Forms and Modifications. There are three general classifications of rubber hydrochloride films, depending on the plasticizers used. The stiffest film is designated as N film, followed by P, then FF, HP, FM, and M.

Properties. The moisture transmission and gas transmission rates vary with the amount of plasticizer used. The less plasticized films are fairly good barriers, but the softer types have very high transmission rates. Resistance is good to most but not all oils and greases, and tests are therefore recommended for products of this type. Acid and alkali resistance is good. Cyclic hydrocarbons and chlorinated solvents will attack rubber hydrochloride. The film softens around 200°F and gets brittle below 0°F. Highly plasticized films for low-temperature applications are very soft and difficult to machine at room temperature. Rubber hydrochloride may whiten or blush when stressed, but this effect disappears in time, more quickly with heat. Gloss and transparency are good, but there may be a brownish cast in thick sections, which increases with age.

Processes. Heat-sealing characteristics of rubber hydrochloride film are excellent, with a broad range of temperatures between 250 and 350°F, and strong seals. Solvent seals also are possible with toluene and a small amount of heat. Dielectric seals can be made, but they take a large amount of power. Printing is usually by flexography for short runs, and gravure for high-volume production. The very soft grades can be troublesome on printing equipment and on wrapping machines. Static buildup also may be a problem on some machines.

Costs. Rubber hydrochloride film is in the middle range of costs, being in a class with the cellulosics. The current price is around $1.10 per pound, but since the yield is fairly good, the cost per thousand square inches figures out to be 4½ cents.

Trade Names. Pliofilm (Goodyear Tire & Rubber Co.); Snug-Pak (Tee-Pak, Inc.).

SARAN (PVDC)

Polyvinylidene chloride has been used as a packaging material since about 1946. The copolymer of vinylidene chloride with vinyl chloride was called Saran by the Dow Chemical Company, and later changed to Saran Wrap to avoid its becoming a generic term. The term "saran" is generally used for any form of polyvinylidene chloride film or coating. It has excellent barrier properties for moisture and gases, but is relatively expensive among packaging films. It is very transparent but has a slight yellowish cast in heavy sections.

Chemistry. Ethylene or acetylene can be chlorinated to yield vinyl chloride:

$$\begin{array}{cc} H & Cl \\ | & | \\ C & = C \\ | & | \\ H & H \end{array}$$

With further chlorination in the presence of calcium oxide, vinylidene chloride is formed, which can be polymerized into long chains:

$$\begin{array}{ccccccccccccc} & H & & Cl & & H & & Cl & & H & & Cl & \\ & | & & | & & | & & | & & | & & | & \\ - & C & - & C & - & C & - & C & - & C & - & C & - \\ & | & & | & & | & & | & & | & & | & \\ & H & & Cl & & H & & Cl & & H & & Cl & \end{array}$$

The dipole of the two chlorine atoms forces the molecule from the normal zigzag pattern to a scalloped configuration that permits close packing into a dense crystalline mass. It is too stiff to be of much use in its pure form, but when it is copolymerized with 30 to 50 percent of vinyl chloride, it gives a flexible material that can be made into a durable film. It is stretched by blowing as it is extruded, and the crystals become highly oriented to yield a very thin, tough film.

Characteristics. Saran is a very soft transparent film that is useful in wrapping food products. It has the best moisture and gas properties of all transparent films, with the possible exception of some very high-priced materials. It has a tendency to cling to itself, which is only partly overcome with silica dusting or slip agents. Saran also stretches rather easily, making it difficult to handle on a machine, and the sharp melting point requires close control of temperatures in the heat-sealing units. Clarity and luster vary from good to hazy, depending on the type of film. The method of manufacture makes saran film highly oriented, and therefore sensitive to elevated temperatures, which cause it to shrink.

Advantages and Disadvantages. The excellent barrier properties of saran film make this a good choice for candies and similar products that must be protected from moisture. Flavors and odors are also maintained to a high degree. It costs two or three times as much as some other transparent films, but good mechanical properties permit its use in very light gauges, which helps the cost situation. The way saran clings to itself makes it very useful for hand wrapping, but causes trouble on automatic equipment. As a shrink film or twist wrap it is excellent for most applications.

Forms and Modifications. Various amounts of plasticizers such as diethyl phthalate give different degrees of softness and cling. Saran film for making bags is sometimes dusted with colloidal silica for ease in opening the bags. With increased slip and less cling there is some loss of transparency. The addition of plasticizers increases the permeability to moisture and gases. One grade of film is specially designed for metallizing, and large quantities are used for Christmas tinsel. Colored film in transparent orange or red is used for packaging cheese and smoked meat. Most saran films shrink when heated, but stabilized films are available that will maintain their dimensions at high temperatures.

Properties. Saran film in 1-mil thickness has a moisture transmission rate around 0.1 g/24 hr/100 sq in., which is the lowest among the commonly used packaging films. The permeability rates for oxygen and other gases also are exceptionally low. Resistance to oils and waxes is good at room temperature, but falls off at elevated temperatures. Impact strength is good under normal conditions, but poor at freezing temperatures. It is not recommended for boil-in-bag packages.

Saran has good resistance to the effects of strong acids and alkalies, except ammonia. It is attacked by halogens, some chlorinated hydrocarbons, ketones, and aromatic ethers, but not by alcohols and aliphatic solvents (except ethers). It is resistant to most petroleum products and turpentine at room temperature. Tetramethylene oxide is a solvent for saran. Shrinkage varies according to type, but can be up to 70 percent in both directions.

Processes. Saran film is difficult to handle on printing and wrapping equipment because of the softness and stretch, but special equipment has been developed for the purpose. Drum-type flexographic presses, for example, have been used successfully for printing on saran. Heat sealing must be done under carefully controlled conditions because of the sharp melting point, and also because of shrinkage in the seal area due to orientation that takes place when the film is made. Sealing bars should be coated with Teflon to prevent sticking. Saran film must be stored in a cool place, as temperatures above 75°F will cause shrinkage over a period of time, and will cause distortion of the rolls and difficulty in handling the film on the machines.

FDA Approval. Most types of saran film are suitable for packaging food products, and large quantities are used for meat, cheese, fish, fruitcake, and fresh produce. Food-grade films have been available since 1946 and are currently being sold for household use in small dispenser packages.

Costs. Because of its high density and low yield, with a fairly high price per pound, saran costs two or three times as much as most other films. At 7 cents per thousand square inches it is about equal in cost to metal foil of equivalent thickness. However, the exceptional barrier properties permit it to be used in lighter gauges than some of the cheaper materials. It is commonly used in thicknesses of 0.0006 and 0.00075 in.

Trade Names. Daran (Dewey & Almy); W. R. Grace & Co.; Rand Rubber Co.; Saran Wrap (Dow Chemical Co.).

STYRENE

Used as an intermediate for the manufacture of synthetic rubber as early as 1925, styrene as a plastic did not become available until the

1930s. It has become one of our most useful materials because of its exceptional clarity and its low cost. It is a very brittle material, however, and until the technique of orientation was introduced in 1958, styrene film could not be used as a free film, but only in laminations. When it is stretched under carefully controlled temperatures, the molecules are rearranged to produce a more flexible film which then becomes suitable for packaging applications.

Chemistry. Ethylene reacts with benzene to form ethylbenzene, which can then be dehydrogenated to make the styrene monomer. Polymerization takes place in the presence of organic peroxides and a small amount of heat to yield polystyrene:

```
       ⬡       ⬡       ⬡
  H    |   H   |   H   |   H
  |    |   |   |   |   |   |
—C——C——C——C——C——C——C—
  |    |   |   |   |   |   |
  H    H   H   H   H   H   H
```

Characteristics. The pure styrene film is brilliantly clear and fairly flexible, but has poor impact strength. It is quite stiff and has a characteristic rattle when it is handled.

Advantages and Disadvantages. The clarity and sparkle of styrene film are outstanding, and its stiffness makes it an easy material to handle on wrapping machines. Barrier properties for moisture and gases are poor, and chemical resistance is very limited; it is therefore generally used only for dry products. Low cost and good optical properties make it very attractive for applications where it is suitable, but poor impact strength, especially at low temperatures, restricts its use to small, lightweight products. Static buildup is a problem in dry weather, causing dust and dirt to collect on the surface of packages on the store shelf, and creating problems on wrapping equipment. It is odorless and tasteless, and approved grades can be used successfully with food products. Flavors and odors are easily lost because of the poor gas barrier properties.

Forms and Modifications. Styrene film and sheet are available as oriented or unoriented material, nearly always without plasticizer. Copolymers with acrylics and rubbers are much tougher than the homopolymer, but with a sacrifice of transparency and at a higher cost. These copolymers are designated in the trade as impact, high-impact, and super-high-impact grades, depending on the proportion of rubber to styrene.

Properties. Optical properties such as haze, gloss, and transparency are the best of all the clear films. The high moisture and gas transmission rates are a disadvantage for some products, but very desirable for fresh produce and baked goods. Styrene is one of the hardest of the plastics and is very susceptible to scratching. Although orientation provides a certain amount of toughness, it is still rather poor in elongation and tear strength. Heat sealing is difficult, impulse-type equipment being the best, but solvent seals and adhesives work very well. Since styrene does not usually contain plasticizers, it ages well and does not require special storage conditions. Moisture absorption is practically nil, and so dimensional stability is excellent.

Processes. Printing and metallizing are readily done on styrene film. With very low moisture absorption and no plasticizers, print registration is good and there is nothing to bleed out and cause problems. Styrene film can be reverse-side printed and laminated so as to "lock in" the decoration, and thus take advantage of its high gloss and sparkle. Static can be a serious problem in handling this film, causing both dirt collection and the clinging of the material to itself. Heat sealing is difficult, and so adhesives are generally used for making bags and for overwrapping. Of the heat-sealing methods, impulse-type equipment works the best. The extreme stiffness of the material helps in handling on fabricating equipment. Styrene film scratches rather easily, and all machine surfaces should be polished or coated to avoid damaging the surface of the film.

FDA Approval. Styrene film complies with FDA regulations for food packaging, provided of course that any additives, release agents, or other components are also acceptable. Since these other materials are seldom used, compliance is not usually a problem.

Costs. Because the specific gravity of styrene is relatively low, the yield is around 26,000 sq in. per lb for 1-mil film. With a price of 45 cents a pound, this gives a figure just under 2 cents per thousand square inches in large quantities for the clear oriented homopolymer film.

Trade Names. Allied Resinous Products, Inc.; Biax (Crane Packing Co.); Bi-Poly-S (Atlantic Refining Co.); Cadillac Plastic & Chemical Co.; Celanese Corporation of America; Dietect A-9 (Dielect, Inc.); Gordon-Lacey Chemical Products Co.; Kardel (Union Carbide Corp.); Ludlow Plastics, Inc.; Nixon-Baldwin Chemicals, Inc.; Polyflex 100 (Monsanto Chemical Co.); Seiberling Rubber Co.; Seilon, Inc.; Southern Plastics Co.; Styroflex (Natvar Corp.); Styrolux (Westlake Plastics Co.); Trycite (Dow Chemical Co.); Visolyte (Elm Coated Fabrics Co.).

SPECIALTY FILMS

A number of plastic films which do not have a broad application in the packaging field may be worthy of mention for unusual requirements.

These are generally higher in cost than the more common materials, but for special applications the higher cost or other limitations might be justified in solving a difficult packaging problem.

Cellulose Acetate-Butyrate. This film is similar to cellulose acetate and cellulose propionate in its properties, but it is somewhat tougher. It often has an undesirable odor, which limits its use for packaging.

Cellulose Nitrate. The mechanical properties of the nitrate film are excellent, but its combustibility makes it too hazardous to be used as a free film for packaging purposes.

Cellulose Triacetate. Similar to the other cellulosics, the triacetate has superior resistance to flexing, but it is not easily thermoformed. It is not affected by most oils, greases, and solvents. Triacetate is generally used for purposes other than packaging.

Chlorotrifluoroethylene. The trade name for the film manufactured by 3M Company is Kel-F, and the Visking Corporation makes a similar material called Trithene. It is a transparent thermoplastic film with a high melting point, and it can be used for surgical materials requiring autoclaving. The moisture and gas transmission rates are very low (10^{-10} g/sq in.), and chemical resistance is excellent. It can be heat-sealed, thermoformed, and printed easily, but surface treatment is necessary for adhesive sealing. Cost is about $1 a square foot.

Edible Film. A transparent starch film made from edible material has been produced by the American Maize Products Company under the name Ediflex, and by National Starch & Chemical Corporation.

Ethylene Butene. Similar to high-density polyethylene, this film has excellent resistance to stress cracking. Impact strength is not quite as good as that of polyethylene. The film has good transparency, but higher moisture and gas permeation rates than polyethylene. It is less stiff and has a softer feel than high-density polyethylene.

EVA Copolymer. Among the copolymers that are used for packaging films, ethylene-vinyl acetate has some interesting properties, which vary according to the different percentages of components that are used. A blend containing 90 percent ethylene will resemble low-density polyethylene, whereas a 70-30 mixture will take on the physical characteristics of gum rubber. Plasticizers are not generally used, and most types have FDA acceptance. Cost of the base resin is around 35 cents a pound.

Fluorocarbon. Best known by the Teflon trademark of Du Pont, this material is remarkable for its nonstick characteristics and a low coefficient of friction. It is a transparent thermoplastic film that can be used over a wide temperature range. It is heat-sealable and has toughness and resistance to most chemicals. The cost is slightly over $10 per pound.

Fluorohalocarbon. Marketed by Allied Chemical Corporation under the name of Aclar, this transparent film performs well at very high and low temperatures. It has exceptional barrier properties and is highly resistant to corrosive chemicals. The price, above $6 per pound, limits its use in packaging.

H-film. Produced by Du Pont, this material has a useful temperature range of −269 to 400°C, and it is infusible and insoluble. It is tough, hard, and flexible, somewhat like polyester film. H film resists gamma radiation up to 10^9 rep, but darkens slightly. Beta rays cause some decrease in impact strength, and exposure to neutron bombardment causes darkening.

Ionomers. These compounds were so named by Du Pont because of ionic bonds in the molecule in addition to the normal covalent bonds that are present in polyethylene. The attraction of sodium, potassium, magnesium, or zinc atoms in the molecule for negatively charged groups provides greater toughness to the film. Surlyn A is transparent and resistant to organic solvents, greases, and oils at room temperatures. Tensile strength and abrasion resistance are good, but tear strength when notched is poor. Special grades can be used for skin and vacuum packaging of food products. Cost is around 50 cents a pound.

Methyl Methacrylate. Intended for outdoor use where weatherability is important, Korad film by Rohm and Haas may have some packaging applications. Optical properties are especially good, and mechanical strength is excellent.

Noryl. This thermoplastic material from General Electric Company is extruded in sheet form 0.005 in. or heavier by Westlake Plastics Company. Its performance does not come up to that of polyphenylene oxide, which it resembles, but it has many of the same characteristics. It is tough and rigid, and has high impact strength. Noryl maintains its strength from −40 to 265°F. With good chemical resistance except to chlorinated and aromatic hydrocarbons, it resists staining and is nontoxic.

Parylene. Film can be produced from Union Carbide's poly-para-xylylene in extremely thin gauges and up to several mils in thickness. It maintains its strength at −330°F, but at temperatures above 175°F it is subject to oxidative degeneration. Parylene film is resistant to attack by organic solvents and does not stress-crack. It is tough under sustained loads, but not under impact. Moisture and gas transmission rates are very low. If a large-volume market develops, the cost is expected to drop to a few dollars a pound.

Polyallomers. Special techniques have been developed for combining polyethylene and polypropylene to give better low-temperature characteristics than the usual copolymers. Other properties fall between those of high-density polyethylene and polypropylene.

Polyethylene Oxide. This film is water-soluble and is used for powder products that can be dropped into water without opening the package. Made by Union Carbide Corporation under the name of Polyox, it can be heat-sealed in the range of 135 to 160°F, indicating that packages may be affected by extremely hot weather conditions. It is resistant to attack by ketones and esters, but is affected by chlorinated hydrocarbons, some acids and alcohols that are not anhydrous, and most organic solvents. Heat and light stability are poor.

Polyethylene oxide film can be heat-sealed or solvent-sealed, but solvents may discolor the seal area. Oriented forms of this material should be useful in shrink packaging, and vacuum forming also is possible. Some types of polyethylene oxide film will machine well, and others are limp and tend to curl. Transparency varies with different resin formulations.

Polyphenylene Oxide. A diphenoquinone known as PPO, this material is offered in sheets 0.005 in. and thicker by Westlake Plastics Company under the trade name Alphalux; it is made from resin supplied by General Electric Company. The useful temperature range is −275 to 375°F, and it is a tough, rigid, and very stable thermoplastic with good chemical resistance except to chlorinated hydrocarbons and some aromatic solvents. Stress cracking is encountered with some organic solvents. Cost of the base resin is above $1 a pound.

Polysulfone. This amber transparent thermoplastic is made from Union Carbide's material by Westlake Plastics Company under the name Thermalux in sheets 0.005 in. and heavier. Impact resistance is not quite as good as that of polycarbonate, but it nevertheless has quite high strength at temperatures from −100 to 300°F. It is abrasion-resistant, odorless, and resistant to acids, alkalies, detergents, and oils, but it is attacked by ketones and chlorinated or aromatic hydrocarbons. Cost is around 10 cents a square foot in 5-mil thickness.

Polyurethane. Available from the B. F. Goodrich Chemical Company under the trade name Estane, this film is transparent, heat-sealable, stain-resistant, and printable, but it discolors in sunlight. Abrasion resistance, elasticity, and resilience give this material some of the qualities of rubber. Impact strength is exceptionally high, moisture and gas barrier properties are good, and resistance to attack by oils and greases is excellent. Polyurethane is affected by strong acids, alkalies, halogens, aromatic hydrocarbons, chlorinated solvents, esters, ketones, and alcohols.

Polyvinyl Alcohol. This is a water-soluble film for packaging dry products that are to be put into water, package-and-all. Manufactured by Mono-Sol Company and Reynolds Metals Company, this film sometimes has a glycerin plasticizer in it. Polyvinyl alcohol film is not affected to any great degree by variations in humidity, but it will dissolve in water in less than a minute. With low humidity the film is more papery, and in damp weather it is inclined to be more stretchy.

Heat seals can be made, but with difficulty because the film clings to itself and sticks to the heating plates. Solvent seals can be made with a dilute water solution of the polyvinyl alcohol. Embossing the film or dusting it with starch improves its slip characteristics. The film also has a tendency to curl when exposed to high-humidity conditions.

It is a tough material, transparent, and resistant to petroleum solvents, greases, and oils, as well as being a good gas barrier, except in the case of water vapor or ammonia. Aldehydes interfere with the solubility of polyvinyl alcohol, and the lower monohydric alcohols have some solvent or swelling action; otherwise the chemical resistance is excellent. When used with detergent powders, the film increases the cleaning power by suspending the dirt when it dissolves.

Polyvinyl Fluoride. Noted for chemical resistance, toughness, and high flex life over a broad temperature range, Teslar is in a price class with the polyester films. It can be heat-sealed, thermoformed, or embossed. Although suitable for greasy, corrosive, or oxygen-sensitive products, it is more often used for other than packaging purposes where outdoor weatherability is a prime requisite.

Silicone. An extremely high permeability rate for moisture and gases is the outstanding property of the silicone film produced by General Electric Research Laboratory. The rates vary with the different gases; for example, oxygen passes through twice as fast as nitrogen. This provides a means of supplying an oxygen-rich atmosphere where this might be desired.

IDENTIFICATION TESTS

It is sometimes necessary to identify a film of unkown origin. Of the several methods that can be used, the simplest and easiest is by visual examination and using the senses of touch, smell, taste, and hearing. The value of such tests is proportionate to the knowledge and experience of the tester, but even the novice can learn something from looking at and feeling the material in question.

Fold it several times to make a number of layers and observe the color and clarity. Is it clear or hazy, watery white or yellowish, crisp and crackly, or soft and sleazy? These qualities are good indicators of the general type of film and will help to narrow down the number of possibilities.

Tearing the film will also help to identify it. The toughness and the way a tear propagates, once it is started, are very good indications of the type of material. Fold it and tear it on the fold. This may be difficult with films having high tensile strength. Then try tearing from a straight edge. Next nick the film and it may be a lot easier to continue the tear, which is characteristic of certain kinds of material.

The way a film burns will give a clue as to its origin. Note whether it burns rapidly, poorly, or not at all. Observe the edge of the film as it burns, and sniff the vapors; and also notice the color of the smoke. Each type of plastic behaves in a different way. (See Table 15.)

Some films give a characteristic green flame with the *copper-wire test.* A clean copper wire is heated and touched to the unknown film, and then put into a flame. A green color indicates a vinyl or rubber type of material.

To determine whether *cellophane* is coated or uncoated, apply water, which will wet out the cellophane if it is uncoated, but will sit on top of the coated film. Acetone will not dissolve the film, although it will remove the coating. Identification of the coating can be made with

TABLE 15 Identification Tests for Plastic Films

Type of film	Visual			Tear test			Burning test												
											Burns slowly with a bead at the edge			Burns poorly					
	Crystal clear	Yellowish clear	Hazy	Easy to tear (notched)	Stretches before tearing	Resists tearing	Burns explosively	Burns like paper	Burns like wax; drips; waxy odor	Burns like wax without dripping; white smoke	Vinegar odor	Rancid butter odor	Marigold odor; black smoke	Sweet odor	Pungent odor	Pungent soapy odor; black smoke	Rubbery odor	Burning hair odor	Green flame with copper wire
Cellophane	X			X				X											
Cellulose acetate	X			X							X								
Cellulose acetate-butyrate	X			X								X							
Cellulose nitrate	X			X			X												
Nylon	X					X												X	
Polyester	X					X								X					
Polyethylene			X		X				X										
Polypropylene			X		X					X									
Polystyrene	X			X									X						
Polyvinyl chloride	X				X	X										X			X
Rubber hydrochloride		X			X	X											X		X
Saran	X					X									X				X

diphenylamine in sulfuric acid, which turns blue with nitrocellulose coatings but not with any other coatings. Pyridine test solution applied to cellophane will indicate a saran coating by turning dark brown.

A more precise method for identifying an unknown film, but requiring some equipment and a little more time, is the measurement of *specific gravity.* Several techniques can be used, but the simplest is to weigh a small amount of the material and put it into a narrow-neck flask. If necessary, cut it into pieces. Add water up to the mark on the neck or level with the top surface. Make sure there are no air bubbles trapped with the film; use vacuum to draw them off if they are present. Weigh the filled flask, and then weigh it filled only with water, at the same temperature. If it is done very carefully, the difference in weight can be converted to specific gravity and compared with Table 16. A 25-cc pycnometer is designed for this type of work and is much more accurate than a flask. Other methods of determining density include floating the film in various liquids of different densities until it sinks. For example, 50 percent methyl alcohol has a specific gravity of 0.92; 44 percent is 0.93; 38 percent is 0.94; and so on.

Another method requiring still more elaborate preparation, but yielding more accurate results, is a *chemical analysis* based on the solubilities of various film materials. Some care is necessary to avoid the hazards of fire, explosion, and toxicity associated with solvents. The film should be cut into small pieces; in the case of sheet material it may be necessary to crush it to increase the solubility rate. The amount of solvent should be at least 10 times the volume of the solid material, and in some cases may need to be 25 times the volume. (See Table 17.)

Probably the most sophisticated technique for identifying plastic films is with *infrared spectroscopy.* This permits an examination of the molecular structure by means of light absorption at various wavelengths. The resulting curve can be compared with charts of known materials, and in

TABLE 16 Densities of Plastic Films

Film	Density	Film	Density
Polypropylene	0.90	Cellulose propionate	1.21
Polyethylene	0.93	Polyurethane	1.24
Polystyrene	1.07	Polyvinyl alcohol	1.25
Rubber hydrochloride	1.11	Cellulose acetate	1.30
Nylon 6/6	1.14	Cellulose nitrate	1.38
Polyester	1.15	Polyvinyl chloride	1.40
Cellulose butyrate	1.18	Cellophane	1.44
Acrylic	1.19	Saran	1.68
Polycarbonate	1.20		

TABLE 17 Solubilities of Plastic Films for Identification

Film	Acetone	Amyl formate	Carbon tetrachloride	Cresylic acid	Cyclohexanone	Dimethyl formamide	Ethyl acetate	Ethyl alcohol	Formic acid	Methyl alcohol	Water	Toluene (boiling)
Acrylic	..	..	I	..	..	..	S	..	..	I		
Cellophane	I											
Cellulose acetate	S											
Cellulose butyrate	S											
Cellulose nitrate	S	S	..	..	..	..	S	I				
Cellulose propionate	S											
Nylon	..	..	I	S	I	..	I	..	S	..	..	I
Polycarbonate	..	..	S	..	..	..	I	..	..	..	..	I
Polyester	I											
Polyethylene	I	..	I	..	..	..	I	..	..	I	..	S
Polypropylene	I	..	I	..	..	..	I	..	..	I	..	S
Polystyrene	S	..	S	..	..	..	S	..	..	I	..	S
Polyvinyl alcohol	..	..	..	..	..	..	..	..	..	..	S	
Polyvinyl chloride	S	..	I	..	S	S	I	..	..	..	..	I
Rubber hydrochloride	..	..	S	..	..	..	I	..	..	I	..	S
Saran	S	..	I	..	S	S	I	..	..	..	..	I

the case of a pure form of the plastic, it will give a reasonably accurate identification. The results can be confused, however, by any additives, coatings, or blending materials that might have been used.

METAL FOIL

Among the flexible packaging materials, metal foil has the most outstanding position. It is the perfect barrier for moisture and gases, far surpassing any paper or plastic material. In appearance it has a luster and color for the most discriminating tastes, and it is widely used for packaging luxury items because of its glamorous appeal. It is fairly high in cost, compared with other flexible materials, bringing about 65 cents per pound. In a 1-mil thickness it will cost about 5½ cents per thousand square inches, but if the thinner gauges can be used, the cost will be brought down quite a bit.

History. Early in the nineteenth century aluminum was first separated from its oxides as a pure metal. The cost was high and it was considered a precious metal. King Christian X of Denmark wore a crown made of aluminum, and Napoleon III had a table service made from this material. For many years the world's greatest scientists tried to find a more economical method of making aluminum. It remained for the son of a Congregational minister, at the age of 22, to show the world how to do it. Charles Martin Hall had worked on this problem while attending Oberlin College in Ohio. Finally on February 23, 1886, in a woodshed behind the family home, with some homemade crucibles and some borrowed batteries, he produced a small amount of pure aluminum. The pellets he produced are preserved in the Smithsonian Institution in Washington, D.C.

In 1898 Ball Brothers started using aluminum covers for the well-known mason jars. The first use of foil was in 1913, for wrapping Life Savers, candy bars, and chewing gum. At the present time over 425 million lb of foil is produced in this country each year, and about half of this is used for packaging. Nearly all of it is aluminum foil; some tinfoil is used, but it is less than 2 million lb per year. About 125 million lb of aluminum foil is used for flexible packaging, 92 million lb for semirigid trays, 15 million lb for barrier material in composite containers and decorative labels, and 5 million lb for cap liners and seals.

A good definition of foil would be: any rolled section of metal less than 0.006 in. in thickness. Metal foil is available in gauges from 0.00017 to 0.0059 in., and in various tempers and surface finishes. The maximum available width is 62 in., except that some of the heavier gauges may be as wide as 66 in. Variation in thickness is held to a commercial tolerance of 10 percent. Flatness of foil, sometimes called *shape,* is difficult to control and becomes a greater problem as the gauge gets thinner. Additional information on aluminum for rigid packages is given in Sec. 7, "Metal Containers."

Alloys. It is possible to make a 99.99 percent pure aluminum foil; such foil is used for capacitors in the electronics industry. It is expensive and does not have the strength necessary for packaging applications. Most "commercial purity" aluminum is around 99.5 percent pure, with about 0.4 percent iron and 0.1 percent silicon making up the balance. A higher iron content increases the bursting strength, which may be an advantage in strip packaging of tablets and capsules, but it reduces the corrosion resistance of the foil. For pie plates and trays 1.25 percent manganese and 0.2 percent copper are often added to provide greater stiffness.

In flexible packaging, alloys 1100 and 1145 are most commonly used. For pie plates and where the heavier gauges are employed for stiffness, alloy 3003 is the most popular. The first digit of these alloy numbers indicates the principal added ingredient; thus the 2000 series contains copper, the 3000 series manganese, the 4000 series silicon, and the 5000 series magnesium. (See Table 18.)

TABLE 18 Aluminum Alloys Used in Packaging

Alloy	Composition percent						
	Al	Fe	Mg	Mn	Cu	Si	Cr
High purity	99.99						
1100	99.0	0.45	. . .	0.05	0.2	0.3	
1145	99.45	0.45	. . .	. . .	. . .	0.1	
2024	93.4	. . .	1.5	0.6	4.5		
3003	98.5	. . .	. . .	1.25	0.25		
5050	98.6	. . .	1.4				
5052	97.25	. . .	2.5	. . .	. . .	. . .	0.25

Temper of Foil. The hardness or softness is dependent upon the composition as well as the treatment of the aluminum foil. In the cold-rolling process it becomes work-hardened and must be annealed for most purposes. The rolls of foil are heated to 650°F for about 12 hr, or until they reach the degree of softness desired. As a result of this heating, the rolling oils are burned off, and the foil is said to be "dry." If half hard or full hard foil is to be used, it should be remembered that not all of the lubricant may be completely burned off and that consequently coating or printing may be more difficult.

As it comes off the rolling mill, the foil is in an extra hard temper (H19) because of the strain hardening that takes place when it is reduced in thickness, and it is necessary to anneal it before it can be used. Heavy foils above 0.002 in. are sometimes only partially softened to an intermediate temper (H25 or H27). For most applications, however, it is necessary to have good dead-folding properties, and a dead soft temper, sometimes called 0-temper, is used.

The tensile strength of aluminum foil is highest when it is in the hardened state. When it is annealed, it loses more than half its strength. Alloy 1100, for example, has a tensile strength of 25 lb per inch of width, per mil of thickness, at the extra hard temper (H19). When it is annealed, this drops to about 12 lb per inch of width.

Characteristics. Aluminum foil is used where the ultimate in protection from moisture or oxygen is required. It is also noted for the attractive appearance of its reflective surface. If it is used for its barrier properties alone, foil is best put on the inside of the package or buried in the walls of the container, to protect it from damage. It is very vulnerable to scratching and abrasion when it is exposed. Lacquer or varnish will help to protect it, and such coatings can be tinted to produce some beautiful effects, so that almost any metallic color can be achieved.

Various matte or embossed textures also are available. Aluminum foil in the softest temper will dead-fold with no springback. This facilitates forming it into any desired shape. Unfortunately, it also makes it wrinkle easily, so that great care must be used in handling, to avoid spoiling its smooth surface. Since aluminum is very ductile, it can be stretched quite a bit if the proper tools are used. Pouches are sometimes made this way for bulky items by forming pockets to conform to the shape of the product. However, aluminum has rather poor tensile strength and tears rather easily. For most applications it is usually laminated to paper, both to reduce cost and to give it the stiffness and tensile strength necessary for handling on packaging machinery.

Aluminum foil is generally smooth and shiny, but it is possible to get embossed or brushed textures which are very attractive. Gauges of 0.001 in. and below are usually produced by pack rolling; that is, two layers are put through the rolls together. This produces a glossy finish on one side and a satin finish where the two sheets are in contact.

Advantages and Disadvantages. The superior barrier properties of metal foil put it in a class by itself. The moisture-vapor transmission rate and the gas transmission rate are zero in the heavier gauges and near zero in gauges below 0.001 in. In appearance foil has a luster and color that add glamour to any package. On the debit side it is higher in cost than the commonly used papers and films, especially because it must often be laminated to paper to make it usable.

Properties. Aluminum foil in gauges above 0.0007 in. is impermeable to moisture and gases. Thinner gauges have pinholes that make the foil permeable to a slight degree. Chemical resistance to solvents and greases is good, but resistance to water is only fair. Resistance to acids, except very mild acids, and to alkalies is poor unless it is protected with a coating of wax or lacquer. Aluminum is unaffected by sunlight or by temperatures up to 550°F. At subfreezing temperatures the tensile strength improves slightly. Reflectivity of bright foil is 88 percent, and for dull embossed foil it is about 80 percent.

The number of pinholes in aluminum foil is considered as negligible in gauges of 0.0007 in. or heavier. The chance of finding one or more pinholes in a square foot of foil is about 15 percent at 0.0007 in. and 8 percent at 0.001 in. These pinholes will range in size from 0.0000001 to 0.00003 sq in. In 100 sq in. of 0.00035-in. foil the total area of all the pinholes will be about 0.00004 sq in. (See Table 19.)

Manufacturing. Aluminum is extracted from aluminum oxide, which is widely distributed in the earth's crust. The process consists of dissolving the aluminum oxide in molten cryolite (Na_3AlF_6) and passing an electric current through the mixture. Pure aluminum settles to the bottom and is drawn off and cast into slabs. The surface of aluminum becomes oxidized almost immediately when it is exposed to the air, and every effort is made to avoid getting this oxide mixed in during

TABLE 19 Moisture-vapor Transmission of Typical Aluminum Foils

Thickness of foil, in.	Grams moisture per 100 sq in. per 24 hr at 100°F, 100% RH
0.00035	0.30
0.0005	0.10
0.0007	0.03
0.001	0

remelting or in the rolling operation; otherwise the particles of oxide would produce pinholes in the finished foil. Another cause of pinholes is porosity of the aluminum due to moisture absorbed from the air while it is in the molten state. Oxygen in the moisture combines with the aluminum, releasing hydrogen, which dissolves in the molten metal. When the aluminum solidifies, this gas comes out, leaving the metal porous. This can be minimized by degassing with chlorine.

The slabs of aluminum are "scalped" before they are rolled, to remove the oxide from the surface. Then they are hot-rolled to reduce the thickness sufficiently for coiling (0.200 to 0.250 in.). Great care must be used to avoid overheating, which would produce oxides that might accumulate on the mill rolls and lead to perforations in the finished foil. The aluminum is annealed once at this stage, and then it is cold-rolled to the required thickness without further annealing.

Surface Finishes

S2S	Two sides bright
B2S	Two sides extra bright
M1S	Matte one side
MF	Mill finish

Each pass through the cold-rolling mill reduces the thickness by one-half. For thickness below 0.001 in. two layers are usually put through together for the final pass. This produces a shiny finish on one side and a satin finish on the other, designated matte one side (M1S), as opposed to bright two sides (B2S). This pairing is not only to increase production but also to help control the thickness. Since the foil is being reduced in thickness, it leaves the rolls faster than it enters and thus burnishes the surface to a mirror finish. A certain amount of lubrication is necessary, depending upon the degree of polish required. Kerosene blended with oleic acid, palm oil, or lauryl alcohol is cooled and sprayed generously over the surface of the foil before it passes through the mill rolls. If the oil is not blended with sufficient slip characteristics, the foil will be marked with a herringbone pattern. On the other hand, if it is too slippery, the surface will not be as brightly polished as it should be.

Adhesives. The least expensive material for bonding aluminum foil to paper is sodium silicate solution, which is widely used for cigarette packages and soap wrappers. Other types of water-soluble and water-emulsion adhesives also can be used. For special applications, casein-latex formulations and resin emulsions may be required, but these are more expensive materials. A drying tunnel must be used before the material can be rolled up, and since the foil is a perfect barrier for moisture, the amount of dryness must be carefully controlled before it is trapped by coiling. If a foil-to-foil lamination is made, it will be necessary to use wax or other solvent-free adhesive because evaporation cannot take place through the foil.

A strike-through lamination is sometimes used to provide a heat-sealable material. In this case a lightweight tissue of about 8 to 10 lb per ream is bonded to foil with up to 20 lb of microcrystalline wax per ream. When heat is applied in fabricating a package, it drives the wax through the tissue to form a bond with the adjacent surface.

Coatings. The materials for coating metal foil include nitrocellulose formulations, which may contain plasticizers and dyes to produce a gold-color foil or other decorative colors. Such lacquer coatings may be used also for scuff resistance, for chemical resistance, or for heat sealing. A primer is sometimes necessary for good adhesion of the lacquer to the foil.

Some of the other coating materials that may be used on foil are polyethylene, ethyl cellulose, methacrylates, chlorinated rubbers and rubber polymers. For protective purposes, the amount of coating that is applied to the foil should be about 1½ to 2 lb per ream (3,000 sq ft), and for severe conditions as much as 5 lb may be required. (See Table 20.)

TABLE 20 Properties of Coated and Laminated Aluminum Foil in Comparison with Paper

Material	Moisture-vapor transmission, g		Flavor retention, days
	Uncreased	Creased	
Greaseproof paper.	150	200	1
0.0007-in foil	0.03	0.42	5
0.0007-in foil/heat-seal lacquer	0	0.01	7
0.0007-in foil/MST cellophane	0	0	21

Decoration. Aluminum foil is usually printed by the flexographic process, although for long runs and high-quality color work, gravure printing is sometimes used. To improve the adhesion of the ink, a primer may be required. Colored lacquers are used to give the appearance of gold, copper, or other metallic colors. Aluminum foil also embosses beautifully, and many attractive patterns are being produced.

Primers. It is difficult to get good adhesion to foil because of its smooth nonabsorbent surface. With some types of coatings it becomes necessary to treat the surface with an adhesion-promoting material. This usually consists of a wash coat of dewaxed shellac or a lacquer with a small amount of phosphoric acid, to improve the wetting characteristics. For water resistance, polyvinyl butyral or other similar vinyl resins may be used in the primer. Printing on foil also may require a primer, depending upon the type of ink used. Oil-base inks in particular must have a vinyl or nitrocellulose primer for good adhesion.

Laminations. In the lighter gauges, foil is not strong enough to be used unsupported. It is frequently laminated to kraft paper for maximum strength and stiffness, although other papers may be used if appearance is an important factor. Sulfite papers give the smoothest surface, but glassine can be used if grease is likely to come in contact with the paper side. Various films also can be laminated to the foil; these include cellophane, polyethylene, polyester, and rubber hydrochloride films.

Forming. Aluminum is quite ductile, and for bulky items it is possible to stretch the foil to form a shallow pocket so that pouches can be made which do not have wrinkles in the seal area. Alloy 1100 is a little better than alloy 1145 for this purpose, and the gauge of the foil usually ranges from 0.002 to 0.003 in. for this purpose. The foil is held on top of a female die by a rubber covered clamp, with sufficient pressure to prevent slippage, and air pressure of 50 psi or less is used to force the foil into the die. Under ideal conditions, a depth of draw of one-fourth the width can be obtained, but one-eighth is a more practical depth.

Heat-seal coatings of vinyl or polyethylene may be used without affecting the forming characteristics, and film laminations with foil can also be shaped easily in a die. Polyester film as a laminant adds strength to the foil and increases the depth of draw that can be made. Other films also may be used for this purpose, but polyester seems to work best.

Foils of 0.005 in. or heavier are usually formed with male and female dies, rather than the air pressure technique that is used with the lighter gauges. Where a hermetic seal is not required, the foil is allowed to slip and wrinkle in the sidewalls without stretching, and much deeper containers can be produced in this way. Although it is possible to get smooth sidewalls on a deep-drawn container with more sophisticated tooling, the cost of the equipment would be much higher and this would be reflected in a higher piece price.

FDA Approval. In 1961 eight manufacturers of aluminum foil collaborated in a study of the residues occurring on the surface of foil that might be considered as food additives under the existing food and drug laws. It was determined by this group that the amount of rolling oil usually remaining on the finished surface of the foil after processing was sufficiently small and that aluminum foil can be "generally recognized as safe" under the Food Additives Amendment of 1958 to the Federal Food, Drug, and Cosmetic Act. Heavy foils tend to have more residues than lighter gauges, and hard tempers more than the softer grades. In nearly all cases, however, the residues are less than 2 mg per sq ft, which would amount to less than five parts per million, even if it all migrated into the contents of the package.

Costs. Heavy-gauge aluminum foil costs 56 cents per pound. With a density (specific gravity) of 2.7, a 1-mil thickness will yield 10,250 sq in. per lb, which makes the cost per thousand square inches about 5½ cents. In the lighter gauges which are generally used for laminating, the cost per pound goes up to 64 cents, making the 0.00035-in. foil about 2.2 cents per thousand square inches. When it is combined with an inexpensive paper, the combination costs around 4 cents per thousand square inches. In terms used by the paper industry, and also in the laminating field, the basis weight of 1 mil (0.001 in.) aluminum foil is 42.1 lb per ream (500 sheets 24 by 36 in.).

Suppliers. Aluminum Company of America; Amax Aluminum Co.; Anaconda Aluminum Co.; Archer Aluminum Division of R. J. Reynolds Tobacco Co.; Consolidated Aluminum Corp.; Fairmont Aluminum Co.; Kaiser Aluminum & Chemical Corp.; Mirro Aluminum Co.; Republic Foil, Inc.; Revere Copper & Brass, Inc.; Reynolds Metals Co.; SCAL GP, Paris; Stranahan Foil Co. Inc.; Western Aluminum Corp.

Section 4

Paper and Paperboard

History 4-1
Structure of Paper 4-2
Wood Fibers 4-2
Pulping 4-3
Papermaking 4-4
Drying 4-5
Pulpwood 4-5
Processing of Pulp 4-5
Groundwood Pulp 4-5
Chemical Pulp 4-6
Semichemical Pulp 4-7
Pulping 4-7
Papermaking 4-7
The Beater 4-7
Machines for Papermaking 4-8
Types of Paper 4-13
Glassine and Greaseproof Papers 4-15
Parchment Papers 4-15
Waxed Papers 4-15
Containerboard 4-16
Chipboard 4-17
Tyvek 4-20
Soluble Paper 4-21
Plastic Papers 4-21

HISTORY

The earliest records indicate that paper as we know it today was first made at Lei-Yang, China, in the year 105. The process was invented by Ts'ai Lun, an official in the court of Ho Ti, Emperor of Cathay. In 751 Muslims captured a Chinese paper mill in Samarkand and forced the workers to reveal the secret process. The Muslims brought paper-making into Spain around 950, and with the beginning of book publishing in the 1450s and the regular publishing of newspapers in 1609, it became an important industry in Europe. The first paper mill in America was built in 1690 by William Rittenhouse on the banks of the Wissahickon Creek in Philadelphia. The methods used up to that period produced one sheet at a time, and it remained for a Frenchman, Nicholas-Louis Robert, to develop a continuous process. His first machine was built in 1799 and was patented in England by the Fourdrinier brothers. Later a cylinder-type machine was invented by John Dickenson and installed near Philadelphia in 1817.

Today the United States leads the world in the use of paper and paperboard, with a per capita consumption of 520 lb per year. There are over 5,000 plants engaged in manufacturing and converting paper and board. The chief source of the raw material in this country is wood pulp, although straw, hemp, cotton, flax, and other materials are used to some extent. About 51 million tons are produced annually, and this tonnage is about equally divided between paper (0.012 in. thick or less) and paperboard (over 0.012 in.). Heavy papers in the range of 0.006 to 0.012 in. are sometimes classified as card stock, but for our purposes we will include them with papers. About half of the paperboard goes into corrugated boxes, and one-fourth into folding and setup boxes. The value of paperboard products made in this country is $5 billion a year, and for all paper products it is over $18 billion.

STRUCTURE OF PAPER

Wood Fibers. Wood is about 50 percent cellulose, 30 percent lignin, an adhesive material principally in the middle lamella (see Fig. 1), and 20 percent carbohydrates such as xylan and mannan along with resins, tannins, and gums. In the conversion of wood into paper, the fibers are separated and regrouped, in more or less random fashion, to form a felted sheet of the desired dimensions and properties. The methods used will depend upon the type of wood and the purpose of the finished sheet. Because wood is a natural material, there is considerable variation in its makeup. Different species of trees and different growing conditions will result in widely variable fiber structures.

"Softwood" is the term used in the paper industry for coniferous or needle-bearing trees, and "hardwood" is applied to deciduous trees, which drop their leaves in the fall. The chief difference from the papermaker's standpoint is fiber length. Hardwood fibers are less than 1/10 in. long, whereas softwood fibers will run up to 1/4 in. in length. (See Fig. 2.) Hardwoods therefore will make a finer, smoother sheet, but not as strong as one made from softwood. The climate in which a tree is grown will also affect its pulping qualities; and northern softwoods have shorter fibers and are more like hardwoods than are the southern softwoods.

Wood fibers have various forms, depending on their source and on their function in the tree. The fibers which are formed early in the growing season (springwood) have much thinner walls than those produced later in the year (summerwood), and the springwood fibers are thus more flexible and more compressible than the summerwood.

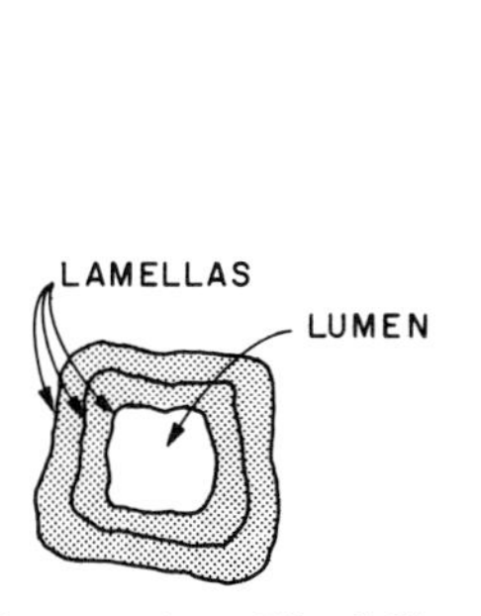

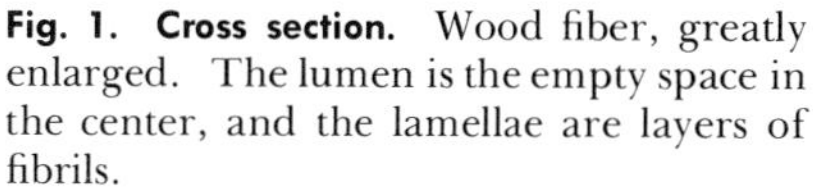

Fig. 1. Cross section. Wood fiber, greatly enlarged. The lumen is the empty space in the center, and the lamellae are layers of fibrils.

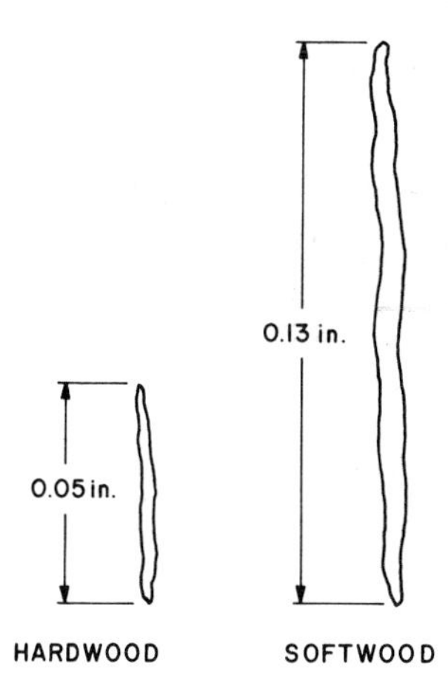

Fig. 2. Wood fibers. Difference in size between typical hardwood and softwood fibers.

Vessels and tracheids are the major types of fibers. Hardwoods have several different kinds of fibers, whereas softwoods have mostly tracheids. A fiber in cross section looks somewhat like a hollow tube of irregular shape. The open center portion is called the lumen, and this is surrounded by a wall made up of layers, or lamellae, of fibrils. (See Fig. 1.) These fibrils, which have diameters of about 0.000001 in., are made up of microfibrils about one-tenth this size and about 0.00005 in. long. These in turn are composed of chains of cellulose molecules, about 3 million in each microfibril, along with short-chain hemicellulose molecules and other polymeric residues.

Cellulose Molecule. Length of chain is from 100 to 3,000 units.

Pulping. The lamellae or layers of fibrils are enclosed by spiral windings called the cambium layer. In the pulping process the fibrils swell and break this outer skin, so that the individual fibrils can extend out and interlock with the hairy surface of other fibers. In papermaking, the water used in the process does not usually go into the crystalline regions of the cellulose molecules, but acts only on the surface of the lamellae. Although some of the short-chain polymers will be dissolved, the principal effect of the water is to get between the lamellae and cause swelling of the fibers. Individual fibers will increase as much as 20 percent in diameter when thoroughly wet, but the length will be only about 3 percent more. The molecular relationship between the water and the fibers is such that the water molecules adjacent to the cellulose are held much more tightly than those which are in a water-to-water relationship. This helps to explain some of the permeability and drying phenomena of cellulose products. (See Fig. 3.)

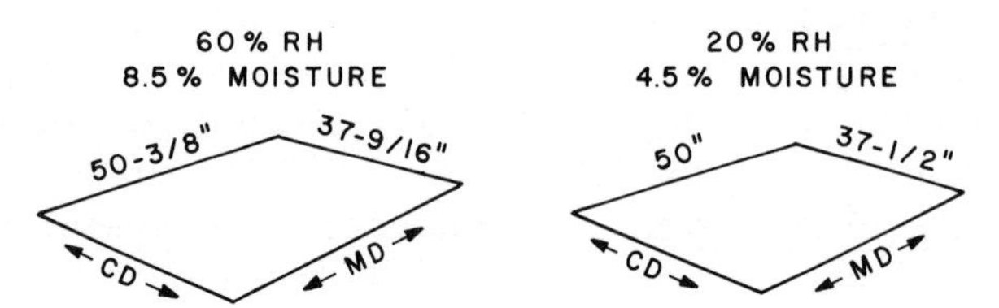

Fig. 3. Shrinkage and expansion. Effect of moisture on paper dimensions. A kraft sheet of the size shown on the left in a damp atmosphere will shrink only slightly in the cross direction, but it will be almost 1 percent smaller in the machine direction when the atmosphere reaches 20 percent RH.

Papermaking. The fibers are suspended in a large amount of water, and as the water is removed, fibers are deposited on a screen which acts as a filter for the remaining suspension. Thin areas, having a higher rate of flow, will cause the fibers to move in their direction, thus tending to give a more uniform sheet. Because coarse fibers, being heavier, will settle more quickly than the finer ones, the side of the sheet toward the wire always has a rougher surface than the top side. (See Fig. 4.) This

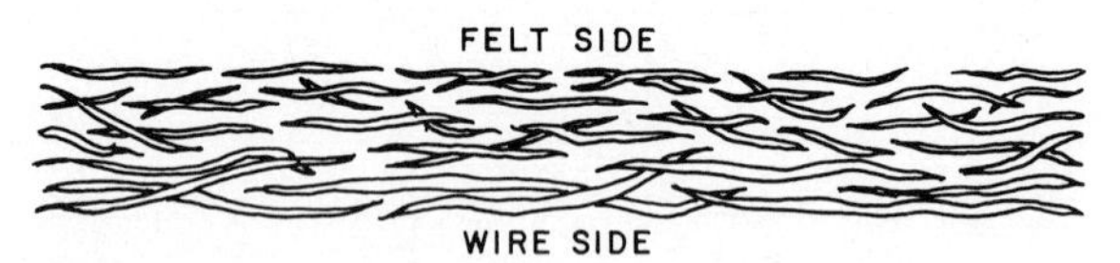

Fig. 4. Surface texture. Coarse fibers settle more quickly than the finer ones; therefore the wire side is always rougher than the felt side.

buildup of fibers on the screen results in a layered structure, with most of the fibers parallel to the surface of the screen, rather than a felted material in which the fibers interweave with other fibers at different levels. Because of this, paper can readily be delaminated into a number of fairly uniform layers. There is also a slight tilt of the fibers toward the wire side in the direction of movement through the machine, caused by the high rate of drainage as the wire passes over the table rolls. Therefore in delamination the cleavage will move to one side or the other, depending on whether it is going in the machine direction or the opposite way. The forces which hold paper together and resist delamination are largely chemical, and where the fibers cross, there is actually a hydrogen bond between them.

Drying. When the slurry is first laid on the screen, the fibers are in suspension, but as the water drains off and the paper reaches a 40 percent solids and 60 percent water ratio, all of this water is within the fibers and none is between them. This is why at this point no more water can be squeezed out but must be driven out by heat. Therefore the function of the entire drier section of the papermaking machine is to dry out the fibers themselves. These fibers shrink as they dry, and if the sheet is very dense, the shrinkage can be considerable. An open, poorly bonded sheet, on the other hand, allows the fibers to slide where they cross each other, and the shrinkage of the finished sheet is greatly minimized. In any case there are strains set up within the fibers at the points where they cross, and anything which affects this strain relationship, such as a change in moisture content, can cause the sheet to expand, contract, curl, or cockle.

Pulpwood. More than half of the pulpwood used in this country for making paper comes from the southeastern and south central states. There is also some from the north central and northwestern part of the country, and a little bit from Canada. Most of this is softwood, but hardwoods, which are from deciduous trees, are being used in increasing amounts, especially for the finer types of paper. Sometimes pulps from different species of trees are blended to provide paper with special characteristics. (See Table 1.)

TABLE 1 Relative Costs of Pulp

Type of pulp	Cost per ton
Groundwood	$ 88
Unbleached kraft	118
Unbleached sulfite	119
Bleached sulfite	134
Bleached kraft	140

PROCESSING OF PULP

Groundwood Pulp. The logs are brought to the pulp mill and the bark is removed and discarded, or used as fuel. Two methods are used to break up the logs to make pulp. Groundwood pulp is made by pressing the logs against a grinder stone while spraying water over the surface of the stone to carry off the ground material. (See Fig. 5.) The stone is cylindrical, and the logs are held against the side, parallel to the axis of

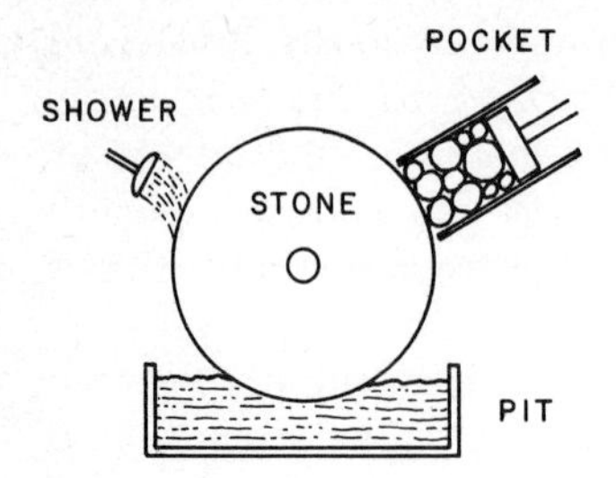

Fig. 5. Groundwood pulp. Made by pushing logs sideways against a grindstone. Water is sprayed on the stone to carry away the fibers and to help keep the stone cool.

the stone, so that the full length of the log is ground at the same time. This method is used only for softwoods, as most hardwoods turn into a useless powder when treated this way. All components of the wood are utilized in this process; that is, the lignin and carbohydrates, which make up half the wood, are not removed. This is in contrast to the chemical process, in which these elements are discarded. About 15 percent of the pulp produced in this country is the groundwood type. It is the cheapest kind of virgin pulp, bringing about $88 a ton, and it is used where low brightness and low strength can be tolerated, as in newsprint and in manila-lined boards. It is not generally used alone in making paper; for example, newsprint is between 50 and 80 percent groundwood, and the balance is made up of chemical pulp for added strength.

Chemical Pulp. The second method of producing pulp is a chemical process rather than a straight mechanical operation. The wood is cooked with chemicals to remove the lignin and carbohydrates, yielding a higher grade of pulp worth $120 to $140 a ton, depending on the type. The oldest of the chemical methods is the *soda process,* discovered in England in 1851, and it is still used for about 2 percent of the pulp produced. In this method caustic soda (sodium hydroxide) and soda ash (sodium carbonate) are used to dissolve the undesirable components of the wood. The soda process is generally used with hardwood pulps.

The *sulfate process,* also known as the *kraft process,* is used primarily on softwoods. Sodium sulfide replaces the sodium carbonate used in the previous method; otherwise the processes are similar. The strongest pulp products are made by this method, and so the name "kraft," which is the German word for "strong," is very descriptive. Sulfate pulp is brown in color and is difficult to bleach, whereas soda pulp is lighter in color and finer in texture, and in strength is somewhere between the groundwood and the kraft pulps. By far the largest percentage of paper pulp is made by the sulfate process.

The *sulfite process* is an acid reaction, in contrast to the soda and the sulfate processes, which are strongly alkaline. The cooking liquor is a solution of calcium or magnesium bisulfite and sulfurous acid. It is generally used on softwoods and yields a light-colored pulp that is stronger than soda pulp but not as strong as kraft. About 15 percent of the pulp produced is this type.

Semichemical Pulp. A combination of chemical and mechanical means also can be used to convert wood into pulp. A soaking in caustic soda or neutral sodium sulfite, to soften the lignin and carbohydrates that bind the fibers together, is followed by grinding in a disk refiner. This method is used mainly with hardwoods and results in a low-cost pulp with most of the lignin retained. Semichemical pulp is difficult to bleach, and it turns yellow when exposed to sunlight. For applications in which strength and stiffness are needed and color is not important, as in corrugating medium, it is often used. About 10 percent of all pulp produced is the semichemical type.

Pulping. The digesters that are used to cook the pulp operate at around 300°F under pressures of about 100 psi. Wood that has been converted into small chips is cooked for several hours and then is blown down into a "blow pit," where it is washed. This is followed by bleaching with one or more chemicals such as calcium hypochlorite, hydrogen peroxide, or chlorine dioxide. Since bleaching reduces the strength of the pulp, it is necessary to reach a compromise between brightness of the finished sheet and its tensile properties. Loss of strength can be minimized to some extent by the choice of chemicals and by bleaching in several stages.

PAPERMAKING

The Beater. The slurry for making paper is composed of 96 percent water and 4 percent solids. This is put into beaters, where the bundles of fiber are broken up and hydration of the fibrils takes place; that is, the fibrils combine chemically and physically with water to become thoroughly "wetted" so that they swell. The cambium layer around each fiber is broken as the fibers swell, and the fibrils open out in a process known as fibrillation. Some shortening or breaking of the fibers is inevitable, but every effort is made to retain as much fiber length as possible.

The Hollander beater, which is still widely used, was invented in Holland around the middle of the seventeenth century. It consists of an elliptical tank with a dividing wall in the center called the midfeather, around which the slurry circulates. (See Fig. 6.) On the bottom of the tank at one end are revolving knives that are set about 0.005 in. apart. After the stock passes between the knives, it travels over a weir called the backfall, which helps to mix the stock, and then around the tank and through the knives again and again until it reaches the proper consistency. A small amount of beating will produce a highly absorbent sheet

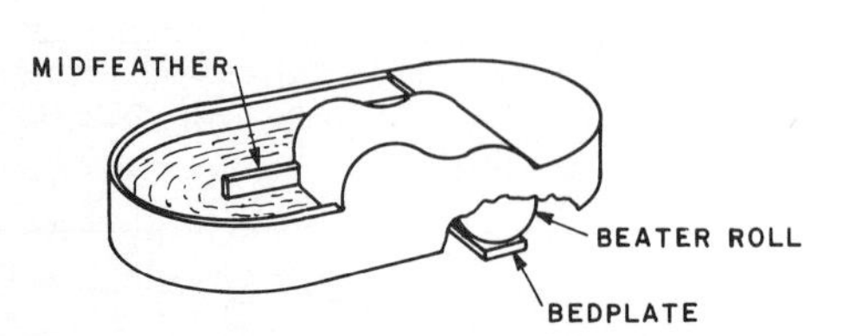

Fig. 6. Hydration of pulp. The Hollander beater is an oval tub with a dividing strip in the center. As the pulp circulates around the midfeather, it passes under a beater roll, which rotates very close to the bedplate, to break open the fibers.

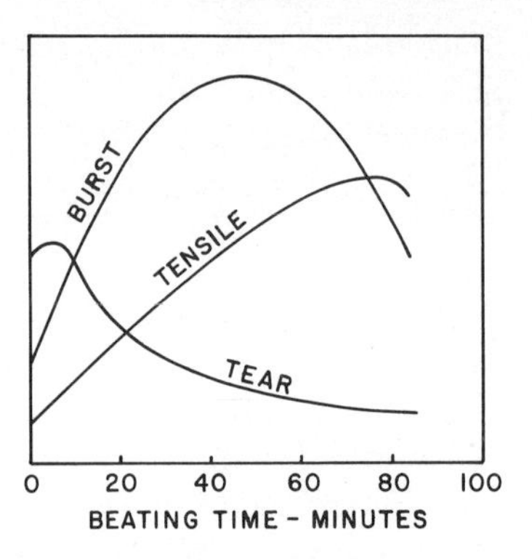

Fig. 7. Beating curves. A small amount of beating makes a very absorbent paper with high tear strength. With more beating the paper becomes more dense, but the tear strength falls off.

with high tear strength but low burst and tensile strength. (See Fig. 7.) With more beating the paper will have higher burst and tensile strength, with a decrease in tear resistance up to a point, but then the burst strength also starts to fall off. A good example is glassine paper, which is carried to almost complete fibrillation.

Sizing materials such as rosin, starch, and papermaker's alum are added in the beater to provide water resistance and ink holdout. Without sizing it would be difficult to print on the paper because the ink would spread and soak into the sheet. The amount of sizing will also affect the behavior of adhesives that may be used later in making packages. Other materials can be added in the beater for color, opacity, stiffness, and similar special properties. These might include titanium dioxide, sodium silicate, diatomaceous earth, casein, wax, and talc. When the stock leaves the beater, it is passed through a refiner such as a jordan. This is a tapered cylinder with rapidly revolving knives set very close together to break up the fibers. From there the mixture of pulp, sizing, and fillers known as the furnish goes into the headbox, ready to start through the papermaking machine.

Machines for Papermaking. The machines for making paper and paperboard are often as long as a city block and several stories high. Some of the newer machines produce paper 30 ft wide at 3,000 ft per min or 800 miles a day, and paperboard 20 ft wide at about half that speed. Two basic machines are used today: the fourdrinier machine and the cylinder machine. (See Figs. 8 and 9.) A third type, which is relatively new, is the Inverform machine, which combines the endless wire of the fourdrinier with the multiple headboxes of the cylinder

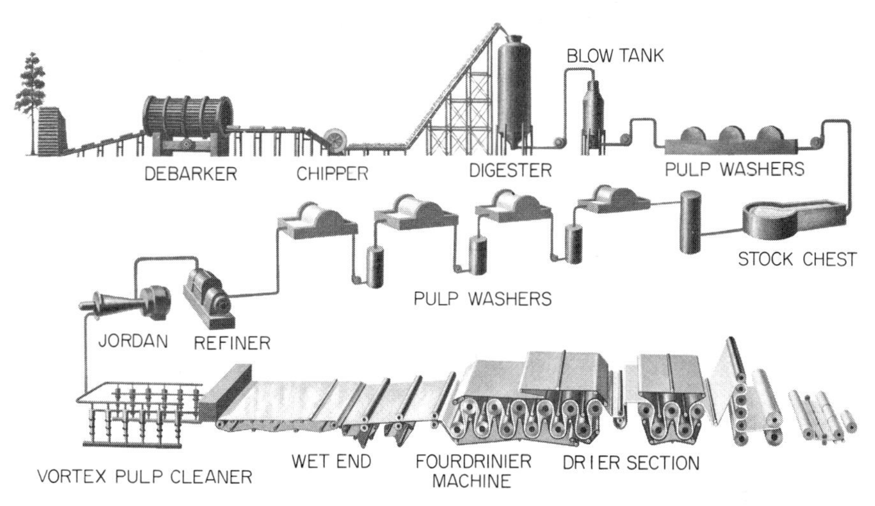

Fig. 8. The fourdrinier machine. The various steps in papermaking are shown here. Logs are chipped and cooked to separate the fibers in the pulp mill, in what is essentially a chemical operation. The fibers are sent to the paper mill, where they are refined by mechanical processes to open up the fibers. A dilute suspension of pulp is spread on a moving screen belt which allows the water to pass through, leaving a mat of wet paper. Pressing, drying, and finishing operations follow.

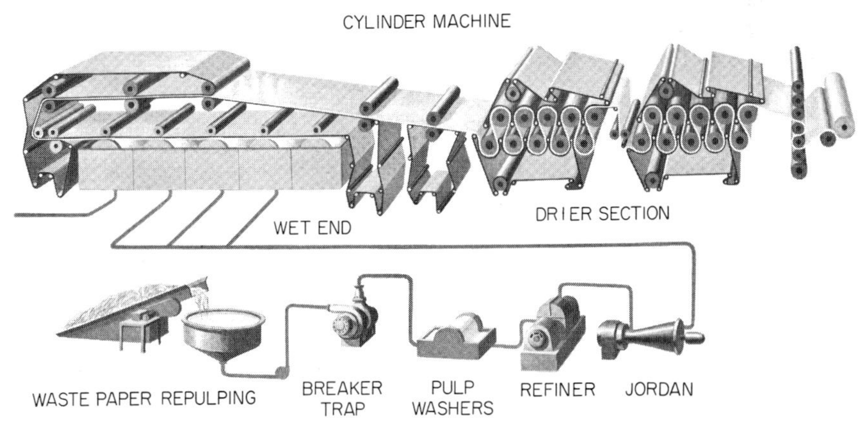

Fig. 9. The cylinder machine. This machine makes heavy grades of paperboard from waste material with a layer of high-grade material on the outside. Screen-covered cylinders pick up a layer of pulp from a vat and deposit it on the underside of a felt blanket. Several layers are built up on the felt and carried into the drier section. The finished paperboard may then be rolled or sheeted.

machine. Thin paper is always made on a fourdrinier, but paperboard above 12 points (0.012 in. thick) can be made on either type of machine. The cylinder machine has a series of six to eight wire mesh cylinders rotating in individual vats. These "wires" carry the pulp upward and deposit it on the underside of a moving felt blanket, which is pressed against the cylinder by a rubber couch roll, each screen adding another layer. Frequently the first and last vats will have a higher grade of pulp, while the vats in between will usually contain waste paper and other reclaimed material blended with a certain amount of virgin pulp. The printing surface which is composed of the highest-quality pulp is called the top liner, and the opposite side is known as the back liner. The plies in between are called the filler. Occasionally an under liner of high-grade pulp is used just below the top liner when the ultimate in quality is sought.

With the fourdrinier machine this combination of materials is not possible because the sheet is formed all at one time from a single vat. The other important difference between these two processes, from a packaging standpoint, is that there is more grain in a cylinder sheet than in a fourdrinier sheet. For example, the stiffness ratio of cylinder kraft linerboard is about 4:1 (machine direction versus cross direction), whereas fourdrinier kraft linerboard has a ratio of approximately 2:1. (See Table 2.)

TABLE 2 Grades of Solid Bleached Sulfate Boxboard in Comparison with Cylinder Board

Type of boxboard	Pounds per caliper point	Stiffness ratio, machine direction to cross direction
Solid bleached	4.1–4.4	2:1*
Low-density bleached .	3.6–4.0	1.76:1
Ultra low density. . . .	3.3–3.7	1.6:1
Cylinder board.	4.0	4:1*

* Solid board is twice as stiff as cylinder board in the cross direction, but about equal in the machine direction for the same caliper. Stiffness increases with caliper; thus 0.024 board is twice as stiff as 0.018 board of the same type.

Whereas the pulp in a cylinder machine is carried upward on a wire mesh roll and deposited on the underside of the felt, on the fourdrinier machine the slurry flows out of the *headbox* through a narrow slot controlled by a long, sharp-edged plate known as a *slice*. It falls on a moving copper-wire screen as it comes up around the breast roll, is carried along over a series of table rolls where some of the water runs out, and then over some suction boxes where more water is removed, so that when the pulp reaches the couch roll at the far end, there is about 20 percent fiber and 80 percent water. (See Fig. 10.) The screen is shaken from side to side about 1/2 in., 200 times per second as it moves along, causing the fibers to be laid in random fashion. A small roll above the wire near the end, called the dandy roll, is used to put in the watermark when one is desired. The pulp leaves the screen at this point and moves unsupported for a very short distance to the press rolls. A woolen felt blanket then carries it between rolls where more of the water is squeezed out.

Fig. 10. Fourdrinier machine. The screen belt can be seen coming up around the breast roll under the headbox in the foreground. The screen moving away from the headbox is carrying a layer of wet pulp received from the slice in the headbox. The water runs through the screen as it travels toward the drier section in the background.

From the press rolls it is carried into the *drier*, where it passes in and out between steam-heated rolls, as many as 150 of them in several stacks, operating at temperatures of 150 to 275°F, until only 4 to 8 percent of the moisture remains. (See Fig. 11.) On some machines a single large roll, about 12 ft in diameter, called a yankee drier, is used in place of the banks of smaller rolls just described, for papers up to about 35 lb per ream.

Surface treatment is sometimes applied while the paper is still on the machine. This consists of dipping the paper into a starch solution and

Fig. 11. Dry end. The drier section of a papermaking machine consists of a long series of steam-heated rolls which evaporate the excess water, so that the paperboard can be rolled up on the roll stand in the foreground.

then squeezing it between rolls. It improves the printing characteristics and makes for a stiffer sheet. The paper can also be *calendered* before it leaves the machine by being ironed through a series of stacked rolls. (See Fig. 12.) This increases the density and smoothness of the surface. As the density is increased, however, stiffness is reduced. This loss of stiffness is proportional to the cube of the caliper reduction. When strength and stiffness are important for packaging purposes, excessive calendering and supercalendering should be avoided.

The variations in processing, along with differences in the starting materials, give an infinite number of different finished sheets. Not only does the same type of paper vary according to manufacturers, but mills within the same company often cannot duplicate each other's products. For consistent results, therefore, it is well to know the source of the papers and boards that are used, and to specify this source rather than depend on specifications that are based upon physical tests alone.

Fig. 12. Calender stack. Supercalender puts a smooth finish on the surface of the paper. As the paper passes between steel or fiber rolls under great pressure, moving from top to bottom of the stack, moisture is added from water boxes in the middle. Various degrees of finish or gloss can be produced, depending on the number of nips and the amount of moisture. Only the bottom roll is powered; the others are held back to cause a certain amount of slippage, which irons the sheet.

TYPES OF PAPER

Two broad categories of paper are being used today: coarse papers and fine papers. Nearly all the papers used for packaging would fall into the coarse paper classification. Fine papers are those which are used for writing paper, bond, ledger, book, and cover papers. The strongest and most useful paper for packaging is that known as natural-color *kraft*. Kraft paper is made from southern softwoods by the sulfate process, in various weights from 16-lb basis weight up to 90-lb basis weight. Coarse papers, but not paperboards or fine papers, are specified by weight of a standard ream 24 by 36 in., usually written 24 × 36—500, which is 3,000 sq ft. Paperboards, however, are measured in pounds per thousand square feet, and range from 26 lb (0.009

in. thick) to 90 lb (0.030 in. thick). Some other ream sizes are shown in Table 3. Comparative costs for different papers are shown in Table 3 in Sec. 13, "Coatings and Laminations," page 13-12.

Kraft paper is sometimes made with no calendering so that when it is made into bags, the rough surface will prevent them from sliding off the pile. More often the paper is given a smooth finish in the calender stack on the paper machine. An MG (machine-glazed) finish can be made on a very large highly polished drying roll.

TABLE 3 Basis Weights for Various Papers

Type of paper	Sheet size, in.	No. of sheets	Total area, sq ft
Paperboard and kraft paper . . .	12 × 24	500	1,000
Bond and ledger paper	17 × 22	500	1,160
Manuscript cover paper	18 × 31	500	1,938
Blotting paper	19 × 24	500	1,583
Cover paper	20 × 26	500	1,806
Tissue paper	20 × 30	500	2,000
Card stock	22 × 28	500	2,139
Wrapping paper	24 × 36	480	2,880
Most coarse papers	24 × 36	500	3,000
Book paper	25 × 38	500	3,299

Creped paper is made by slowing down the press rolls in relation to the wire speed so that the paper builds up and crepes on the press roll. The angle of the doctor blade, which scrapes the paper off the roll, also helps to put the crepe in the paper. Up to 300 percent stretch in one direction can be put in by this method. Extensible paper is made by carrying a web of wet paper between rubber blankets which are under tension, and allowing the blankets to relax as they travel through the paper machine. The paper made in this manner will stretch as much as 7 percent in the machine direction (MD) and 5 percent in the cross direction (CD); it is very useful for multiwall bags and other applications requiring impact strength. *Bleached* and *semibleached krafts* are the strongest white papers, although they lose some of their inherent strength in the bleaching process. *Wet-strength* papers are made by adding melamine formaldehyde or urea-formaldehyde resin to the paper stock. After curing with heat, which takes but a few minutes, or at room temperature over a couple of weeks, paper treated in this manner will retain 20 to 40 percent of its dry tensile strength when saturated with water.

Glassine and Greaseproof Papers. Both of these types of papers are made by prolonged beating of the pulp in preparation for the paper machine. The appearance and properties of these papers are due entirely to the working of the pulp, and not to any additives. Glassine is the same as greaseproof paper, with the exception that it has been heavily calendered to give it the glasslike surface from which it derives its name; that is to say, glassine is made from greaseproof paper, by taking it through one additional step. The term *greaseproof* as applied to paper refers to a class of material rather than its properties. Like glassine, it does have a high resistance to grease, fats, and oils, as its name implies; it is not, of itself, however, moistureproof, although it does provide an excellent surface for moistureproof coatings, such as wax and lacquer, which may be applied as a top coating, and for laminations with the greaseproof paper sandwiched between two sheets. There are various other materials that can be incorporated in this way for special purposes; for example, plasticizers for softness and pliability, release coatings for sticky products, antioxidants to retard rancidity, and mold inhibitors to prevent the growth of molds, yeast, and fungi.

Parchment Papers. Although greaseproof paper is sometimes called mechanical parchment or greaseproof parchment, the term "parchment" is usually taken to mean vegetable parchment. This type of paper is made by passing a web of waterleaf, which is a high-quality unsized chemical pulp, through a bath of sulfuric acid, after which it is thoroughly washed and dried on conventional papermaking driers. The acid attacks the surface of the cellulose fibers, forming a jellylike amyloid, similar to laundry starch, which fills the interstices between the fibers, and yielding a material with high tear strength which has the unique property of being stronger when it is wet than when it is dry. It is available in basis weights ranging from 15 to 120 lb, but the most commonly used weights are 27, 35, and 45 lb. It has good grease resistance, especially in the heavier weights, and has excellent wet strength even in boiling water. It has a fiber-free surface and is odorless and tasteless. Vegetable parchment is not a good barrier for gases, except possibly in the heavier weights, unless it is coated with a material for that purpose. (See Sec. 13, "Coatings and Laminations.")

Waxed Papers. Almost any type of paper can be waxed, and the question of which basic sheet to use generally comes down to a matter of whether it needs to be a high grade of sulfite paper for food packaging or a lower grade for a less demanding application. There are several ways of incorporating the wax in the paper. Either the wax can be added in small amounts as a sizing in the papermaking process, or it can be applied to the finished sheet as a dry wax or a wet wax treatment.

The basic material is paraffin wax, which has a melting point between

115 and 165°F, blended with one of the following: microcrystalline wax ranging from 130 to 190°F; polyethylene in the range of 195 to 255°F; or petrolatum, which melts between 105 and 125°F. Waxed paper is one of the lowest-cost moisture barriers, and it offers good grease resistance and heat-sealing characteristics as well, making it very useful for packaging such things as food, soap, tobacco, and similar products.

When paper is coated with wax and is carried over heated rollers so that the wax is kept fluid and soaks into the paper, it is called *dry wax paper.* If the hot rolls are omitted, the wax solidifies on the surface to produce what is called *wet wax paper.* The wax can be applied either to one side or to both sides of the paper. Generally, dry wax paper does not offer as much protection from moisture as does the wet wax paper. *Wax-sized papers,* in which the wax is added at the beater during the papermaking process, have the least amount of wax and therefore give the least amount of protection.

Containerboard. Two types of containerboard are used in making corrugated board for boxes: linerboard and corrugating medium. Containerboard is the largest single grade of paper that is currently being made in the industry. The *corrugating medium* is made only in a 26-lb basis (0.009 in. thick), and it is usually produced from hardwoods by the sulfate process, although some strawboard is still being made. (See Fig. 13.) The *linerboard* is made from southern pine and other softwoods generally on a fourdrinier machine and is called kraft. Linerboard is made in a range of weights from 26 lb per 1,000 sq ft (0.009 in. thick) to 90 lb per 1,000 sq ft (0.030 in. thick). Some linerboard is being made on a cylinder machine with a filler of waste paper and liners of kraft paper; this is called *jute.* (See Sec. 5, "Corrugated Fibreboard.")

Fig. 13. Surface of paper. Scanning electron photomicrograph of the surface of corrugating medium on the felt side. Magnification is 60X. Composition of the paper is semichemical 80 percent, kraft 4 percent, refined kraft screenings 11 percent, and machine broke 5 percent. (*U.S. Forest Products Laboratory*)

Chipboard. In the manufacture of chipboard, old newspapers and other scrap papers are put through a beater with various sizing materials and fillers. This furnish is then put through the regular papermaking process to produce various kinds of paper and boxboard. When this material is made into lightweight papers, it is known in the trade as *bogus.* This type of paper, which is sometimes indented to give it more cushioning, is widely used to protect glassware and other fragile articles in shipment. Although bogus paper can be used also as wrapping paper, it has very little tensile strength and is most often used merely as a separator or crumpled and stuffed into void spaces as a filler. In order to produce *folding boxboard,* certain bender materials must be added to provide the necessary folding characteristics, and usually a top liner of bleached virgin pulp is put on one side in the papermaking machine. Sometimes the back liner also is made of a better grade of pulp with a higher chemical pulp content, in contrast to the mechanical pulps which make up the major part of the filler; this is called *newsback.* A white liner can be put on the back as well as the front, if this is desired, by using bleached pulp, blended to the brightness needed. Folding boxboards are made in various thicknesses from 13 point (0.013 in.) up to 53 point (0.053 in.) although it is difficult to crease boards for folding boxes that are heavier than 32 point. The vast majority of this type of board ranges from 16 to 24 points in thickness. (See Table 4.)

TABLE 4 Gauges and Densities of Folding Chipboard, Lined and Unlined

Thickness, in.				Weight per 1,000 sq ft, lb
Finish No. 1	Finish No. 2 (most common)	Finish No. 3	Finish No. 4	
0.042	0.040	0.038	0.036	144
0.038	0.036	0.034	0.032	131
0.035	0.033	0.031	0.029	120
0.031	0.030	0.028	0.026	111
0.029	0.028	0.026	0.024	103
0.027	0.026	0.024	0.023	96
0.025	0.024	0.022	0.021	90
0.023	0.022	0.021	0.019	85
0.021	0.020	0.019	0.018	80
0.020	0.019	0.018	0.017	76
0.019	0.018	0.017	0.016	72
0.018	0.017	0.016	0.015	69
0.017	0.016	0.015	0.014	65

Grades of Bending Board

Grades of Bending Board
Chipboard
Cracker-shell board
Mist board for suit boxes
Single manila-lined chip
Bleached manila-lined chip
Patent-coated
Machine clay-coated
Solid bleached sulfate for foods
Extra strength for bottle carriers

Standard Gauges of Folding Boxboard, in Inches

0.011	0.020	0.032
0.012	0.022	0.034
0.013	0.024	0.036
0.015	0.026	0.038
0.016	0.028	0.040
0.018	0.030	

Trade practice permits a tolerance of ±5 percent in caliper or ±7.5 percent in weight.

Standard Finishes for Boxboard

No. 1. *Rough dry.* Low density, rough surface, light weight, high yield per ton.
No. 2. *Medium water.* More dense than No. 1, suitable for ordinary printing. Recommended for most packaging purposes.
No. 3. *Smooth water.* Greater density than No. 2 and smoother, for high-quality printing.
No. 4. *Very smooth.* Highest density with a slick surface, and lowest yield per ton.

Heavier-weight boxboard that is used for setup boxes is made of *nonbending chipboard.* Some of this type of board is too thick to wrap around the drier rolls in the papermaking machine; it is called *wet machine board* because it is taken off the machine wet. This heavyweight board is used mostly for such things as book bindings and shoe board. The thickness of nonbending chipboard is usually specified by *regular number,* which is the number of 25- by 40-in. sheets (1,000 sq in.) in a 50-lb bundle. Thus the lower the number, the heavier the board. Regular number should not be confused with *count,* which is the actual quantity of sheets of any given size in a 50-lb bundle, rather than a "standard area" sheet. Equivalent thicknesses in points of caliper are given in Table 5.

The *brightness* of boxboard is given in reflectance values as a ratio of comparison with magnesium oxide taken as 100. Brightness is highly desirable in a packaging material; in the case of boxboard it is a result of the quality of the furnish and the bleaching method, along with any fillers and coatings that may be used. The figures in Table 6 can be thought of as percentages of a dead white.

There is an increasing demand for *solid bleached sulfate* for critical applications such as butter cartons, bacon wrappers, and frozen food packages. This type of boxboard is used also if high-speed machines require a strong board for consistent operation, and in cosmetic packages, for which it is used to suggest quality. Solid bleached sulfate, sometimes called bleached kraft, is higher-priced than other boxboards,

TABLE 5 Gauges and Densities of Nonbending Chipboard

Regular number in a 50-lb bundle	Thickness, in.				Weight per 1,000 sq ft, lb
	Finish No. 1	Finish No. 2 (most common)	Finish No. 3	Finish No. 4	
35	0.065	0.062	0.058	0.053	206
40	0.057	0.054	0.051	0.046	180
45	0.051	0.048	0.045	0.041	160
50	0.046	0.043	0.040	0.037	144
55	0.041	0.038	0.036	0.033	131
60	0.038	0.035	0.033	0.030	120
65	0.035	0.032	0.030	0.028	111
70	0.032	0.030	0.028	0.026	103
75	0.030	0.028	0.026	0.024	96
80	0.028	0.026	0.024	0.022	90
85	0.026	0.024	0.023	0.021	85
90	0.024	0.022	0.021	0.019	80
95	0.023	0.021	0.020	0.018	76
100	0.021	0.020	0.019	0.017	72
110	0.019	0.018	0.017	0.016	65
120	0.017	0.016	0.015	0.014	60

TABLE 6 Brightness of Typical Boxboards

Type of boxboard	Percentage of reflectance*
Single manila-lined chip.	42–46
Bleached manila-lined chip	56–60
White patent-coated news	66–70
Super white patent-coated news . .	71–75
Machine clay-coated news	74–78
Brush clay-coated (off machine). . .	78–80

* Magnesium oxide as 100.

but because of its superior strength it can be used in a lighter caliper. As a rule of thumb, solid sulfate can be used 2 points lighter than bending chipboard and will perform just as well. (See Table 7.) Folding cartons made from solid board have a greater tendency to bulge, however, because of less stiffness in the machine direction, that is, less grain than cylinder board in the horizontal direction of the box. Scores hold well with aging, and there is less creep or fatigue under compression. Ultra low density with 80 percent of the weight of regular density board retains 88 percent of its stiffness. (See Table 2.)

TABLE 7 Relative Costs of Different Types of Boxboard

Type of boxboard	Cost per ton
Plain chipboard	$109
Filled newsboard.	116
White lined cylinder board, 60 brightness	142
White lined cylinder board, 70 brightness	158
White patent-coated newsboard	158
Clay-coated board, 80 brightness	167
Solid bleached sulfate	227
Cast coated gloss finish	250
High-gloss finish.	320

Tyvek. A new material that has recently become available is spun-bonded high-density polyethylene. Made by Du Pont under the trade name Tyvek, it is being used in place of paper for special applications. In appearance it resembles a slick paper with good whiteness and exceptional strength. It has no grain, and it does not shrink and expand with changes in humidity. It is lint-free and is resistant to staining and to mold and mildew growth. Chemical resistance is excellent (see "Polyethylene" in Sec. 8, "Plastics," page 8-23), and it is said to resist the passage of bacteria into sterile packages made from this material.

Cost is higher than for paper, with tag and label stock selling for about 18 cents per square yard. Most inks suitable for paper or film printing can be used on Tyvek without any special treatment, but some solvents such as xylene and toluene can cause swelling, which will affect registration. There are two types available at the present time. Tyvek 10 series is a stiff material to be used in place of paper. Tyvek 14 series is soft and drapable and can be used for garments. The difference is in the bonding: Tyvek 10 series is bonded throughout, whereas Tyvek 14 series is bonded only at scattered points, giving greater fiber mobility and consequently better flexibility and tear strength in the sheet.

Properties of Tyvek 1073

Thickness, mils	8
Basis weight, oz per sq yd	2.2
Melting point, °F	275
Tear strength, Elmendorf lb	1.2
Fiber size, denier	1
Elongation, percent MD	26
Mullen test, psi	175
WVTR, g per sq m	850
Porosity, sec per 100 cc air	13

Soluble Paper. A new type of paper which dissolves in water is being offered by the Gilbreth Company, Philadelphia, Pennsylvania, under the trade name Dissolvo. It has the appearance and texture of bond paper, and it can be printed and fabricated in much the same way. It is not affected significantly by high humidity, but it dissolves quickly in water. At the time of writing, it had not been approved by the FDA for food use.

Offset printing requires that a minimum of water be used and that the ink be slightly tacky. Otherwise the standard methods of converting paper can be used. Standard thickness is 5 mils, but heavier weights can be supplied. A special type, Dissolvo-A, is soluble in a 2 to 5 percent alkaline solution, but does not dissolve in water that contains no alkali.

Plastic Papers. There have been several attempts to develop synthetic papers, particularly in Japan, where the sources of cellulose for paper are very limited. In addition to Tyvek, which is discussed above, an impact styrene sheet is being produced in two forms by Japan Synthetic Paper Company. These are designated as Q-Kote, which is coated on both sides, and Q-Per, which is uncoated but solvent-etched on the surface. A pigmented high-density polyethylene material is being made by two companies: AcroArt, by the Mead Corporation, Dayton, Ohio; and Polyart, by Bakelite-Xylonite Ltd., London, England.

Bursting strength, tear strength, and folding endurance are somewhat better than for coated paper of the same thickness, as shown in Table 8. Stiffness is considerably less than for paper made from cellulose, and this may cause some problems on printing and packaging equipment. These synthetic papers will not change dimension with changes in humidity, and their wet strength is equal to that in the dry state. They will not grow mold or mildew, and they are greaseproof and moistureproof to a high degree. Chemical resistance far surpasses paper for most purposes. They take printing well, but drying temperatures must be kept low as the softening point of polystyrene is around 175°F.

TABLE 8 Properties of Synthetic Papers

Type of paper	Cost per lb	Cost per 1,000 sq in.	Caliper, mils	Burst factor	Tear factor, MD*	Elongation percent, MD*	Folding endurance, MD*	Taber stiffness, CD* g-cm	Brightness
No. 1 enamel paper	$0.25	$0.039	2.9	18	60	2	150	3.3	84
Q-Kote 150	0.50	0.076	3.6	30	50	11	13,000	1.7	86
AcroArt.	0.72	0.122	3.9	21	460	200	30,000	1.6	91
Tyvek	1.35	0.185	5.0	175	. . .	28	90,000		

* MD = machine direction; CD = cross direction.

An unpigmented high-density polyethylene film is being made in England by M.G.S. (Plastics), Ltd., of High Wycombe, Buckinghamshire, under the name of Tissuthene. A similar product is being made by C. K. Addison & Co., Ltd., of Louth, Lincolnshire under the Finoplas label. It is also reported that a very thin film of this material is being sold in France under the name of Freshpak film.

In a thickness of 1/2 mil, this oriented film resembles tissue paper, while at a thickness of 1 mil it has some of the characteristics of glassine paper. The present price of 40-gauge material is $1.80 per ream for 20- by 30-in. cut sheets. This is about 20 percent less than the cost of an equivalent amount of greaseproof paper.

Section 5

Corrugated Fibreboard

History 5-1
Board Construction 5-2
General 5-4
Manufacturer's Joint 5-4
Corrugations 5-4
Combining 5-5
Flute Selection 5-5
Stacking Strength 5-6
Dimensions 5-8
Scoring Allowances 5-8
Fitting to Folding Cartons 5-10
Manufacturer's Joint 5-10
Design Considerations 5-11
Printing 5-11
Coating 5-14
Costs 5-15
Closing and Sealing 5-16
Testing 5-18

HISTORY

The most common type of shipping container being used commercially today is the corrugated box. The first patents for making corrugated paper were recorded in England in 1856. In the United States the first patents were granted to A. L. Jones in 1871 for an unlined corrugated sheet for packing lamp chimneys and similar fragile objects. The first to use a box made of double-lined corrugated board was a cereal manufacturer, who in 1903 got acceptance in the official freight classification for this type of shipping container. By the end of World War I about 20 percent of the boxes were corrugated or solid fibreboard, and 80 percent were of wood construction. By the end of World War II these figures were reversed, and 80 percent of all shipments were being made in fibreboard boxes. Now over $2 billion worth of corrugated containers are being produced in about 1,200 plants in this country. About half of these are sheet plants, which buy combined board from other plants and do only the printing and cutting.

The proper name for a fibre shipping container, if you want to be precise, is a *box* rather than a carton or a case. The most frequently used style of box is the *regular slotted container,* generally referred to as an RSC, in which all the flaps are the same length and the outer flaps meet in the center. (See Fig. 1.) It is made from a single piece of fibreboard and is shipped flat to the user's plant. There are many other styles of boxes which will be discussed further on, but the RSC type is the mainstay of the box business.

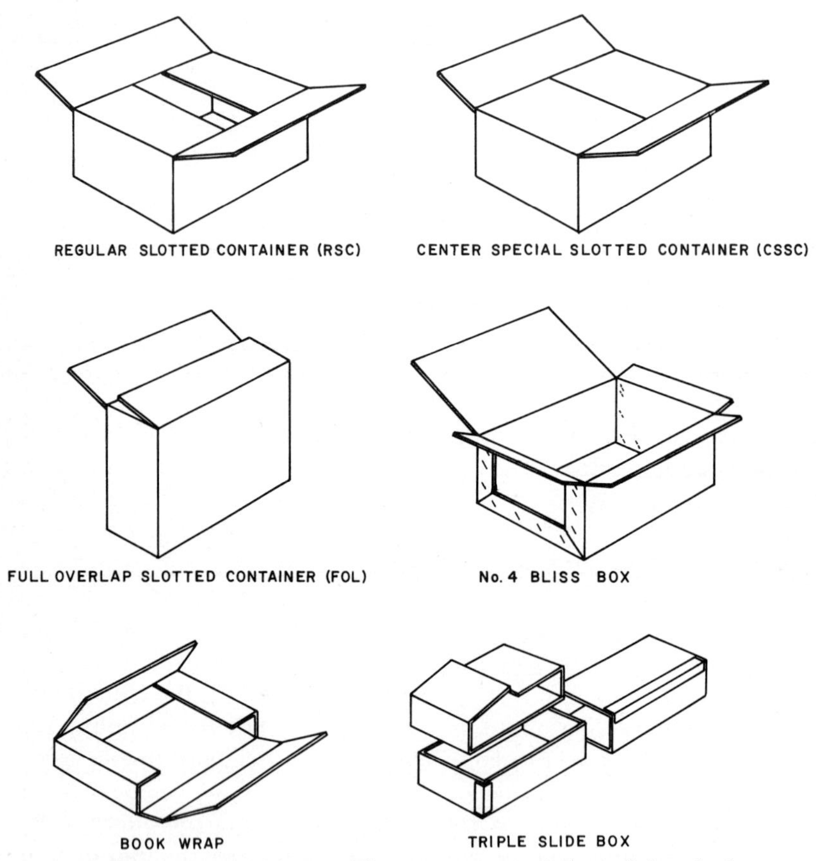

Fig. 1. Styles of corrugated boxes. There are many variations of these basic types.

BOARD CONSTRUCTION

Corrugated fibreboard is either single-faced, that is, a flat sheet of paper to which has been glued another sheet of corrugated paper (or "medium" as it is known), or double-faced, in which case a flat sheet is glued to both sides of the corrugated medium. It is also possible to get double-wall or triple-wall board by alternating additional layers of corrugated and flat sheets. The flat sheets, or facings, will vary in thickness according to the strength required, as shown in Table 1, but the corrugated medium is nearly always 0.009 in. thick and weighs 26 lb per 1,000 sq ft.

TABLE 1. Grades of Linerboard

Weight per 1,000 sq ft, lb	Caliper, in.	Bursting resistance per sq in., lb
26	0.009	65
33	0.012	80
38	0.014	90
42	0.016	100
47	0.018	110
69	0.024	135
90	0.030	175

The standards which we use for containerboard are largely the result of the work done by the railroads, through their Classification Committee, in attempting to reduce damage claims. The minimum specifications which they developed have been written into the DOT regulations and adopted by most of the other carrier groups. This has established the 9-point corrugated medium as the standard throughout the corrugated industry. It has also made the 42-lb linerboard almost universal for all but the most unusual packing requirements. (See Table 2.)

TABLE 2. Corrugations

Flute	Height of medium, in.	Combined height, in.	Number of flutes per foot
A	0.167	3/16	36
B	0.089	1/8	51
C	0.130	5/32	42
E	0.036	1/16	96

Both outer facings are usually the same weight, but occasionally they are "unbalanced." Such an unbalanced sheet has a tendency to warp and may be difficult to handle, either in the box maker's plant or at the point of assembly, and for this reason it is to be avoided unless there is a very good reason for using it.

GENERAL

Manufacturer's Joint. The ends of the box blank can be joined in several ways. The edges may be butted and the joint covered with a suitable tape, or an extended flap can be joined to the adjacent panel by waterproof glue or by metal stitches or staples. The flap may be inside or outside, although it is usually inside. It may be an extension of the end panel and fastened to the side, or vice versa. If stitches are used, they should be 2½ in. apart, although the distance can vary; the applicable regulations should be consulted for individual cases. (See Fig. 2.)

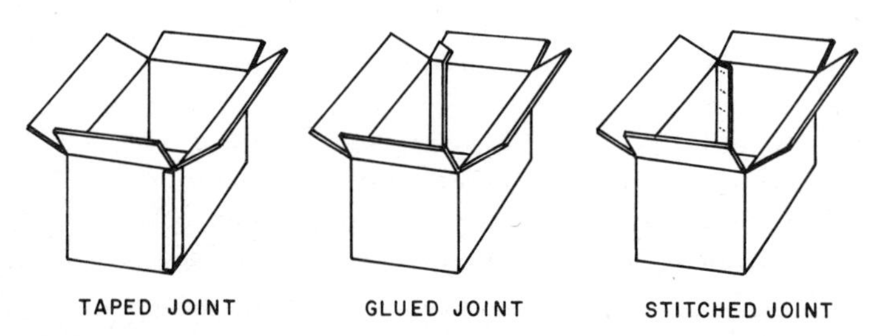

Fig. 2. Manufacturer's joints. The three main types are shown. The glued joint can be made with or without the glue lap extending onto the top flap, and with the glue lap inside or outside.

Corrugations. Normally the direction of the corrugations in a box is vertical, to provide the maximum stacking strength. The interior corrugated parts generally have the corrugations vertical also, although a liner will sometimes have the corrugations in the horizontal direction to withstand the shock from sliding down chutes and from being carried on conveyors, or from humping of freight cars. Actually the difference in the strength of vertical and horizontal corrugations is not very great; reference to the figures in Table 3 shows that B-flute board really has more stacking strength horizontally.

TABLE 3 Stacking Strength of Horizontal Corrugations or Stacking on Side

A-flute horizontal = 80 percent of A-flute vertical
B-flute horizontal = 120 percent of B-flute vertical
C-flute horizontal = 90 percent of C-flute vertical
E-flute horizontal = 150 percent of E-flute vertical

The corrugated medium may be made of kraft, bogus, semichemical, or strawboard. (See Sec. 4, "Paper and Paperboard," for more information on these materials.) Linerboards are generally fourdrinier or cylinder kraft, or a filled sheet made from scrap paper, called jute. In 1949 about 28 percent of all linerboard was jute, and by 1959 less than 14 percent. There is also a shift from cylinder kraft toward fourdrinier kraft, so that the ratio is now about 40:1 in favor of the latter. Cylinder kraft has more grain than fourdrinier, and therefore greater stiffness and tensile strength in the machine direction, but less in the cross direction (flute direction) than fourdrinier. The stiffness ratio of cylinder kraft linerboard is about 4:1 (machine direction versus cross direction), whereas fourdrinier kraft linerboard is said to be more square, with a ratio of approximately 2:1.

Combining. The process of adhering the flat sheet, known as the liner, regardless of which side it is on, to the corrugated sheet, which is called the medium, is called combining. (See Fig. 3.) The machine which is used for the purpose is called a combiner. On the end of the machine are cutting and scoring wheels and a cutoff knife, so that the sheets can be delivered cut to the proper length and width, with creases perpendicular to the corrugations. These creases in the conventional RSC box are the horizontal scores which form the hinges for the top and bottom flaps. The vertical scores are made in a subsequent operation. The adhesive which joins the peaks of the corrugations to the linerboard is usually a solution of sodium silicate or a starch paste, although for special purposes other adhesives may be used. The adhesives used for combining corrugated board are described in Sec. 9, "Closures, Applicators, Fasteners, and Adhesives," page 9-30.

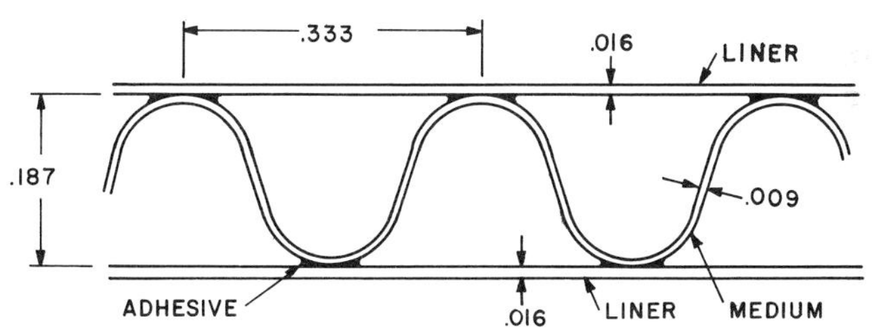

Fig. 3. Basic construction. Structure of corrugated board, showing typical dimensions for 200 test, A-flute board.

FLUTE SELECTION

Whether A-, B-, C-, or E-flute is used depends upon the type of contents, particularly its fragility, density, and self-supporting characteristics. If top-to-bottom compressive resistance is important, as in the case of nonsupporting products stacked to a great height in the warehouse, A-flute is the proper choice. Fragile articles also will receive better cushioning from A-flute fibreboard, except in cases of high density that may indicate a higher flat crush value. (See Table 4.) For greater crush resistance, B-flute with more lines of contact between the corrugated medium and the facing is a better choice. It also has greater strength at the score line, where canned foods have a tendency to tear out, and better end-to-end crush resistance. For very small boxes, B-flute folds more easily and makes a neater-looking package. For interior parts, A-flute is usually more serviceable because of its greater thickness and better cushioning properties, but the density of the contents must be taken into account to prevent complete collapse of the corrugations under impact.

TABLE 4 Flat Crush Values

Flute	Flat crush, psi
A	40
B	57
C	50
E	140

A compromise between A and B which is growing in popularity is a C-flute construction. This gives reasonably good stacking strength and a fair amount of stiffness. For average types of loads it is a good choice, and often is easier to get because of its wide use. A combiner can usually be set up to make only two types of flutes on any machine, and as new plants are built, a decision must be made as to which two it should be. Most of the new equipment is being built to make B- and C-flute, and it is becoming more difficult to find sources for A-flute. For special applications, E-flute has some interesting values. It provides an excellent printing surface, works well in automatic equipment, even folding carton machines in some cases, and takes up less space. Not all suppliers can furnish E-flute, but it is becoming more available as the demand increases.

STACKING STRENGTH

A rule of thumb for long-term storage is to use one-fourth of the compressive strength of a corrugated box as a safe load. For example, if an empty box with flaps sealed was found to collapse under a force of 800 lb in a compression tester, a stacking load of 200 lb on the bottom box in the stack would be the maximum for normal warehouse conditions. A more accurate method would be to calculate the fatigue factor for the length of time the material is expected to remain in storage. To this result should be applied a factor for moisture, depending on the climate and the season. This assumes that the contents is not self-supporting and that the box must support the entire load.

To determine the stacking strength, turn to Fig. 4 and find the value corresponding to the perimeter of the box (length + width inside × 2), the Mullen test, and the type of flute. The figure on the chart is the load that could be supported for a minute or two. Multiply by the percentage figure selected from Table 5, corresponding to the expected storage

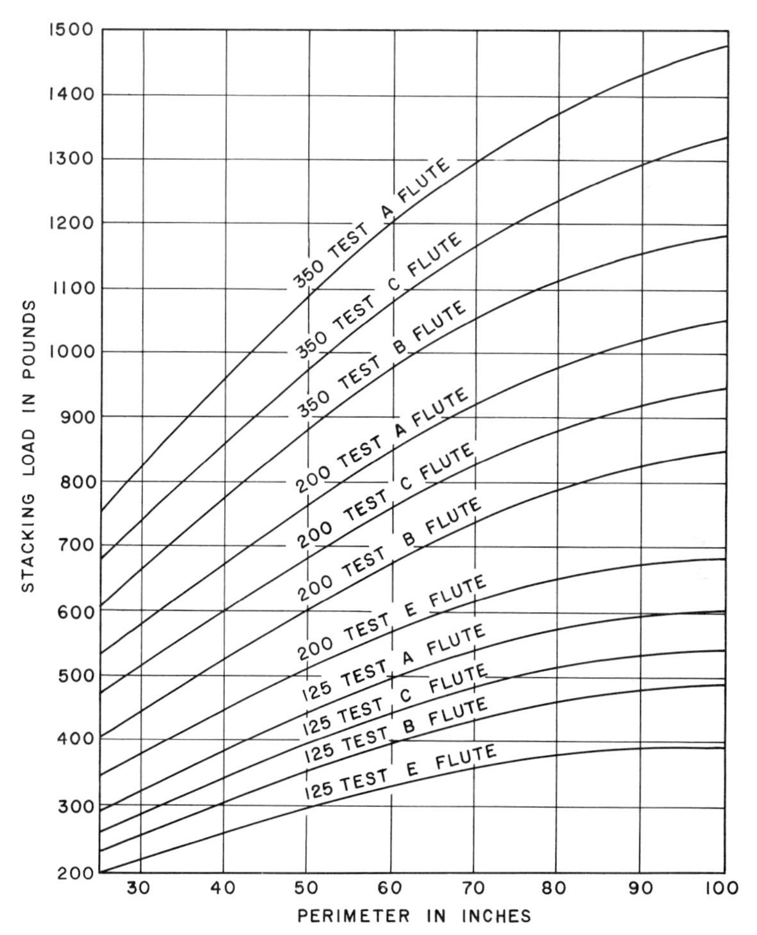

Fig. 4. Compressive strength. Normal stacking strength of regular slotted corrugated boxes. Although this is an oversimplification of a complex set of calculations, it is accurate enough for most purposes. For tall narrow boxes, or large amounts of printing, deduct 10 percent. Double-wall board will be 50 percent stronger than the figures indicated for single-wall board of the same test. Other factors given in Tables 3, 4, and 5 also can be applied to the values shown on the chart.

TABLE 5 Fatigue Factors

Duration of load	Stacking strength, percent
Short term	100
10 days	65
30 days	60
100 days	55
1 year	50

TABLE 6 Humidity Factors

Humidity, percent RH	Stacking strength, percent
Dry	100
25	90
50	80
75	65
85	50
90	40

period. Then choose the highest humidity condition in Table 6 that is likely to occur during this storage period, and multiply by this factor. This will give the maximum safe load that can be supported by the bottom box in a stack, assuming of course that the box is properly made.

DIMENSIONS

The size of a corrugated box is always given in terms of the inside dimensions, with the longer dimension of the opening given first, followed by the shorter dimension of the opening, and finally the depth, or distance between the top and bottom openings. Unless it is otherwise stated, the box maker will always assume that this is what is intended. Sometimes the direction of the corrugations is indicated by underlining the dimension which is parallel with it; otherwise it is assumed that the corrugations are vertical, that is, parallel with the last dimension. (See Fig. 5.)

SCORING ALLOWANCES

When corrugated fibreboard is scored and folded at right angles, the center line of the sheet will intersect the score line; that is, half the

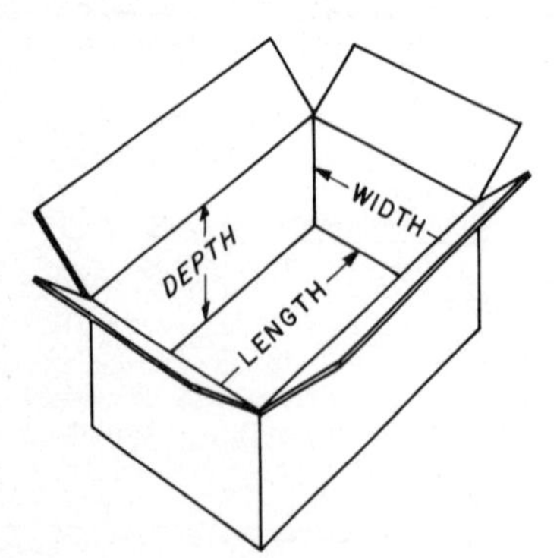

Fig. 5. Size designation. Dimensions are always understood to be inside dimensions. The first dimension always indicates the length, the second the width, and the third the depth.

thickness of the board will be on one side of the score line and half will be on the other side. (See Fig. 6.) Therefore, when the sheet is folded into a rectangle to form a box, the inside dimensions will be less than the score-to-score dimensions by an amount equal to one thickness of board. Likewise, the outside dimensions will be greater than the score-to-score dimensions by one thickness of board. Starting with the inside dimensions that are desired, the settings for the scoring wheels on the printer-slotter can be calculated by adding a certain amount according to the thickness of the corrugated board being used. (See Fig. 7.) If the result is an odd fraction in thirty-seconds or sixty-fourths of an inch, it should be carried to the next higher sixteenth, as box makers do not work closer than 1/16 in. It should also be noted at this point that the dimensional tolerance that is accepted in the trade for the finished box is ±1/16 in. If specifications are written with closer tolerances than this, costs may be affected.

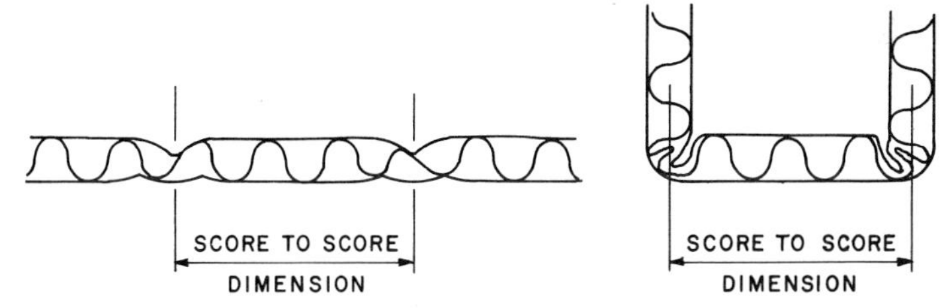

Fig. 6. Allowance for scoring. This is based on the thickness of the corrugated board. For a given inside dimension, the scores should be one thickness farther apart. When the board is folded on the score lines, the scoring dimension falls in the center of the wall that is formed.

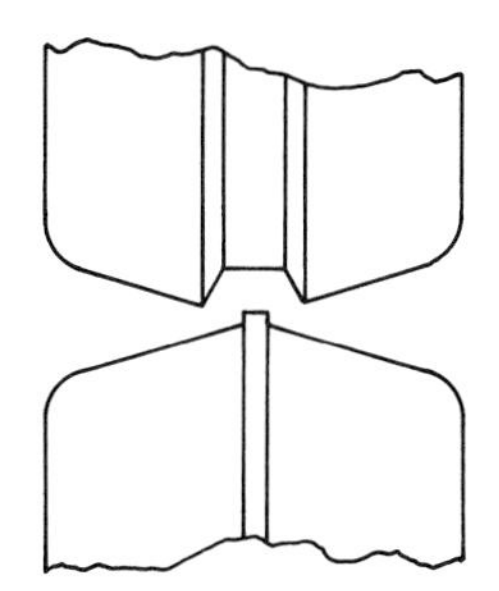

Fig. 7. Scoring wheels for corrugated boxes. The face or "point" on the male wheel on the bottom, which makes the score on the inside of the box, is 0.0625 in. wide. The space between the points of the female wheel is 0.3125 in., although this may vary for different grades of board. There are other styles of wheels, such as "five point" for double-wall board, single V against a flat wheel sometimes used on printer-slotters, and double V used on combiners. The type shown makes the strongest box when used for either vertical or horizontal scores.

The allowances for the horizontal scores of a box are a little more complicated than for the vertical scores. Since the small flaps go inside the large flaps in an RSC box, allowance is made for one top flap and one bottom flap thickness in addition to one thickness of board as before. Thus, the inside height is less than the score-to-score dimension by three thicknesses of board. For calculations the thickness of the various types of board is given in Table 2. If the result is an odd fraction, it can be adjusted up or down to give the desired fit. For good stacking with a dense load it is usually better to work on the tight side.

FITTING TO FOLDING CARTONS

Using the score-to-score dimensions of a folding carton, it is possible to determine the approximate inside dimensions of a corrugated box to hold a multiple of these, at least sufficient for cost purposes. Before starting production, however, these figures should be checked with actual samples.

Allowances for Folding Cartons

Front-to-back	+ 3/32 in. × no. of pieces + 1/16 in.	= ID of box
Side-to-side	+ 3/64 in. × no. of pieces + 1/16 in.	= ID of box
Top-to-bottom	+ 3/32 in. × no. of pieces	= ID of box

MANUFACTURER'S JOINT

The ends of the box blank are brought together and joined by what is called the box manufacturer's joint. There are three methods for making the joints commonly used: taped joints, glued joints, and wire-stitched joints. (See Fig. 2.) The cost is about the same for all three. The tape that is used for making a taped manufacturer's joint is different from the tape that is applied to the flaps in closing the box, and it is considerably stronger. It is usually a lamination of two plies of heavy paper with asphalt or other waterproof material in between. Fibers of sisal or other reinforcing material are laid crosswise in the asphalt. A taped joint makes a box that is smooth inside and out, but it is not as strong as one with a stitched or glued joint. The glued joint is the strongest of all, in most instances. If tape is used for the joint, it will cover a portion of two panels which might otherwise be printed, and the layout for the printing must take this into account.

One disadvantage of a glued joint or a stitched joint is that it will have a bump with a sharp edge where it overlaps. Although the corrugations are crushed at this point in the manufacturing process, the extra thickness will sometimes cause excessive abrasion of the contents. If this is a problem, the lap can be put on the outside, but then it has a tendency to get snagged on conveyor rails or the side frame of trucks and railcars. Another disadvantage of stitches is that they will sometimes rust; if they are expected to be exposed to moisture, they can be specially coated to resist corrosion. It might be well to specify this added protection in any case, as the extra cost is not significant. The lap must be at least $1\frac{1}{4}$ in. wide, and the stitches should not be more than $2\frac{1}{2}$ in. apart in accordance with Rule 41 of the Uniform Freight Classification. The first and last stitch should not be more than 1 in. from the horizontal score lines. The size of the stitching wire should be not less than 0.028 by 0.100 in. in cross section.

DESIGN CONSIDERATIONS

The most economical box for a given cubic contents has the proportions 2 : 1 : 2; that is, the length is twice the width and the height is equal to the length. This uses the least amount of board to enclose a given amount of space, using the RSC construction, and while it does not take into account the cost of material used for the manufacturer's joint, it is a satisfactory guide for all but the smallest boxes. A container which is twice as long as it is wide has the further advantage of interlocking in a stack to form a more stable pile; each box can be placed at right angles to the ones below it. A perfect cube is just about the poorest shape for warehousing because it cannot be interlocked.

If a choice can be made in the arrangement of the contents, it is well to give some thought to the proportions of the box in terms of maximum utilization of space in storage and shipment. The width of conveyors, size of truck bodies and railcars, and pallet patterns all have some bearing on the design of the optimum box shape. A standard 42- by 48-in. pallet ideally should take the container in an interlocking pattern with not more than 1-in. overhang and with a minimum of open space. (See Fig. 8.)

PRINTING

It is possible to print slotted boxes in one or two colors from rubber dies very economically, in the same operation which makes the slots and vertical scores. The machine which performs this operation is known as a printer-slotter. A third color, however, may increase the cost considerably, since most printer-slotters are limited to two colors, and more than this number will require running the box blanks through the machine twice. It is even possible to use halftones and process colors, but these require a long run to justify the complex setup that is necessary for such printing.

The printing dies are made of fairly soft rubber. If the rubber is too hard, it will crush the corrugations and weaken the box. Various typefaces are available, and logotypes or special designs may be engraved directly in the rubber or made up as woodcuts from which the rubber is

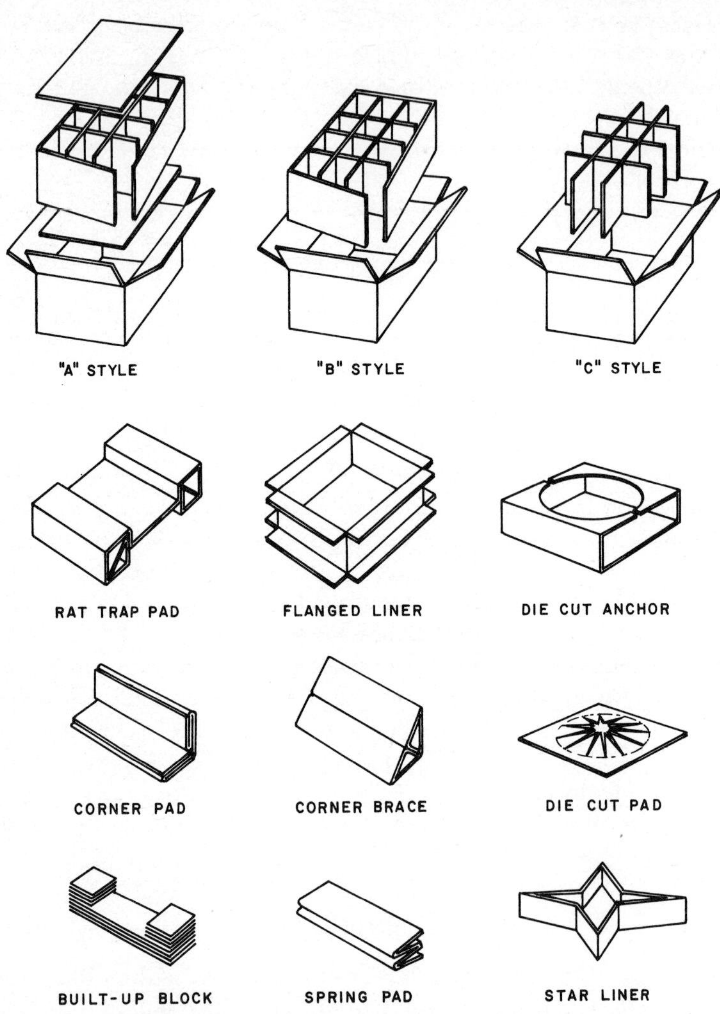

Fig. 8. Interior packing parts. The A, B, and C style designation is used in the glass industry for reshippers. Various types of pads, blocks, and braces are made of corrugated board to secure and cushion fragile items.

molded. Since the printing dies are flexible and the surface of the fibreboard is uneven, it is best to use an open-face style of type or design, because small openings in the dies tend to fill up and it is difficult to keep the printing sharp and clean. Lightface block letters without serifs give the most satisfactory results. Avoid designs with colors that butt or overlap, as trapping and register are more difficult than with metal type; colors on corrugated sometimes lap or miss as much as $^3/_{32}$ in. Large areas of solid color will often smear and look unsightly, especially red and orange tones.

Keep the printing at least ½ in. away from horizontal score lines, as the depression made by the scoring wheels of the combiner will not take the ink properly. It is best also to stay away from the vertical scores, but this is not as critical. If there is printing in this area, the ink may be picked up on the scoring wheels of the printer-slotter and transferred along the score line.

In laying out the copy to be printed on the box, identification of contents should be repeated on all four sides of the box in bold type not less than 1 in. in height. This information can be carried also on one of the top flaps along with the handling and storage cautions, but the other top flap should be left blank for consignee's name and address, or other shipping data. (See Fig. 9.) The quantity and size of contents can be placed in the upper left corner of each panel, with the stock number in the upper right corner. In the center should be the brand name, variety, and product name. Manufacturer's name and address can best be put on the bottom of each panel. The information that should be put

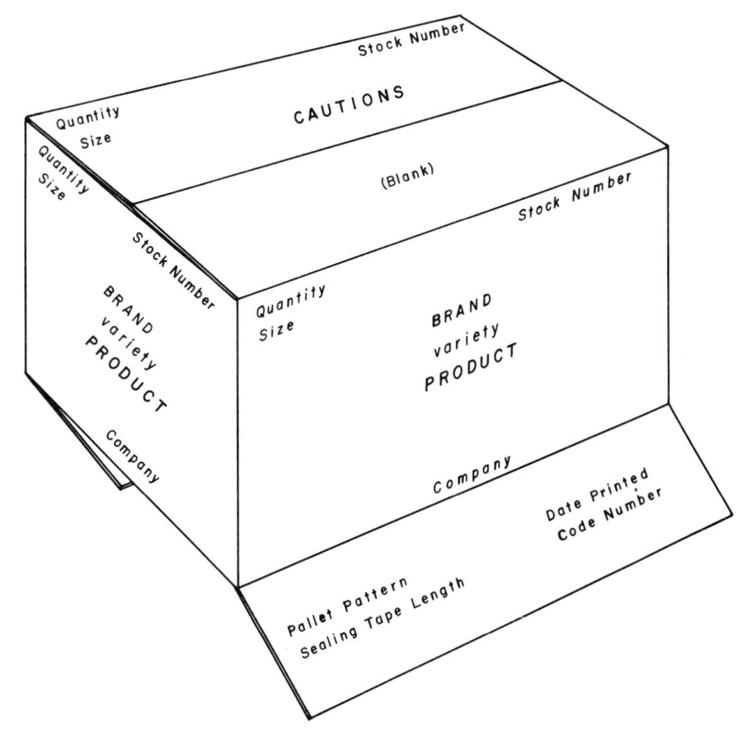

Fig. 9. Printing. Shipping containers should provide essential information on all sides, as well as the top. The extra cost is insignificant. The arrangement shown is based on surveys among wholesalers and retailers in the food and drug trades for preferred locations of quantity and size designations. One top flap is left blank for name and address of consignee and other shipping information.

on the bottom flaps with the box maker's certificate includes raw material code for the box, date of manufacture of the box, length of long tape and short tape for sealing the box if tape sealing is used, and pallet pattern diagram.

COATING

Various types of coatings can be applied to corrugated board to minimize abrasion against finished surfaces of appliances or furniture, or to provide waterproof or water-resistant properties to the containers. Other reasons for coating might be to improve appearance, to protect the printing, to provide an easily cleaned surface, to give grease resistance, or to improve mechanical strength under high-moisture conditions.

One technique that is used is called curtain coating. In this process flat blanks that have already been slotted, scored, and printed are passed through a continuous stream of molten material that is falling from an extrusion slot or weir in the top of the machine, through an opening in the conveyor table, into a reservoir where it is collected and sent back up through the system again. (See Fig. 10.) A corrugated blank moving

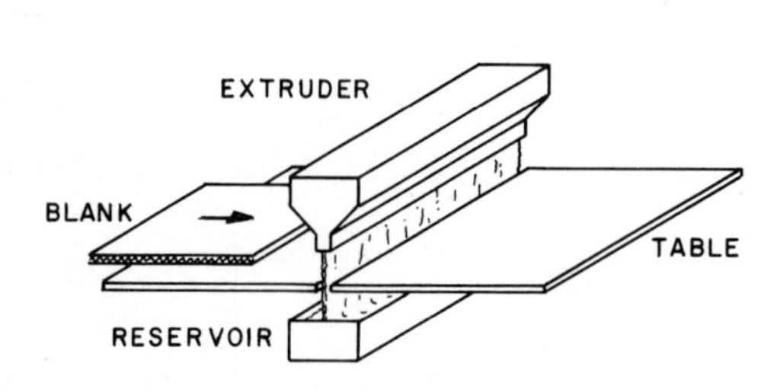

Fig. 10. Hot-melt application. Curtain coater for applying waterproofing to box blanks. A blend of wax and plastic flows down from the extruder in a continuous sheet and is collected in the reservoir at the bottom, to be reheated and sent back up to the extruder. A box blank, passing from left to right through this curtain, receives a uniform coating on the top surface and over the edges.

along on a conveyor at about 1,000 ft per min passes through this falling curtain and gets a uniform coating on the top surface and over the edges. Maximum sheet size is about 5 by 7 ft. It is possible to restrict the coating to a strip and thus save material where the design does not require coating all the way to the edge, but there is a tendency in that case for the coating to be heavier near the edges of the strip.

Materials used for curtain coating generally are made up of about 60 percent paraffin, with the remainder consisting of microcrystalline waxes, ethylene-vinyl copolymers, and petroleum resins. They cost about 20 cents per pound, and they can be coated as thin as 5 lb per 1,000 sq ft (less than 0.001 in. thick).

COSTS

It is possible to figure the cost of a corrugated box with a fair amount of accuracy, sufficient at least for purposes of design. The market price of kraft linerboard has been fairly steady at $127.50 per ton for the past few years, and the figures shown are based on that price. Each change of a dollar in the price per ton will affect the figures in Table 7 by about 0.10 if it is desired to update these prices.

TABLE 7 Cost of Corrugated Fibreboard

Bursting resistance (Mullen test)	Linerboard facings, basis weight	Cost per 1,000 sq ft
Single wall:		
Nontest	26/26	$14.00
125	33/26	14.55
150	33/33	15.10
175	42/33	15.50
200	42/42	16.00
250	69/42	18.70
275	69/69	21.45
300	90/69	24.00
350	90/90	26.65
Double wall:		
Nontest	26/26/26	20.35
200	33/26/33	22.90
275	42/26/42	24.20
350	42/42/42	25.40
400	69/42/69	27.85
500	69/90/69	35.20
600	90/90/90	40.00

To calculate a typical RSC, it is necessary only to measure the total area in the box blank; for example, a box 1 by 1 by 1 ft will have a total of 4 sq ft in the side panels, and 4 sq ft in the top and bottom flaps, for a total of 8 sq ft of board area. From Table 7 we find that 10,000 boxes made from 200 test board will cost $1,280 for the board alone:

$$8 \text{ sq ft} \times 10{,}000 \text{ boxes} \times \$16/1{,}000 \text{ sq ft} = \$1{,}280 \text{ per lot}$$

To this we add the extra charges from Table 8.

Board	$1,280
Setup charge	25
Stitch joint 12 in. × 0.45	54
Printing 1 color, 4 panels: $1 + $2 + $10	13
Total	$1,372 per lot

TABLE 8 Additional Charges

Chargeable item	One-time charges	Repetitive charges	
	Setup charges	Per 1,000 boxes	Per 1,000 sq ft
RSC box	$25.00	$6.00	
Flap cutting (CSSC)	4.00	. . . .	$0.25
Die cutting	. . . .	. . . .	2.50
2-in. paper tape	. . . .	0.45/in.	
3-in. paper tape	. . . .	0.80/in.	
2-in. cloth tape	. . . .	0.80/in.	
3-in. cloth tape	. . . .	1.10/in.	
Glue or stitch joint	. . . .	0.45/in.	
1-color printing	1.00 + 0.50/panel	1.00	
2-color printing	4.00 + 1.00/panel	1.75	
Solid color	. . . .	3.00	2.20
Paraffin coating	7.50	. . . .	{2.00, 1 side 3.00, 2 sides
Tear tape	35.00	0.15/in.	
Slit scores	8.00	3.50	
Hand holes (2)	3.00	2.00	
Vent slots (3)	2.00	2.00	
Pad or sheet	7.00	2.50	
Scored sheet	8.00	3.00	
Horizontal corrugation	. . . .	2.50	
L or W under 4 in.	. . . .	3.00	
L and W under 4 in.	. . . .	6.00	
Depth under 3½ in.	. . . .	3.00	
Antiskid coating	4.75		
Mottled white	. . . .	. . . .	0.90
Bleached white	. . . .	. . . .	2.40
½- mil PE lamination	. . . .	. . . .	6.00
Double slide box	20.00	4.00	
Triple slide box	28.00	3.67	
Partitions, assembled	10.00	7.00	
Corner-cut tray	20.00	11.00	

CLOSING AND SEALING

The most economical way to seal boxes if labor is not an important factor is with adhesive. (See Table 9.) At least 50 percent of the area of contact must be firmly glued in accordance with Rule 41 of the Uniform Freight Classification. Gluing makes a very strong package when properly done, but it has the disadvantage of making the box difficult to open, and sometimes dirt will sift in where the flaps meet. Three strips of tape at top and bottom will make a dust-free seal which is strong and easy to open by the user. (See Fig. 11.) Tape must be made from 60-lb kraft paper at least 2 in. wide. It must cover the entire length of the

TABLE 9 Sealing Costs for a Typical Corrugated Box 21 by 16 by 14 In.

	Glue seal 0.25¢/sq ft	Three-strip dust seal 0.1¢/ft	One-strip reinforced tape 0.5¢/ft	Preformed staples 0.1¢ each
Material	0.5¢	1.3¢	2.2¢	3.8¢
Labor	1.2	3.0	0.9	1.6
Total	1.7¢	4.3¢	3.1¢	5.4¢

joint, and it is recommended that it extend onto the end panels about 1½ in. The end seams also must be completely covered, and it is preferable to turn the corner and run onto the flap score about 2½ in. on each side. A single strip of reinforced tape at top and bottom may be substituted for the three strips of kraft tape, provided it is 3 in. wide and meets the strength requirements of Rule 41.

Stitches made from coiled wire or preformed staples may be used to close the box securely. This is a clean, dry operation which makes a strong package that is easily opened at its destination. There is some

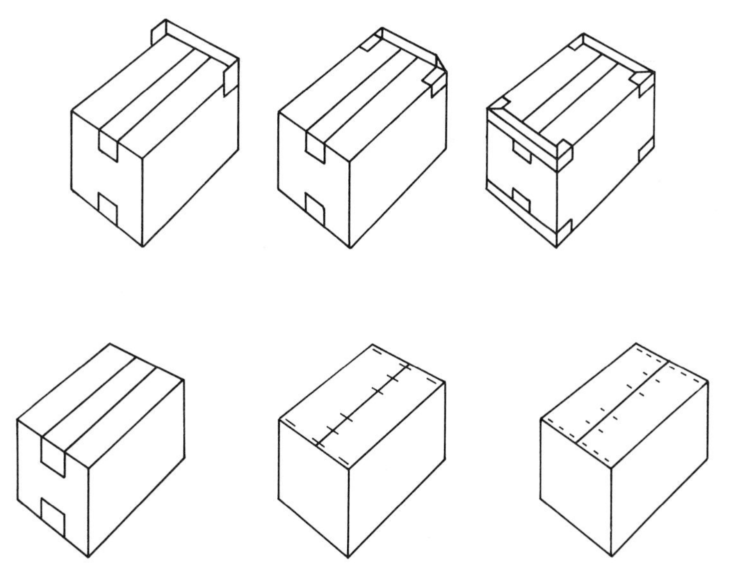

Fig. 11. Closure methods for corrugated boxes. Three strips of gummed tape are applied at top and bottom as shown in the sequence at the top, to make a dustproof seal. A single strip at top and bottom is adequate, provided special reinforced tape is used. Wide-crown staples are shown in the center picture, and regular metal stitches are illustrated on the right.

risk of damaging the contents unless a corrugated pad is provided at top and bottom. Stitches must be not more than 2½ in. apart on each side of the center seam where it overlies the inner flaps, and across both ends 1 to 1½ in. from the edge of the flap. Staples which are at least 1¼ in. wide may be placed across the center seam not more than 5 in. apart where it overlies the inner flaps. The wire used for such staples is defined in Sec. 7(b) of Rule 41.

TESTING

A new design for a corrugated box should be thoroughly tested before going into full production. Various tests can be used, some of which are described in Sec. 18, "Test Methods." The choice of tests to be made will be determined by the type of contents that is to be packed. If it is fragile, a drop test will be very important. For rail shipment in carload lots, a drop of 12 to 18 in. on a couple of edges and the normal bottom should be adequate. Parcel post shipment would call for a much more severe test, in the range of 30 to 48 in. on several corners, edges, and flat surfaces.

If scuffing is likely to be a problem, vibrating for about ¾ hour should give an indication of what may happen in a typical domestic shipment. A revolving drum test for about 30 falls, or an incline-impact test from the fourth zone (9 mph) also can be used. Unless the contents is self-supporting, it is recommended that a stacking load be applied for several weeks if time will permit.

Section 6

Glassware

History 6-1
Chemistry of Glass 6-3
Colored Glass 6-5
Colors for Screening 6-6
Mechanical Properties of Glass 6-6
Annealing 6-8
Design Considerations 6-8
Analysis of Fractures 6-9
Manufacturing 6-10
Labeling 6-13
Coatings 6-13
Luster Decorating 6-15
Design and Molds 6-15
Finishes 6-19
Tolerances 6-19
Defects 6-22
Critical Defects 6-23
Major Defects 6-26
Minor Defects 6-26
Quality Control 6-27
Tubing Products 6-27

HISTORY

Glass is one of the oldest materials known, and a material similar to glass of volcanic origin was used for arrowheads during the Bronze Age. In the first century Pliny wrote of sailors who used blocks of soda from their cargo to make a fireplace on the beach and discovered that the soda fused with the sand to form glass. It is believed that the Egyptians were using glass as early as 3000 B.C. The blowpipe is known to have been used in Sidon during the first century B.C. for making hollow glass articles, and by the third century of the Christian era articles of glass were in fairly common use in Roman households. The commercial success of the Venetian glassblowers in the sixteenth century is well known, and the term "flint glass" refers to the very pure silica in the form of flint used by these artisans of Venice.

Around 1870 the shop system of making glass containers was introduced in this country. This consisted of a team of seven men: three skilled workmen and four boys. The first man on the team would take a lump of molten glass out of the furnace on the end of his pipe and would partially blow it. A boy called the mold tender then would open the mold while the blower put the glass into the mold and inflated it with pressure from his mouth. The pipe was then cracked off and the glass was taken by the "snapping-up boy" to the finisher, who shaped the neck and the lip with tools. The "carrying-in boy" then took the ware to the lehr, while a "cleaning-off boy" prepared the blowpipe for the next gathering. Two blowers would alternate in filling the mold, or one would preblow and swing while the other did the final blowing. Since the neck was made last, it was called the finish and even today, although the machines make it first, the top of a glass container is still called the finish.

The milk bottle was invented in 1884 by a physician who was concerned about the delivery of safe, clean, fresh milk. Mechanization was first introduced on a large scale in 1892, and this permitted some of the skilled workers to be replaced by unskilled operators. In 1896 the Atlas Glass Works was using semiautomatic machines in regular production for wide-mouth jars. The first fully automatic machine was designed and built by Michael J. Owens in 1903, while working in the Toledo plant of Edward D. Libbey. In this machine the blank mold was dipped down into the tank and the molten glass sucked up into the mold, while the bottom was trimmed off, to give the correct amount of glass to make one bottle. The parison was then transferred to another mold for final blowing. By 1917 there were 200 of these machines in operation around the country. The Hartford-Empire Company was perfecting their IS (individual section) blow-and-blow machine during this period, and the principle of dropping a gob of glass into the mold from the top started to replace the Owens method of feeding from the bottom in most of the glass plants. By 1925 the production on the two types of machines was nearly equal. The efficiency of the IS machine has been further increased with double-gobbing and triple-gobbing techniques. Lynch blow-and-blow machines and Miller press-and-blow machines are of the rotary type, in contrast to the IS machines, which are stationary. All three are gob-fed from an overhead tank.

The Lynch machine is somewhat slower than the IS machine, and for long runs it is at a disadvantage. However, it can be changed over from one size to another quickly, often in less than an hour, and it reaches peak efficiency after a changeover much quicker than the IS machine. The GMI (Glass Machinery, Inc., Indianapolis, Indiana) machine was designed specifically for wide- and semiwide-mouth containers. It is a press-and-blow type of machine which is a modification of the IS machine and provides good distribution of glass. As much as 10 percent less glass can be used, because of the better distribution.

Starting with a solid blank, there is no baffle seam, and consequently there is less thermal shock breakage than with containers made by the blow-and-blow method. There are limitations in design, however, because the blank must be stripped from the parison, and at least a 2° taper must be provided, from the neck ring down.

About 35 billion glass containers, worth about $1.5 billion, are produced in this country each year. The industry has been set back in recent years by strikes and price increases, which have accelerated the swing from glass to plastic containers by many packing companies. The broad range of types and sizes of glass containers, one of the chief advantages of this form of packaging, is being reduced by the elimination of slow-moving items. Only in the beverage business is there any significant growth. The nonreturnable bottle is approaching an annual figure of 10 billion, five times the use of returnables at this writing.

CHEMISTRY OF GLASS

Glass is not a crystalline material in the usual sense of the word. Although the molecular structure is believed to be orderly in the shorter ranges (2 to 20 Å), it is probably disordered in terms of long-range crystal lattices. The constituent crystallites are very small, ranging in size from 0.1 to 1 μ. Since by definition a crystal is a strict repetition of identical units, and glass does not conform to this description, it is more realistic to consider it as a congealed liquid. Its structure is more dependent upon its thermal history than upon the chemical composition, but it is worth noting that the less sodium oxide, the stronger the glass, although only by a small percentage. (See Table 1.)

The principal ingredients of glass are sand, soda ash, and limestone. The sand is almost pure silica, and the soda and lime are usually in the form of the carbonates $NaCO_2$ and $CaCO_2$. There are also some traces of other materials such as lead, which gives clarity and brilliance to the

TABLE 1 Composition of a Typical Bottle Glass

Element*	Percentage
Oxygen	60
Silicon	24
Sodium	10
Calcium	4
Aluminum	2

* The elements are usually present in the form of their oxides: SiO_2, Na_2O, CaO, and Al_2O_3.

product but is inclined to yield a relatively soft grade of glass. Alumina (Al_2O_3), however, is often used to increase the hardness and durability.

The formulation of the glass can be adjusted for specific purposes. Thus if resistance to chemical action is especially important, less sodium and more aluminum compounds can be used; if alumina supplies one-eighth of the alkali content, the glass will be highly resistant to chemical attack. Magnesia also is a valuable addition for chemical durability, but it has a tendency to form "flakes" in solutions.

To reduce seeds and blisters, fining agents such as salt cake (Na_2SO_4) and arsenic (As_2O_3) are often used. Since lower temperatures are possible by adding certain ingredients, and machine operators like to work with glass below 1200°, the glass is usually formulated in this way for better machinability. If breakage from thermal shock is a problem, it is possible to use high-silica, low-alkali glasses, but these may lead to cord problems. Cords are strains that are not relieved by annealing.

In the manufacture of glass the sand, soda ash, and limestone are mixed with *cullet,* which is broken glass to help in the melting, and fed into the furnace at the end. The soda ash melts first, acting as a solvent for the sand, which would otherwise require a much higher temperature to become molten. A gas flame coming from the side above the melted glass keeps the material in the tank between 2700 and 2900°F. Currents are formed in the molten glass by the bubbling of the escaping gases, which mixes each batch of fresh materials uniformly with the material already in the furnace. The steady flow toward the opposite end, where it is drawn off for the machines, also helps to generate currents which aid in the mixing. (See Fig. 1.)

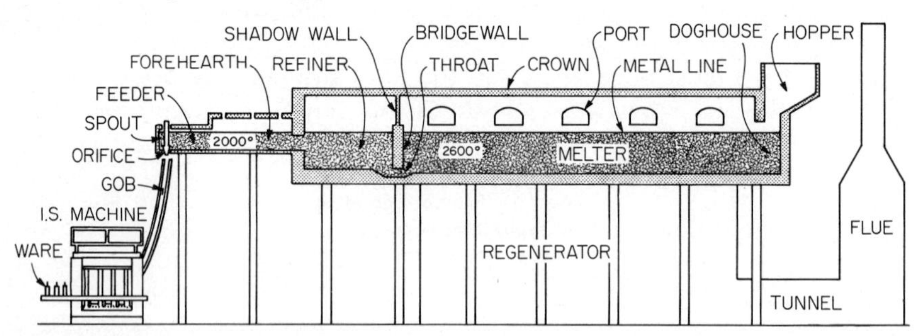

Fig. 1. Glass tank for melting container glass. Gas flames from the side ports go across the surface and into the regenerator on the opposite side. Every 20 min the direction of firing is reversed, to take advantage of the heat that accumulates in the regenerators on each side. The molten glass is mixed by currents that develop in the melter. Materials are charged in at the hopper end and move from right to left in the illustration, and then pass through the throat into the refiner. Impurities that float on top are held back by the bridge wall. Several forehearths lead off from the refiner, where the temperature is reduced and the formula adjusted. The glass then passes into the feeders, where it is pushed through an orifice and cut off by shears into uniform gobs, ready to be formed into containers.

Although glass is a fairly inert material and is used to contain strong acids and alkalies, as well as all types of solvents, it has nevertheless a definite and measurable chemical reaction with some materials, notably water. The sodium is rather loosely combined with the silicon, and it is leached from the surface by plain water. In 1 year distilled water which is stored in flint glass will pick up 10 to 15 ppm of NaOH, along with traces of the other ingredients of the glass. The addition of about 6 percent boron to form a borosilicate glass reduces this leaching action so that only about ½ ppm would be dissolved in 1 year.

When glassware is stored for several months where fluctuations of temperature and humidity cause condensation of moisture from the air, salts are dissolved out of the glass; this is called blooming. Since the moisture tends to run down the inside as well as the outside of the ware, it is common to find a heavier deposit near the bottom. The bloom sometimes interferes with labeling and decorating operations, but it can be removed with an acid wash. It is best to work with freshly made ware whenever possible, to avoid such complications.

If the alkali in the glass may affect the contents, soaking in plain water will remove most of it from the surface. Adding some dilute acid or heating the water will speed up the process. A special sulfur treatment is offered by some manufacturers, and the sodium sulfate which is formed on the surface is much more resistant to attack by water and by acids. It does not, however, make the glass any more resistant to alkaline solutions. The borosilicate glass, designated as type I in the *United States Pharmacopoeia*, is substantially more resistant to attack by alkalies than regular glass, although resistance varies with the kind of alkaline material. It is about 10 times as resistant to acids. (See Table 2.)

TABLE 2 Types of Glass for Pharmaceuticals

Glass type	Material	Size	Alkalinity,* ml
Type I.	Borosilicate glass	. . . .	1.0
Type II	Treated soda-lime glass	Up to 100 ml	0.7†
		Over 100 ml	0.2†
Type III.	Soda-lime glass	. . . .	8.5
NP (nonparenteral).	Soda-lime glass	. . . .	15.0

* Maximum amount of 0.02 *N* sulfuric acid required to neutralize water that has been autoclaved in contact with 10 g of powdered glass.

† Type II glass is tested with 100 ml of water that has been autoclaved in contact with the treated surface of the glass.

Colored Glass. Some colors of glass are readily available for containers, such as amber, green, and opal. (See Table 3.) Full tanks of these colors are maintained at several glass plants. Other colors are available only if the orders are very large. A method has been developed for

TABLE 3 Coloring Agents Used in Glass

Color	Agent
Red	Cuprous or cupric oxide, cadmium sulfide
Yellow.	Ferric oxide, antimony oxide
Yellow-green . .	Chromic oxide
Green	Ferrous sulfate, chromic oxide
Blue.	Cobalt oxide
Violet	Manganese
Black	Iron oxides in large amounts
Opal	Calcium fluoride
Amber	Carbon and sulfur compounds

adding the coloring material in the forehearth, so that much smaller orders of special colors can be processed and made available to many small users who could not justify a full tank of special glass. The forehearth is shown in Fig. 1.

Colors for Screening. In addition to the decorative use of colored glass, certain colors are effective in protecting a product from the effects of sunlight by screening out some of the harmful rays. Amber glass is required by the USP to screen out light in the range of 2900 to 4500 Å. Flint takes out the rays at 2900 to 3200 Å (far ultraviolet), and emerald green is effective at 4000 to 4500 Å (visible violet-blue). (See Fig. 2.)

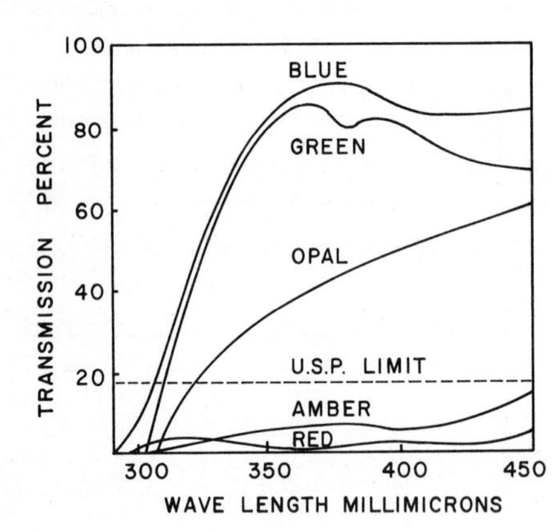

Fig. 2. Light transmission of glass. Colored glass will protect the contents of a bottle from light in varying degrees, depending on the color. In the critical ultraviolet region, only amber and red glass are really effective.

Mechanical Properties of Glass. The strength of glass has very little relationship to chemical composition, one type of glass being very much like another. The condition of the surface, however, has a great deal to do with its tensile properties. The theoretical tensile strength of glass is greater than 2×10^6 psi. The effective strength of annealed glass is between 3,000 and 8,000 psi. Tempered glass, which has a high compressive stress in the outer layers, with tension in the interior, of the type used for automobile windows, will have strength between 12,000 and 40,000 psi. New techniques of strengthening glasses chemically have produced strengths up to 200,000 psi. A smooth surface on a bottle, just as it comes from the mold, will withstand stresses of 100,000 psi or more. When the surface becomes scratched or bruised even the slightest amount, however, more than half the strength is lost. By the time it reaches the store shelf, it may be down as low as 3,000 psi. An understanding of "notch sensitivity" of brittle materials may help in explaining this great loss of strength. Stresses are concentrated at the tip of a notch in any material, but with ductile materials the tip of the notch yields, the sharpness of the notch is blunted, and the stress concentration is reduced so that the fracture does not go any farther. In fact, most ductile materials "strain-harden" at the tip of the notch sufficiently to offset the increased stress at that point. With brittle materials such as glass, however, the stresses frequently exceed the strength of the material at the tip of the notch. A crack develops, and failure is instantaneous. It is known that microcracks exist on all glass surfaces, as a result of handling in manufacturing and distribution, in amounts varying from 1,000 per square centimeter on fine plate glass to over 50,000 per square centimeter on commercial-grade glass. Although they may be small, their stress-concentrating effect can be considerable if they are sharp enough. This effect is proportional to the square root of the ratio of the notch depth to the notch tip radius. Thus a notch only 1 μ deep could cause a concentration factor of 100 if it were atomically sharp.

Various coatings can be sprayed on the ware as it comes from the annealing oven, in order to preserve as much of the original strength of the glass as possible. The coatings do not add strength to the glass directly, nor do they combine chemically with the glass. They act merely as lubricants so that contact between bottles, or between bottle and machine, is much less likely to damage the surface of the glass. (See Tables 4 and 5.)

TABLE 4 Properties of Coatings for Glass Containers

Property	Silicone	Metallic oxides	Wax	Resin	Sulfur
Insoluble in hot or cold water	Yes	Yes	Yes	Yes	No
Wettable after application	No	Yes	Yes	Yes	Yes
High film strength	No	Yes	Yes	Yes	No
Easy to apply	Yes	No	Yes	Yes	Yes
Visible on surface of glass	Yes	Iridescent	Yes	Yes	No
Difficulty in labeling	Yes	No	No	No	No
Protects when wet	Yes	Yes	Yes	Yes	No
Nontoxic	Yes	Yes	Yes	Yes	Yes

TABLE 5 Lubricity of Coatings for Glass Containers

Coating	Coefficient of friction	Max. angle of repose
Silicone oil. . . .	0.26	14°
Polyethylene. . .	0.27	15°
Glyceride	0.28	16°
Polyoxyethylene.	0.29	17°
Stearate	0.30	18°
Metallic oxide . .	0.40	24°
Uncoated glass .	0.84	40°

Annealing. If glassware is allowed to cool too quickly, strains will be set up in the glass and will make it more susceptible to breakage. For this reason, bottles and jars are always put through an annealing *lehr,* where the temperature is raised to 1000°F and held for 15 min. They are then cooled very slowly to room temperature, which relaxes any strains which may have developed from contact with the molds and conveyors. Occasionally there are strains, known as *cords,* which are not relieved by the annealing process. These are caused by poor mixing in the melting tank, resulting in a variation in the composition of the glass. This produces a weak container which may break spontaneously. Such strains can be detected with a polariscope, by cutting sections and examining the colored bands produced by the polarized light; they are usually considered critical defects. The definition of critical defects is given under "Defects," page 6-23.

DESIGN CONSIDERATIONS

The shape of a container will have a very important influence upon its strength. A sphere is the strongest shape, a cylinder is next, and a rectangular shape is the poorest from an engineering standpoint. The flat panels of a rectangle will yield more readily than convex panels, particularly if the stress occurs near the middle. Therefore, the designer should provide ridges or beads to take the shock in the center. These ridges can be further strengthened by making them knurled instead of straight. Knurling or stippling is used also on the base to reduce the area of contact with conveyors in the glass plant, thereby minimizing thermal shock. Although these principles apply particularly to square shapes, they can be used also for rounds, ovals, and other styles of containers. As much as 50 percent can be added to the strength of a bottle by designing it so as to direct the shocks where they will do the least harm. (See Fig. 3.)

The corners of a glass container tend to be the thinnest part because the glass has to stretch farthest when the corners are blown. The thinness can be minimized to some degree by the shape of the *parison,* which is the partly formed glass in the blank mold before it is transferred to the blowing mold. It is not possible, however, to get perfectly uniform distribution of the glass, and the shoulder and heel are usually the thinnest parts. The bottom corners are especially vulnerable because the containers are usually dumped on the filling line conveyor so that they are right side up; this often bruises the bottoms. If the inside surface chips below the parting line, which is generally about 1/8 in. up from the bottom, it is called a "light bottom" and it is likely to break at this point on the production line or in shipment if it is dropped on a hard surface.

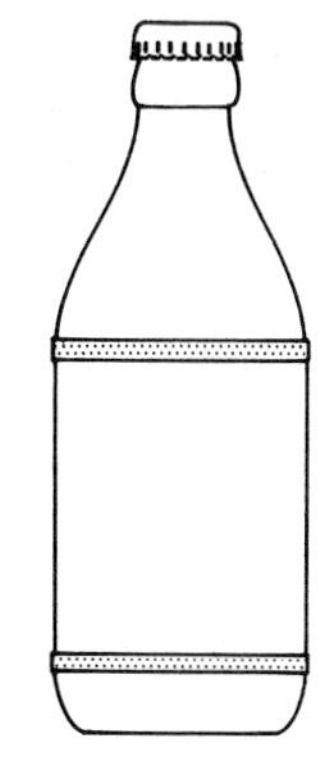

Fig. 3. Shock bands. Breakage is reduced by providing raised areas at the shoulder and base to take the impact stresses. Stippling these bands will help to limit the effects of bruising in this area.

A source of damage which does not occur very often, but which is peculiar to certain types of products, is a chipping of the inside surface at the center of the bottom panel. When heavy syrups or pureed foods are dropped, especially if vacuum-packed, they develop a vacuum bubble at the bottom. The bubble forms and collapses rather quickly, but in doing so it concentrates sufficient hydraulic force at one point to break the surface, which becomes pitted and consequently weakened. See Table 6 for weight of glass in bottles of various sizes.

Analysis of Fractures. The failure of a glass container is usually the result of thermal shock or impact stresses. It is sometimes desirable to know the cause of the damage, and by examining the fragments it is possible to learn a great deal about the stresses which caused the break.

TABLE 6 Glass Allowance in Mold Design

Capacity, fl oz	Weight in av oz, rounds	Weight in av oz, odd shapes
1	1 1/2	1 7/8
2	2 1/2	3
3	3 1/4	4
4	4	5
6	5	6
8	6 1/2	8
12	8	10
16	10	12 1/2
32	16	20

Although it takes some training and experience to make a precise evaluation, it might be useful to know some of the principles involved.

Glass fails only in tension, never by compression. Therefore we can reconstruct the container from the pieces and read a history of why it failed, from the shape of the fragments. A fracture moves out from the point of impact in a series of waves. The broken surface nearest the point of impact is smooth, shiny, and flat, indicating that it separated rapidly. Farther away it is angular and rough, especially near the edges, where the break occurred more slowly. Still farther away the surface again becomes smooth and flat. (See Fig. 4.)

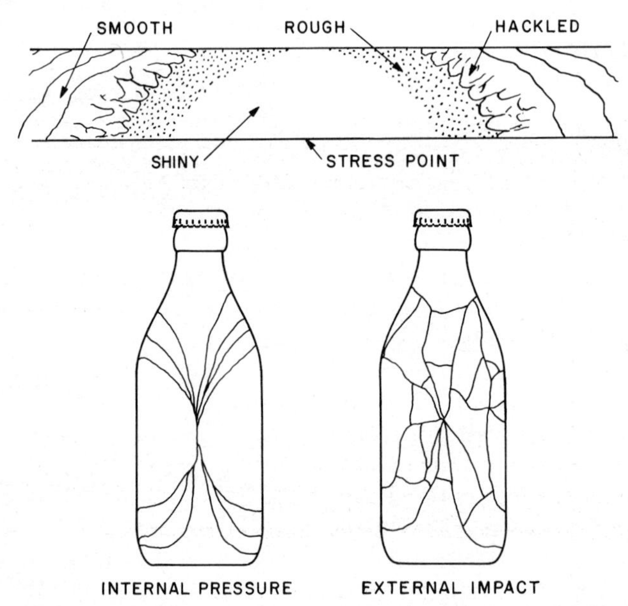

Fig. 4. Analysis of glass failure. Fragments of glass can reveal considerable information about the severity and type of stress that caused the failure of a container. A mirrorlike surface indicates a rapid fracture, usually at the starting point. A rough surface occurs where the fracture proceeded at a slower rate. The shape of the pieces will show the type of shock; thin slivers usually indicate an outward force, and short chunks indicate a blow inward.

MANUFACTURING

A mixture of sand, soda ash, limestone, and cullet or broken glass is fed into a furnace which is 60 ft long and 40 ft wide. The molten glass is about 4 ft deep, and is kept at around 2700°F by gas flames across the top surface. (See Fig. 1.) The tank is built high above the floor and is lined with firebrick. At one end is a "throat" or narrow opening below

Fig. 5. Glass container manufacturing. A glass plant, with four feeders coming from the refiner in the upper right. An IS machine is in the foreground, with a conveyor carrying the hot ware into the lehr at the lower left. (*Cobelcomex*)

the surface which allows the glass to flow into a smaller refining chamber while it holds back the impurities that are floating on the surface. From there it flows into a feeder, where the temperature is about 2000°F. (See Fig. 5.) An orifice in the bottom of the feeder allows the glass to flow out to where it is cut into *gobs* by shears. These gobs go down a chute into a *blank mold,* where they are forced by compressed air into the neck ring at the bottom, which makes the threaded or other type of finish. At the same time the body is given the general shape that it will ultimately have, but in reduced size since it is not completely filled out. At this point it is called a *parison.* The shape of the blank mold is very important in getting good distribution of the glass in the final piece. The parison is held by the neck ring mold and carried by an arm into the final mold, where it is blown out to its finished form. It is then picked up by the neck ring with tongs and automatically placed on the conveyor that takes it through the annealing lehr. (See Fig. 6.)

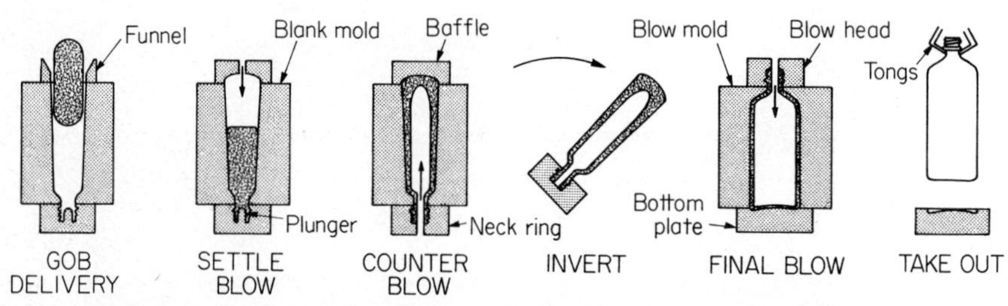

Fig. 6. Glassblowing operations. A schematic diagram of the successive operations performed by an IS machine in making a bottle by the blow-and-blow process. The gob of hot glass comes down a chute into the blank mold and is sucked down over the plunger by vacuum. The baffle comes down on top, the plunger is withdrawn, and compressed air is introduced to blow the "metal" to the shape of the blank mold. The mold opens slightly to allow the surface of the parison to reheat, by flow of heat from the interior. The parison is inverted to bring the neck ring on top, and the blow mold closes around it. Compressed air forces the glass out against the blow mold, giving the container its final shape. The mold opens and tongs pick up the bottle by the neck ring, placing it in a programmed location on the conveyor, which carries it through the annealing lehr.

The timing of the opening and closing of the molds, and the various stages of blowing and transferring, are all controlled by cams mounted on a revolving drum at the back of the machine. These are adjusted at the start of a run by a mechanic who checks each adjustment by examining the pieces as they come out of the mold. The inspectors at the other end of the lehr are examining each piece and sorting the good from the bad. At first they may keep 20 percent and discard 80 percent of the bottles produced, and the rejects go back into the tank as cullet. Over a period of several hours the adjustments are made so that eventually about 90 percent of the output can be shipped to the customer.

Labeling. The pristine surface of glass as it comes from the mold is very reactive, and it quickly combines chemically with the moisture in the air to form OH hydroxyl groups so that the surface is negatively charged. This means that the surface is readily wetted with water, alcohol, and other substances having similar OH groups in their chemical formulas. Likewise, NH_3 amine groups, such as are found in dextrin or casein molecules, will readily adhere to the glass. Thus the glass surface can be said to be hydrophilic in character, and as long as the adhesives are neutral or alkaline, a good bond will result from the chemical action with the OH molecular groups on the surface of the glass. The adhesive should be used in as thin a film as possible, to avoid residual stresses in the adhesive as it dries, which would weaken the bond. Opal glass is difficult to label because the fluoride ions cause the surface to be hydrophobic. This makes it difficult to get a good wetting action. Silica and quartz glass are difficult to label for the same reason. Borosilicate glass is not quite so bad, but the acidic nature of boron compounds can cause some problems.

The moisture which condenses on the surface of a glass container can sometimes interfere with labeling. If bottles are brought from a cold warehouse to the filling line, which is at room temperature, dew will form on the surface. This may be only a small amount which is barely perceptible, and yet it can prevent the adhesive from sticking to the surface of the glass. The solution, of course, is to allow enough time for the ware to warm up before putting it into the filling line.

Coatings. To preserve the inherent strength of the glass, it is necessary to avoid scratching the surface, as much as possible. One way is to coat the surface, to minimize the abrasion from bottle against bottle and from bottle against the metal parts of the filling machine. One of the best coatings, from a lubrication standpoint, is silicone. It makes the surface water-repellent, however, and may cause some problems in labeling.

Polyethylene also is an excellent coating, but it is difficult to work with unless it is applied in the form of an emulsion so that it forms a discontinuous film which is suitable for labeling. It is also possible to oxidize the surface of the polyethylene coating with special treatments that will facilitate bonding with adhesives. Polyethylene glycol and the stearate of polyethylene glycol are much easier to handle than plain polyethylene. The chief disadvantage is that they are water-soluble and therefore are not very permanent. Oxides and chlorides of iron and tin also are used for this purpose.

Beverage and some food filling lines operate with the glass containers wet on the outside. Under this condition most of the surface treatments are ineffective. Silicone is one exception and is quite durable even when thoroughly wet. Returnable bottles that are repeatedly washed with caustic solutions, however, cannot be expected to retain the surface coating for very long. Silane has been combined with polyvinyl alcohol to produce a coating that is somewhat resistant to water. The silane acts to provide a good chemical bond with the glass surface, and curing of the film gives it the inertness to water.

Coatings used on glassware for food or internal medication should be made from materials that are listed as safe by the Food and Drug Administration, even though they do not get on the interior surfaces of the container. Regulations (Federal Food, Drug, and Cosmetic Act §409(C)(1), §72 Stat. 1787, 21 USC 348 (C)(1) §121) permit the use of resins or polymeric coatings formulated from these materials if the extraction level is below 50 ppm, based on the volume of the container. For stearates they will allow 1 ppm, and for green soap the limit is $^1/_{10}$ ppm.

To test the effectiveness of the surface treatment, two treated containers are rubbed against each other under controlled conditions. There should be no damage under a 10-lb load with movement so as to continuously expose new glass surface to static drag damage, at a rate of 0.02 in. per sec.

Coated glass is as much as three times as strong as uncoated glass, under impact breaking stresses, after it has been through a typical filling line. Pristine glass with a breaking stress value of 9 in.-lb will drop to about 3 in.-lb after abrasion on a filling line, whereas the treated ware will continue to give values of 9 in.-lb or better after abrasion.

Line breakage is reduced to a tenth or less of that found before surface treatment, under actual production conditions. This is considerably more than the test results would indicate and is explained by the fact that it is the weakest bottles that get broken; a slight improvement affects a high percentage of the weak bottles. Therefore the improvement becomes very significant.

Filling lines for beer bottles run as high as 1,230 bottles per minute, and for baby food in glass jars almost as fast. Breakage is a serious problem at these speeds, not only because of the loss of production, but also from the risk of getting fragments into the product. The importance of coatings in this kind of operation cannot be overemphasized.

If an overflow filling system is used, as with pickles, olives, and cherries, there is a risk that the coating on the bottles may be washed off and become concentrated in the system, as the material is recirculated. Silicones would not be as much of a problem, for example, as stearates, because they are less soluble.

Polyoxyethylene monostearate, often called just stearate, costs about 50 cents per pound. It is diluted at the rate of 1 lb in 800 gal of water.

Labeling may be a problem if dextrin adhesives are used. Better results will be obtained with casein adhesives on glassware that has been treated with stearate. The Food and Drug Administration has approved the use of stearate as a bottle coating, provided the contamination of the interior of the container is below 1 ppm. Stearate treatment can be washed off and is therefore considered nonpermanent.

Potassium palmitate, known as green soap or soft soap, is made from palm oil; it costs about 21 cents a pound. It is diluted at the rate of 1 lb in 120 gal of water. Labeling may be a problem on Pneumatic Scale labelers, which apply the glue to the glass instead of to the label. The glue is not picked up as readily by the glass surface if it is treated with green soap. With other types of labelers there should be no problem, as nearly all types of adhesives are compatible with the soap treatment, except some hot melts. Since the coating is soluble in water, it is classified as nonpermanent. The Food and Drug Administration has approved the use of green soap as a bottle coating if the contamination on the interior is below 1/10 ppm.

Luster Decorating. Organic compounds of tin, bismuth, iron, titanium, and other metals dissolved in hydrocarbons or chlorinated hydrocarbons can be applied to glassware to provide surface coloration. The pieces are then fired at 1000°F in an oxidizing atmosphere to provide a metal oxide film. The luster can be applied by spraying, dipping, silk-screen printing, and several other techniques. Some luster coatings closely resemble colored glass, they are used for short runs if coloring the glass in the tank would not be practical. They have resistance to abrasion and to detergents. Available colors are pink, copper, orange, turquoise, and light and dark blue. These colors cannot be mixed to produce other colors because the results do not follow the usual rules and the effect is usually disappointing. Two different coats with a firing in between produce attractive results, especially if the base coat is gold or platinum. Treating the second coat, while it is still wet, with ethyl acetate produces a marbled effect. Colored coatings of the epoxy type provide a certain amount of surface lubricity and thus reduce breakage on the production line and in transit; they provide also the decorative effect of the color and the protection of the product from light. Epoxy coatings carry an up-charge of 30 to 50 cents per gross.

DESIGN AND MOLDS

A design for a new container is first worked out on paper, with the necessary allowances for headspace, weight of glass, and other such factors. The amount of headspace may be determined by the customer, depending on where the fill point is desired, or based on tests at 0 and

120°F to see that breakage will not occur under the most adverse weather conditions. A pharmaceutical fill is usually halfway up the shoulder, whereas a cosmetic fill is to the base of the neck. (See Fig. 7.) Since the mold determines the outside dimensions of the container, the space taken by the walls must be added to the desired capacity. The specific gravity of glass is 2.6 in relation to water taken as 1.00, or 163 lb per cu ft. (See Table 6.)

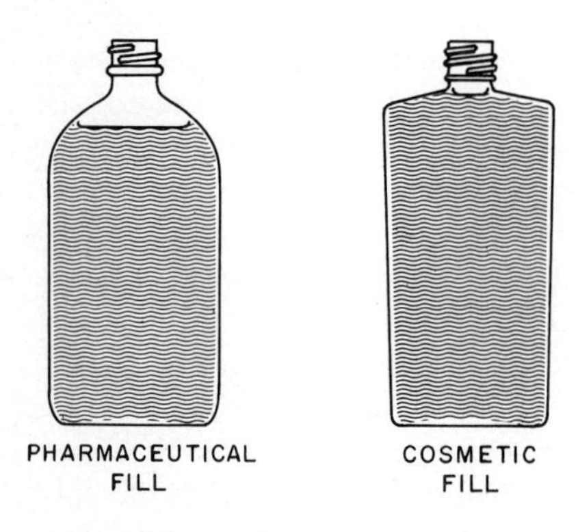

Fig. 7. Fill points. Pharmaceutical products are usually filled to a point halfway up the shoulder. Cosmetics and toiletries often are filled to the base of the neck, for a better appearance.

Molds are made of cast iron. They consist of a bottom plate, a body mold which is divided vertically into two halves, and a neck or finish mold which also is split into two parts. The finish molds are generally interchangeable with other shapes of body molds, and it is often possible to substitute finish molds which are a size larger or a size smaller, or of a different type, for use with the same body molds. This permits some flexibility when using stock molds for short runs. The cost of a set of molds for a private mold design will be between $3,000 and $7,000, depending upon the size and the complexity of the design.

The bottom plate usually carries the manufacturer's symbol or, in the case of private molds, the customer's name. The mold number is put into the bottom plate to identify which mold of a set made a particular bottom. In this way defects can sometimes be traced to a fault in one of the molds. A plant number also may be added if the manufacturer has several plants, and sometimes the date of manufacture is indicated by the last two digits of the year. (See Figs. 8 and 9.)

It is good practice to stipple the bottom plate, at least that portion which will be in contact with the conveyor belt, to minimize the thermal shock of contact with the moving metal floor of the annealing oven. A minor disadvantage of stippling the bottom is that dirt will more likely be picked up on the filling line, getting into the crevices of the stippled surface.

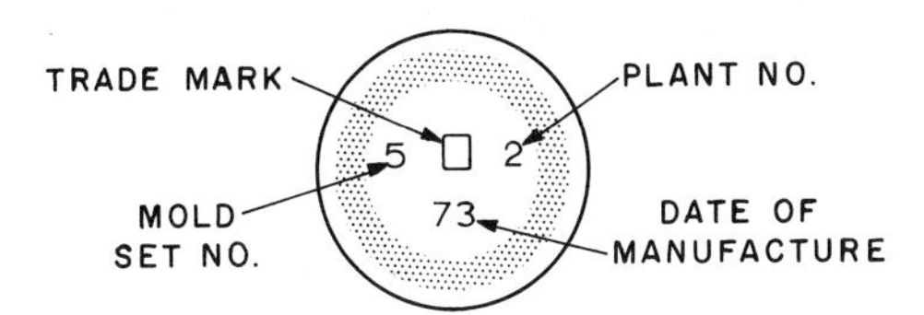

Fig. 8. Bottom marks on bottles and jars. Identifying symbols and numbers are sometimes put in the bottom plate of the mold for a glass container. These marks are not standardized, but the arrangement shown is typical. Some of the trademarks used by the various glass plants are shown in Fig. 9.

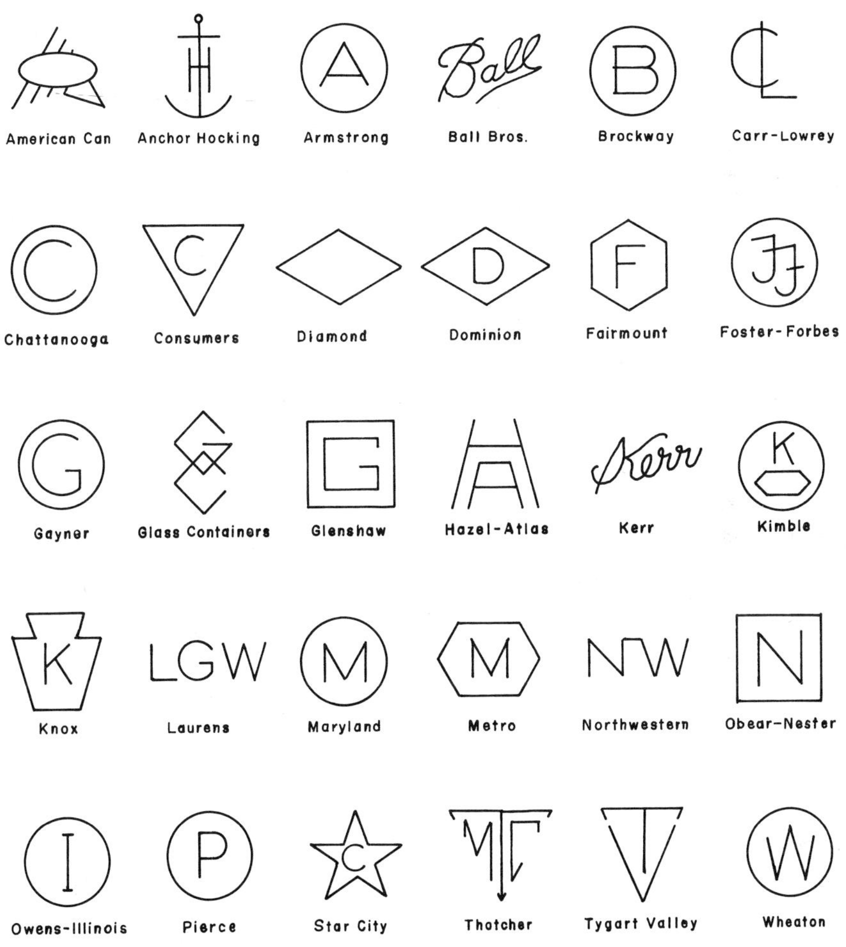

Fig. 9. Trademarks of glassware manufacturers. The bottom plate of a mold sometimes carries a symbol to identify the manufacturer. This chart may be helpful in locating the source of a bottle or jar from a sample.

Body molds are made in two halves, which are divided in a vertical plane. In a square or oblong shape the parting line is usually placed on diagonally opposite corners so that the mark made by the joint will be less noticeable. A raised band or bar at the shoulder and the base to provide contact points on the filling line will greatly increase the durability of the container, if it can be worked into the design. (See Fig. 3.) It should be remembered that as the mass of the bottle increases, the energy of impact increases in direct proportion. What is even more significant, with the trend toward higher filling speeds, is that the impact energy increases as the square of the velocity. That is, at three times the speed there is nine times the shock. Many decorative effects can be obtained with sunken panels, flutes, stippling, and similar devices. Sharp corners and edges should be avoided, however, because they are difficult to make and will lead to a high incidence of defects in the finished ware. Sharp convex edges in the mold, for example, tend to cause hot checks or pressure checks in the glass. Acute concave angles require high temperature and pressure to force the glass into the corner, possibly causing cocked necks, hollow necks, or other problems.

The shape of the body should be considered in terms of filling and labeling equipment. An oval bottle may shingle if the radius at the sides is too sharp. (See Fig. 10.) If the width or thickness at the top is different from that at the base, the bottles may topple when they are pushed together by the conveyor. (See Fig. 11.)

Fig. 10. Design fault. A poorly designed oval bottle may jam on a conveyor because of shingling, as shown toward the right.

Fig. 11. Design fault. A bottle that is wider at the top than at the bottom can cause problems on the production line. The movement of the conveyor belt may cause tipping, as shown toward the right.

Because placing labels accurately may be a problem with certain shapes, it is often wise to put an indexing dimple or button in the bottom which can be sensed by the machine and oriented to the proper position. (See Fig. 12.) Insofar as making the bottle or jar is concerned, a sphere would be the easiest to blow, and a cylinder would be next. The farther we depart from these ideal shapes, the more difficult it becomes to fill out the mold, the more glass it takes, and the greater is the chance of having thin corners.

Glassware which is wider at the top than at the bottom, such as tumblers, ashtrays, and some wide-mouth jars, are made by pressing. Most ointment jars and cosmetic wide-mouth ware are made by the press-and-blow method. By far the greatest number of glass containers however, are made by blowing.

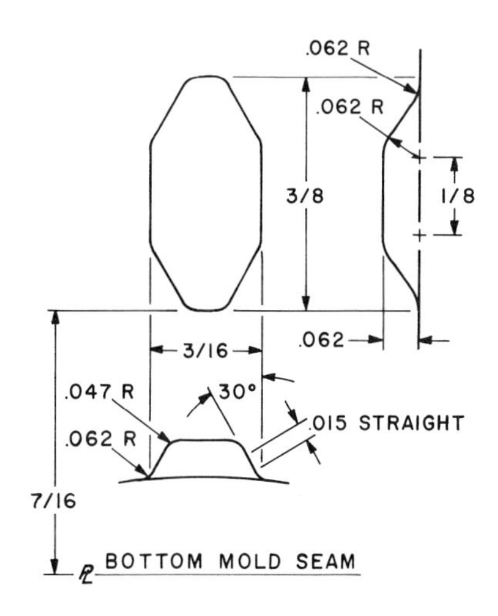

Fig. 12. Locating bar. For placing labels in register, or to avoid mold seams when decorating, a small raised button is molded in the side near the bottom of a cylindrical glass container. This is used to orient the bottle or jar automatically on the production line.

Finishes. The part that takes the closure is called the finish because in the days of handmade bottles the neck was made last. Today the automatic machines make this part first, but it is still called the finish. Containers can be generally classified as "wide-mouth" and "narrow-neck" ware. As the name implies, wide-mouth containers have an opening almost as large as the body of the container. Tablet jars and ointment pots are examples of this type. Liquids are usually put into the narrow-neck variety, for ease in pouring. (See Fig. 13.)

The neck may be threaded to take a screw cap, or it may have an interrupted thread to take a lug-type cap, sometimes known as a quarter-turn closure. Friction covers are often used on food containers; for instance, on jelly glasses they preserve the smooth edge for after use as tumblers. (See Table 7.)

Tolerances. The Glass Container Manufacturers Institute (GCMI) has established limits which are generally accepted as reasonable tolerances by most manufacturers. Allowance has been built into these ranges not

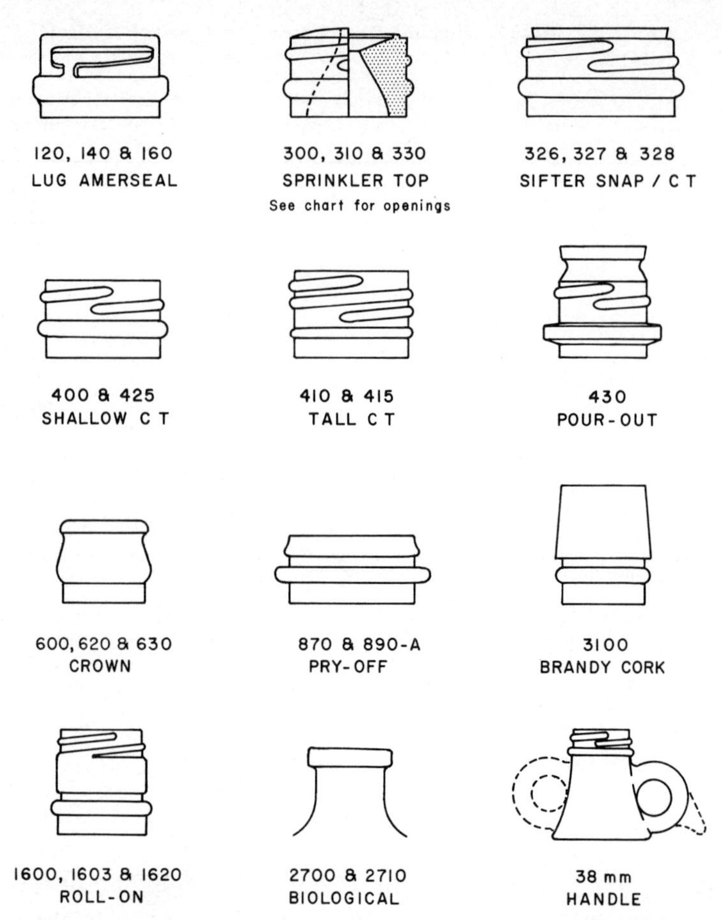

Fig. 13. Glass finishes. A few of the great variety of finishes for bottles and jars are shown. Others are described in Table 7. Standard dimensions for the most popular types of finish are given in Table 9.

TABLE 7 Standard Glass Finishes

Series no.	Description	Series no.	Description
120	2-lug Amerseal quarter-turn finish	285	Short goldy and slip cap catsup finish
140	4-lug Amerseal quarter-turn finish	300	Sprinkler finish with flat top
160	6-lug Amerseal quarter-turn finish	310	Sprinkler top drain-back finish with sealing ring
200	Combination large goldy and CT (continuous thread) catsup finish	320	Sprinkler finish with raised center
225	Combination small goldy and CT catsup finish	326	Pour-out snap cap CT combination
235	Short goldy and CT catsup finish	327	Snap cap CT combination
250	Combination large goldy and slip cap catsup finish	328	Sifter snap cap CT combination
275	Combination small goldy and slip cap catsup finish	330	Sprinkler top drain-back finish without sealing ring
		400	Shallow CT finish
		401	Wide sealing surface on 400 finish

TABLE 7 Standard Glass Finishes (Continued)

Series no.	Description
405	Depressed threads of 400 finish at mold seam
410	Medium CT concealed-bead finish
414	Brockway pour-out CT
414-B	Snap cap CT combination
415	Tall CT concealed-bead finish
416-11	Tamperproof jug handle 1 ear
416-12	Tamperproof jug handle 2 ears
418-A	Pour-out on one side
425	Shallow CT sizes 8 to 15 mm
430	Pour-out CT
440	Unpopular sizes of 400 finish
445	Deep S CT finish
450	Deep CT mason finish
456	Deep skirt mason CT, 456-910 with bail eyelet
460	Home canning jar finishes
462	Freezer jar CT finish
471	Wide sealing surface on 400 finish
600	Beverage crown finish
610	Crown finish for steinie beer bottle
620	Crown finish for stubby beer bottle
630	Crown finish for cone-top beer bottle
640	Crown finish for series 166 beer bottle
650	Crown finish for nonreturnable beer bottle
660	Crown finish for nonpressure bottle
700	Crown finish larger sizes
870	Anchor vacuum side seal pry-off
890-A	Anchor vacuum top and side seal pry-off
900	Repeal finish 28-400 with wide transfer bead
910	Bail eyelet used with G-450 finish
1100	Kork-N-Seal finish
1110	English Kork-N-Seal finish
1125	Kork-N-Seal tamperproof finish
1150	Kork-N-Seal pouring lip finish
1240	Vacuum lug-style finish
1300	National Seal Co. Duplex finish
1310	National Seal Co. Uniplex finish
1337	Roll-on pilferproof finish
1500	Jelly glass side seal finish
1525	Anchor side seal finish
1550	Anchor side seal with bead
1600	Roll-on finish
1603	Roll-on Hytop finish
1620	Roll-on pilferproof finish
1624	Roll-on pilferproof finish
1640	Roverseal finish
1705	White side seal pry-off finish
1710	Side seal pry-off, IS machine finish
1712	Side seal pry-off, Owens or Miller machine finish
1715	White snap-on pry-off finish
1720	White deep side seal pry-off finish
1725	Wide mouth snap-on pry-off finish
1740	Side seal pry-off baby food jar finish
1751	White twist-off vacuum seal
1754	White deep twist-off vacuum seal
1880	Large goldy finish
1885	Small goldy finish
2000	Vacuum top seal, threaded reseal finish
2300	Upressit finish
2600	Anchor pry-off finish
2700	Spun-on finish
2710	Biological and aerosol finish
2800	Cup cap finish
2900	Jigger seal finish
3100	Brandy cork finish
3105	Short brandy cork finish
3110	Cognac cork finish
3120	Square bead cork finish
3130	Gin cork finish
3135	Tapered bead cork finish
3140	Champagne cork finish
3150	Vermouth cork finish
3160	Sweet wine cork finish
3170	Medium brandy cork finish
3190	Standard wine cork finish
3200	Constricted cork finish
3210	Constricted cork double bead finish
3250	Pharmaceutical cork finish
3320	Guardian 30-400 large bead finish
3340	Guardian 30-400 small bead finish
3350	Guardian 30-400 tall finish
3380	Guardian 30-400 double spin finish
3390	Guardian 38-400 double spin finish
3750	Side or top seal jelly glass beaded edge finish
3761	Owens side seal finish
3767	Owens Vapak side seal finish

Chart for 300, 310, 320, and 330 Sprinkler Finish Openings

	A	B	C	D	E	F	G	H	I	J	K	L	M	N
Opening	3/32	7/64	1/8	9/64	11/64	3/16	13/64	7/32	15/64	1/4	17/64	9/32	19/64	5/16
Filling tube	3/64	1/16	5/64	3/32	1/8	9/64	5/32	11/64	3/16	13/64	7/32	15/64	1/4	17/64

Example: 20-400-310K finish has 17/64-in. opening (1/32-in. tolerance) for 7/32-in. maximum diameter filling tube.

only for the process capabilities, but also for a certain amount of growth, that is, enlargement of the mold due to wear. The tolerances are quite liberal, and it is possible to work somewhat closer without penalty. (See Tables 8 and 9.) If it is necessary to work still tighter, it can be done, but there will be a premium for more frequent replacement of the molds and a higher rejection rate by the inspectors at the cold end of the lehr.

TABLE 8 Standard Tolerances for Glass Containers

Height, in.*	Tolerance either way, in.	Capacity, fl oz*	Tolerance either way, fl oz	Weight, av oz*	Tolerance either way, av oz
Below 4 1/4	1/32	Below 1/8	1/100	Below 1 3/8	1/16
4 1/4–8 1/2	3/64	1/8–1/2	1/64	1 3/8–2	3/32
8 1/2–12	1/16	1/2–1	1/32	2–2 1/2	1/8
12–15	5/64	1–2	3/64	2 1/2–3	5/32
15–20	3/32	2–3 1/4	1/16	3–3 3/4	3/16
20 and up	1/8	3 1/4–4 1/4	5/64	3 3/4–6	1/4
Diameter, in.*	Tolerance either way, in.	4 1/4–5	3/32	6–7 1/2	5/16
		5–6	7/64	7 1/2–9 1/2	3/8
		6–8	1/8	9 1/2–14	1/2
Below 1	1/64	8–10	5/32	14–17	5/8
1–2 1/4	1/32	10–12	3/16	17–20	3/4
2 1/4–3	3/64	12–16	7/32	20–24	7/8
3–4 1/2	1/16	16–20	1/4	24–28	1
4 1/2–5 3/4	5/64	20–29	5/16	28–35	1 1/4
5 3/4–6 3/4	3/32	29–37	3/8	35–40	1 1/2
6 3/4–7 3/4	7/64	37–46	7/16	40–48	1 3/4
7 3/4 and up	1/8	46–57	1/2	48–54	2
Width and thickness, in.*	Tolerance either way, in.	57–75	5/8	54–60	2 1/4
		75–95	3/4	60–72	2 1/2
		95–115	7/8	72–88	3
Below 1 5/8	1/32	115–140	1	88–112	4
1 5/8–2 3/4	3/64	140–165	1 1/2	112–160	5
2 3/4–4	1/16	165–192	2	160–224	7 1/2
4–5	5/64	192–256	3	224–288	10
5–6	3/32	2–3 gal	4	288–352	12 1/2
6–7	7/64	3–5 gal	6	352 and up	15
7–8 3/4	1/8	5 gal and up	8		
8 3/4 and up	5/32				

* The tolerances apply to dimensions up to but not including the second figure in the range.

DEFECTS

There are six broad classifications of glass defects: (1) checks, (2) seams, (3) nonglass inclusions, (4) dirt, dope, adhering particles, or oil marks, (5) freaks and malformations, and (6) marks. These vary in importance, depending upon their severity, but the type of defect will usually determine its seriousness. Defects can be grouped as *critical*—those that are hazardous to the user and those that are completely unusable because they are freaks or did not completely fill the mold; *major*—those that materially reduce the usability of the container or its contents; and *minor*—those that do not affect the usability of the container, but detract from its appearance or acceptability to the customer. (See Fig. 14.)

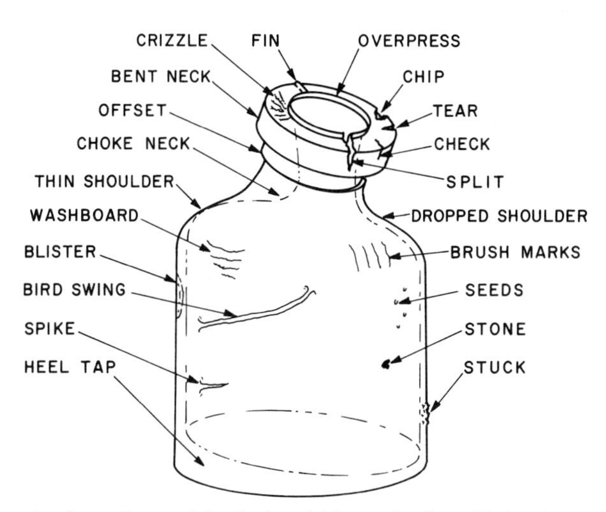

Fig. 14. Defects in glass. Some of the faults which may be found in bottles and jars. Many of these are very rare, but others will be found more frequently.

Critical Defects

Bird-swings and Spikes. Long thin strands inside the bottle that would probably break off when the bottle is filled.

Overpress. A rim extending up from the inside edge of the finish, sometimes sharp.

Filament. Hairlike string inside the bottle.

Split. An open crack starting at the top of the finish and extending downward.

Check. A small, shallow surface crack. Cold checks are usually wavy, and hot checks are generally straight lines. Groups of checks at the top of the finish are called a crizzled finish. Pressure checks occur near the seam; they may be small or run the full length of the bottle. Bruise checks often show up at the shoulder or the heel areas. Mold checks are deep and run from the bottom up the sides. Disk checks occur on the bottom from contact with the transfer disk in the molding operation. Panel checks are those found on flat areas of the bottle.

Unfilled Finish. Wavy top surface or a dip, usually right above the start

TABLE 9 Standard Finish Dimensions for Glass Containers

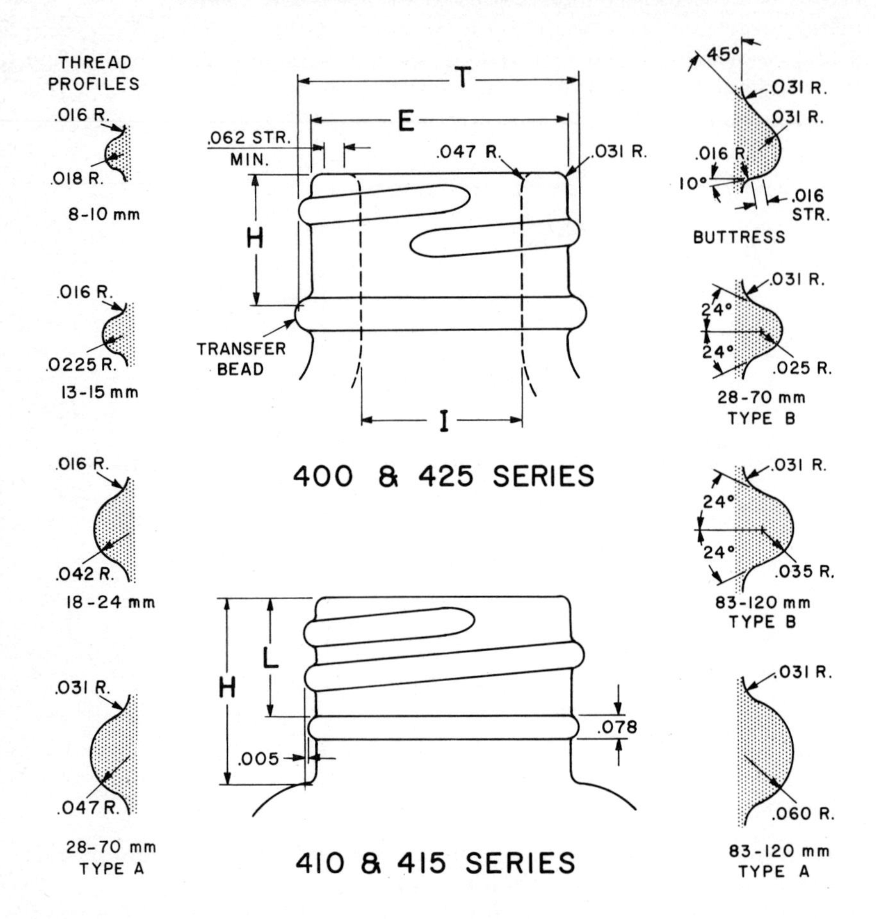

ALL RADII ARE MAX.

Type	Series	Millimeter size	T min.–max.	E min.–max.	H min.–max.	I min.
Shallow continuous thread	425	8	0.348–0.358	0.298–0.308	0.252–0.262	0.114
		10	0.398–0.408	0.343–0.353	0.260–0.275	0.176
		13	0.499–0.514	0.439–0.454	0.285–0.300	0.176
		15	0.566–0.581	0.506–0.521	0.285–0.300	0.176
	400	18	0.684–0.704	0.600–0.620	0.349–0.364	0.325
		20	0.763–0.783	0.679–0.699	0.349–0.364	0.404
		22	0.842–0.862	0.758–0.778	0.349–0.364	0.483
		24	0.920–0.940	0.836–0.856	0.378–0.393	0.516
		26	0.984–1.009	0.890–0.915	0.378–0.393	0.535
		28	1.063–1.088	0.969–0.994	0.378–0.393	0.614

TABLE 9 Standard Finish Dimensions for Glass Containers (Continued)

Type	Series	Millimeter size	T min.–max.	E min.–max.	H min.–max.	I min.
		30	1.102–1.127	1.008–1.033	0.378–0.398	0.653
		33	1.240–1.265	1.146–1.171	0.378–0.398	0.791
		35	1.329–1.364	1.235–1.270	0.378–0.398	0.875
		38	1.441–1.476	1.347–1.382	0.378–0.398	0.987
		40	1.545–1.580	1.451–1.486	0.378–0.398	1.091
		43	1.619–1.654	1.525–1.560	0.378–0.398	1.165
		45	1.705–1.740	1.611–1.646	0.378–0.398	1.251
		48	1.853–1.870	1.741–1.776	0.378–0.398	1.381
		51	1.933–1.968	1.839–1.874	0.378–0.408	1.479
		53	2.032–2.067	1.938–1.973	0.378–0.408	1.578
		58	2.189–2.224	2.095–2.130	0.378–0.408	1.735
		60	2.302–2.342	2.208–2.248	0.378–0.408	1.853
		63	2.421–2.461	2.327–2.367	0.378–0.408	1.972
		66	2.539–2.579	2.445–2.485	0.378–0.408	2.090
		70	2.696–2.736	2.602–2.642	0.378–0.408	2.247
		75	2.873–2.913	2.779–2.819	0.378–0.408	2.424
		77	2.995–3.035	2.901–2.941	0.457–0.487	2.546
		83	3.228–3.268	3.108–3.148	0.457–0.487	2.753
		89	3.466–3.511	3.346–3.391	0.505–0.535	2.918
		100	3.892–3.937	3.772–3.817	0.567–0.597	3.344
		120	4.674–4.724	4.554–4.604	0.655–0.685	4.131
Medium continuous thread	410	18	0.684–0.704	0.600–0.620	0.501–0.526	0.325
		20	0.763–0.783	0.679–0.699	0.532–0.557	0.404
		22	0.842–0.862	0.758–0.778	0.563–0.588	0.483
		24	0.920–0.940	0.836–0.856	0.624–0.649	0.516
		28	1.063–1.088	0.969–0.994	0.686–0.711	0.614
Tall continuous thread	415	13	0.499–0.514	0.439–0.454	0.430–0.455	0.218
		15	0.566–0.581	0.506–0.521	0.535–0.560	0.258
		18	0.684–0.704	0.600–0.620	0.595–0.620	0.325
		20	0.763–0.783	0.679–0.699	0.720–0.745	0.404
		22	0.842–0.862	0.758–0.778	0.815–0.840	0.483
		24	0.920–0.940	0.836–0.856	0.935–0.960	0.516
		28	1.063–1.088	0.969–0.994	1.060–1.085	0.614

of the thread. More rarely the thread or the transfer bead is not filled out.

Freaks. Odd shapes and conditions that render the container completely unusable. Bent or cocked necks are a common defect.

Poor Distribution. Thin shoulder, slug neck, choke neck, hollow neck, heavy bottom, and slug bottom are terms used to describe an uneven thickness of glass. A slug is a heavy spot, most often found at the parting line.

Finish Marks. Lines on the sealing surface, sometimes called shear marks.

Soft Blister. A thin blister usually found in or near the sealing surface. It can, however, show up anywhere on the bottle.

Cracks. Partial fractures generally found in the heel area, but sometimes occurring at the shoulder.

Cord. A strain which is not relieved by annealing.

Pinhole. Any opening causing leakage. It occurs most often in bottles with pointed corners.

Major Defects

Chipped Finish. Pieces broken out of the top edge in the manufacturing process.

Stone. Small inclusion of any nonglass material.

Rocker Bottom. A sunken center portion of the bottom.

Mismatch. One half of the finish may be shifted to the side or upward of the other half, or the finish may be "set over" from the rest of the container.

Fin. A seam on the top surface or down the side at the parting line.

Out-of-round finish. A pinched, flattened, or oval finish.

Flanged Bottom. A rim of glass around the bottom at the parting line.

Minor Defects

Sunken Shoulder. Not fully blown, or sagged after blowing.

Tear. Similar to a check, but opened up. A tear will not break when tapped, as a check will.

Washboard. A wavy condition of horizontal lines in the body of the bottle.

Hard Blister. Deeply embedded blister that is not easily broken.

Dirt. Scaly or granular nonglass material. Also oil, carbon, dope, rust, graphite, or other foreign substance.

Heeltap. Heavy glass on one side of the bottom.

Mark. A brush mark is composed of fine vertical laps. Oil marks are the result of oil accumulation on the molding equipment. Carbon marks come from the feeder or from mold dope.

Droplet. A small projection of glass, usually near the parting line of the finish, caused by the neck rings being chipped.

Neck Ring Seam. Bulge at the parting line between the neck and the body.

Long Neck. Stretched-out neck resulting from bottle being too hot when picked up.

Stuck. If bottles touch while still soft, they may stick and, in pulling apart, leave a rough spot.

Seam. The seam from the blank mold may not correspond with the mark made by the joint of the finishing mold. This is not a defect unless it is very large or unsightly.

Seeds. Small bubbles in the glass.

Cold Appearance. Also called cold mold. It is a wavy condition on the surface of the bottle.

Wavy Bottle. An irregular surface on the inside.

QUALITY CONTROL

At least 100 samples should be taken at random from the lot, preferably 300 or more to ensure that critical defects are not missed. Samples should not be taken from cases that show shipping damage. (See Sec. 19, "Quality Control.") If critical or major defects exceed the limits, they should be rechecked. If they still appear to be excessive, the lot should be rejected. For minor defects that are beyond the limits, the lot is sometimes accepted with a warning to the supplier.

TUBING PRODUCTS

Glass tubing is made by drawing the molten glass through a die, mounted on the furnace, in a horizonal direction until it is cool enough to cut to length. It is supported on rollers as it leaves the die, and every few feet there are more rollers to keep the soft tubing from sagging. After it has traveled 100 ft or more, it is nicked and snapped off in lengths of approximately 10 ft, by automatic equipment. Then it is put into a sorting machine which gauges the outside diameter at several points along its length, and rolls it into one bin or another, according to its diameter.

From this tubing can be made a variety of containers such as vials, bottles, and ampuls, as well as applicator rods, pipettes, and syringe barrels. Type I glass is frequently used for these items because it is more resistant to chemical action, and because it is generally more uniform and therefore can be more consistently worked by the machines which make these articles from tubing. However, type II glass and type III glass are available, and they are lower in cost.

Small bottles of 30 cc or less can be made from tubing and will have certain advantages over the usual molded bottles. The wall is more uniform and therefore can be made thinner. This gives better clarity and complete absence of black specks and other defects resulting from

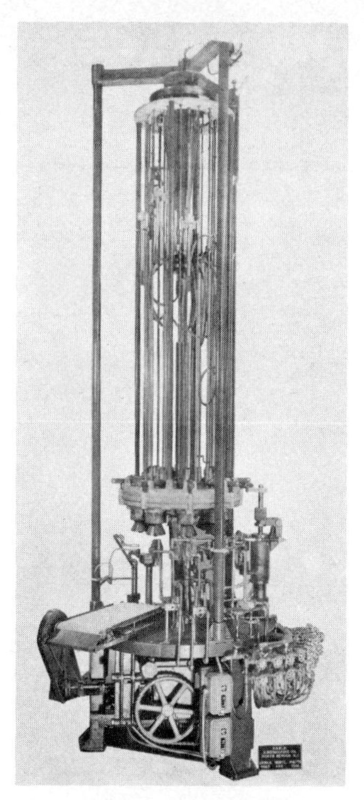

Fig. 15. Tubing products machine. Vials, ampuls, and bottles are made on machines of this type. Ten-foot lengths of glass tubing are placed in the turret and fed down as the machine rotates. Gas flames heat the glass, and forming tools cut and shape the piece into a finished container. (*Kahle Engineering Co.*)

mold dope. As much as a one-third reduction in weight can be made by a change from molded to tubing containers, in the small sizes. There are no molds seams, so that decoration of the bottles is much easier. Delivery is often 7 or 8 weeks instead of up to 32 weeks for blown ware. (See Fig. 15.)

Section 7

Metal Containers

Metal Cans 7-1
History 7-1
Manufacture 7-2
Tin-free Steel 7-6
Linings 7-8
Decoration 7-9
Fabrication 7-9
Joints and Seams 7-12
Shaped Cans 7-13
Equipment 7-14
Aluminum Cans 7-14
Aerosol Cans 7-16
Friction-plug Cans 7-18
Screw-top Cans 7-18
Seamless Cans 7-20
Steel Drums and Pails 7-21
History 7-21
Regulations 7-21
Manufacturing 7-21
Linings 7-21
Types of Drums 7-22
Steel Pails 7-23
Collapsible Tubes 7-24
History 7-24
Materials 7-25
Manufacturing 7-26
Decoration 7-27
Linings 7-29
Closures 7-29
Sealing 7-30
Aluminum 7-31
Aluminum Trays 7-32

METAL CANS

History. The quest for a better container, which led to the development of the tin can, was probably started by Napoleon's offer of 12,000 francs for a method of preserving food to sustain his army. This award was claimed by a Parisian chef and confectioner, Nicholas Appert, in 1809 for his new process of canning. A year later Peter Durand, an Englishman, invented the "tin canister," and when William Underwood migrated from England to Boston in 1817, he brought the beginnings of the United States canning industry with him. For a long time the cans were made slowly and laboriously by hand. Both ends were soldered to the can, with a hole about an inch in diameter in the top. After filling the can through this hole, a tinplate disk was soldered in place.

The "sanitary-style" open-top can was introduced in the early 1900s, and this type of construction made it possible to develop high-speed equipment for making, filling, and closing these cans. The original handmade cans were produced at the rate of five or six per hour, but modern machines can produce 500 to 1,000 per minute. Over 50 billion cans are made each year in this country, and this takes about 5 million tons of steel. Looking at it another way, Americans open 175 million cans every day. (See Table 1.)

TABLE 1 End Use of Cans Based on Tonnage of Steel Used

Product	Percentage*	Product	Percentage*
Beer	18.0	Motor oil	3.6
Vegetables	16.0	Meat and poultry	3.4
Fruits	13.8	Paint products	3.3
Miscellaneous foods	11.0	Fish	2.5
Miscellaneous nonfoods	6.0	Aerosols	2.2
Dairy products	4.3	Shortening	2.0
Coffee	4.3	Antifreeze	1.0
Pet food	4.0	Baby food	0.6
Soft drinks	4.0		

* To convert to number of base boxes used annually, multiply by 1.2 million.

Manufacture. The name "tin can" is a misnomer, since these containers are 99½ percent steel, with only a very thin coating of tin on both sides. (See Fig. 1.) The steel is usually about 0.010 in. thick, although larger cans may require heavier gauges in proportion to their size. The metal that is used for the ends is generally of a stiffer temper than that which is used for the body, and the can maker uses the highest temper he can work with, in order to get the strongest can for the lowest cost.

The tin coating is only about 0.00001 in. thick, but it is sufficient to protect the base metal from rust and corrosion, except under extreme conditions. It protects the steel not only by covering the surface, but also by an electrochemical action with the steel. It also alloys with the iron compounds that are on the surface of the sheet. However, the protective value of the tin without any additional coatings is somewhat limited. When exposed to atmospheric conditions of high humidity, rust will develop rapidly. On the inside of a sealed can, oddly enough, the polarity of the two materials changes and there is considerable protection against perforation by the electrochemical action of the tin with the steel.

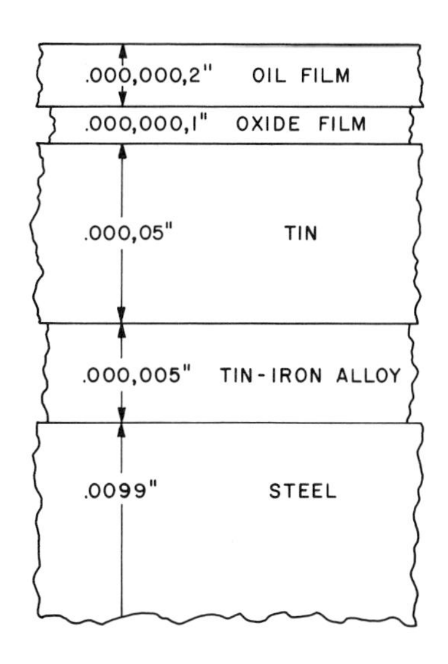

Fig. 1. Composition of tinplate. The relative thicknesses of the various layers that make up tinplate are shown greatly enlarged.

Most steel is made in an *open-hearth furnace,* but a more economical method, developed in Europe and known as the *basic oxygen process,* has recently come into use. It cannot be used with low-cost scrap however, and *electric furnaces* are being used for this purpose. Mini-steel plants based on these electric furnaces are springing up around the country to convert old automobiles and other scrap iron into sheet stock for metal containers.

Another technique that was developed in Europe and will cut cost and improve the quality of steel for packaging is the *continuous casting* process. This method produces a thick ribbon of steel instead of ingots, and its yield is higher. This process used in conjunction with *vacuum carbon deoxidizing* produces a cleaner and more uniform steel, which is important for high-speed can making.

The temper of the steel is selected according to the amount of working that has to be done to it; for example, sardine cans, which are deep-drawn, would be made from T1 temper, which is quite soft. At the other end of the scale is T6 temper, which is very hard. (See Table 2.)

Type MR. Most widely used steel for containers. Residual elements are not limited, except phosphorus, which is kept at a low level.

Types L and LT. Used for tinplate that is intended for packing corrosive food products. Residual elements such as phosphorus, silicon, copper, nickel, chromium, and molybdenum are limited to the least practicable amounts. Type LT has the same composition as type L that has been tested for corrosion resistance.

Type MC. This steel is rephosphorized to provide extra stiffness. It is

TABLE 2 Tempers of Tinplate

Temper	Rockwell hardness 30-T scale	Description	Applications
T1	46–52	Soft for deep drawing	Nozzles, spouts, and closures
T2	50–56	Moderate drawing where some stiffness is required	Rings and plugs, pie pans, closures, shallow-drawn and specialized can parts
T3	54–63	Shallow drawing, general purpose with fair degree of stiffness	Can ends and bodies, large-diameter closures, crown caps
T4	58–64	General purpose where increased stiffness desired	Can ends and bodies, small screw closures
T5	62–68	Rephosphorized steel to resist buckling	Can ends and bodies for noncorrosive products
T6	67–73	Rephosphorized steel for maximum stiffness	Very stiff applications

used in can ends for noncorrosive products such as beer. This grade of steel is being replaced to a large extent by 2CR, described below.

Type D. This nonaging type of steel is aluminum-killed (deoxidized) and processed to improve the drawing characteristic. It is used for deep-drawn parts or where freedom from fluting or stretcher strains is required.

The recent introduction of a stiffer grade of steel, known as double cold reduced plate, or 2CR, has permitted lighter gauge weights to be used for cans for beer, motor oil, and citrus juices. Three tempers are available in 2CR. These are designated DR-8, DR-9, and DR-10 in order of increasing strength and hardness. Tin coatings are available in various weights on 2CR, in bright, matte, and satin finishes, just as they are for regular tin plate. It is reported that beer cans made from 2CR cost about 87 cents per thousand cans less than those made of regular tinplate.

The standard of measurement for tinplate in the can industry is the base box. This is the equivalent of 112 sheets 14 by 20 in. or 31,360 sq in. of plate (62,720 sq in. of plated surface). A base box provides enough material for almost 400 cans of No. 2 size, which are 3$\frac{7}{16}$ in. in diameter by 4$\frac{9}{16}$ in. high. (See Tables 3 and 4.)

The original method for applying tin was by hot dipping, and this is still used to some extent, but more than 95 percent of the tinplate is now being made by the electrolytic process. Electrolytic tinplate is normally sold in a bright finish in which the tin coating has been melted to a lustrous finish. It is available also in a matte finish, in which the tin is not melted; it is often used in this form for crown caps, for its attractive

TABLE 3 Can Maker's Standard Base Weights for Tinplate*

Weight, lb	Thickness, in.	Weight, lb	Thickness, in.
55	0.0061	95	0.0105
60	0.0066	100	0.0110
65	0.0072	107	0.0118
70	0.0077	112	0.0123
75	0.0083	118	0.0130
80	0.0088	128	0.0141
85	0.0094	135	0.0149
90	0.0099		

* Tolerance for individual sheets ±10 percent.

appearance. A satin finish can be produced also by applying the tin coating before the final cold reduction. Other forms of sheet metal used for containers include bare mild steel designated CMQ for can-maker quality; this is usually called black plate because some of the early production was covered with black iron oxide. It is used for packaging powders, oils and greases, and other noncorrosive products. Sometimes it is enameled or chemically treated, but since these procedures are about as costly as tinplating, there is not always a real saving in its use. Terneplate, which gets its name from the French word for "dull," is steel which has been hot-dipped in a mixture of 80 to 90 percent lead and 10 to 20 percent tin; it is designated SCMT for special-coated manufacturing terne, which carries no specific weight of coating. For most purposes the thinnest coating that will cover the metal is used. This is about 1.30 lb per base box, although heavier coatings can be supplied if needed. Terneplate cannot be used in contact with foods because of the toxicity of the lead, but for nonfood products it has many applications, being less costly than tinplate.

TABLE 4 Typical Base Weights of Tinplate for Cans

Name	Dimensions*	Use	Weight of base box, lb	
			Body	Ends
No. 1	211 × 400	Soup	90	95
No. 2½	401 × 411	Vegetables	95	100

* Figures given are for the outside dimensions of the sealed can. The first figure always indicates the diameter (e.g. 401), and the second the height (e.g. 411). In each case the first digit designates whole inches, and the second and third digits designate sixteenths of an inch. Thus 401 × 411 means the diameter is 4¹⁄₁₆ in. and the height is 4¹¹⁄₁₆ in.

The usual hot-dipped coatings for the canning industry would average 1.00 to 1.50 lb per base box. During World War II the supply of Straits tin from Malaysia was cut off. Although some tin was available from Bolivia and Africa, it was not as well suited for the purpose. In order to utilize more fully the limited amount of tin that was available, a concentrated effort was made to develop the process of electroplating the tin onto the steel. This process had been used up to that time on a very limited basis, but in the wartime emergency the government induced the manufacturers to install the necessary equipment. The tin can be put on more uniformly by this process, with as little as one-third as much being used for a satisfactory coating for certain purposes. In performance, No. 100 (1.00 lb per base box) electrolytic tinplate has been said to be equal to No. 125 hot-dipped tinplate, although it is not clear whether this statement is based on pot yield or on the actual coating thickness.

A recent refinement called "differential coating" permits a different thickness of tin to be put on each side of the sheet. A common example is a 1.00-lb coating on the inside, where the greatest protection is needed, and 0.25 lb on the outside. The savings of a differential coating over a hot-dipped coating, in terms of No. 2½ cans, would be about $1.24 per thousand cans. Lower tin coatings are also available, such as No. 75/25 and No. 50/25. (See Table 5.)

TABLE 5 Comparative Prices of Various Coating Weights of Conventional Tinplate (per base box)

Black plate	$ 8.50
0.25/0.25	9.40
1.00/0.25	10.00
1.25/1.25 common coke	12.60

Tin-free Steel. Until recently the principal material for can making has been tinplate. A base box of tinplate contains from 50 cents to $2 worth of tin, and much of the effort to reduce the cost of metal containers has been toward finding a substitute for this tin coating. Three types of tin-free steel have come into use in recent years: (1) chemically passivated steel, which has a phosphate-chromate film on the surface; (2) chrome-coated steel with a layer of chromium oxide on top of the metallic chromium; and (3) aluminum-coated steel. There are some problems, however, in drawing or ironing these substitute materials without damaging the surface, and in getting solder to flow and stick to the joints. Two new methods have been developed for fastening the seams, on a production basis: cementing and welding.

Conventional tinplate resists corrosion not only by the protective layer of tin on its surface, but also by a cathodic reaction which minimizes oxidation at any pinholes or bare spots that may be present. A tin coating also prevents the iron from being dissolved in certain beverages and food products, which would acquire an undesirable iron taste. Tin-free steel does not have this advantage, and it requires complete coverage of the coating to be effective. Fortunately, the adhesion of enamels to tin-free steel is superior to their adhesion to tinplate.

The earliest tin-free steel was made with a phosphate-chromate type of passive film on the steel surface. This has had only limited applications in the food and beverage field. Although it is used to some extent in packaging beer and citrus juices, it has not been very successful in containing carbonated beverages. It is being used at the present time for large drums and pails for many industrial products.

Annual Use of Steel for Containers

Tinplate	123,679,000 base boxes
Tin-free steel	11,971,000 base boxes
Black plate	10,019,000 base boxes

One of the most promising new materials for cans is a chromium-coated steel developed in Japan and improved in this country. Chromium coating alone is not satisfactory, because of staining with certain food products. Chemical passivation alone does not provide the corrosion resistance and the enamel adhesion that are required. However, a combination of *metallic chromium* followed by treatment with solutions to produce chromium oxide on the surface has given excellent results. The metallic coating is extremely thin, being only 0.2 μin. thick, or about 1/100 the thickness of tin on No. 25 tinplate. This is about 20 atoms in thickness. The *passivation treatment* involves chromium salts such as CrO_3 or $Na_2Cr_2O_7$, which put an oxide coating over the metallic chromium, which is an additional 0.2 μin. in thickness. This combination is considered to be equal in corrosion resistance to No. 50 electrolytic tinplate, and although it will not withstand attack by strong acids or alkalies, it is generally satisfactory with weak acids such as citric acid, lactic acid, or acetic acid, as well as with weak alkalies. It cannot be soldered, but it can be welded or cemented, using nylon adhesives. It cannot be double-seamed because chromium is very brittle and would crack in the forming process. Therefore the side seams of cans must be lap seams and not lock seams.

Aluminum-coated steel is another development that offers lower-cost materials for containers. Two methods are used for applying the aluminum to the steel: hot dipping for heavy-gauge stock for pails and drums, and vapor coating for can making material. In the hot dipping process a strip of cold-rolled steel passes first through a cleaning and heat-treating chamber, and then is submerged in a molten bath of aluminum, alloyed with 8 percent silicon. This yields a coating weight of 3½ lb (0.001 in. thick) per base box on each side, which is about 20 times the thickness of tinplate. Vapor coating produces a much thinner layer of aluminum on the steel and is more practical for small containers.

Aluminum-coated steel cannot be soldered, but it can be welded with special fluxes, and cemented side seams also can be used to fabricate containers. It is highly resistant to atmospheric corrosion, and it is well suited for dry products. Some food products, however, are much more corrosive to aluminum than to tinplate, and extensive testing is recommended before adopting this material for other than dry products. The enamels required for aluminum-coated steel are very different from those used for tinplate, but they are just as effective in most cases.

Welded side seams for containers made from tin-free steel are suitable for most purposes, and they provide a neat appearance on the outside, with only a 9/32-in. strip of bare metal exposed. *Cemented side seams* using nylon adhesives are adequate for beverages, but are not recommended for processed foods. The outside appearance of the cemented cans is excellent, with no bare metal exposed. The nylon adhesive that is used for these containers is strongest when it is in shear, rather than in tension, and so a lap seam is better for this purpose than a hooked seam.

Linings. There are various methods of treating the inside of these containers to protect the metal from corrosion, and to avoid contamination of the product by the can. Special coatings, called enamels by the can makers, are usually applied to the tinplate in the flat, and then baked in gas-fired ovens. These enamels must be formulated to suit a particular product, and sometimes even different brands of the same product may require different coatings.

Some of the general types of interior *enamels* have the following uses: An oleoresin coating known as berry and fruit enamel is used for packing berries or fruits that are apt to be discolored by tin salts. Oleoresins, or a C enamel which is an oleoresin pigmented with zinc oxide, are used for vegetables and meats that would form dark sulfide compounds in the presence of tin. Phenolics, or a special combination of oleoresin, phenolic, and zinc oxide which is called seafood enamel, are used for packing fish. Citrus enamel is an oleoresin, or a phenolic that has been modified to withstand attack by the citrus acids. Beer can enamel is usually a double vinyl coating, or a vinyl over epoxy or butadiene coating, to prevent cloudiness and to avoid a tinny taste. Off flavors are really due to dissolved iron, rather than to any tin that might be present. For difficult products, acrylics are sometimes used as a lining material, as well as epoxies, which are rather high in cost, and polybutadiene, which is inclined to be brittle at the required thickness. Acrylics can stand high baking temperatures, and they are often used with meat release agents. Enamels are generally specified in terms of the amount in milligrams which will cover 4 sq in. A typical coating for beer cans, for example, would be 4 on this basis.

Corrosion of the inside of the container is usually either of two general types. One is a surface *corrosion* which proceeds gradually but does not penetrate very deeply. This is fairly easy to control by following good practices in storage and in the processing operations. The other general type of corrosion is the pitting or perforating type, in which the effect is localized and often difficult to see. It may not affect the product, but can be troublesome by causing leakers. It most often occurs at the surface line of the product, but can be above in the vapor phase area, or below in the liquid phase, depending upon the nature of the product.

Some of the pack components which can aggravate a corrosive condition are copper even in small amounts, such as might be picked up from pipes or mixing vessels; sulfur, which is used in foods as a bleaching agent; pectin; phosphoric acid; and some colorants such as azo-type dyes. Materials which act as corrosion *inhibitors* and sequestering agents include certain types of sugar and caramel, ascorbic acid, some enzymes such as glucose-oxidase-catalase, stannous ion (in the amount of 5 to 10 ppm), and calcium or sodium ethylene diamine tetraacetic acid, which is used to pick up traces of iron.

Decoration. The outside of the can may be labeled or it may be lithographed. Paper labels provide the lowest cost with the greatest flexibility. Particularly in the packing of processed foods it is convenient to have the cans undecorated and to add suitable paper labels for the different foods as they are packed. If the can is lithographed, this is usually done in the flat, before the sheet is formed into a can. (See Fig. 2.) A base coat of a paintlike material, generally light in color, is rolled on and dried by being passed through an oven. Prints of several different colors can be applied to get excellent reproduction of the finest artwork. Embossing can be used to heighten the decorative effect for luxury items.

Fabrication. The sheets, which have been enameled on one side and printed on the other, are 1 to 2 ft wide and 2 to 3 ft long. They are cut into strips which are the width of one can, on a machine called a slitter.

The slit strips are cut into body blanks and fed into a body maker, which is the first machine in an automatic can line. In the body maker the corners are notched to avoid the extra thicknesses of metal where the side seam is curled into the end pieces. (See Fig. 3.) The opposite

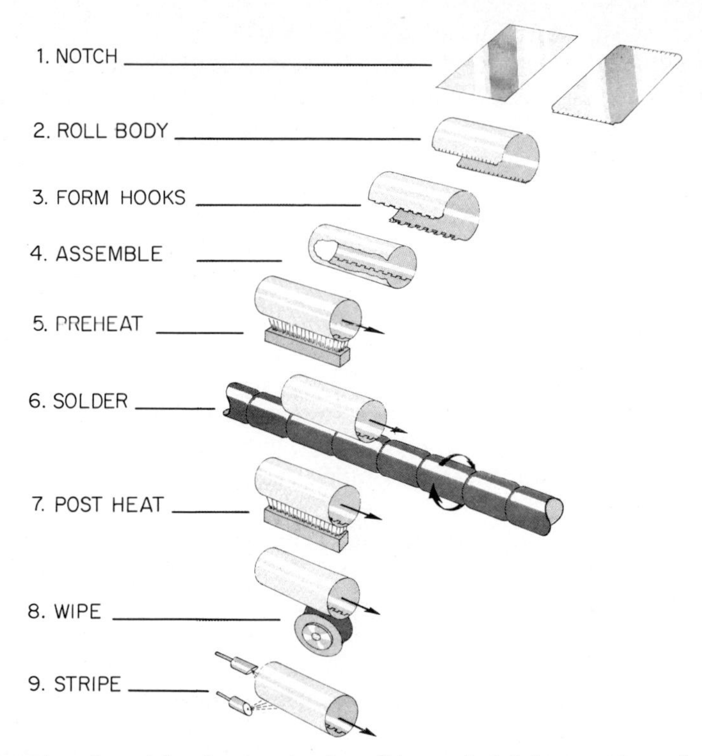

Fig. 2. Body maker. After the sheet has been lithographed, it is cut to size and notched. In the case of aerosol cans the side seam is made with a series of hooks, as shown, to withstand the internal pressure. Solder is applied to the side seam, and a protective coating is sprayed along the joint. (*American Can Co.*)

edges which will form the longitudinal seam are bent to form hooks. Flux is put on the hooks, and they are bumped together to make the side seam. The next machine is the side seamer, by which the joint is soldered and the excess solder on the outside is wiped off. Lacquer stripes are sometimes applied to one or both sides of the seam. The bodies are next transferred to a flanger, by which the edges are flared out to receive the end piece.

The plate for the can ends is sheared into scrolled strips for the most economical layout. If they were cut in a straight line, there would be triangular sections of waste material along the edges between the circular ends. In order to reduce this waste space, the edge is sheared in a wavy line, and the amount that is saved becomes a part of the adjoining strip.

The scrolled strips then go into a punch press, where the ends are blanked out, concentric beads are formed for profile strength, and the edge is curled. Sealing compound is then flowed into the curl and

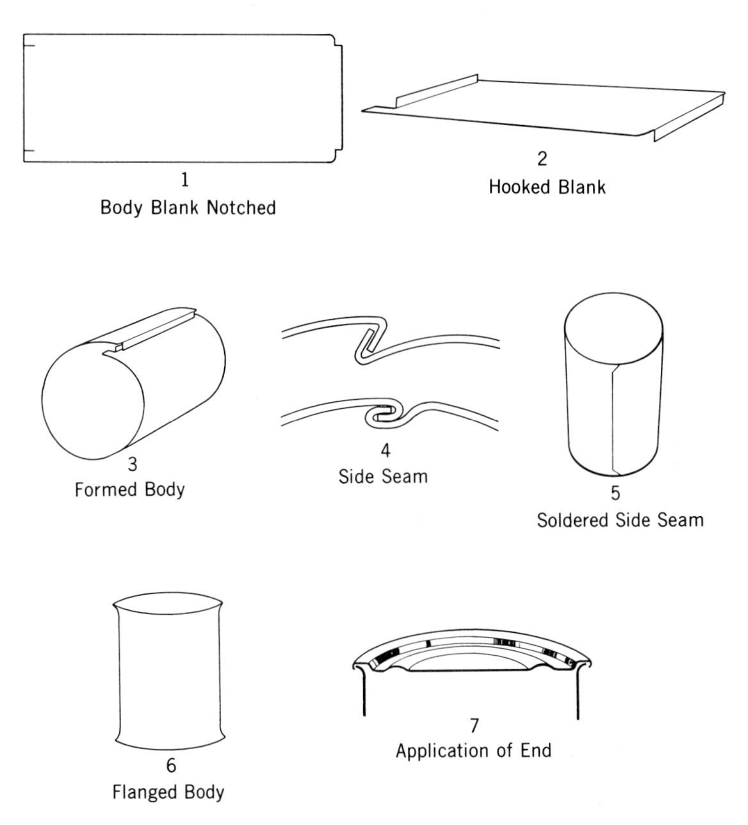

Fig. 3. Can fabrication. The body blank is notched to eliminate some of the extra thicknesses of metal where they overlap. The edges are hooked and bumped together, and then soldered. The end pieces are placed over the flared ends of the body, and the edges rolled together by a crimping machine. (*American Can Co.*)

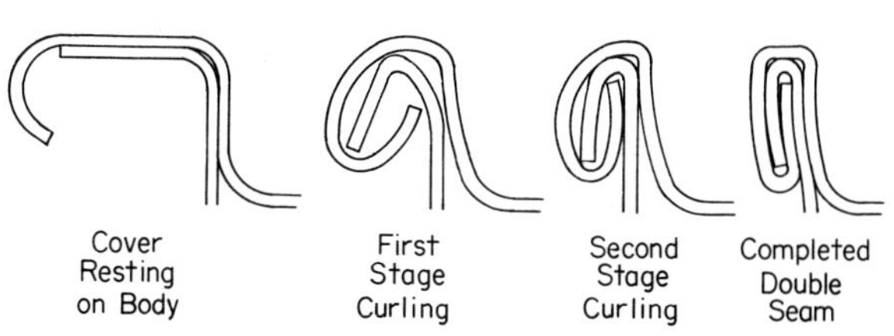

Fig. 4. Double seam. The various stages of attaching the end piece to the body are shown enlarged. Sealing compound between the two pieces was omitted for clarity.

allowed to dry. The compound is applied from a nozzle while the can end is revolved in a holding chuck. Two complete turns are made to ensure an even distribution of the compound. The ends and bodies are joined together in a machine called a double-seamer, and sent on to a can tester to be checked for leaks. (See Table 6 and Fig. 4.)

TABLE 6 Approximate Cost of Various Sizes of Cans

Size	Price per thousand
4-oz mushroom can	$ 26.55
8-oz vegetable can	27.25
16-oz light fruit can	36.55
16-oz dark fruit can	38.20
46-oz juice can	81.25
6½-lb tomato puree can	141.50

Joints and Seams. The side seam is usually soldered to make a tight joint on cans that must withstand internal pressure, such as beverage cans or aerosols. It is necessary to avoid decoration in the area near the joint, so that bare metal will be exposed to take the solder. If the cans have an internal coating, the joint usually is "striped" on the inside in a subsequent operation.

Various types of *solder* are used, depending on the kind of product to be packed. They may contain lead, tin, silver, or antimony in various proportions. For very corrosive products, pure tin may be specified. The most common alloy composition is 97.5 percent lead and 2.5 percent tin. The soldering temperature is between 725 and 750°F.

Silver added to tin or tin-lead solders will increase their resistance to creep. The phenomenon of *creep*, or plastic flow, is the very slow movement of the solder when the can is under constant stress, as in the case of some aerosol containers. Eventually the solder bond may break, resulting in failure of the can, even though the original breaking strength was probably above 400 psig of internal pressure. Tests on solders should extend over a 2-year period to determine resistance to creep and the effects of the product on the solder.

The side seam can be cemented instead of soldered, for some products. A major portion of the cans for lubricating oil and some coffee and pet food containers are being made from black plate or chemically treated steel plate with *cemented side seams*, as are some cans for citrus juices, antifreeze, and detergents. Cemented side seams permit all-around lithography, with no bare strip required at the joint for soldering. Cans that are under pressure, such as beer cans, in which the pressure on the shelf is around 35 psi, and which in processing may reach 85 psi, are being made with cemented side seams. A thorough test program should be followed before using cemented side seams for cans that are under pressure.

The ends of the cans have a *gasket compound* flowed in around the circumference, so that when they are curled into the flared end of the body they form a hermetically sealed container. The sealing compound is usually made from pure latex, synthetic rubber, or a soft plastic material. Cans are usually delivered to the packer with one end seamed on and the other end shipped separately. After filling, the second end is seamed on by means of closing machines which can apply vacuum or inert gas, if required, during the seaming operation. Some of these machines can operate at speeds of more than 1,000 cans a minute.

Cans are commonly shipped to the packer in paper bags, or "carriers." Large users will take them in bulk pallets, which consist of layers of cans separated only by chipboard and are entirely overwrapped with paper. Automatic equipment is available for depalletizing the cans in the packer's plant. Reshippers also are used to some extent; that is, the same corrugated box that brings the cans to the packer's plant is used to pack the finished products. (See Table 7.)

Shaped Cans. It is possible to get can bodies that are formed to provide attractive designs. These have been made in the 211 × 414 size

TABLE 7 Standard Sizes of cans

Name	Diameter and height*	Name	Diameter and height*
No. 1 (picnic)	211 × 400	No. 2 vacuum	307 × 306
211 cylinder	211 × 414	No. 95	307 × 400
Pint olive	211 × 600	No. 2	307 × 409
7Z pimiento	300 × 206	Jumbo	307 × 510
	300 × 308	No. 2 cylinder	307 × 512
8Z mushroom	300 × 400	Quart olive	307 × 704
No. 300	300 × 407	No. 1¼	401 × 207.5
No. 1 tall	301 × 411	No. 2½	401 × 411
303	303 × 406	No. 3 vacuum	404 × 307
303 cylinder	303 × 509	No. 3 cylinder	404 × 700
No. 1 flat	307 × 203	No. 10	603 × 700
Kitchenette	307 × 214		

* Figures given are for the outside dimensions of the sealed can. The first figure always indicates the diameter, and the second the height. In each case the first digit designates whole inches, and the second and third digits designate sixteenths of an inch. Thus 401 × 411 means the diameter is $4\frac{1}{16}$ in. and the height is $4\frac{11}{16}$ in.

for the beverage trade, but other sizes are possible if there is sufficient volume to justify the high tool costs. The production rate for these shaped metal containers is much slower than for the standard cylindrical cans, and the filling and handling equipment must be reduced in speed. Consequently the cost of these cans is nearly double that of the standard can; they are generally recommended only for premium products that can absorb the extra cost.

Equipment. The cost of setting up a can filling line will depend, of course, on the type of product, operating speed, and other factors. But to take one example, a line for carbonated beverages operating at 100 cans a minute would require a capital investment of about $50,000, with another $8,000 for installation. As the operating speeds go up, the equipment must be more sophisticated and the costs are correspondingly higher. (See Table 8.)

TABLE 8 Cost of Equipment for Filling Carbonated Beverages at Different Speeds

Equipment	Line speed for 12-oz cans		
	175–315 cans per min	350–500 cans per min	500–750 cans per min
Pallet unloader	.....	$ 7,500	$ 7,500
Empty can feeder	$ 1,600	6,500	11,000
Empty can washer	2,150	3,250	3,250
Can filler	22,500	28,500	45,000
Closing machine	23,000	47,500	54,000
Filled can warmer	12,000	14,000	17,500
Filled can drier	3,250	3,250	3,250
Filled can weigher	5,500	5,500	5,500
Filled can marker	1,100	1,100	1,100
Packer	12,000	35,000	65,000
Conveyors	22,000	27,500	32,500
Carbo-cooler, synchrometer, deaerator, refrigeration	50,000	70,000	105,000
Installation	16,000	22,000	34,000

Aluminum Cans. First used for cans in 1959, aluminum has increased its share of market very significantly in recent years. Since aluminum cannot be easily soldered, however, its use has been restricted mostly to those products that can be put into cans with cemented side seams, such as citrus juices and motor oil. Of the beverage cans, 10 percent, or about 2½ billion, are being made of aluminum, and about 80 percent of the tinplate beverage cans have aluminum ends with pull tabs. Aluminum weighs only about one-third as much as steel of the same gauge, but since it is less strong, heavier gauges must be used. Because the aluminum companies have increased their production capacities in recent years, they have reduced the price to make it competitive with tinplate. Nearly 400 million lb of aluminum is now being used for cans each year in this country.

Comparative Shipping Weights of 6-oz Citrus Juice Cans

Paper/foil composite body, aluminum ends	34 lb per thousand
Paper/foil composite body, tinplate ends	47 lb per thousand
Tinplate body, tinplate ends	67 lb per thousand

Light weight and resistance to atmospheric corrosion make aluminum very attractive as a container material. When it is used properly, it is very useful and should continue to occupy an increasingly important place in packaging. Since it can be impact-extruded, drawn, drawn and ironed, or extruded and ironed, aluminum is sometimes used to make one-piece aerosol cans, on a type of equipment similar to that used for collapsible tubes. (See Fig. 5.) Aluminum alloys such as 5082 and 5182, which contain 1 to 3 percent magnesium, are most often used for cans. These alloys are corrosion-resistant, and they can be used in the full hard H19 temper. For more information on alloys and tempers see "Metal Foil" in Sec. 3, "Films and Foils," page 3-54.

Fig. 5. Drawing and ironing. Can bodies are made without seams by deep-drawing sheet metal. The several steps in this process are shown. (*American Can Co.*)

The limitations of aluminum must be recognized, however, if we are to avoid the pitfalls of using it without sufficient knowledge and adequate testing. Although it is highly resistant to atmospheric effects, there are some products which are much more corrosive to aluminum than they would be to tinplate. Special coatings must be developed, and these are usually quite different from the enamels that are applied to tinplate containers. Since can openers work better on tinplate than on aluminum, and many of them are equipped with magnets to retain the lid, aluminum cans are being made with one end of tinplate, and the opposite end embossed "Open Other End." These are being used primarily for orange juice concentrate.

Aerosol Cans. The same materials that are used for metal food containers can be used for aerosol cans. There are certain conditions, however, which are peculiar to these pressurized containers and must be taken into account. The stress on the side seams from the internal pressure, for one thing, will cause the solder to creep. As was mentioned under "Joints and Seams," page 7-12, the solder bond may break after a long period of time, causing leakage or a blowout. Silver or antimony added to the composition of the solder will improve the creep characteristics.

Another effect of the pressure is distortion of the can, especially in the dome section, if the gauge or the temper of the metal is not suitable. Where an overcap is a snap fit with the dome of the can, this distortion can become a problem. Elongation in the countersink area of the dome, for example, can be as much as 0.020 or 0.025 in. as a result of going through the hot bath which is used for testing on the production line. For this reason the water should not be allowed to overheat, as the pressure will increase considerably with a small rise in temperature, and this puts an excessive strain on the cans. This is covered more fully in Sec. 11, "Aerosols."

Interior coatings for aerosol cans must withstand the effects of the propellant as well as the product. Some ingredients tend to strip the enamel lining away from the metal, and pieces of the lining may then get into the solution and eventually into the valve, causing the valve to become inoperative. If this occurs after large quantities have been sent out into the market, the result can be a near catastrophe. Such things as shave creams and detergents which are made from sodium lauryl sulfate have been known to do this. A different effect is sometimes caused by products which contain such things as starch and polyvinyl alcohol; this is the pitting of the enamel and eventual perforation of the can. This condition is hard to detect, and test packs must be examined very carefully under a strong glass, especially at the liquid surface line. Interior coatings should be thoroughly tested over an extended period of time before large-scale production is undertaken.

TABLE 9 Standard Dimensions for Aerosol Containers

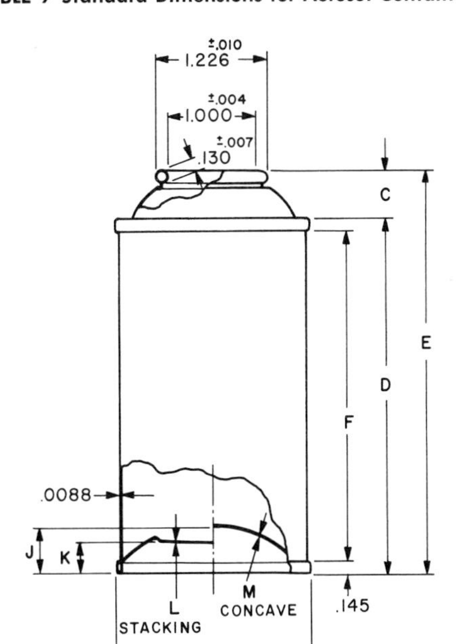

Nominal size*	B ±.031	C ±.016	D ±.031	E ±.047	F min.	J ±.015	K ±.015	L min.	M min.	Capacity, fl oz†
202 × 200	2.188	0.396	1.990	2.386	1.700	0.400	0.315	0.0130	0.0120	3.6
202 × 214	2.188	0.396	2.865	3.261	2.575	0.400	0.315	0.0130	0.0120	5.2
202 × 314	2.188	0.396	3.865	4.261	3.575	0.400	0.315	0.0130	0.0120	7.0
202 × 406	2.188	0.396	4.365	4.761	4.075	0.400	0.315	0.0130	0.0120	7.9
202 × 509	2.188	0.396	5.552	5.948	5.262	0.400	0.315	0.0130	0.0120	10.0
202 × 700	2.188	0.396	6.990	7.386	6.700	0.400	0.315	0.0130	0.0120	12.7
202 × 708	2.188	0.396	7.490	7.886	7.200	0.400	0.315	0.0130	0.0120	13.6
207.5 × 413	2.484	0.815	4.802	5.617	4.512	0.452	0.388	0.0130	0.0120	11.4
207.5 × 509	2.484	0.815	5.552	6.367	5.262	0.452	0.388	0.0130	0.0120	13.4
207.5 × 605	2.484	0.815	6.303	7.118	6.013	0.452	0.388	0.0130	0.0120	15.2
207.5 × 701	2.484	0.815	7.053	7.868	6.763	0.452	0.388	0.0130	0.0120	16.9
211 × 407.5	2.703	0.798	4.458	5.266	4.168	0.500	0.435	0.0140	0.0123	10.0
211 × 413	2.703	0.798	4.802	5.600	4.512	0.500	0.435	0.0140	0.0123	14.1
211 × 510	2.703	0.798	5.615	6.413	5.325	0.500	0.435	0.0140	0.0123	16.5
211 × 604	2.703	0.798	6.240	7.038	5.950	0.500	0.435	0.0140	0.0123	18.3
211 × 612	2.703	0.798	6.740	7.538	6.450	0.500	0.435	0.0140	0.0123	19.7
211 × 713	2.703	0.798	7.802	8.600	7.512	0.500	0.435	0.0140	0.0123	22.8
300 × 709	3.015	0.798	7.552	8.350	7.262	0.582	. . .	. . . .	. . . .	26.9

* Figures given are for the outside dimensions of the assembled can. The first figure always indicates the diameter, and the second the height. In each case the first digit designates whole inches, and the second and third digits designate sixteenths of an inch. Thus 211 × 413 means the diameter is $2^{11}/_{16}$ in. and the height is $4^{13}/_{16}$ in.

† Overflow capacity without valve.

The three main types of enamels used in the interiors of aerosol containers are vinyls, phenolics, and epoxies. The vinyls have good flexibility for fabrication, but they are rather poor in process resistance, and the chemical resistance will vary according to the product that is being packed. Epoxies are generally good in all respects, but they are more expensive than the vinyls or the phenolics. Perhaps the most useful of the enamels are the phenolics; although they are brittle and tend to be troublesome for the can manufacturer, they process very well and have good chemical resistance to a wide variety of formulations.

It must be emphasized that there is no substitute for an adequate test program to select the interior coating for aerosol cans. Tests should be made not only at room temperature for the expected shelf life of the product, but also at elevated temperatures. It is easy to be misled by accelerated tests, which are not equivalent to actual storage conditions. Time is an important ingredient in the end result, and 2 years is not too long a time to continue to look for trouble. It is well also to keep in mind that any subsequent change in the product, no matter how insignificant it may seem, should call for a new series of tests. Even a change of suppliers for the same ingredient may upset the compatibility balance, and a change in the process that alters the amount of air in the product has been known to cause problems. (See Table 9.)

Friction-plug Cans. The familiar paint can is an example of a friction plug container. It is made in sizes from 1/4 pt to 1 gal. There are single friction plugs and double friction plugs. The former are used mainly for powders and pastes, while the latter are better for liquids, giving a much tighter seal.

When this can is used as an outer shipping container, the cover is tacked in several places with solder, or special metal clips can be used to meet the requirements of Postal Manual Regulation No. 125.222. The clips cost 1 to 2 cents each, depending on size and quantity; four clips are required for each can up to and including the quart size, and six clips for 1/2 and 1-gal cans. (See Table 10.)

Screw-top Cans. Containers with screw closures are made in an oblong shape, sometimes designated as F style (first used for Flit), in 1/4-pt up to 2-gal sizes; in cone-top rounds from 3 oz to 1 qt; and in squares, which are usually limited to a 5-gal size. (See Fig. 6.) A wide variety of special spouts and applicators, made to fit the screw threads, are stocked by the can manufacturers.

The size of the closure, which is listed in inches and fractions, refers to the outside of the threads on the container, and not the cap. These sizes bear no relationship to the screw threads on glass containers, which are based on millimeter sizes. Since there are no standards in the industry for thread profiles or helix angles, the caps from one manufacturer are

TABLE 10 Approximate Cost of Friction-plug Cans

Size	Price per thousand			
	Without bail*		With bail*	
	Not lined	Enamel-lined	Not lined	Enamel-lined
1/4 pt	$ 55.80	$ 57.50		
1/2 pt	66.40	68.50		
1 pt	78.25	80.40		
1 qt	107.70	111.00		
1 gal	243.60	251.70	$252.00	$269.00

* A bail is a wire handle fitted into ears that are soldered on each side of the can.

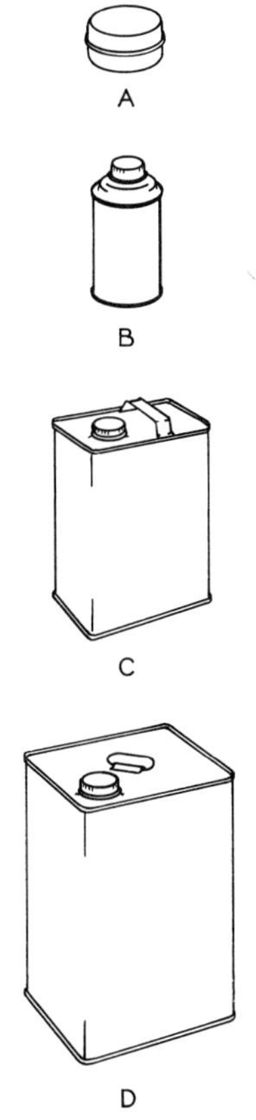

Fig. 6. Types of metal containers. (*A*) Seamless can with slipcover. (*B*) Cone-top round can with screw top. (*C*) Oblong gallon container with strap handle. (*D*) Five-gallon square can with ring handle.

not likely to fit the containers from another. The caps and containers must be purchased at the same time from the same source, to be sure of a good fit. (See Table 11.)

Seamless Cans. Shallow cans with slipcovers are made by blanking and drawing tinplate to the proper size, and hemming or curling the edge. The sheet metal can be attractively lithographed before it is formed, with due allowance for the distortion that takes place in forming. Examples of this type of container are ointment tins, shoe polish boxes, and typewriter ribbon containers. (See Table 12.)

TABLE 11 Approximate Cost of Screw-top Cans

Size	Price per thousand		
	Round	Oblong	Square
3 oz, 5/8-in. cap	$ 65.80		
4 oz, 1-in. cap	69.50		
1/4 pt, 1-in. cap	100.25	$ 67.10	
1/2 pt, 1-in. cap	105.90	88.40	
12 oz, 1-in. cap	118.75		
1 pt, 1-in. cap	128.20	100.90	
1 qt, 1-in. cap	162.40	139.10	
1/2 gal, 1 1/4-in. cap	.....	245.90	
1 gal, 1 3/4-in. cap	.....	287.90	
2 gal, 1 3/4-in. cap	.....	470.80	
5 gal, 1 3/4-in. cap	.....	.....	$722.75

TABLE 12 Approximate Cost of Round Seamless Slipcover Cans

Capacity, fl oz	Diameter and height, in.	Price per thousand
1/16	7/8 × 1/4	$ 9.60
1/8	1 1/8 × 1/4	10.75
1/4	1 1/4 × 1/2	16.10
1/2	1 1/2 × 5/8	17.90
3/4	1 3/4 × 3/4	23.25
1	2 × 3/4	23.50
1 1/2	2 1/8 × 7/8	31.75
2	2 3/8 × 7/8	32.40
2 1/2	2 1/2 × 1	45.90
3	2 3/4 × 1 1/8	47.00
4	3 × 1 1/4	52.80
6	3 1/2 × 1 3/8	70.40
8	3 3/4 × 1 1/2	81.65

STEEL DRUMS AND PAILS

History. In 1902 the Standard Oil Company started using its Bayonne steel barrel as a substitute for the wooden barrels it had been using. It was bilged and shaped like its wooden counterpart. The straight-sided steel drum, first introduced in 1907, was made of 16-gauge terneplate, with soldered side seams and soldered chime seams. The need to reduce cost soon resulted in the development of the 55-gal drum made from 18-gauge steel with welded seams. By 1920 hand painting was replaced by automatic painting and baking of enamel coatings. Annual sales of steel shipping containers now are over $200 million, about half of which is for the chemical processing industry.

Regulations. Light shipping drums for shipment of nonregulatory products, also known as single-trip drums, are specified in Rule 40 of the Uniform Freight Classification. Single-trip drums for regulatory products, sometimes referred to as dangerous articles, and returnable drums are covered by Tariff No. 23 for railroads, also known as George's Tariff as it was issued by T. C. George for the Bureau of Explosives and approved by the Department of Transportation (DOT). Similar regulations were issued by the American Trucking Associations, Inc., as Tariff No. 14 and by the Airline Tariff Publishers, Inc., as Tariff No. 6-D.

Manufacturing. Both cold-rolled and hot-rolled steels are used for making drums. Although cold-rolled is cleaner and easier to coat with resin, hot-rolled is sometimes preferred for certain types of treatment. Acid pickling, detergent scrubbing, sand blasting, and shot blasting are frequently used to prepare the steel for fabrication. The bottoms and tops are blanked out of the sheets. Bodies, however, are sheared to size. They are then rolled in a machine until the edges come together to form a cylinder, and the joint is butt-welded. The slag caused by the welding operation is taken off and the joint is flattened so that it is flush with the rest of the body. Flanges, beads, and rolling hoops are then rolled into the shell, where necessary. The lining material is sprayed on the different parts separately, and baked on. The head and bottom then have a seam compound flowed in around the edges and are next joined to the body by spinning the edges together to form a five-layer chime. The outside is then painted, silk-screened, or lithographed. (See Table 13.)

Linings. Some resistance to corrosion is provided by phosphatizing, which is a chemical treatment of the surface of the steel. For better protection or for products that need a better coating, phenolics are widely used. They are very effective against acids, which are the most troublesome materials to package in metal. They are low in cost, but they are brittle, and their resistance to alkalies is generally poor.

TABLE 13 Steel Sheet Thicknesses

Gauge no.	Nominal thickness, in.	Minimum thickness, in.	Weight, lb/sq ft*
12	0.1046	0.0946	4.375
14	0.0747	0.0677	3.125
16	0.0598	0.0533	2.500
18	0.0478	0.0428	2.000
20	0.0359	0.0324	1.500
22	0.0299	0.0269	1.250
24	0.0239	0.0209	1.000
26	0.0179	0.0159	0.750
28	0.0149	0.0129	0.625

* The approximate tare weight of a 55-gal drum can be calculated by assuming that 24 sq ft will be used for a tight-head drum. For example, an 18-gauge drum will weigh about 48 lb (actually a few ounces over this) and an open-head drum will weigh about 2 lb more.

Vinyls are very good with alkaline materials; they are flexible and reasonable in cost. Epoxies have good resistance to alkalies and fair resistance to acids. They are flexible but tend to become brittle with age when in contact with certain products, and they are a bit higher in cost than the others. It is important that these lining materials be properly cured; if they are undercured, they will not have the proper chemical resistance, and if they are overcured, they tend to be brittle. Other coatings which are available but not often used are saran, Teflon, vinylidene chloride-acrylonitrile copolymer, styrene-butadiene, neoprene, and plastisols. Polyethylene is widely used, but usually in the form of a loose piece which is shaped to fit closely to the interior of the drum.

Types of Drums. Steel drums are made in tight-head or open-head styles. Tight-head drums usually have two flanged openings: a 2-in. and a ¾-in. opening placed at opposite sides in the head of the drum, or one in the head and the other in the middle of the body, or the end of the body, on the opposite side. The openings are threaded and fitted with screw plugs and rubber or asbestos gaskets. Sizes given refer to U.S. Standard Pipe Thread sizes. The small opening is for venting, and the large opening can be used for attaching a spigot. (See Table 14.)

Open-head drums have a loose cover with a gasket in the channel which is formed around the edge. (See Fig. 7.) This makes a tight seal against the rim of the drum. A split locking ring made of 16-gauge steel is provided to draw the cover down tightly on the drum and hold it in place for shipment. The ring is secured by a toggle lever which is wired

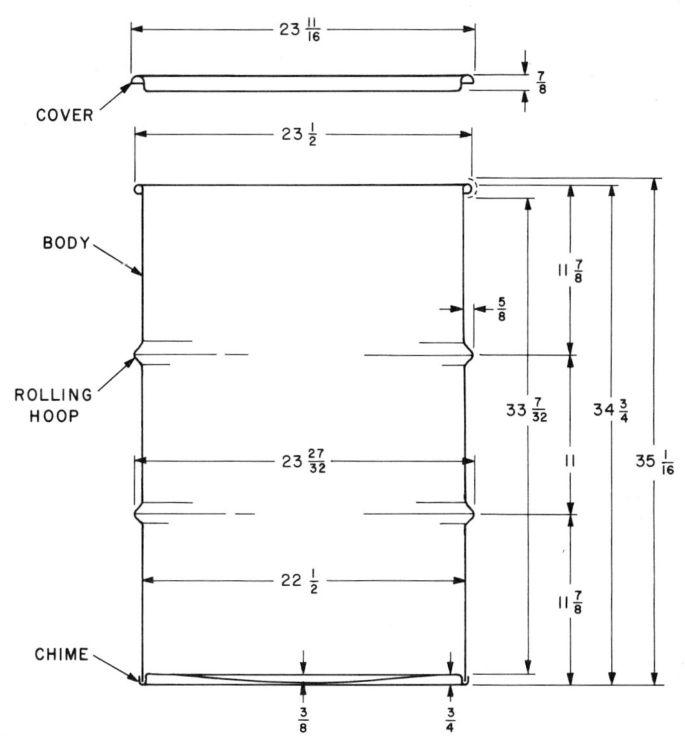

Fig. 7. 55-gal removable-head drum. Dimensions are shown for a standard Rule 40 universal drum.

down, or by a nut and bolt which is sometimes fitted with a locknut. The toggle lever is more expensive, but if the drum will be opened and closed frequently, it is worth the added cost. Rolling hoops are formed by pressing out the side wall at two equally spaced points. Although originally intended to provide for easy rolling, they also serve to greatly stiffen and strengthen the drum.

Steel Pails. Many sizes and styles of pails are made, some with bails or handles and some without, both straight-sided and tapered, tight-head or open-head, single-bead, double-bead, necked-in top, and ducked-in bottom, to name a few. The common scrub bucket is often used as a shipping container when it is fitted with a tight cover. Pails come in a range of sizes from 1 to 10 gal. Several types of closures, spouts, and other fittings can be supplied to suit the requirements of the product.

Most of the interior coatings that are used on drums can be applied also to pails, and the same facilities for exterior decoration are available for the smaller containers. Thickness of metal is left up to the customer,

TABLE 14 Approximate Cost of Unlined* Drums and Pails (per unit)

1 gal, lug cover	$ 0.52
2 gal, lug cover	0.67
5 gal, lug cover	1.00
10 gal, lug cover	2.15
16 gal, lug cover, 22-gauge	2.50
30 gal, tight head, 24-gauge	3.10
30 gal, tight head, 20-gauge	4.80
30 gal, open head, 19-gauge	5.60
30 gal, stainless steel, 16-gauge	69.00
45 gal, tight head, 22-gauge	4.00
55 gal, used, tight head, 18-gauge	4.50
55 gal, rebuilt, tight head, 18-gauge	4.75
55 gal, new, tight head, 18-gauge	6.55
55 gal, open head, 18-gauge	6.95
55 gal, galvanized, tight head, 16-gauge	12.55
55 gal, ICC 5, tight head, 14-gauge	15.40
55 gal, ICC 5A, tight head, 12-gauge	30.00
55 gal, aluminum, tight head, 12-gauge	40.00
55 gal, stainless steel, tight head, 16-gauge	86.10

* Linings cost about 70 cents per coat (if pigmented 80 cents per coat) in a 55-gal drum.

and will depend on how severe the service is expected to be. Small sizes are commonly made from 26-gauge (0.0179 in.) steel, and the larger ones most often are made from 24-gauge (0.0239 in.) material. The trend is toward lighter gauges, however, and 5-gal 28-gauge straight-sided and nesting pails with a 26-gauge lug-style cover are now authorized by motor carriers and by the railroads. Nesting pails are gaining in popularity at the expense of the straight-sided containers because they occupy one-third the storage space required for straight-sided pails when empty.

COLLAPSIBLE TUBES

History. An American portrait painter, John Goffe Rand, looking for a better container for his oil colors, extruded the first metal tube of tin in 1841. The first manufacturing plant in this country was established in 1870, and up until the turn of the century only two additional companies were formed for this purpose. Around 1895 toothpaste was offered to the public in tubes, and it was such a great success that other semifluid products were tried in these packages.

There has been a continued growth until today nearly 1½ billion tubes are produced each year, worth about $70 million. Toothpaste, shaving cream, and cosmetics account for about 63 percent of the volume; medicinal products about 21 percent; and household and industrial products about 16 percent. A small percentage is used for such food products as cake icings, cheese, and sundae toppings, but general acceptance in the food field has been slow. Europeans seem to find wider use for tubes as food containers; jams, jellies, fish pastes, and meat spreads are more often seen in shops abroad than in this country.

Metal Tube Production
Annually in the United States

Dentifrices	650 million
Pharmaceuticals	400 million
Household products	240 million
Cosmetics	95 million
Shaving creams	10 million
Food products	5 million

Of the metal tubes produced in 1968, 60 percent were aluminum, 25 percent were lead, 10 percent were tin, and 5 percent were tin-lead combinations.

The collapsible tube is an attractive container that permits controlled amounts to be dispensed easily, with good reclosure, and adequate protection of the product with a minimum risk of contamination of the remainder while dispensing, because it does not "suck back." It is light in weight and unbreakable, and the nonrefillable units lend themselves to high-speed automatic filling operations. They have a smooth surface that is easily decorated in multiple colors with good fidelity, in high gloss colors, with embossed shoulders, and with attractive closures and special fitments for many applications.

Materials. Any ductile metal that can be worked cold is suitable for collapsible tubes, but the most common ones are tin, aluminum, and lead. Tin is the most expensive, at least in the larger sizes, and lead is the cheapest. Very small tubes can sometimes be made more cheaply out of tin because it is more easily worked than the other metals, and the cost of material becomes less significant as the size gets smaller. (See Table 15.)

TABLE 15 Base Cost of Metals

Tin	$1.65 per pound
Aluminum	0.26 per pound
Lead	0.13 per pound

The tin that is used for this purpose is alloyed with about 1/2 percent of copper for stiffening. When lead is used, around 3 percent of antimony is added to increase hardness. Aluminum work-hardens when it is formed into a tube, and must be annealed to give it the necessary pliability. Aluminum also work-hardens in use, sometimes causing tubes to develop leaks. *Lead* is used only for nonfood products such as shoe polish and adhesives, and should never be used alone for anything taken internally because of the risk of lead poisoning. To reduce cost, or when tin is in short supply, it is possible to *laminate* tin and lead to form a sandwich, with lead in the middle and tin on the outside. When slugs are cut from sheets that are made this way, and extruded into tubes, the proportion of the two metals remains fairly constant. Thus, if the slug was composed of 3 percent tin on one side and 7 percent on the other, with the other 90 percent being lead, the walls and the shoulder of the finished tube would have about the same distribution. The heavier coating of tin should be put on the inside, in contact with the product, and the lighter skin on the outside to give protection from atmospheric corrosion and a good base for decoration. *Alloys* of tin and lead also can be used to reduce cost, but again they should never be used with foods or internal medication. There is some competition from plastic collapsible tubes and laminated tubes. (See "Collapsible Tubes" in Sec. 8, "Plastics," page 8-74.) The laminates are fairly new, and after some unsuccessful attempts to get a toehold in the tube market, it appears that they may at last be a practical process. The cost is still much higher than for other materials, and the order quantities must be up in the millions.

Manufacturing. The metal for tubes is received as pigs and is cast into slabs preparatory to rolling into sheets. Slugs are punched from these sheets, about the size of coins. They are then tumbled and lubricated, and sent to the presses to be made into tubes. The method of making tubes is by impact extrusion. (See Fig. 8.) With this technique, a slug of metal is put into a shallow female die. The male die is the shape of the inside of the tube and is somewhat longer than the tube that is to be produced. When the die closes on the slug with considerable pressure, the metal squeezes out between the two parts, hugging the inner die as it moves rapidly out around it. When the die separates, the tube is blown off with compressed air. Subsequent operations, such as cutting the thread, reaming the orifice, trimming to length, decorating, and lining, are then performed on another machine. Aluminum tubes are annealed after trimming, in ovens at temperatures between 900 and 1200°F for a short time. After decorating, the tube is carried through an oven to dry the coatings, and then the closure is applied.

The end opposite the closure is left open to the full diameter for filling, and the tubes are packed into nested boxes to preserve the

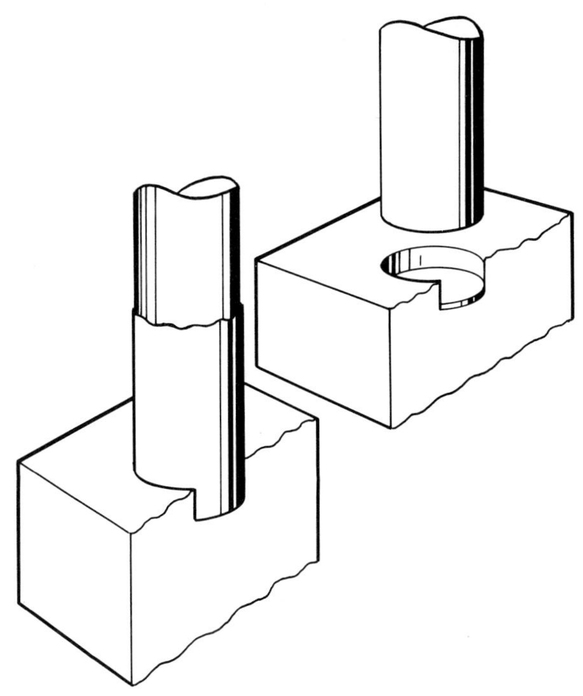

Fig. 8. Impact extrusion. A slug of aluminum is shown in place in the lower half of the die on the right. When the die closes, the metal is forced up around the upper part of the die as shown on the left. This method is used to make collapsible tubes. (*American Can Co.*)

cylindrical shape of the body. They are easily damaged at this stage and must be packed in sturdy boxes and handled rather carefully. (See Table 16.)

Decoration. The shoulder can be embossed in the extrusion die with a design or legend, but it cannot be printed easily. Embossing of the side walls of tin or lead tubes also is possible, within limitations, by special tools for this purpose. The enamel coating does not extend onto the shoulder, but a matching color can be sprayed on. Sometimes clear lacquer is used to prevent tarnishing of the metal, particularly with lead tubes. A brushed finish that is very attractive can be applied to the shoulder of aluminum tubes. Most tubes have an enamel base coat applied to the side wall by roller coating. Up to five colors can then be printed simultaneously by the offset process. Best results are obtained if the colors do not have to be registered too closely. The base coat is the lightest color, usually, with the printing inks in the darker tones. Protective coatings can be applied over the ink, but with good gloss inks they are not usually necessary.

TABLE 16 Standard Dimensions of Metal Tubes

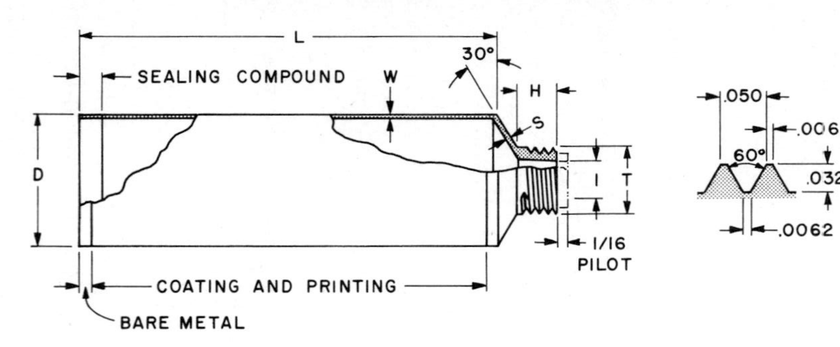

Neck size*	I Orifice ID ±0.010″	T Thread OD ±0.005″	H Neck height ±1/64″	D Outside diameter ±0.005″	S Shoulder thickness min.–max.	W Wall			L Max. length ±1/32″	Capacity at max. length, cc.
						Aluminum max.	Tin min.–max.	Lead min.–max.		
No. 12	0.188	0.312	3/16	0.375	0.006–0.014	. . . .	0.0027–0.0033	. . . .	2	2.77
				0.500	0.009–0.021	. . . .	0.0027–0.0033	0.0040–0.0050	3	7.39
				0.625	0.011–0.023	0.0048	0.0032–0.0038	0.0045–0.0055	4	14.79
No. 16	0.250	0.375	7/32	0.750	0.012–0.026	0.0048	0.0031–0.0039	0.0046–0.0056	4 3/4	27.73
				0.875	0.014–0.028	0.0050	0.0036–0.0044	0.0049–0.0061	5 1/2	43.87
No. 20	0.313	0.438	7/32	1.000	0.014–0.030	0.0050	0.0038–0.0046	0.0051–0.0063	6	66.54
				1.125	0.015–0.031	0.0055	0.0040–0.0050	0.0053–0.0067	6 1/2	82.33
No. 28	0.438	0.563	7/32	1.250	0.016–0.032	0.0055	0.0045–0.0055	0.0058–0.0072	6 7/8	111.83
				1.375	0.016–0.034	0.0060	0.0045–0.0055	0.0058–0.0072	7 1/8	151.57
				1.500	0.017–0.035	0.0060	0.0050–0.0060	0.0063–0.0077	7 1/4	171.65

* The neck number is the diameter of the orifice in sixty-fourths of an inch.

The preparation of the final artwork for printing should be left to the tube manufacturer because the necessary allowance for wrapping the printing plates around the printing roll is best done by him. A rough sketch with color separation and typewritten text is all that needs to be furnished.

The base coat can be formulated, if desired, so that it is easily peeled off. This is useful for prescription items if it is desirable that the druggist remove the identity of the product and replace it with the prescription information furnished by the doctor.

Linings. If the product is not compatible with bare metal, the interior can be flushed with wax-type formulations or with resin solutions, although the resins or lacquers are generally sprayed on. Care must be taken to prevent clogging the orifice with the lining material, especially in the case of ophthalmic, mastitis, or other elongated nozzle tips. A tube with an epoxy lining will cost about 25 percent more than the same tube uncoated. (See Table 17.)

Wax linings are most often used with water-base products in tin tubes, and phenolics, epoxies, and vinyls are used with aluminum tubes giving better protection than wax, but at a higher cost. Phenolics are most effective with acid products; epoxies protect better against alkaline materials.

TABLE 17 Approximate Cost of Metal Tubes

Size	Price per gross
5/8″ × 2 1/4″ aluminum, 0.004″ wall, 0.018″ shoulder . . .	$4.00
3/4″ × 3 1/4″ tin, 0.004″ wall, 0.018″ shoulder	7.90
1″ × 6″ aluminum, 0.0045″ wall, 0.024″ shoulder	5.50
1 3/8″ × 6 1/4″ aluminum, 0.005″ wall, 0.026″ shoulder . .	7.40
1 1/2″ × 6 1/2″ aluminum, 0.0055″ wall, 0.028″ shoulder. .	8.00

Closures. Screw caps are available in various styles under such descriptive names as fez, mushroom, reverse taper, flowerpot, and eye tip. Some caps are made of metal, but the large majority are of plastic, such as phenolic, urea, and polyethylene. Any of the usual colors that are available in bottle closures—and they are almost limitless—can be used for tube closures. This subject is covered in greater detail in Sec. 9, "Closures, Applicators, Fasteners, and Adhesives."

Applicators, which can be either integral or separate parts of the tube, come in a variety of shapes, including pile and vaginal pipes, ophthalmic tip, nasal tip, mastitis tip, and screw-eye tip. (See Fig. 9.) Tubes can be made with a blind opening, for volatile products or where sterility is

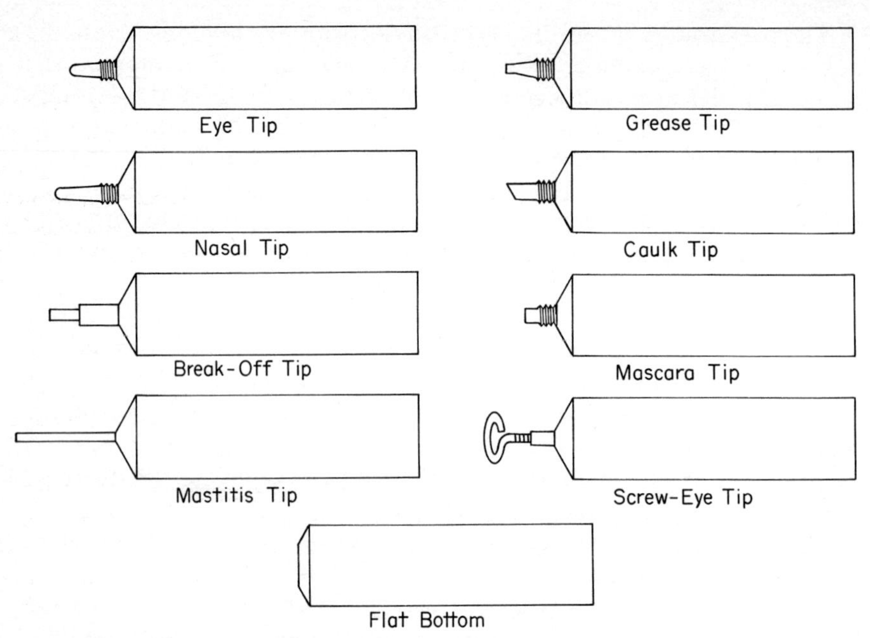

Fig. 9. Styles of tips. Collapsible tubes are available with a variety of openings for different purposes.

important, and a screw-eye or a pointed cap can be provided to puncture the tip when it is ready to be used. Tubes with plastic necks are available; with these the product, as it is dispensed, does not accumulate the gray spots and streaks of metal particles and oxides that are the residue of the tube fabrication processes and the grinding action of the cap on the metal threads. Some single-use containers have neither neck nor cap; they are opened by tearing off the folded end of the tube. Break-off tips also can be used for the same purpose.

Sealing. The tube is filled through the bottom, and the end is then flattened and folded to form a seal. A folded piece of stiff metal, called

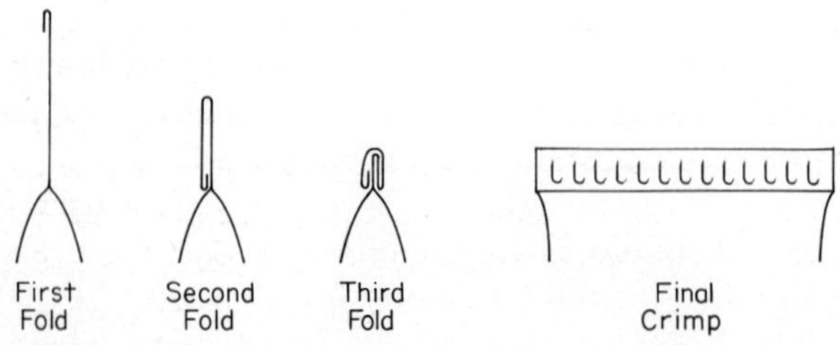

Fig. 10. Bottom closure for tubes. Sequence of folds in closing the bottom of a metal tube is shown.

a clip, is sometimes placed over the end and tightly closed to keep the tube from leaking when pressure is applied for dispensing. More often, however, a clipless closure is made by folding the metal several times and crimping it. (See Fig. 10.) Heat-setting sealants can be applied as an inside or outside coating near the tube end. When heat is applied after closing, the coating softens and flows to make a more positive seal. One coating for this purpose is known as Westite cement. A welded seal also is possible with a patented method recently developed.

ALUMINUM

The use of aluminum in packaging can be separated, for purposes of discussion, into two categories: flexible packaging and rigid packaging. Metal foils for bags and pouches are discussed in Sec. 3, "Films and Foils." Taking the subject of rigid containers at this point, aluminum alloys for this purpose are primarily of the non-heat-treatable aluminum-magnesium type. They are available in both annealed and strain-hardened tempers. The composition of these alloys varies according to the intended use, with up to 5 percent magnesium, 1 percent manganese, and traces of iron, silicon, zinc, chromium, copper, and titanium. (See Table 18 in Sec. 3, "Films and Foils.")

As the forming characteristics and resistance to corrosion are improved, the yield strength usually goes down, and heavier gauges are then required to do the same job. Cold working causes these alloys to strain-harden, and a 50 percent reduction in size by rolling results in about 30 percent higher yield strength. This will drop off with time, losing perhaps 5 percent in a year owing to an age-softening phenomenon. When enamels or other coatings are baked, there is also a slight drop in yield strength. The extra stiffness of this work-hardened material is used to advantage in making container bodies and ends. The rolling process that is used to make the sheet material also results in what is called "preferred orientation," which shows up in the form of ears around the rim of a formed piece, usually 90° apart. This can be controlled to some degree by changing direction during rolling, and with annealing between passes.

For deep-drawn containers and impact-extruded cans or tubes an alloy with good workability is necessary. The annealed forms of Nos. 1100 and 3003 are among the best for this purpose. Shallow-drawn parts, such as can ends, take alloys with less ductility, and they can be of a medium temper so that the flat panel which does not get worked will have sufficient stiffness. Alloy 5052 is one of the most popular for this use, and a temper of H34 is about right for average conditions. For can bodies a harder temper around H19 would be more suitable, in thicknesses of 0.008 to 0.020 in. (See Table 18.)

TABLE 18 Properties of Aluminum Alloys*

Alloy and temper	Tensile strength, psi	Yield strength, psi	Elongation, percent
1100-0. . .	13,000	5,000	35
1100-H12 .	16,000	15,000	12
1100-H14 .	18,000	17,000	9
3003-0 . .	16,000	6,000	30
3003-H12 .	19,000	18,000	10
3003-H14 .	22,000	21,000	8
3003-H19 .	38,000	34,000	3
3004-0 . .	26,000	10,000	20
3004-H34 .	35,000	29,000	9
3105-0 . .	18,000	7,000	25
5052-0 . .	28,000	13,000	25
5052-H34 .	38,000	31,000	10
5082-H19 .	57,000	54,000	4
5086-H19 .	61,000	57,000	4
5154-H19 .	55,000	52,000	3

* From William A. Anderson, New Alloys for Rigid Containers, *Modern Packaging,* vol. 40, no. 7, March 1967, p. 134.

Commercial purity aluminum, known as alloy 1100, contains not less than 99.0 percent aluminum, with small amounts of copper, iron, and silicon, and traces of other metals. The best corrosion resistance is obtained at the lowest iron and copper levels. Super-purity aluminum (99.99 percent) has the best corrosion resistance of the pure aluminums. The addition of 1¼ percent manganese to make alloy 3003 increases the corrosion resistance; this alloy is used for cooking utensils. Alloy 5052, with 2½ percent magnesium, is a stronger material but has slightly less resistance to uniform corrosion, although it is more resistant to pitting than 3003.

Aluminum Trays. For frozen food and baked goods, shallow trays made from thin-gauge aluminum are finding widespread use. They are

TABLE 19 Approximate Cost of Aluminum Trays

Size, in.	Price per thousand
3 × ½ tart	$ 6.13
4 × 2 casserole	22.63
8 × 1 pie plate	26.85
9 × 7 × 1 TV dinner	45.52

available in round, square, oblong, and some odd shapes like sliced pie, in a myriad of sizes. (See Table 19.) Most have crinkled sides, although smooth sides can be made. For smooth sides the tool costs are much higher, making the piece price greater. There are three general types of closures: a vertical flange, which is folded over a flat board cover; a horizontal flange, to which a cover is heat-sealed; and a horizontal flange with a hood cover which is crimped under the flange.

Covers are of three kinds: disk of cardboard, sometimes with a transparent window; a flat hood of aluminum with a downward flange which can be curled under the rim of the tray; and the raised hood for products which are higher than the tray.

Section 8

Plastics

General 8-2
History 8-2
Industry Statistics 8-2
Costs 8-2
Chemistry 8-3
Additives 8-4
Copolymers 8-4
Processes 8-4
Materials 8-5
Selection of Materials 8-5
Decoration 8-6
ABS (Acrylonitrile-Butadiene-Styrene) 8-7
Acetal 8-9
Acrylic 8-10
Cellulose Acetate 8-11
Cellulose Propionate 8-13
Ethyl Cellulose 8-14
Nylon 8-15
Phenolic 8-17
Polycarbonate 8-19
Polyethylene, Low-density 8-21
Polyethylene, High-density 8-24
Polypropylene 8-26
Polystyrene 8-29
Polyurethane 8-30
Polyvinyl Chloride 8-33
Urea 8-35
Chemical Properties Defined 8-36
Application of Plastic Materials 8-37
Costs 8-37
Clarity 8-37
Stiffness 8-37
Water-Vapor Transmission 8-37
Gas Permeability 8-38
Chemical Resistance 8-39
Mar Resistance 8-39
Temperature Range 8-39
Warpage 8-40
Impact Strength 8-40
Tear Strength 8-40
Elongation 8-40
Copolymers 8-40
Plasticizers 8-41
Additives 8-42
Fillers and Reinforcements 8-44
Processes 8-45
Injection Molding 8-45
Design 8-46
Molds 8-49
Molding Machines 8-51
Compression Molding 8-53
Blow Molding 8-56
Design 8-58
Molds 8-60
Machines 8-62
Processes 8-65
Materials 8-65
Coatings 8-71
Testing 8-71
Rotational Molding 8-72
Slush Casting 8-73
Collapsible Tubes 8-74
Thermoforming 8-76
Design 8-77
Molds 8-77
Processes 8-79
Materials 8-83
Foamed Plastics 8-84
Polystyrene Foam 8-85
Urethane Foam 8-86

GENERAL

History. In 1843 Dr. Montgomerie reported that gutta percha was being heated and formed by hand into knife handles by the natives of Malaya. Cellulose nitrate was synthesized in 1835 by J. Pelouze. A patent was granted in 1854 to J. Cutting, of Boston, covering the use of gum camphor as a plasticizer in collodion.

J. L. Baldwin was granted U.S. Patent No. 34,344 on February 11, 1862, for "Molds for Making Daguerreotype Cases" in plastic material. It is known that molds for this purpose were in use 10 years earlier than this. The materials used were gutta percha or shellac mixed with various fillers, humectants, and plasticizers.

Dr. John Wesley Hyatt, an Albany, New York, printer, seeking a material to replace ivory for billiard balls, obtained a number of patents for celluloid between 1868 and 1870. Dr. Leo Hendrik Baekeland of Yonkers, New York, recognized the importance of the reaction between phenol and formaldehyde in 1907 and started the plastics industry as we know it today.

Industry Statistics. About 20 percent of the plastics manufactured in this country, or about 1 million tons, finds its way into packaging materials each year. The largest portion of this is for films and coatings, which use 600,000 tons per year. Containers account for 350,000 tons, closures 60,000 tons, and adhesives about 45,000 tons. (See Table 1.)

TABLE 1 Plastics Used in Packaging in the United States Annually

Polyethylene	800,000 tons
Polystyrene	180,000 tons
Vinyls	85,000 tons
Polypropylene	33,000 tons
Cellulosics	20,000 tons
Thermosets	20,000 tons

Costs. A comparison of various plastics with an equivalent volume of various metals yields the following cost figures per cubic inch: polystyrene costs 0.56 cents, low-density polyethylene 0.35 cents, high-density polyethylene about 0.85 cents, polyvinyl chloride 0.75 cents, methacrylate 2.4 cents, cellulose acetate about 2.4 cents, nylon 3.8 cents, and polycarbonate about 4.6 cents; of the metals, aluminum is about 2.2 cents, steel about 2.8 cents, die-cast zinc 4.3 cents, and brass 13.4 cents.

Chemistry. Carbon atoms have four points of attachment to each other or to atoms of other materials. Thus, we can write the formula for methane as CH_4, indicating that four atoms of hydrogen are attached to one atom of carbon to form a molecule of methane. The connecting points are equally spaced around the carbon atom, 109°28′ apart and not in the same plane. When carbon atoms hook onto each other, they form chains which have a zigzag pattern:

$$
\diagup CH_2 \diagdown CH_2 \diagup CH_2 \diagdown CH_2 \diagup
$$

The shortest of these chains are called *monomers.* These monomers will add onto one another to form very long chains called *polymers.* When the chains are grouped together in a random pattern like a pile of straws, they are said to be amorphous. If they are packed nearly parallel in an orderly arrangement, the material is more crystalline and therefore has a higher density and is stiffer and tougher. Polyethylene, for example, can have a specific gravity anywhere from 0.900 to 0.980, depending upon the degree of crystallinity. This type of plastic is easily melted and re-formed; it is classified as *thermoplastic.*

Sometimes these chains have side chains, for example, polypropylene:

$$
\begin{array}{ccccccccc}
-CH_2- & CH & -CH_2- & CH & -CH_2- \\
 & | & & | & \\
 & CH_3 & & CH_3 &
\end{array}
$$

Such side chains interfere with the close packing of the molecules and result in lower density of the material. (See Fig. 1.) Other combinations of atoms go to make up the various plastics, such as polyvinyl chloride $-CH_2-CHCl-$ or polyvinylidene chloride $-CH_2-CCl_2-$, but they all have one thing in common—the characteristic of "condensing" into very long chains under the right conditions. If the side chains become connected, or *cross-linked,* the plastic takes on some of the properties of the *thermosets;* that is, it becomes more difficult to remelt.

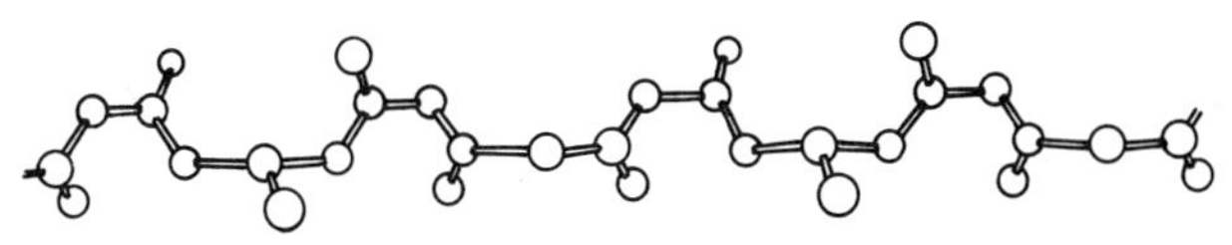

Fig. 1. Molecule. A chain of carbon atoms with hydrogen or other atoms attached may have side branches as shown. The four connecting points around each carbon atom are equally spaced in all directions, resulting in a spiral configuration of the chain.

Additives. Some plastics are combined with plasticizers, antioxidants, slip agents, UV light degradation inhibitors, and fillers of various kinds. Vinyls are normally rigid, but with the addition of plasticizers they become soft and pliable, and their usefulness is greatly extended. Polypropylene would have a very short life if it were not for the antioxidants that are added during the processing. Glass fibers used as fillers will double the strength of most plastics; other fillers are used to add stiffness or to reduce the cost of the finished piece.

Copolymers. It is possible to combine different plastics and get certain desirable properties that are not available in the homopolymers. Such combinations as styrene-acrylonitrile and ethylene-vinyl acetate are very useful additions to the growing family of plastics. They have a toughness and resistance to stress cracking that make them especially suitable for use under difficult shipping conditions. Blends of two or more mutually soluble resins will sometimes take on the character of a plasticized resin, in that they are more flexible, stronger, and tougher than they would otherwise be, and they do not have the problems of exudation, migration, and leaching which are associated with plasticizers.

Processes. The technique of *injection molding* is used for *thermoplastic* materials, that is, those materials which melt under heat after they have been processed. The molds are expensive, but if there is sufficient volume, it usually gives the lowest piece price. *Compression molding* is required for *thermosetting* materials which soften under heat in the mold, but which cure in the process and thereafter cannot be remelted. *Extrusion* is another method of forming plastics; it consists of pushing the softened plastic through a die to form rods, tubes, and other continuous shapes. *Blow molding* is a combination of extrusion, injection molding, and blowing to produce hollow articles. Extrusion and blowing in a continuous operation are also one of the techniques used for producing film. Another method of making *film* and *sheet* material is extrusion through a long narrow slit in a die, allowing it to chill and "set up" in the air or in a cold water bath. The difference between film and sheet is merely in thickness: anything under 0.010 in. is designated as film, and above that it is called sheet. Film *casting* consists of pouring the fluid material onto a chilled roll or steel belt, and after it has cooled peeling it off and winding it on rolls. *Calendered* film is made by a similar process, by forcing the molten plastic between rollers.

A method of *thermoforming* film or sheet material is heating it to the softening point and then forcing it against a mold by pressure or vacuum. When it is cool, it is taken from the mold, and it retains the shape into which it has been formed. The molds are inexpensive, and it is possible to get very thin sections so that the amount of plastic used is reduced to a minimum. *Foamed* or *expanded* plastics have some interesting properties. Structurally they provide considerable stiffness with a minimum of material, and they are excellent thermal insulators. They may be molded and formed into many shapes for attractive and durable products.

In a process called *slush casting*, resins are mixed at room temperature with nonsolvent types of plasticizers, along with any fillers, stabilizers, or colorants that may be desired, to form pastes or viscous fluids called plastisols. These are poured into a warm mold, and the excess is poured out, leaving a coating on the inside of the mold. The temperature of the mold is then raised to effect a cure of the plastisol.

MATERIALS

Selection of Materials. The choice of plastics to be used for a particular application will be governed by a number of considerations. First the environmental conditions of packing, storage, and shipment should be taken into account. The material must have the physical characteristics to protect the contents against mechanical damage and against atmospheric conditions. It is necessary to study the tensile and impact strength, permeation rates, and chemical compatibility with the product or with other components of the package. The information given in condensed form in Table 7 can be used as a starting point. For more detailed information, the next few pages describe more completely the various plastics used in packaging. If this is not sufficient, it will then be necessary to consult textbooks on the subject.

Next we must look at the form in which the material is required. If it is to be used as a wrapping material, it must be available as a film. For liquids or powders the container will require a secure closure, which can be obtained with bags, blown bottles, pouches, blisters, and tubes. A rigid container that must provide certain mechanical functions may require injection or compression-molded parts. Each of these forms of plastic places limitations upon the type of material that can be used. Cost is always an important factor, and when used in conjunction with the listing of physical properties in Table 7, it will aid in the selection of the optimum material for the purpose.

If the product comes under FDA regulations, there are limitations on the materials that can be used. The regulations on packaging for toiletries, cosmetics, and foods are fairly simple, but pharmaceutical packaging is fraught with all sorts of problems. The only plastics that are approved by FDA for rigid packages for pharmaceuticals are polyethylene and polystyrene, and to a limited extent polyvinyl chloride. The drug manufacturers must guarantee that the efficacy of their products is not affected by the containers, and this can be difficult.

On the basis of physical requirements of the package, a process of elimination may help in choosing the right plastic. For example, if *elasticity* is needed, then consider polyethylene, vinyl, polypropylene, polyurethane, or very thin sections of some of the stiffer materials. For *high-temperature* requirements the choice is limited to acetal, nylon, polycarbonate, and any of the thermosets. If *impact strength* is necessary, some thought can be given to the cellulosics, olefins, nylon, polycarbonate, or vinyl; but the impact strength shown in the tables of physical properties does not always correlate with actual experience, and samples should be tested thoroughly before the final selection is made.

Color is no problem with any of the thermoplastics, and almost any hue or tint can be had, but with thermosetting resins the phenolics are limited to dark colors. If *transparency* is desired, use one of the cellulosics, polystyrene, phenoxy, vinyl, or acrylic. For moisture resistance or retention of aqueous products, the olefins and saran are best. Chemical resistance varies among the different plastics; it is discussed further on in this section under the particular materials. Do not overlook the synergistic effect of combinations of chemicals which separately would not be a problem.

If *grease* resistance or retention of oily products is required, the vinyls and cellulosics should be considered. The protection of *odors* and *flavors* is particularly difficult, and although the gas permeation rates shown in Table 7 will give some guidance, careful tests over a long period of time are very important. The problem here is that often a selective loss of some of the aromatics will result in a disagreeable odor or flavor. A further discussion will be found under the individual materials in the pages that follow.

Tests should be performed at the extremes of temperature which the package is expected to undergo in commercial use, as well as at room temperature. Such tests should be continued for several weeks and preferably much longer. Conditions of testing should simulate actual service environment of motion, impact, storage positions, light, temperature cycles, and humidity as there are some strange interactions between plastic packages and their contents. A plastic closure, for instance, may behave very differently under tension on a bottle from when it is immersed in the product for test purposes. For more discussion see "Application of Plastic Materials," page 8-37.

Decoration. Various printing processes are used for decorating plastics. Flexographic printing is used for films, offset printing for bottles, jars, and collapsible tubes, and silk-screen printing for small quantities of rigid parts or for luxury items where an embossed effect is desired. If a bright gold is needed, the hot-stamping process offers a good method, and this technique can be used for colors as well. (See Figs. 2, 3, 4, and 5.)

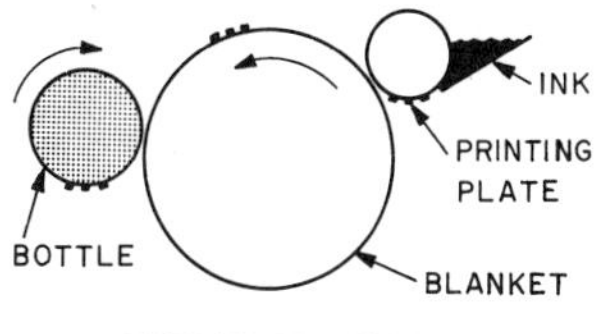

Fig. 2. Offset printing. This method of decorating is inexpensive but does not give good coverage.

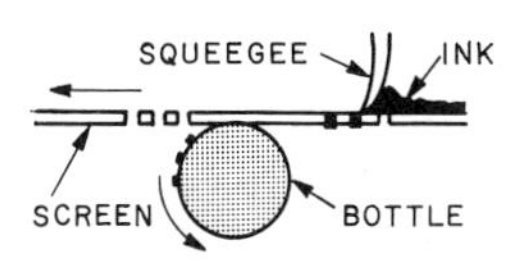

Fig. 3. Silk-screen printing. For solid coverage at a higher cost, this process is preferred for luxury goods. It does not provide fine detail or halftones.

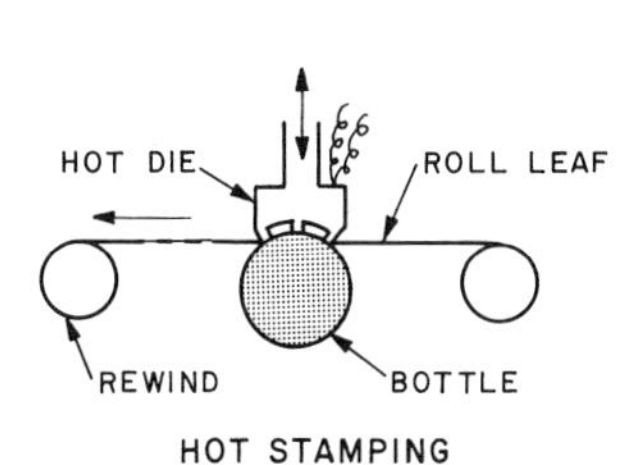

Fig. 4. Hot stamping. For bright gold and strong colors this method is best. It is not as clean and sharp as other types of printing.

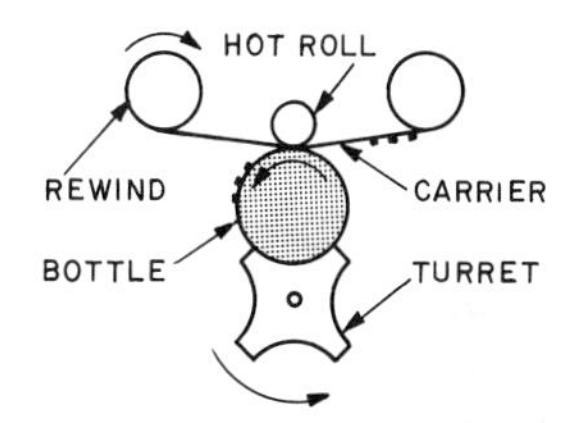

Fig. 5. Heat transfer decorating. For long runs with several colors, this technique is the most economical. Its disadvantage is the halo of clear lacquer that surrounds the print.

ABS (ACRYLONITRILE-BUTADIENE-STYRENE)

Considerable study is being given to combinations of plastics, either as copolymers in which the different types of molecules bond together forming chains of mixed composition, or as blends in which there is no chemical reaction but merely a mixing of the different materials. One of the more successful combinations of this type has been ABS, more properly called a terpolymer because of the three rather than the usual two components.

Chemistry. The three constituents of ABS have the basic structures:

$$
\begin{array}{cc} H & H \\ | & | \\ C & = C \\ | & | \\ H & CN \end{array}
\qquad
\begin{array}{ccccccc} H & & & & & & H \\ | & & & & & & | \\ C & = & C & - & C & = & C \\ | & & | & & | & & | \\ H & & H & & H & & H \end{array}
\qquad
\begin{array}{cc} H & H \\ | & | \\ C & = C \\ | & \\ C_6H_5 & \end{array}
$$

These combine to form:

$$
\begin{array}{ccccccccccccccccc} & H & & H & & H & & H & & H & & H & & H & & H & \\ & | & & | & & | & & | & & | & & | & & | & & | & \\ - & C & - & C & - & C & - & C & = & C & - & C & - & C & - & C & - \\ & | & & | & & | & & & & & & | & & | & & | & \\ & H & & C_6H_5 & & H & & & & & & H & & H & & CN & \end{array}
$$

Copolymers and terpolymers can vary in the proportions of the different constituents from which they are made, and the chemical and physical properties will differ accordingly. In the case of acrylonitrile, butadiene, and styrene, it is the butadiene portion that has the greatest influence on the type of plastic that results. As the amount of butadiene goes up, the impact strength increases, but the tensile strength goes down. The molecular chain may have a regular pattern, but more likely will have a random configuration:

—Bu—Bu—St—Ac—St—St—Bu—Ac—Ac—St—

Characteristics. ABS is a translucent thermoplastic of exceptional toughness. It has a slight yellowish cast and the surface finish is glossy. Because ABS has good hardness and stiffness, thin-walled containers can be made from this material. It is a medium-cost plastic, and while it is usually thought of as an engineering material for such things as automobile parts, ABS is finding increasing use in packaging applications, particularly for thermoforming. Plasticizers are almost never used with ABS, as there is no need for them.

Advantages and Disadvantages. Impact resistance and toughness are the chief advantages of ABS. It is not transparent, and a good white color is not possible because of the yellowish color of the base resin. The surface finish is glossy but not as sparkling as, for example, styrene. Cost is in the medium range, but the stiffness and light weight of ABS permit the use of less material than in the case of some of the lower-priced plastics.

Properties. ABS is highly resistant to scuffing, marring, and staining. Impact strength is equal to, or better than, impact styrene, and in a falling-dart test it is exceeded only by nylon and polyethylene. In tensile and flexural strength, only nylon and acetal are better, among the common plastics. Cold flow under load is minimal, even at elevated temperatures. Chemical resistance of terpolymers is not as good as for the components separately, and ABS is no exception. It is soluble in ketones, aldehydes, esters, and some chlorinated hydrocarbons. It is resistant to some hydrocarbons, but others cause softening and swelling. Some chemicals which ordinarily do not affect ABS may, under load, cause stress cracking. This can happen with some alcohols, oils, or greases. ABS is resistant to alkalies and weak acids, but is attacked by strong acids.

Processes. ABS can be calendered, extruded, blow-molded, injection-molded, and thermoformed. It is necessary to predry ABS before processing. Injection molding is rapid and easy at 400 to 500°F and 10,000 to 20,000 psi, with no mold release agents needed, no degradation of the resin, and very little warpage. ABS takes decoration readily, without any surface preparation.

Costs. The light weight of ABS, 1.10 density, makes the cost per cubic inch fairly reasonable. The current price is around 42 cents per pound, which makes the cost per cubic inch about 1.58 cents.

Trade Names. Abson (B. F. Goodrich Chemical Co.); Cycolac (Marbon Chemical Div., Borg-Warner Corp.); General Tire & Rubber Co.; Royalite (United States Rubber Co.); Sullvac (O'Sullivan Rubber Corp.).

ACETAL

This material is noted for its rigidity and high heat resistance, and it is well suited to the blow molding of aerosol bottles. Acetal is strong, is dimensionally stable, and has good impact resistance.

Chemistry. The acetals are diethers of alkylidene glycols, and contain two ether-oxygen atoms attached to the same carbon atom:

$$\begin{array}{c} \quad\ \ O-R \\ \quad\ \ | \\ R-CH \\ \quad\ \ | \\ \quad\ \ O-R \end{array}$$

When this is polymerized, we get $-O-CH_2-O-CH_2-$ in long chains containing more than a thousand carbon atoms.

Characteristics. Acetal is exceptionally stiff and strong, with good barrier properties and high solvent resistance. It is translucent in its natural state, but may be colored white or tinted to any desired hue.

Advantages and Disadvantages. It is one of the few plastics that is stiff enough to withstand the pressure in an aerosol container. It is rather high in cost as compared with other materials that are generally used for blow-molded bottles.

Properties. Impact strength is equal to that of high-impact styrene, and tensile strength compares favorably with that of nylon. Acetal is resistant to some weak acids and weak alkalies, but is attacked by strong acids, oxidizing agents, and some strong alkalies. It has good resistance to most of the organic solvents, with the exception of acetone, acetate, methyl alcohol, and methylene chloride.

The relative permeation losses for thin-walled containers over a normal shelf life are given here for comparison purposes, and not as a basis for design. Water loss is about 5 percent, alcohol around ½ percent, and methyl salycylate near ¾ percent. Propellant 11 can be considered as zero, propellant 11/12 about ¾ percent, and propellant 114 around 2 percent. Not more than 35 percent of propellant 11 should be used in a formulation because it may cause distortion of the container above 120°F. With thick walls in a container the permeation losses mentioned above will become insignificant. Carbon dioxide and nitrous oxide have high permeation rates and are not recommended as propellants.

Processes. Because acetal is highly crystalline, mold temperatures must be above 200°F and sometimes as high as 270°F to produce a smooth, glossy surface. Acetal can be heat-sealed or adhesive-bonded, provided the surface is acid-etched first. It can be blow-molded, injection-molded, spin-welded, or extruded with standard equipment, and the processing temperatures are not critical. The resin is available as a homopolymer or a copolymer, with internal or external lubricants, ultraviolet light stabilizers, or with various fillers if desired.

Costs. The moderately high price of acetal should be considered in the light of its high density. The cost per cubic inch is around 3⅓ cents, putting it in the high-price category, along with nylon and polycarbonate.

Trade Names. Celcon (Celanese Plastics Co.); Delrin (E. I. du Pont de Nemours & Co., Inc.).

ACRYLIC

This is a crystal-clear thermoplastic material with good optical properties, but without the brilliance of styrene. It was first introduced in this country around 1937 and is best known by the trade names Lucite and Plexiglas. It is a tough plastic and in thin sections is quite flexible.

Chemistry. The term "acrylic" is applied to a group of homopolymers and copolymers based on methyl methacrylate. In its pure state this material is very hard and stiff, and it is usually combined with a small percentage of a softer plastic like ethyl acrylate. When it is polymerized, it forms long chains:

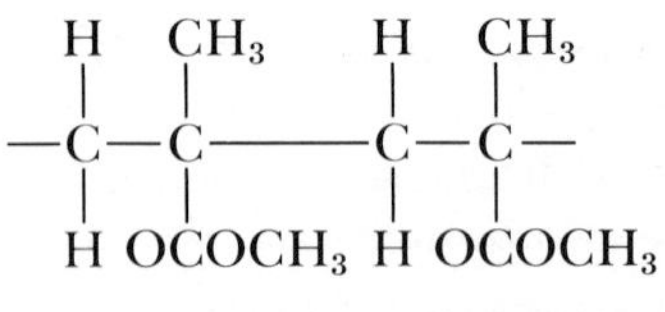

$$\begin{array}{ccccccccc} & \mathrm{H} & & \mathrm{CH_3} & & \mathrm{H} & & \mathrm{CH_3} & \\ & | & & | & & | & & | & \\ - & \mathrm{C} & - & \mathrm{C} & \text{———} & \mathrm{C} & - & \mathrm{C} & - \\ & | & & | & & | & & | & \\ & \mathrm{H} & & \mathrm{OCOCH_3} & & \mathrm{H} & & \mathrm{OCOCH_3} & \end{array}$$

Characteristics. Acrylic is noted for excellent transparency and good rigidity. It is used in place of glass for lighting fixtures and for windowpanes where vandalism is a problem. Although high in cost, it is used where toughness and resistance to oxygen and light are required.

Advantages and Disadvantages. Acrylic has a rather low softening point (150°F), has a tendency to cold-flow, and is easily scratched, depending on the formulation. It develops quite a static charge that attracts dust and dirt almost as badly as styrene.

Properties. Although acrylic is resistant to petroleum derivatives, it is attacked by the lower alcohols, chlorinated hydrocarbons, oxidizing acids, ketones, esters, and aromatic solvents. It will not support fungal growth, and it is not affected by strong acids or alkalies. Shrinkage continues for several weeks after molding and can amount to as much as 0.005 in. per in.

Copolymers. Acrylic copolymers and multipolymers have better melt strength and temperature stability for blow molding, but they sometimes whiten when bruised.

Processes. Acrylic can be molded, cast, extruded, or thermoformed, but because it is slow to heat up and slow to cool, it requires a long cycle time. Because it is easily machined and cemented, by using chloroform or carbon tetrachloride, it is useful for preparing models of packages.

FDA Approval. This material is widely used in transparent covers for food containers, vessels for beverages, and beverage dispenser pumps.

Costs. Acrylic is in the middle price range, but because of its fairly low specific gravity, its cost on a volume basis is down around 2¼ cents per cubic inch. The processing cost may be a factor in the total cost figures because, as mentioned above, cycle times tend to be longer than for most other plastics.

Trade Names. Lucite (E. I. du Pont de Nemours & Co., Inc.); Pacrosir (Società Italiana Resine); Plexiglas (Rohm & Haas Co.).

CELLULOSE ACETATE

A crystal-clear thermoplastic material, cellulose acetate has toughness and is easily processed. Plasticizers are generally added to improve the impact strength. It was first used for photographic film in 1912 and later served as a coating material for the fabric on airplanes. Other information will be found in Sec. 3, "Films and Foils."

Chemistry. Raw cellulose is combined with acetic acid and acetic anhydride by means of a catalyst and a solvent. This yields a clear solution of cellulose triacetate (62 percent combined acetic acid). It is hydrolyzed with water and hydrolyzing agents and then dried to produce cellulose acetate flake. The percentage of combined acetic acid and the length of the molecular chains determine the physical properties of cellulose acetate:

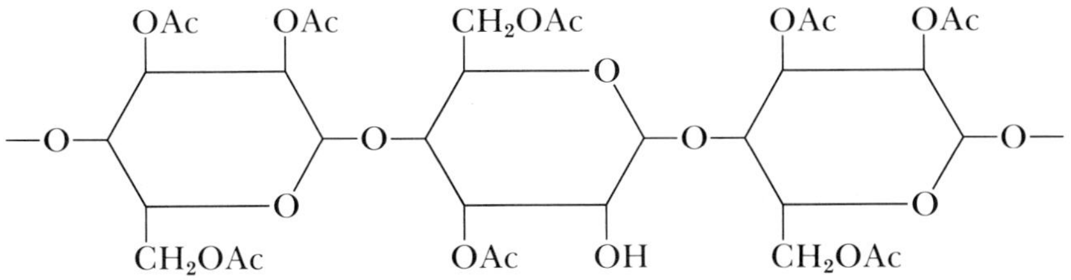

OAc OAc CH_2OAc OAc OAc
O
—O— —O— —O— —O—
O O
CH_2OAc OAc OH CH_2OAc

Since there are three OH groups attached to each ring in the cellulose chain, the maximum substitution possible would be 62½ percent. In the molecular diagram shown there is one OH that has not been replaced by an acetyl (OAc) group. Therefore it is 58.4 percent acetylated in this particular case.

Characteristics. Cellulose acetate is noted for its clarity, toughness, and ease of fabrication. It is stronger than high-density polyethylene under impact, for example, but it is not quite as strong as cellulose propionate.

Advantages and Disadvantages. It is resistant to abrasion, but scratches will show up as they do on any transparent material. The slightly yellowish cast of cellulose acetate can be masked with suitable dyes, and discoloration by sunlight can be prevented with a stabilizer such as 0.01 percent tartaric acid. It has fairly good heat resistance, but becomes brittle at low temperatures.

Properties. Cellulose acetate is resistant to oils and greases, but is attacked by strong acids, alkalies, alcohols, ketones, esters, and chlorinated hydrocarbons; and it tends to swell when exposed to moisture. The effect of these materials depends somewhat on the percentage of combined acetic acid in the acetate. As the percentage goes up, solubility decreases. Cellulose acetate may craze and discolor when exposed to light, oxygen, and water at the same time. Impact strength improves in proportion to the amount of plasticizer used; the most commonly used plasticizer is diethyl phthalate.

The barrier properties of cellulose acetate against moisture and gases is only fair. Tensile and impact strengths are good, but the tear strength of films is poor. (See Table 2.)

Processes. Cellulose acetate can be compression-, transfer-, and injection-molded, extruded, blow-molded, thermoformed, drawn, swaged, and solvent-cast. It must be predried in most instances before it is processed, to avoid bubbles that will develop from the absorbed mois-

TABLE 2 Strength of Unplasticized Cellulose Acetate

Percentage of combined acetic acid	Tensile strength, psi
52.9	10,800
54.8	11,000
56.3	11,300
58.9	14,100
60.7	16,100

ture. Static can be a problem in handling cellulose acetate, and static eliminators may be necessary for some types of operations.

FDA Approval. Cellulose acetate, in specific formulations, is acceptable to the Food and Drug Administration. It is important that plasticizers and other additives, as well as mold release compounds, also be approved materials.

Costs. With a price around 50 cents a pound, cellulose acetate would be considered a medium-priced plastic. Since it has a very high specific gravity, however, the price on a volumetric basis becomes quite high at 2.34 cents per cubic inch.

Trade Names. Bexoid (B. X. Plastics, Ltd.); Lumarith (Celanese Corporation of America); Plastacele (E. I. du Pont de Nemours & Co., Inc.); Sicaloid (Mazzucchelli Celluloide S.p.A.); Tenite I (Tennessee Eastman Co.); Vuepak (Monsanto Chemical Co.).

CELLULOSE PROPIONATE

A patent was issued in 1931 for a method of synthesizing cellulose propionate, but it was not produced on a commercial scale until 1945, when an economical process was developed for making propionic acid from natural gas.

Chemistry. Cellulose propionate is made by treating cellulose with propionic acid and anhydride, or with a mixture of acetic and propionic acids and anhydrides, in proportions to yield the desired properties in the final product. A catalyst such as sulfuric acid is generally used. A typical mixed ester would be 15 percent acetyl and 33 percent propionyl.

Characteristics. Similar to cellulose acetate, but with twice as much impact strength, propionate has good transparency and is easily fabricated. It will absorb moisture and swell, but not so much as cellulose acetate.

Advantages and Disadvantages. Cellulose propionate develops good flexural strength with only about half the amount of plasticizer required for the other cellulosics. Retention of plasticizers is better, and there is a wider choice of materials that can be used for this purpose than in the case of cellulose acetate. Its moisture absorption is the lowest of all the cellulosics, and there is no odor from the propionic acid such as occurs with the butyrates.

Properties. A tensile strength of 9,100 psi and elongation of 15 percent make cellulose propionate a tough material. Specific gravity is around 1.24. It is attacked by strong acids and alkalies, alcohols, ketones, and esters.

FDA Approval. Certain compounds of cellulose propionate are acceptable to the Food and Drug Administration and may be used for food packages.

Costs. Propionate is more expensive than acetate, the current price being around 62 cents per pound. On a volumetric basis, a cubic inch will cost about 2.71 cents.

ETHYL CELLULOSE

This material has been available for a long time, having been discovered by an Austrian physician, Dr. L. Lilienfeld, in 1916. It is stable at elevated temperatures and is often used in hot-dip coatings, hot lacquers, and strippable coatings. It is thermoplastic and generally contains some plasticizers.

Chemistry. Ethyl cellulose is an ether in which ethyl groups —OC_2H_5 have replaced the hydroxyl groups —OH of the cellulose molecule:

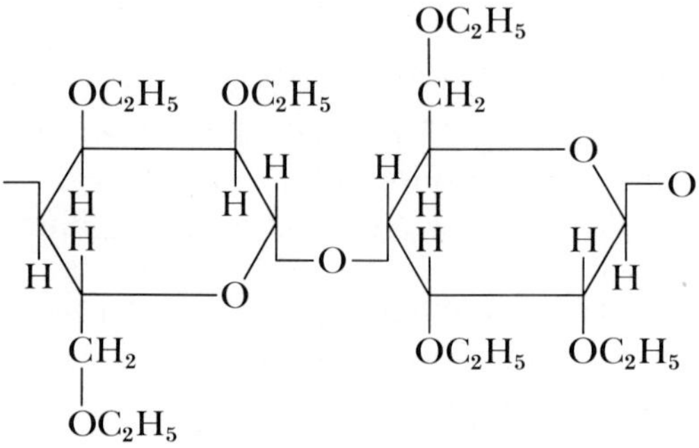

OC_2H_5
OC_2H_5 OC_2H_5 CH_2
H H O
H H H O
H H H H
H O H
O
CH_2 OC_2H_5 OC_2H_5
OC_2H_5

While a complete etherification is possible, yielding a triethyl derivative, the commercial product averages 2¼ to 2½ ethoxyl groups per glucose unit, with the lower figure giving the greater toughness and lower solubility, but poorer compatibility with plasticizers and other additives.

Manufacture. Wood pulp or cotton linters are treated with a 17½ percent solution of sodium hydroxide, in a manner similar to the process for cellophane. The alkali cellulose is then treated with ethyl chloride to form ethyl cellulose. This is washed to remove sodium chloride, and plasticizers are added.

Characteristics. Ethyl cellulose is a clear thermoplastic material with good toughness and impact strength. It is a poor barrier for moisture and gases, and chemical resistance is not good. It is fairly high in cost and has not been used extensively in packaging.

Advantages and Disadvantages. A softening temperature around 110°F and a high water absorption which causes swelling are drawbacks to the use of ethyl cellulose in packaging. It has a tendency to cold-flow or deform under stress, and this limits its use in closures. It is highly permeable to moisture and gases, and it has poor resistance to organic solvents. It is colorless, odorless, and tasteless. It is difficult to heat-seal unless the surface is activated with solvents. It can be decorated easily, but requires a special type of ink.

Properties. Ethyl cellulose has good toughness and impact strength. Tensile strength falls and extensibility rises with increasing temperature. At low temperatures the flexural strength is outstanding. There is no degradation up to about 375°F. The deep-draw properties exceed those of almost any other plastic used in packaging, when properly plasticized. Ethyl cellulose has good resistance to greases and oils such as butter and olive oil. Resistance to alkalies and weak acids is good, but strong acids may cause decomposition. It ages well and is not seriously affected by sunlight. Barrier properties for water vapor and gases are poor, and it is soluble in most solvents except aliphatic hydrocarbons, glycols, and water.

FDA Approval. Some formulations of ethyl cellulose are covered for certain food applications in accordance with section 121.1087 of the *Federal Register,* volume 27, 1962, page 4915. Specifications are given also in the *National Formulary XII,* 1965, page 164.

Costs. The specific gravity of ethyl cellulose varies, depending on the amount of etherification; it is around 1.15. With a base price of 72 cents per pound, the cost of a cubic inch is about 3 cents—rather high for packaging purposes.

Suppliers. Dow Chemical Co.; Hercules Inc.

NYLON

In the early 1930s W. H. Carothers of the Du Pont Company discovered the process for making nylon. The name nylon applies to a whole series of long-chain polyamides which vary according to the starting ingredients from which they are made. The first commercial product was toothbrush bristles made from nylon 6/10 in 1938. Hosiery yarns made from nylon 6/6 were first introduced in 1939. More than 50 different kinds of nylon are now being produced commercially.

Chemistry. Nylon (polyamide) is made from a dibasic acid combined with a diamine. Since there are many dibasic acids and many different amines, there are a great many kinds of nylon. The type of acid and amine that is used is indicated by an identifying number; thus nylon 6/10 has six carbon atoms in the diamine and ten in the acid. When a straight-chain dibasic acid $HOOC—(CH_2)_x—COOH$ combines with a straight-chain diamine $H_2N—(CH_2)_y—NH_2$, we get nylon $NH(CH_2)_y—NHCO(CH_2)_xCO$.

Different types of nylon are sometimes copolymerized. For example, nylon 6/6 and nylon 6/10 can be combined to give a copolymer with a lower melting point than either of the homopolymers, and with good transparency. It often happens, however, that there is a sacrifice of some desirable properties when different types are combined in this way. There are other ways of varying the properties of nylon, as with plasticizers and fillers, to produce a confusing array of materials. Over 100 formulations are now available for use in packaging, but for most purposes nylon 6/6 or nylon 6 can be used. The latter is polycaprolactam $NH(CH_2)_5CO$, which accounts for the single-number designation. Nylon 6/6 is somewhat stiffer, creeps more slowly, and yields at a higher stress level than nylon 6.

Characteristics. Nylon is a tough and fairly stiff thermoplastic material. It is translucent with a yellowish cast, and it has a high softening temperature, so that it can be used for sterilizable items. The low coefficient of friction makes it useful for moving parts, as in aerosol valves. It is a high-priced material as packaging materials go, but its superior impact strength makes it a very useful plastic.

Advantages and Disadvantages. Although it is transparent in thin films, in molded parts it is opaque. It is difficult to get a good white color because discoloration in processing gives it a yellowish cast. It will absorb moisture, and changes in humidity can cause appreciable amounts of swelling or shrinkage. It is tasteless, odorless, and nontoxic.

Properties. Tensile strength is good, and yield strength is exceptionally high. Elongation is comparable to that of polyethylene, but strength is about three times as much. Because notch sensitivity is low, nylon is useful for intricate molding. It is resistant to alkalies and dilute acids, but not to strong acids or oxidizing agents. Formic acid and phenol will dissolve nylon and can be used as cements. It is compatible with greases and oils and holds up under abrasion quite well. Additional information is given in Sec. 3, "Film and Foils."

Processes. Nylon has a sharp melting point, which makes processing temperatures very critical. It must be perfectly dry when it is molded; since it can absorb as much as 8 percent water, this can cause splay in the finished pieces if it is not removed. Normal molding temperature is around 550°F although thin sections may require a higher temperature. Nylon flows easily, permitting sharp corners and thin walls to be molded, but it also causes flash at the parting line if the mold is not carefully matched.

Nylon can be machined like brass, and standard saws or other tools can be used without modification. Special adhesives are available, but phenol with 12 percent water works quite well, and a solution of equal parts of resorcinol in ethyl alcohol is also widely used. These particular cements, however, should not be used for food packages.

FDA Approval. Nylon can be used for all food products, with the exception of milk and dry milk products, for which approval has been temporarily withheld until it is determined whether the cleanability standards of the Public Health Service are met. Nylon 6, nylon 6/6, nylon 6/10, nylon 11, and certain copolymers are cleared by FDA subject to limitations on extractables.

Costs. The current price of nylon is around 90 cents a pound. Its density is 1.14, making the cost of a cubic inch about 3.71 cents.

Trade Names. Allied Chemical Corp.; Belding Chemical Co.; Chemstrand Div. Monsanto Chemical Co.; Foster Grant Co., Inc.; Nypel (Nypel Corp.); Ultramid (Badische Anilin- & Soda-Fabrik AG); X-tal (Spencer Chemical Div. Gulf Oil Co.); Zytel (E. I. du Pont de Nemours & Co., Inc.).

PHENOLIC

One of the earliest plastics to come into broad commercial use was phenol-formaldehyde. Dr. Leo Hendrik Baekeland of Yonkers, New York, is credited with the first practical application of this material in 1907, and the trade name Bakelite has since become a household word. Phenol-formaldehyde had been known as a laboratory curiosity more than 50 years earlier, but Baekeland learned how to control the reaction with catalysts. He also introduced the use of fillers to improve the physical properties of the finished product.

Thermosets account for about 25 percent of the total plastics market, and thermoplastics for the remaining 75 percent. Among the thermosetting plastics, phenolics have the widest acceptance because they are the lowest in cost and the easiest to fabricate. They are limited in color range, but they have good dimensional stability and high heat resistance, which make them useful for many applications. In packaging they are most often used for caps on bottles. Over 200 million lb of phenolic molding compounds is used each year in this country for all purposes.

Chemistry. Phenol and formaldehyde are combined in the presence of an alkaline catalyst to form complicated molecules that have not been clearly defined. In theory the product of the reaction is dihydroxydiphenylmethane:

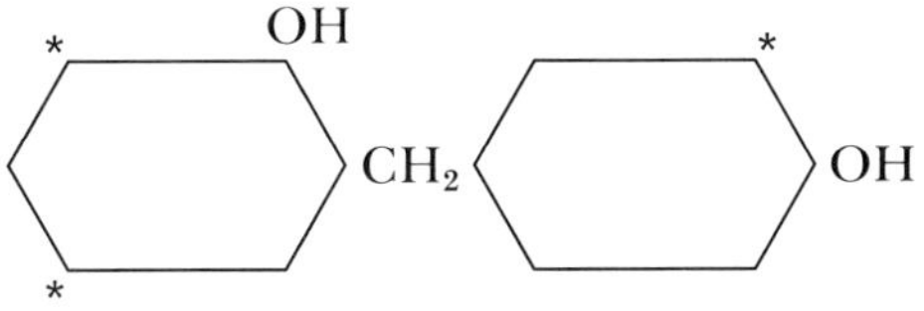

```
 *        OH                *
   ________           ________
  /        \         /        \
 <          > CH2   <          > OH
  \________/         \________/
 *
```

Condensation is believed to occur through CH_2 groups attached at the points marked with an asterisk.

There are three stages in the formation of the final product, known as the A, B, and C stages. The first reaction yields a low-molecular condensate, the intermediate phase gives a fusible but insoluble material, and the final product is infusible and insoluble.

Fillers. The properties of phenolics depend to a large extent upon the filler material that is used. Wood flour improves impact resistance and reduces shrinkage. Cotton flock gives better strength, and rag fibers and clippings are still better. Asbestos and clays will improve the chemical resistance. A long list of other materials can be used for special purposes; some are fine powders, others are crystalline or fibrous in nature.

Characteristics. There are different grades of phenolic molding compounds known as general purpose, heat-resistant, impact, nonbleeding, and special purpose. They are available only in dark colors, usually black or brown. In price phenolics are among the lowest of the packaging plastics, and they are hard, strong, and resistant to most chemicals. Phenolics are used where a hard, rigid piece is needed and where dark colors can be tolerated.

Advantages and Disadvantages. Low cost, rigidity, heat and chemical resistance, strength, and creep resistance are the outstanding properties of the phenolics. Color limitation is the chief drawback, although coatings and platings are available at a premium price. For bottle closures and fitments the phenolics are an excellent choice. They can withstand the torquing forces of the capping machines, and they will maintain a tight seal over a long period of time without loosening.

Properties. The phenolics are resistant to some dilute acids and alkalies and are attacked by others, especially oxidizing acids. Organic acids and reducing acids do not usually have any effect. Strong alkalies will decompose phenolics. The bleedproof type is generally resistant to organic solvents of all types. (See Table 3.)

Processes. There are three methods of molding phenolics: compression molding, transfer molding, and injection molding. The oldest technique is compression molding. It is slow and costly, but for small quantities it may well be the most economical. With the transfer

TABLE 3 Fillers for Phenolics

	Impact strength, ft-lb/in. of notch	Flexural strength, psi	Specific gravity	Heat distortion, °F	Shrinkage, in./in.	Tableting	Molding temperature, °F	Molding pressure, psi
Wood flour	0.28	10,000	1.40	290	0.006	Good	350	8,500
Mica	0.34	9,000	1.80	310	0.002	Good	325	4,000
Asbestos	0.35	8,000	1.75	400	0.004	Good	310	3,500
Flock	3.00	11,500	1.40	320	0.004	Fair	350	4,500
Cord	6.00	12,000	1.39	320	0.003	Poor	330	4,500
Glass fiber	17.00	25,000	1.90	600	0.001	Fair	300	4,000

process, a preform is made and preheated. Gases are allowed to escape, and the material is then put into the finishing mold. More heat is applied to make it "kick over" into the curing phase. The newest machines are the injection molding type, very similar to those used for thermoplastic molding. They are generally limited to small parts, whereas a compression press can handle pieces that weigh 8 lb or more. The principal difference is that the molds must be heated for curing the thermosetting material, whereas thermoplastic molding requires cooling of the mold. It is claimed that injection molding produces more uniform parts at about 50 percent less cost than compression molding, and around 25 percent less than transfer molding for certain parts. Because mold costs and setup costs are higher, it is better suited to long runs. There are also limitations to the materials that can be injection-molded, because of a tendency to clog in the machine. Shrinkage factors are different for the different processes, and the molds must be designed accordingly. Shrinkage for transfer molding, for instance, is greater than for compression molding. Additional information is given under "Compression Molding," page 8-53.

Costs. Phenolics will vary in price, depending on the fillers used. A typical figure would be 20 cents per pound. With a density around 1.40, a cubic inch would cost just about 1 cent. This is a little bit higher than styrene or polyethylene, but still a very low-cost material. The price of a finished piece, however, will be determined by the method of molding as well as the material cost. Compression molding is generally more expensive than injection molding, although there are some rotary-type machines that are very efficient. When comparing the cost of thermosets with thermoplastics, be sure to include the molding costs.

Trade Names. Bakelite (Union Carbide Corp.); Durez (Hooker Chemical Corp.); Fiberite Corp.; General Electric Co.; Mesa (Allied Chemical Corp.); Plenco (Plastics Engineering Co.); Rogers Corp.

POLYCARBONATE

Polycarbonate was discovered in the early 1950s by Dr. Daniel W. Fox while working for General Electric Company. At about the same time the Farbenfabriken-Bayer AG of Germany also produced small quantities of this material. Mobay Chemical Company, partly owned by Farbenfabriken-Bayer, is one of the manufacturers of polycarbonate in this country.

The use of polycarbonate in packaging has been somewhat limited because of its high cost, but where toughness and a high softening temperature (270°F) are required, it serves very well. It is suitable for food packaging, being odorless and nonstaining, but resistance to alkalies is poor.

Characteristics. Polycarbonate has good clarity with a slight yellowish tinge, excellent machinability, and unusual ductility (over 100 percent). It is a high-priced material, but for special applications it has exceptional mechanical properties and is sometimes substituted for metal. It absorbs moisture but does not swell significantly. Polycarbonate must be absolutely dry when it is processed, or bubbles and silvery flow marks will show up in the finished piece. It has very little cold flow or deformation under load.

Properties. Polycarbonate is resistant to dilute acids, oxidizing or reducing agents, salts, oils and greases, and aliphatic hydrocarbons. It is attacked by alkalies, amines, ketones, esters, aromatic hydrocarbons, chlorinated aliphatic hydrocarbons, and some alcohols. It will dissolve in methylene chloride, ethylene dichloride, dioxane, and cresol, and the first two are useful for bonding polycarbonate parts: the first for small quick-setting joints, and a 50-50 mixture with the second for large areas and longer open times. Use the minimum of solvent, and heavy clamping pressure, for at least 20 min.

Polycarbonate is not stained by coffee or most food products. It will stress-crack under certain conditions, and it is subject to crazing under tension in the presence of chemicals such as carbon tetrachloride, but not when it is under compression. Built-in stresses may have the same effect. Annealing is not usually effective in relieving these stresses. The heat resistance of polycarbonate is high, making it suitable for sterilization and food processing. It has a temperature range beyond any of the common thermoplastics (270 F), and sub-zero performance is also superior. Moisture pickup will affect dimensions about 1/10 percent for a moisture increase of 1/2 percent.

Processing. For molding or extruding, the polycarbonate must be absolutely dry; otherwise silvery streaks, splays, chicken tracks, or air bubbles may spoil the appearance. Even when the material is dry at the start, if it is not protected from the atmosphere, within 10 min it can pick up enough moisture to affect production. Not only will the finished piece not look as good, but the material will not have the toughness it should have, resulting in brittleness failure under impact. Film is made by solvent casting or extrusion, the former giving better optical properties although at a higher cost.

Polycarbonate can be heated rapidly with little risk of thermal degradation. For thermoforming, the sheet should be dried at 270°F for at least 2 hr. Cold forming of polycarbonate is possible, and bending, stamping, cold heading, drawing, coining, and rolling have been done successfully. Cold rolling can be carried out with successive passes of 10 percent up to a total of 50 percent reduction. Bending can be accomplished without heat by overbending; for a right-angle bend the amount of springback is about 15°.

Cementing of polycarbonate can be done with methylene chloride or with special adhesives made for the purpose. A solution of 5 percent polycarbonate in methylene chloride may be used, but a higher concentration than this may result in bubbles in the joint. For large joints and higher open time add up to 40 percent ethylene dichloride to slow down the rate of evaporation. Other bonding methods include heat sealing, press welding, spin welding, hot-air welding, and ultrasonic welding.

Heat sealing of polycarbonate requires temperatures around 410°F and can be done on thin sections with settings of 450°F and 2 sec dwell time. Seal strength is around 10 lb per in. Dielectric sealing is not feasible because of the low power factor. Press welding of thick sections consists of heating the edges 3 or 4 sec against a hot plate (660°F) and pressing them together. For spin welding the parts must have circular joints. A shallow tongue and groove is preferred for easy alignment. When one piece is spun with a peripheral speed of about 50 ft per min and pressed against the mating part which is held fixed, with a pressure of about 300 psi, friction produces a molten zone in a fraction of a second, which becomes a weld when the rubbing stops.

FDA Approval. The approval of certain polycarbonate resins was published in the *Federal Register,* May 22, 1963, page 5083. If molded parts are made from approved resins, and if they meet the requirements set forth in the above regulation, they can be used in contact with meat, with milk and milk products, and with other foods without FDA review.

Costs. Polycarbonate has a specific gravity of 1.2 and a price of $1.05 per pound. The cost per cubic inch figures out to be about 4½ cents. For packaging purposes this is on the high side, but since the impact strength is almost five times as much as for the common packaging plastics, it is possible to design parts with thinner cross sections and keep the cost down.

Trade Names. Lexan (General Electric Co.); Merlon (Mobay Chemical Co.).

POLYETHYLENE, LOW-DENSITY

Developed by M. W. Perrin and J. C. Swallow of Imperial Chemical Industries, Ltd., in England during the early 1930s, polyethylene reached commercial importance during World War II. In 1939 ICI licensed Du Pont, who started to produce polyethylene for wire and cable insulation. Union Carbide Corporation developed their own methods of making polyethylene in 1940, followed by I. G. Farbenindustrie in Germany in 1942. Over 3½ billion lb is now produced in this country each year, about three-quarters of which is of the low-density type. There is continued expansion in the industry, with oil and gas companies playing a major part because of their supply of ethane gas, which can readily be made into ethylene. A greater proportion of polyethylene finds it way into packaging applications than any of the other plastics. These applications are: film and sheet 44 percent, injection molding 20 percent, coating 14 percent, blow molding 3 percent, and nonpackaging 19 percent.

Chemistry. Polyethylene is essentially a straight-chain compound:

$$\begin{array}{ccccccccccccccc} & H & & H & & H & & H & & H & & H & & H & \\ & | & & | & & | & & | & & | & & | & & | & \\ - & C & - & C & - & C & - & C & - & C & - & C & - & C & - \\ & | & & | & & | & & | & & | & & | & & | & \\ & H & & H & & H & & H & & H & & H & & H & \end{array}$$

with hundreds, and sometimes thousands of carbon atoms in each chain. There may be side branches, and even some cross-linking between these groups, but excessive branchiness interferes with the close packing of the molecules and results in a soft greasy material. Less branching, on the other hand, allows the molecules to pack together and produce a crystalline structure which is stiffer, denser, and less permeable. (See Fig. 6.)

Cross-linking is usually undesirable as it makes processing the polyethylene more difficult. In some cases, however, the cross-linking is purposely brought about by adding an organic peroxide, to provide greater abrasion resistance, chemical resistance, or good weathering characteristics. There are dozens of different grades of polyethylene, and the variations are due either to the different lengths of the molecules (not only the greatest length but the proportions of different lengths), to the branchiness, or to the amount of cross-linking. Low-density polyethylene is made from ethylene gas under pressures of around 50,000 psi and temperatures of about 300°F, with small amounts of organic peroxide as a catalyst. For this reason it is sometimes called high-pressure polyethylene.

Fig. 6. Polyethylene crystals. Long-chain molecules form dendritic (branched) crystals, which are folded back every 100 Å or so, making lamellae which are about 150 layers thick. Chilling of the surface causes spherulites to nucleate very close together. The lamellae which grow from these nuclei propagate perpendicular to the surface, since they are restricted by neighboring spherulites, forming a transcrystalline region which is about 15 μ in thickness. (*Journal of Applied Physics*)

Characteristics. The flexibility, moisture protection, low cost, and light weight of polyethylene are its most outstanding attributes. It has good transparency in thin sections, but in thick-walled containers it is translucent and waxy in appearance. Polyethylene is nearly odorless and tasteless for most applications, but it should be carefully checked when it is used with foods.

Advantages and Disadvantages. The softness of polyethylene makes it useful as a wrapper for garments and baked goods, and as a material for squeeze bottles, but it makes it almost useless for rigid containers. Flexible packages made of polyethylene film are difficult to open, because of the way it stretches without breaking. Stress cracking can be a problem in the presence of certain chemicals such as detergents. Flavor and odor problems are more prevalent than with polypropylene.

Properties. Polyethylene is highly resistant to most solvents, but at temperatures above 140°F it is attacked by some aromatic hydrocarbons. Oils and greases sometimes cause a container to become sticky on the outside, making it necessary to check carefully before using polyethylene with this type of product. Mineral oil, for instance, will go through a low-density 1½-mil film in 4 or 5 days. Polyethylene is unaffected by acids and alkalies, with the possible exception of hot concentrated nitric acid. (See Table 4.)

Polyethylene is a good barrier for moisture, but it allows the passage of gases rather readily. (See Table 5.) It is a good dielectric, but this makes it difficult to seal by dielectric means because of the large amount of power required. Softening temperature is around 210°F for low-density material, and up to 260°F for the higher densities. The surface of

TABLE 4 Weight Loss in 1 Year from a 4-oz Polyethylene Bottle with 1¼-mil Wall

Contents	Low density, percent	High density, percent
Acetone	40	4
Alcohol	7	1
Amyl acetate	26	4
Benzene	900	150
Carbon tetrachloride	800	80
Ether	900	140
Ethyl acetate	60	9
Peppermint oil	12	2
Turpentine	78	9
Water	0.4	0.1
Wintergreen	15	2

TABLE 5 Permeability of Gases through 100 Sq In. of 1-mil Polyethylene in 24 Hr

Gas	Low density, cc	High density, cc
Carbon dioxide	1,940	340
Hydrogen	1,275	318
Nitrogen	175	50
Oxygen	440	110

polyethylene is nonpolar, which means that it is difficult to get adhesives or inks to stick to it. Treatment with flame or corona discharge is necessary before it can be printed. (See Figs. 11 and 12 in Sec 3, "Films and Foils," page 3-31.)

Odors and flavors are sometimes lost through polyethylene; this can be a partial loss in which certain fractions move out more rapidly than others, causing some odd results.

Processes. Film is made by casting onto a chilled roll or by extrusion as a tube or flat film. The tube is inflated by air pressure, which orients the molecules, making a tougher film. (See Fig. 9 in Sec. 3, "Films and Foils," page 3-29.) Tubing is usually extruded upward, although it can be made downward when it is desired to quench it in a water bath to improve transparency. Polyethylene is produced in many forms such as sheet, film, coatings, extrusions, and moldings. It was one of the first materials used for blow molding of bottles, and in its high-density form it is still the leading material for this type of container. Polyethylene films are in wide use for bread wrappers and bags for textile products. As a coating material it is used on butter cartons, bacon wrappers, and milk containers. The sheet material is being thermoformed for margarine tubs and frozen food containers, among other things.

FDA Approval. Polyethylene itself is acceptable for packaging foods and drug products, provided that no unacceptable additives or mold release agents are used in the manufacturing processes.

Costs. The price of polyethylene is around 10 cents per pound, which puts it among the lower-cost materials. The low specific gravity of this plastic makes the cost even more favorable. On a unit-of-volume basis polyethylene has a value of about 0.35 cent per cubic inch. Film sells for around 1 cent per thousand square inches per mil of thickness.

Trade Names. Alathon (E. I. du Pont de Nemours & Co.); Dylan (Sinclair-Koppers Co.); Chemplex Co.; Columbian Carbon Co.; Dow Chemical Co.; Eastman Chemical Products, Inc.; Enjay Chemical Co.; Fortiflex (Celanese Plastics Co.); Gulf Oil Corp.; Monsanto Chemical Co.; National Petro Chemicals Corp.; Phillips Petroleum Co.; Rexall Chemical Co.; Union Carbide Corp.; U.S. Industrial Chemicals Co.

POLYETHYLENE, HIGH-DENSITY

In the middle 1950s Prof. Karl Ziegler of Germany and several others at about the same time, notably Phillips Petroleum Company, developed a process for making polyethylene by using a catalyst of titanium tetrachloride and triethyl aluminum. This process works at normal atmospheric pressure and room temperature; the product is sometimes called low-pressure polyethylene.

About one-fourth of all the polyethylene produced is the high-density type, and most of this is used for packaging. The approximate percentages are: blow molding 48 percent, injection molding 28 percent, film and sheet 6 percent, and nonpackaging 18 percent.

Chemistry. The molecular structure of high-density polyethylene is essentially the same as for the low-density material, with this main difference: there are relatively few side branches, and so the molecules are permitted to line up nearly parallel and close together. Thus the crystallinity can range up to the 95 percent level, providing a hard, stiff, impermeable material.

Characteristics. This low-cost, moderately flexible plastic is used to a large extent for blow-molded bottles. It is stiffer and has better barrier properties than low-density polyethylene. (See Tables 4 and 5.) Clarity is poor except in cast films, in which the rapid chilling reduces the crystal formation. The surface finish is also inferior to that of low-density polyethylene. It is translucent in its natural state and can be tinted with any opaque color. Polyethylene is essentially tasteless and odorless.

Advantages and Disadvantages. The low density and low cost per pound make this one of the most useful of plastics for packaging food products and household items. It is a good barrier for moisture, but relatively poor for oxygen and other gases. It will stress-crack in the presence of some products such as detergents, unless it is formulated with other resins to minimize this tendency. The addition of 15 percent polyisobutylene, for example, will provide stress-crack resistance. Sometimes the product will absorb oxygen from the headspace and cause the bottle to distort or "collapse." This reaction can be minimized or concealed with proper design of the bottle. The higher the density, the stiffer the material, which means that a thinner wall section can be used. With bleach bottles, for example, the stiffest type of polyethylene can be used because liquid bleach does not cause stress cracking.

Properties. Most solvents will not attack polyethylene, and it is unaffected by strong acids and alkalies, with the possible exception of hot concentrated nitric acid. It is a good barrier for moisture, but gases pass through rather readily. Odors and flavors are sometimes lost differentially; that is, certain fractions of the perfume or flavoring oils transpire more rapidly than others, leaving an odd taste or smell that is undesirable. By substituting high-density for low-density polyethylene, the weight can be reduced by 40 percent or more without affecting rigidity, based on a comparison of tensile moduli. The surface of polyethylene is nonpolar and therefore must be treated with a gas flame or corona discharge before printing or using adhesives. (See Figs. 11 and 12 in Sec. 3, "Films and Foils," page 3-31.)

Processes. Because of the stiffness of high-density polyethylene as compared to the low-density type, it is preferred for thin-walled containers. The blow molding field is a large user of this material, and so is the thermoforming industry. It is sometimes extruded as a stiff tubing or other shape, but it is not often made into film because of its poor transparency. Injection molding of closures and similar parts is fairly common, and rotational molding is used for large containers such as carboys.

FDA Approval. High-density polyethylene meets the requirements of the government for plastics in contact with food, under Regulation 121.2501, provided that there are no additives or mold release agents that do not comply.

Costs. The cost of high-density polyethylene, a relatively low 16 cents a pound, makes this material very useful for packaging foods and related products. The cost per cubic inch is around 0.6 cent. Film costs about 1¼ cents per thousand square inches per mil of thickness.

Trade Names. Alathon (E. I. du Pont de Nemours & Co.); Alkathene (Imperial Chemical Industries, Ltd.); Allied Chemical Corp.; Ampacet (American Molding Powder & Chemical Co.); Blapol (Blane Chemical Corp.); Carag (Shell International Chemical Co.); Chemplex Co.; Dow Chemical Co.; Fortiflex (Celanese Plastics Co.); Hi-Fax (Hercules, Inc.); Hostalen (Farbwerke Hoechst AG); Lupolen (Badische Anilin- & Soda-Fabrik AG); Monsanto Chemical Co., Petrothene (U.S. Industrial Chemicals Co.); Phillips Petroleum Co.; Riblene (A.B.C.D.), Stamylan (Verenigd Plastic-Verkoopkantoor N. V.); Super Dylan (Sinclair-Koppers Co.); Super Modulene (Muehlstein & Co.); Union Carbide Corp.

POLYPROPYLENE

The discovery in 1954 by Prof. Giulio Natta, of the Milan Polytechnic Institute, of a sterically regular structure in a polymer was largely responsible for the polypropylene resin which we know today. He coined the term "isotactic" to describe the corkscrew arrangement of the short branches on the molecular chain. Along with these regularly ordered molecules there is produced a small amount of irregular structures forming a sticky gummy substance which he called "atactic" material. Over 30 million lb of polypropylene is now being produced in this country each year by about 10 different companies. Compared with polyethylene, it is only one-tenth as much, but its growth has been more rapid.

Chemistry. Although similar to polyethylene in some respects, polypropylene has a more complicated molecular structure. When propylene gas molecules $CH_2{=}CH{-}CH_3$ combine to form long chains in the presence of suitable catalysts, the CH_3 side groups usually follow a regular pattern:

$$
\begin{array}{cccccc}
-\underset{|}{CH}- & CH_2- & \underset{|}{CH}- & CH_2- & \underset{|}{CH}- & CH_2- \\
CH_3 & & CH_3 & & CH_3 &
\end{array}
$$

which allows the molecules to line up nearly parallel and pack together in a crystalline pattern. (See Fig. 1.) About 5 percent of the molecules do not conform to this neat arrangement, but make a soft sticky substance which is an undesirable by-product. The amount of this amorphous substance in the polypropylene resin determines some of its properties. Some additives are necessary to prevent the rapid degradation of the plastic from the effects of oxygen and light.

Characteristics. The density of polypropylene at 0.90 makes it the lightest of all plastics. It is not transparent except in film form, which can be made crystal clear by rapid chilling, with a good surface finish. Polypropylene has the interesting property of forming a hinge in thin sections, which seems to become stronger with use. One of the greatest shortcomings of polypropylene, however, is its brittleness at low temperatures. In its purest form it is quite fragile around 0°F and must be blended with polyethylene or other material to give it the impact resistance which is required for packaging.

Advantages and Disadvantages. The light weight and low cost per pound make this a very useful plastic for packaging purposes. It is widely used for screw caps, and because of its resilience it can be designed as a "linerless" closure. As a film it has excellent clarity, although low-temperature brittleness may cause some splitting in winter. Its high melting point makes it suitable for boilable packages and for sterilizable products. It is difficult to make the film into bags with tight seals because of its sharp melting point and because the plastic strings out in hairlike threads from the hot knife. Impact resistance of the film is not good, and it is not recommended for bags to hold sharp objects.

Properties. Polypropylene has good resistance to strong acids and alkalies, and it is unaffected by most solvents at room temperature, except the chlorinated hydrocarbons. At elevated temperatures benzene, xylene, toluene, and turpentine are unsatisfactory, and strong nitric acid will react with the polypropylene. It resists oils and greases, and it does not stress-crack under any conditions. Polypropylene is a fairly good barrier to moisture and gases. For printing or using adhesives it is necessary to flame-treat the polypropylene because the surface is nonpolar.

Nucleated polypropylene contains additives which promote the rapid formation of crystallites, which group together to form spherulites. These in turn grow out from the nucleus like a sunburst until they touch one another and can grow no farther. By providing many nuclei, the spherulites are kept small, thus giving the material higher impact strength and greater elongation, but lower tensile strength. When film is cast onto a chill roll, no spherulites are formed because of the very rapid cooling. With injection molding, on the other hand, there is crystallization which contributes greatly to the shrinkage problems: packing the mold with excess pressure will make up only for that part of the shrinkage which is due to temperature change, and not for that which is due to crystallization. Nucleated resins keep the crystals small so that there is improved clarity and less shrinkage, which helps to reduce warpage and sink marks. (See Fig. 7.)

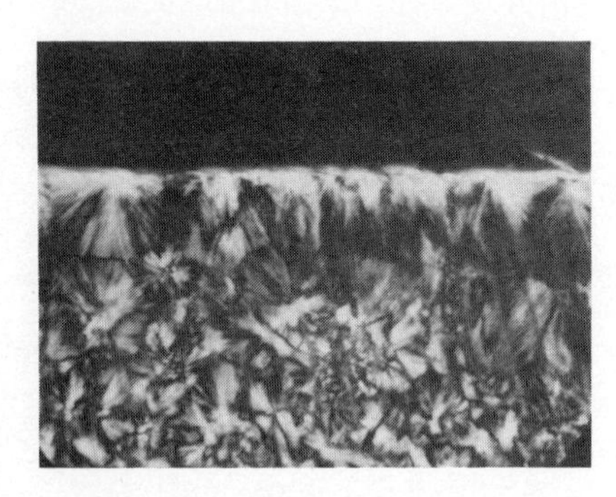

Fig. 7. Polypropylene crystals. Rapid cooling produces large numbers of small spherulites, improving the transparency of thin films. These spherulites grow perpendicular to the surface because they are limited by adjacent spherulites from going in any other direction. This causes a transcrystalline region at the surface about 12 μ thick, which can be seen in the microphotograph. (*Journal of Applied Physics.*)

Processes. Thermoforming, blow molding, injection molding, and extrusion can all be used with polypropylene. Cycle times are long because polypropylene is slow to heat up and slow to cool.

FDA Approval. There are some types of polypropylene which meet the specifications and requirements of Regulation 121.2501, provided that additives or mold release agents which do not comply are not used.

Costs. The light weight of polypropylene, coupled with a price of around 19 cents a pound and rigidity which permits thin walls, makes this one of the most promising of the new plastics for packaging purposes. The cost per cubic inch is around 0.6 cent, and film costs are about 2½ cents per thousand square inches per mil of thickness.

Trade Names. Bexphane (B. X. Plastics, Ltd.); Dynafilm (U.S. Industrial Chemicals Co.); Escon (Enjay Chemical Co.); Luparen (Badische Anilin- & Soda-Fabrik AG); Olefane (Avisun Corp.); Pro-Fax (Hercules Inc.); Propathene (Imperial Chemical Industries, Ltd.); Udel (Union Carbide Corp.).

POLYSTYRENE

A material called "styrol" was synthesized by E. Simon in 1839, but it was not used commercially until 1925, when the I. G. Farbenindustrie of Germany introduced it as an intermediate in the manufacture of synthetic rubber. It was first produced as a plastic in this country in the 1930s. Today it is one of the most widely used plastics because of its low cost.

Chemistry. Benzene is combined with ethylene gas to form ethylbenzene. This is dehydrogenated to make the styrene monomer. When this is polymerized, or joined together in long chains, we have the plastic:

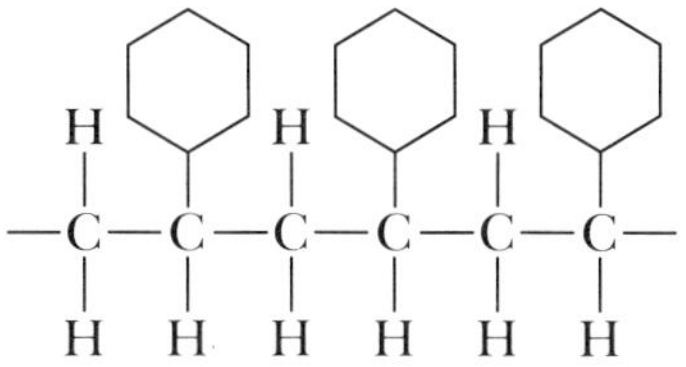

```
    H  ⬡  H  ⬡  H  ⬡
    |  |  |  |  |  |
  —C—C—C—C—C—C—
    |  |  |  |  |  |
    H  H  H  H  H  H
```

By controlling the rate of polymerization with catalysts and processing conditions, it is possible to make short chains or long chains with molecular weights between 40,000 and 220,000. This will affect the processing characteristics of the material, and the choice depends on whether it is to be used for film, extrusion, or injection molding.

Characteristics. General-purpose styrene is crystal clear and very hard and brittle; it is the cheapest thermoplastic on a per pound basis. The surface finish is excellent, and it has a brilliance without equal among plastics. It is attacked by many chemicals which cause it to craze and crack, and so it is generally used only with dry products.

Advantages and Disadvantages. This crystal-clear material is as brilliant as glass, and like glass it is also very brittle. When it is not subjected to any great stress, it is the best-looking and cheapest plastic that can be used. There is a slight tendency to shrink with age, and in strong sunlight it will discolor. When polystyrene is in contact with some solvents, or even their fumes, it will craze and become cloudy. If it is placed against other plastics, notably vinyl, it may absorb the plasticizers and become discolored. Polystyrene will build up a static charge easily, and thus has a great affinity for dust and dirt.

Properties. Polystyrene has a low melting point (190°F) and cannot be used for hot foods or other high-temperature applications. Elongation under stress is nil and therefore impact strength is poor. Styrene is resistant to acids and alkalies, except strong oxidizing acids. It is not affected by the lower alcohols or glycols, but is attacked by higher alcohols, esters, ketones, and aromatic and chlorinated hydrocarbons. Polystryrene is not a good barrier for moisture or for gases. It is odorless and tasteless and can be used with foods.

Copolymers. Many of the disadvantages of general-purpose polystyrene are overcome by combining with acrylic and rubber compounds. Although clarity is sacrificed in some cases, the resulting copolymer has greatly improved mechanical properties. This is covered more fully on page 8-40.

Fabrication. Polystyrene works easily in injection molds and has a low shrinkage rate, so that there is very little warpage or few sink marks. For blow molding it is necessary to avoid trimming operations, because of the brittleness of styrene. It works well in thermoforming and is widely used for food containers. Film is usually oriented or stretched to give it the required impact strength. Foam polystyrene is available in a number of forms; for more information see "Polystyrene Foam," page 8-85.

FDA Approval. Some polystyrene resins can be used with foods, provided that unacceptable materials are not introduced as coatings, adhesives, or mold release agents.

Costs. The base price for general-purpose natural-color polystyrene in large quantities is about 14½ cents per pound, and the cost per cubic inch is under 0.6 cent. The medium-impact grade is 2 cents higher, high-impact is 3 cents above the base price, heat-resistant high-impact is 4 cents higher, and super-high-impact is 8 cents more. Color costs an additional 3 cents.

Trade Names. Bextrene (B. X. Plastics, Ltd.); Carinex (Shell International Chemical Co.); Dylene (Sinclair-Koppers Co.); E-Z Flow (Solar Chemical Corp.); Fostarene (Foster Grant Co.); Kardel (Union Carbide Corp.); Lorkalene (Etablissements Kuhlmann); Luran (Badische Anilin- & Soda-Fabrik AG); Lustrex (Monsanto Chemical Co.); Restirolo (Società Italiana Resine); Vestyron (Chemische-Werke Huls AG).

POLYURETHANE

The application of the polyurethanes in packaging has two forms: solid and foam. The solid material as a film or as a molding material has unusual abrasion resistance, along with elasticity and resilience that put it in a class with rubber for some purposes. As an example of these properties, polyurethane heel lifts for women's shoes last seven times as long as nylon lifts.

The polyurethane foams were available around 1945, but they did not come into widespread use until 10 years later. Today over 250 million lb of flexible and rigid polyurethane foams is used in this country per year.

Chemistry. Isocyanic ester R—NCO is formed by the reaction of a diisocyanate such as tolylene diisocyanate with a polyester such as diethylene glycol. This ester, however, is being replaced to a large extent by the less expensive ethers, which are made by a process in which the diisocyanate is reacted with polyoxypropylene. A small amount of water is used in the reaction to combine with some of the diisocyanate to form carbon dioxide. This gas forms bubbles in the gel to make the foam. Catalysts such as dibutyl-tin laurate will accelerate the foaming, and surface-active agents like silicone oil help to keep the pores small, along with blowing agents such as methylene chloride to supplement the carbon dioxide. The polymerization yields a complicated structure containing, among other things, linkages of urea with ethane, and thus we get the name urethane for this group of compounds.

Characteristics. Polyurethane foams are available in a variety of densities and flexibilities, determined by the type of polyol which is combined with the diisocyanate. If it is highly cross-linked, it will result in a more rigid foam than one that has a low cross-linked density. Another factor that has an important effect on the type of foam produced is the gel strength during the blowing phase, which may result in closed cells that give a stiff material, or which may permit ruptured cells that yield a more resilient foam. The amount of water also controls the amount of CO_2 generated, and consequently the density of the foam.

Advantages and Disadvantages. Polyurethane foam can be tinted, but in its natural state it is a creamy white. It quickly turns to a yellowish brown when exposed to light, however, and for this reason it is usually colored to cover up this change. If it is foamed in place, it is important to provide good ventilation because toxic vapors are produced by the reaction.

Properties. Polyurethane foam is odorless and resistant to oxidation, oils, greases, and fungi. It is affected by strong acids and alkalies,

TABLE 6 Maximum Recommended Load for Flexible Polyurethane Foam

Density per cu ft, lb	Dead load, psi
1½	0.12
1¾	0.18
2	0.24
2½	0.36
3	0.42
4	0.54
5	0.60

halogens, aromatic hydrocarbons, chlorinated solvents, esters, ketones, and alcohols. For cleaning up the equipment that is used in processing, methylene chloride can be used as a solvent.

Polyurethane foam will adhere to any surface that is free of wax and oil. If it is foamed against a cold surface, a dense skin will be formed, but if this is undesirable, the surface can be preheated. Polyurethane is considered an efficient cushion, and it will regain up to 90 percent of its original thickness after long standing under load. (See Table 6.) The polyethers have a livelier recovery than the polyesters. The *cushion factor*, that is, the ratio of maximum stress to the energy absorbed, is around 4, which is considerably better than curled hair with a factor of 6. Polyurethane foams will withstand extremes of temperature from −50 up to 250°F with very little change, although they tend to stiffen somewhat as they go below −10°F. (See Sec. 17, "Cushioning.")

Processes. Urethane foam can be made in large slabs and then skived or sliced into various thicknesses. It can be slit, die-cut, hot-wire-cut, or sawed to the finished size. It can also be molded into intricate shapes for special internal packing. Polyurethane can be foamed in place by mixing two liquid components together and pouring directly into the shipping container. The products being packed must be protected from adhesion by the foam, as urethane sticks to almost any surface; also the container must be liquid-tight. The products can be enclosed in a polyethylene bag, or the foam can be put into two or three bags around the products. Contaminants containing sodium or potassium ions will spoil the reaction, and if water is used, it should be deionized. The pressure developed in a mold or container is usually less than 5 psi.

Costs. Foamed polyurethane costs about 1/10 cent per cubic inch. While this is higher than some other cushioning materials, it has greater efficiency and is more easily worked and assembled, and less expensive molds can be used than, for example, for polystyrene foam, which costs only half as much. When comparing costs, keep in mind that corrugated fibreboard weighs 10 lb per cu ft, and curled hair weighs around 4 1/2 lb per cu ft. Foamed polyurethane on a weight basis costs around 50 cents per pound, and in slab form it costs about 15 cents a board foot.

Trade Names. Arothane (ADM Chemicals); Chem-o-thane (Chemical Products Corp.); Chempol (Freeman Chemical Corp.); Expandofoam (Armstrong Cork Co.); Isofoam (Isocyanate Products, Inc.); Lux-Foam (Morristown Foam Corp.); Nopofoam (Nopco Chemical Co.); Stafoam (Dayco Corp.); Stanfoam (Standard Plastics); Thermothane (Thiokol Chemical Corp.); Unifoam (Wm. T. Burnett & Co.); Uralane (Furane Plastics, Inc.).

POLYVINYL CHLORIDE

Sometimes called simply vinyl or PVC, polyvinyl chloride was introduced commercially in 1927. It can be used as a rigid material, or it can be plasticized to a soft flexible sheet that makes an excellent substitute for leather or rubber. Although it is not so popular in this country as in some foreign markets, the annual production in the United States is over $2\frac{1}{4}$ billion lb. About 20 percent of this goes into film and sheeting, for household goods and garments as well as packaging, one-third is used for molding and extrusion of such things as phonograph records, pipe, and wire coatings, and about 10 percent is used for paper coating and treatment. Plasticizers can make up as much as 50 percent of a vinyl formulation, while antioxidants and colorants are usually only a fraction of a percent, and heat or UV stabilizers can run as high as 2 to $2\frac{1}{2}$ percent. Rigid PVC always has a heat stabilizer in it: usually one of the organotins (organic tin compounds).

Chemistry. The vinyl molecule is characterized by the double bond between two carbon atoms $CH_2{=}CH{-}$, and when this combines with chlorine, we have vinyl chloride $CH_2{=}CHCl$. This is made by passing acetylene gas through hydrochloric acid. At this stage it is called a monomer and is in the form of a liquid. When it is polymerized, it becomes polyvinyl chloride:

$$
\begin{array}{ccccccccccccc}
 & \mathrm{Cl} & & \mathrm{H} & & \mathrm{Cl} & & \mathrm{H} & & \mathrm{Cl} & & \mathrm{H} & \\
 & | & & | & & | & & | & & | & & | & \\
- & \mathrm{C} & - & \mathrm{C} & - & \mathrm{C} & - & \mathrm{C} & - & \mathrm{C} & - & \mathrm{C} & - \\
 & | & & | & & | & & | & & | & & | & \\
 & \mathrm{H} & & \mathrm{H} & & \mathrm{H} & & \mathrm{H} & & \mathrm{H} & & \mathrm{H} &
\end{array}
$$

Characteristics. In its natural state PVC is crystal clear and stiff but has poor impact resistance. It can be made to any degree of softness with plasticizers. For more information see "Plasticizers," page 8-41. Various stabilizers, antioxidants, lubricants, or colorants may be incorporated, and PVC is seldom used in its pure form.

Advantages and Disadvantages. Polyvinyl chloride is an inexpensive tough, clear material that is easily processed. It must not be overheated, however, as it starts to degrade at about 280°F and the degradation products are very corrosive. PVC yellows when exposed to heat or ultraviolet light, unless a stabilizer is included by the resin supplier. From the standpoint of appearance, the best stabilizers are the tin compounds, but most of these cannot be used with foods or internal medication. At this writing only certain dioctyl-tin mercaptoacetate and maleate compounds have been approved in the United States. These additives have a slight odor which is noticeable in freshly blown bottles. Other stabilizers that are acceptable for this purpose give a yellowish cast, or else they make the plastic hazy. As with any transparent plastic, scratches will show up badly if it is abused in production or shipment.

Properties. PVC is an excellent barrier for oils, alcohols, and petroleum solvents, but the plasticizer, if any is used, will be extracted by solvents. This action usually leaves the plastic hard and stiff, but sometimes the effect is not immediately apparent, because the solvent either softens the plastic or replaces the plasticizer. Later, when the solvent evaporates, the full effect is realized.

Polyvinyl chloride retains odors and flavors quite well and is a good barrier for oxygen. Rigid PVC is a fairly good barrier for moisture and gases in general, but plasticizers will reduce these properties. It is affected by aromatic hydrocarbons, halogenated hydrocarbons, ketones, aldehydes, esters, aromatic ethers, anhydrides, and molecules containing nitrogen, sulfur, or phosphorus. Water loss from a 4-oz bottle is about 2 percent per year. PVC is not affected by acids or alkalies, except some oxidizing acids, but the plasticizers may be hydrolyzed by concentrated acids and alkalies. Oxidizing and reducing agents have no effect, nor does chlorine. Impact resistance of PVC is poor, especially at low temperatures.

Processes. Blow-molded bottles of PVC are being made in large quantities, especially in Europe. As new compounds are developed that are acceptable to FDA, it is expected that these plastic containers will replace glass bottles to a large extent. For more information on this topic, read the discussion under "Blow Molding," page 8-56.

Film is made by casting or extrusion, and PVC film can be oriented to provide an excellent shrinkable material. This is covered more fully in Sec. 3, "Films and Foils." Injection molding of rigid polyvinyl chloride has increased rapidly since the reciprocating screw type of molding machine made it possible to minimize the problem of thermal degradation. Sheet material can be thermoformed quite readily and is used a great deal in blister packaging. Sometimes an unpleasant odor will develop from oxidation of the plasticizer; if PVC is used in a confined space or with food products, it should be tested for a week or more under extreme climatic conditions.

FDA Approval. The use of PVC for foods or medicines depends on the type of additives used. Some of the best plasticizers from a practical standpoint are too toxic for food packaging, and most of the stabilizers containing tin compounds are not acceptable. With careful selection of plasticizers, lubricants, stabilizers, and impact modifiers, it is possible to get a plastic that meets all the requirements, but usually at some sacrifice in appearance and processing characteristics.

Costs. Among the very transparent materials, PVC is the cheapest except for polystyrene. At 1/2 cent per cubic inch it is a fairly low-cost packaging material. The price of 10 cents per pound is very attractive, but since it has a rather high specific gravity, comparison with other plastics should be on a unit-volume basis. Off-grade material has been known to sell for as little as 8 cents a pound, and the amount and type of plasticizers also have a lot to do with the price.

Trade Names. Air Reduction Corp.; Allied Chemical Corp.; American Chemical Corp.; Atlantic Refining Co.; Borden Co.; Carina (Shell International Chemical Co.); Cary Chemical Co.; Diamond Alkali Co.; Elvax (E. I. du Pont de Nemours & Co.); Geon (B. F. Goodrich Chemical Co.); Hostalit (Farbwerke Hoechst AG); Irvinil (Great American Plastics Co.); Kenron (Sechrist Chemicals, Inc.); Marvinol (United States Rubber Co.); Opalon (Monsanto Chemical Co.); Pantasote Co.; Rucoblend (Rubber Corp. of America); Vinoflex (Badische Anilin- & Soda-Fabrik AG); Vygen (General Tire & Rubber Co.).

UREA

A thermosetting resin, urea was developed around 1930. It is a hard translucent material that takes coloring well, and it is widely used for closures and cosmetic cases. They are more expensive to produce from this material than from some of the thermoplastics, but the heat resistance and other fine properties of urea make it suitable for such premium items. Beautiful colors are obtainable because the translucency gives a brightness and depth of color similar to opal glass.

Chemistry. The term "urea" is short for urea-formaldehyde, which indicates the two main ingredients used in the manufacture of this plastic. When these two materials react in the presence of an alkaline catalyst $NH_2—CO—NH_2 + HCHO$, a monomer is formed: $NH_2—CO—NH—CH_2OH$. When this is combined with a latent catalyst and heated in the mold, an acid which catalyzes the polymerization is produced.

Characteristics. Urea plastic is available in an unlimited range of colors. In the case of pink and orchid shades, however, and to a less extent yellow and orange, it is difficult to get a good color match, and these colors have a tendency to fade with time. In the past, the colors for urea were standardized by the U.S. Department of Commerce as MUP (molded urea plastic) standards, known familiarly as "mup colors." Urea is a hard, brittle material which is odorless and tasteless; it has good gloss and a pleasing translucency. It does not build up static electricity, which would attract dust.

Advantages and Disadvantages. Compared with phenolics, urea plastics are quite expensive. The range of colors is so much better, however, that there is little choice if appearance is critical. Being a thermosetting plastic, urea can withstand high temperatures without softening, but it will char at about 390°F. It will absorb water under very wet conditions, but this does not seem to have any serious effect on the plastic.

Properties. Urea is not affected by any of the organic solvents, but it is affected by alkalies and strong acids. It has good resistance to all types of oils and greases. Although urea can withstand elevated temperatures, it cannot be steam-sterilized. There will be shrinkage in the parts up to 0.003 in. per in. after molding, in addition to the shrinkage that occurs while it is in the mold.

Processes. Since urea is a thermosetting resin, it can be processed only by transfer molding, injection molding, or extrusion. Cold spots in the mold will cause warpage, as will excess moisture in the material. Other defects are streaking at the weld lines, blistering due to undercure, overcure, or trapped gas, and "orange peel" surface caused by poor flow in the mold.

Costs. Among thermosets urea is in the middle range of costs. With a unit-volume cost of 1¾ cents per cubic inch, it is considerably higher than the phenolics, but not as high as melamine. Among the thermoplastics it is in a price class with acrylic.

Trade Names. Arodure (ADM Chemicals); Beetle (American Cyanamid Co.); Kaurit (Badische Anilin- & Soda-Fabrik AG); Resfurin (Resine Sintetiche ed Affini S.p.A.); Scarab (British Industrial Plastics Ltd.); Siritle (Società Italiana Resine); Sylplast (FMC Corp.); Synvarol (Synvar Corp.).

CHEMICAL PROPERTIES DEFINED

Where properties of plastics are shown in the individual discussions on pages 8-7 to 8-36, the resistance to certain types of chemicals is mentioned. For those who may not be familiar with all the terms used, the following is a partial list of some typical materials in each category:

Hydrocarbons:

Benzene	Turpentine
Hexane	Kerosene
Toluene	Gasoline
Xylene	Naphtha

Chlorinated hydrocarbons:

Carbon tetrachloride	DDT
Chloroform	Ethylene dichloride

Ketones:

Acetone	Methyl ethyl ketone

Esters:

Ethyl acetate	Methyl salicylate
Amyl acetate	Tricresyl phosphate

Essential oils:

Camphor oil	Orange oil
Cinnamon oil	Peppermint oil
Citronella oil	Spearmint oil
Eucalyptus oil	Turpentine oil
Lemon oil	Wintergreen oil

APPLICATION OF PLASTIC MATERIALS

To determine whether plastics are suitable for a particular packaging application, or to help in choosing between the different polymers, there are a number of useful criteria. These are based on test methods that have been developed over a long period of time, with the usual requirement of standardized tests that they be reproducible and that they can be duplicated in any well-equipped laboratory. These tests cannot always be directly related to service conditions, however, and they must be interpreted in terms that are applicable to the packaging problem at hand. It should also be kept in mind that the figures in Table 7 apply to laboratory tests under very specific conditions. Any change in those conditions, as for instance a change in temperature, will alter the results. For this reason the figures given in other texts may not always agree with the values given in this book.

Costs. Table 7 shows the current figures at the time of going to press. While there may be fluctuations in the market prices, the relative positions on the cost scale are not likely to change, and these figures can be used for rough calculations. To do close figuring, it will be necessary to get the latest prices from a supplier.

Clarity. Since many packaging applications require that the contents be visible, the relative transparency of the different plastics is of considerable interest. The values shown in Table 7 are in terms of a fairly thick sheet. A thin film, of course, will be much more transparent than a heavy section. Surface luster varies among the different plastics, but this was not taken into account in making up this tabulation. The transparency ratings are based strictly on the transmission of light.

Stiffness. The wall thickness of a container is often dependent upon the stiffness of the plastic. As shown in Table 7, high-density polyethylene is much stiffer than low-density polyethylene. Even though the high-density material is more expensive, a container can be made more cheaply with it because the thinner walls take less material.

Water-Vapor Transmission. The water-vapor transmission rate (WVTR), sometimes called moisture-vapor transmission (MVT), is the measure of the gain or loss of water through the walls of the package. In actual storage this will vary according to the season of the year, the average humidity being considerably higher in the summer than in the winter. Higher temperatures will also accelerate passage of moisture

TABLE 7 Properties of Plastics for Packaging

Plastic	Cost in dollars			Clarity	Water vapor transmission[a]	Gas permeability[b]			Dust attraction	Printability[c]
	Per lb	100 cu in.	1,000 sq in., 1 mil			O_2	N_2	CO_2		
ABS	0.42	1.58	0.03	Translucent						
Acetal	0.65	3.34	. . .	Translucent	1.9	. . .	High	. . .	Low	F
Acrylic	0.51	1.65	. . .	Transparent	. . .	. . .	. . .	. . .	High	
Cellulose acetate	0.50	2.34	0.05	Transparent	150	117	40	1,000	Low	E
Cellulose propionate	0.62	2.71	0.05	Transparent	150	. . .	High	. . .	Medium	
Ethyl cellulose	0.72	3.00	0.04	Transparent	10	2,000	600	5,000	Low	
Nylon	0.90	3.71	0.09	Translucent	19	25	. . .	160	Medium	
Phenolic	0.20	1.01	. . .	Opaque	. . .	. . .	. . .	. . .	Low	
Polycarbonate	1.05	4.54	0.09	Transparent	11	300	50	1,000	Medium	E
Polyethylene, high-density	0.16	0.55	0.02	Translucent	0.3	600	70	450	High	F
Polyethylene, low-density	0.10	0.35	0.01	Transparent	1.3	550	180	2,900	High	F
Polypropylene	0.25	0.81	0.02	Translucent	0.7	240	60	800	High	G
Polystyrene, general-purpose	0.15	0.57	0.02	Transparent	8	310	50	1,050	Very high	E
Polystyrene, high-impact	0.18	0.60	. . .	Opaque	. . .	. . .	. . .	. . .	High	E
Polyurethane	1.20	3.95	. . .	Translucent	0.6	. . .	. . .	. . .	Low	
Polyvinyl chloride, rigid	0.10	0.50	0.03	Transparent	4	150	. . .	970	High	E
Urea	0.32	1.74	. . .	Translucent	. . .	. . .	. . .	. . .	Low	

[a] g loss/24 hr/100 sq in./mil at 95°F, 90% RH.
[b] cc/24 hr/100 sq in./mil at 77°F, 50% RH; ASTM D1434-63.
[c] E = excellent; G = good; F = fair; P = poor.

through the plastic. The figures in Table 7 are for thin films, but for heavier gauges there is a direct relationship which is almost the exact inverse of the thickness. With those materials that contain plasticizers and other additives, the transmission rate may be different from the figures shown, usually higher, as these values are based on unplasticized resin. Pressure differences do not have any effect upon the rate of transmission.

Gas Permeability. The rates shown in Table 7 for the transmission of gases through the various plastics are based on thin films. The rate goes down almost proportionately to the increase in thickness; that is, twice the thickness gives about half the rate. The permeation rates for the gases shown are independent of pressure, but temperature can be an important factor, especially for those materials with the lowest rates. Combinations of gases act independently and transpire through the plastic as though they were alone.

Chemical resistance			Temperature range, °F	Mar resistance[d]	Warpage[e]	Stiff-ness[f]	Impact strength[g]	Tear strength notched[h]	Brittle-ness[i]
Acids	Alkalies	Solvents							
G	G	F	−65–215	100	0.006	300	6.2	. . .	60
F	F	G	−40–250	120	0.022	410	1.4	. . .	15
E	E	F	to 150	120	0.004	430	0.4	. . .	5
F	P	G	−15–140	60	0.005	200	2.5	15	40
F	P	P	−30–200	70	0.004	200	6.0	25	80
F	E	P	−70–250	80	0.006	. . .	4.0	20	30
F	E	G	−100–200	110	0.010	200	1.2	75	100
F	F	G	to 250	120	0.010	1,000	0.5	. . .	1
G	P	F	−210–270	118	0.006	340	3.0	25	75
E	E	E	−20–250	38	0.040	150	10.0	30	100
G	G	G	−70–180	112	0.030	10	20.0	100	400
E	E	E	0–275	90	0.020	200	1.0	25	300
G	E	F	−80–175	120	0.004	. . .	0.3	. . .	1
G	E	F	−55–200	75	0.004	. . .	8.0	. . .	60
P	P	G	to 190	60	0.009	. . .	. . .	. . .	500
E	E	F	−50–200	45	0.002	. . .	8.0	90	20
F	F	G	to 170	150	0.010	. . .	0.4	. . .	1

[d] Rockwell hardness, R scale; ASTM D785-51.
[e] Mold shrinkage, in. per in.
[f] Flexural modulus, psi ÷ 1,000; ASTM D790.
[g] Izod impact strength, ft-lb per in., notched; ASTM D256-54T.
[h] Elmendorf, g/mil; ASTM D689-62.
[i] Elongation at breaking point in percent of length; ASTM D882-61T; lowest figure is most brittle.

Chemical Resistance. The effect of chemicals varies with each plastic, sometimes causing swelling and softening, or a stickiness on the surface. At other times the result is a weakening which leads to stress cracking, or it may be a stiffening effect due to removal of the plasticizer. Table 7 gives the net effect of usefulness as a packaging material without regard to the type of deterioration.

Mar Resistance. Packages are subjected to many kinds of hazards in manufacture, on the filling line, and in shipment. The resistance of the various plastics to scratching, scuffing, denting, bruising, and abrasion is summed up in Table 7 under the heading "Mar resistance." The appearance of a container at the point of sale will depend also on whether it is transparent or opaque, since clear materials tend to show up defects more obviously than do the colored plastics.

Temperature Range. The variations in temperature that a package is likely to encounter in this country are between −20 and 120°F. Winters get quite cold in the North Central states, and it is possible for packages in transit to be exposed to low temperatures for short periods of time. The other extreme might occur in the Southwest in the middle of summer, if a freight car is exposed to the sun or if packages are displayed in a store window in direct sunlight; and heat may affect packages stored near a radiator. The figures in Table 7 give the practical limits, with the assumption that the package may be under stress from stacking loads or tight closures or abusive handling at these extremes.

Warpage. The tendency of rigid plastic containers to warp is a function of shrinkage in the mold. Thick sections shrink more than thin sections, and parts remote from the gate in the mold are affected more than those close by. Shrinkage will vary also with mold temperatures and pressures, but basically it is a function of the molecular structure of the polymer.

Impact Strength. There are various ways of checking impact resistance of plastics. The figures shown in Table 7 are based on the Izod test, using notched pieces which are struck with a swinging weight. This is not the same as dropping a filled container on a hard surface, but it is a more reproducible test and it gives a reasonably good comparison for packaging purposes. It should be noted that variations in thickness may not give the expected differences in impact strength because of "skin effect." This is more pronounced in some materials than in others. The resistance of a plastic material to damage by shock is a function of tensile strength in conjunction with its elongation under stress and its flexural stiffness. The ability of a package to withstand this type of abuse can be determined with certainty only by tests with the actual container and contents.

Tear Strength. Although it is a combination of tensile, shear, and elastic properties, the tear strength of a plastic film or sheet can be very accurately measured. It is quite important to have this information to determine not only the processing characteristics, but also the shipping qualities of the package, and the ease with which the customer can open the package for use. The figures in Table 7 are based on the Elmendorf test.

Elongation. The amount of stretch in a plastic material is a measure of its ability to conform to an irregular surface, and also its ability to absorb stresses without breaking. On the negative side, it is the resistance of a packaging material to being opened easily by the consumer if it is a wrapper or a blister type of package.

Copolymers. The combining of two monomers into one polymer chain is called copolymerization. The proportions are not usually equal; the major portion is called the *base monomer,* and the smaller amount the *comonomer.* The properties of the combination will depend on several things. The proportions of the component materials are the most important factor, but the type of comonomer and the way in which it is processed also contribute to the character of the final product. For example, butadiene and styrene might be combined in a regular pattern: —Bu—St—Bu—St—Bu—St—; or in a more random configuration: —Bu—Bu—St—Bu—St—St—St—Bu—.

Also possible is the combination of three materials, called a terpolymer, as in the case of acrylonitrile-butadiene-styrene, commonly referred to as ABS, which is a rugged material used for appliances, automobile parts, and many similar applications. Styrene-acrilonitrile, known as SAN or by the trade number C-11, is a tough plastic that is used for closures if plain styrene is not good enough. Ethylene-vinyl acetate, sometimes called EVA, has a rubbery quality that makes it a prime candidate for cap liners.

The chemical and physical properties of copolymers will vary according to the proportions of the components. For example, as the amount of butadiene goes up, the impact strength increases, but the tensile strength goes down. In some cases there may be a loss of transparency and luster, but this is a small price to pay for the improved strength characteristics. A disadvantage of copolymers is that their chemical resistance may be much less than that of their various components used separately. Also if butadiene or other elastomers are used, white blush marks may occur in transparent copolymers as a result of impact, crushing, bending, or built-in stresses. Copolymers which contain ethylene generally have poor resistance to high temperatures.

Plasticizers. Stiff plastics can sometimes be made softer and more pliable by the addition of oily liquids called plasticizers. These materials are dissolved in the plastic during processing, and they have the effect of separating portions of the long molecular chains, allowing them to slip and slide against each other. Typical of these plasticizers are the high-boiling esters of various 8-carbon alcohols such as dibutyl phthalate, dioctyl phthalate, and tricresyl phosphate.

There is certain information about plasticizers that may be of importance to the packaging engineer. They tend to exude under long standing or high temperature, and they may migrate from one plastic part to another, as, for example, from a vinyl to a styrene if they are in contact. The plasticizers may also be leached out by solvents or by a liquid product which acts as a solvent. The relationship between the resin and the plasticizer may vary with temperature, and some plastic materials become stiff and brittle when cold.

Some additives which serve as plasticizers may also be stabilizers or lubricants at the same time. Most of the plasticizers, however, simply improve the flexibility of the plastic. They are probably the most important additives for plastics that are used in packaging. Over a billion pounds of plasticizers are used in the plastics industry each year. Thousands of different plasticizers have been developed, and new ones are being introduced all the time, but there are fewer than 100 that have real commercial value and only about half that number are used to any extent in packaging. Prices vary with the material, but the major types are selling for around 15 cents a pound. By far the greatest use of plasticizers is in polyvinyl chloride, accounting for about 90 percent of the production; polyvinyl acetate is second, accounting for around 4 percent.

Plasticizers have three main functions: to lower the processing temperature and avoid decomposition of the polymer, to modify the processing characteristics, and to make the finished product more flexible. They accomplish these functions by acting as a lubricant to allow the molecules to slide over one another freely, or by acting as a partial solvent for the resin. In the second case they break some of the polymer-to-polymer bonds in the molecule and replace them with polymer-to-plasticizer bonds.

The most important class of plasticizers are the *phthalates.* Typical of this group is dioctyl phthalate, known in the trade as DOP, which accounts for about 30 percent of all plasticizer production. High-purity types of DOP meet FDA requirements for food packaging materials. The chief drawback of the phthalates is poor resistance to oils. As much as one-third will migrate into an oily product. Secondary plasticizers which sell for as little as 6 cents a pound are sometimes used as extenders for the more expensive plasticizers. These are usually oily or waxy types of petroleum derivatives.

For low-temperature flexibility, the *adipates,* such as dioctyl adipate, are often used. They have the disadvantage of being somewhat volatile. Among the *phosphates* one of the earliest, and still used to a large extent, is tricresyl phosphate (TCP); it has poor low-temperature characteristics, is very toxic, and is fairly high in cost. It is used with cellulosics, vinyls, alkyds, and phenolics. It is often combined with phthalate plasticizers to improve the processing characteristics of polyvinyl chloride.

To reduce migration and extraction by petroleum products, *polyesters* which are not highly polymerized, and therefore are viscous fluids, can be used, but they are in the higher-price class. Their chief advantage is to extend the useful life of household items, and they are of less importance in packaging. Chlorinated straight-chain *hydrocarbons* are used for the same general purpose, and they too are seldom used in packaging plastics.

Additives. Plastics for packaging may have other ingredients added.

Epoxy-type plasticizers, derived from soya bean oils and from fatty acids, serve as stabilizers against heat and light, as well as being plasticizers. Some of these have FDA approval and can be used for food packaging. They are in the medium-price class at about 27 cents a pound. A typical *internal lubricant* would be a fatty acid glyceride, and a typical *external lubricant* would be a montanic acid ester. The purpose of the internal lubricant is to decrease the friction between plastic molecules and thereby the melt viscosity during processing. External lubricants, on the other hand, migrate to the surface and reduce friction with the processing equipment.

The action of light on plastics varies with the different compounds. The shorter wavelengths are much more destructive than the longer ones, and most of the damage comes from the ultraviolet part of the spectrum, between 3000 and 4000 Å. For protection against these UV rays, and against heat, the best *stabilizers* from an appearance standpoint are the organotins (organic tin compounds) such as di-*n*-octyl-tin mercaptide at a level of 1 to 1½ percent. (See Table 8.) Most of these are not acceptable for contact with foods or internal medication, however, and a more suitable material might be 1 percent aminocrotonic acid ester, with about 1½ percent calcium-zinc stearate. This gives a yellowish tint to the polyvinyl chloride, which may not be objectionable for some applications. The function of the heat stabilizers is to scavenge the hydrogen chloride that is liberated as a degradation product. They are nearly always used in conjunction with antioxidants. The amount of UV stabilizer used in a formula must depend on the thickness of the finished part; thin sheets and films require a great deal more than heavy sections of molded parts, because of the greater area of exposure. These are expensive materials, most of them being $4 or more a pound.

TABLE 8 Toxicity of Stabilizers

Material	LD_{50} level,* mg
Diphenyl thiourea	1,500
Di-*n*-octyl-tin mercaptide	2,000
Di-*n*-octyl-tin-dilaurate	6,000
Phenylindole	6,000
Zinc stearate	6,000
Aminocrotonic acid ester	6,000
Epoxidized soya bean oil	6,000

* Acute oral toxicity is expressed as the *L*ethal *D*ose for *50* percent of the test animals in milligrams per kilogram of body weight (LD_{50}).

Of special importance in packaging are the *antistatic agents,* because the tendency to attract dust on store shelves has been a deterrent to greater use of plastics in this field. When two materials are in contact, the electrons on the surface atoms intermingle and may move from one material to the other. The nature of the plastic will determine the degree to which this takes place; styrene is the most active, followed by acrylic and polyethylene. Pressure and friction will increase the movement of electrons. When the two materials are separated, the one that has lost electrons is positively charged, and the one that has gained electrons is negatively charged. With sufficient moisture in the air there is an ionization of dissolved substances which neutralizes or bleeds off these electrons. The purpose of most antistatic agents is to absorb moisture from the air for this purpose. One of the simplest ways to do this is with a dip or spray of a 1 percent solution of a dishwashing detergent. Unfortunately it is effective only for a few weeks. Other antistatic agents such as polyethylene glycol are hydrophilic materials, which can be mixed into the resin at the rate of about ½ percent; they have the effect of slowly exuding onto the surface and attracting moisture over a much longer time than surface treatments. They are only partially effective, however, and resin suppliers are constantly searching for better antistatic additives. The best antistats, unfortunately, are not approved by FDA for use with food products or internal medication.

Fillers and Reinforcements. Various materials can be used as fillers in plastics. Both thermosetting and thermoplastic resins can incorporate these fillers with beneficial results, although they are used more consistently with the thermosets. Cotton, talc, wood flour, glass fibers, and dozens of other materials can be used. They are chosen for different reasons: to reduce cost if the filler is cheaper than the resin; to add stiffness to a plastic that is too flexible; or to increase the tensile strength with tough fibrous materials. Shrinkage in the mold is also greatly reduced, minimizing any tendency for warpage.

In theory the amount of resin should be just enough to coat the surface of the filler particles. In actual practice the filler is less than this amount, more nearly 50 percent of the total mixture. When glass fibers are used as fillers, or, as the suppliers prefer to call them, reinforcements, their surface must be treated with a coupling agent such as silane (aminopropyltriethoxysilane) to get good adhesion to the plastic. These glass fibers are 0.0004 to 0.0006 in. in diameter, and ⅛ to ⅜ in. long for soft plastics; ½ in. for styrenics and other hard materials. The shortest fibers give almost the same tensile strength as the longer ones, but the impact strength suffers when ½-in. fibers are used. (See Table 9.) The cost per pound may be doubled, but the greatly improved mechanical properties of the glass-filled material permit the use of thinner sections.

TABLE 9 Strength of Fiber-glass-reinforced Thermoplastics

Thermoplastic	Tensile strength, psi		Impact strength, ft-lb per in.	
	Unreinforced	Reinforced	Unreinforced	Reinforced
Nylon	11,800	30,000	0.9	3.8
Polycarbonate	9,000	20,000	2.0	4.0
Polyethylene	3,300	8,000	. . .	4.5
Polystyrene	8,500	14,000	0.3	2.9
Polypropylene	5,000	8,000	1.8	3.5

Even more exotic fibers can be used with astounding results, but the costs are even more astounding: upward of a thousand dollars a pound for boron or sapphire fibers.

PROCESSES

INJECTION MOLDING

The most widely used method of making plastic parts is injection molding. With this technique the plastic is melted and forced under high pressure into a heavy steel mold. After a few seconds the plastic chills and hardens, taking the shape of the mold as it sets up. The mold then opens up into two parts so that the piece can be ejected, and the mold is closed again to repeat the cycle. The mold usually consists of two main parts: the *cavity*, or female half, which forms the outside of the piece, and the *force*, or male portion, which shapes the interior of the part. (See Fig. 8.) There may be only one cavity, in which case the mold makes one piece at a time, or there may be as many as 16 or 24 or more cavities, depending on the size of the piece. The greater the number of cavities, the higher the cost of the mold, but the lower the cost of the pieces being produced.

The hot plastic enters the mold through a center channel called a *sprue*. From the sprue there are *runners* to carry the molten material to each of the cavities. These runners are about 1/4 in. in diameter and get smaller as they get closer to the cavities. Where the plastic enters a cavity, there is a restricted opening called a *gate*, which is usually 0.020 to 0.030 in. in diameter. When the parts are ejected from the mold by the *knockout pins*, they are connected by a spiderlike plastic piece which comes out of the runners and sprue, and must be separated by breaking or cutting the gate. The scrap material can be reground and put back into the hopper, so that it is not wasted. The complete set of parts with

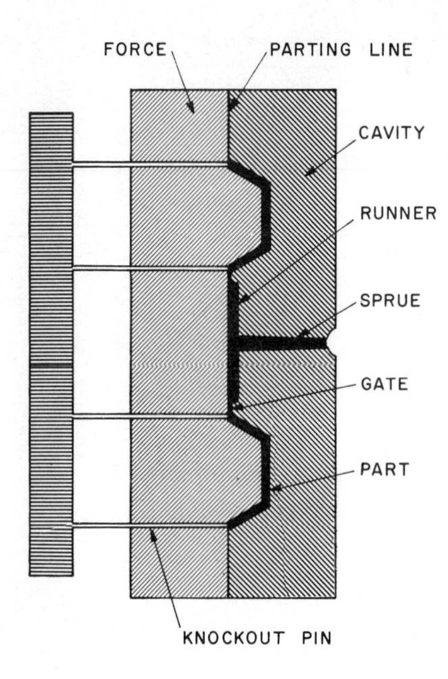

Fig. 8. Injection mold. The molten plastic is forced into the mold under high pressure. After the plastic chills and hardens, the force separates from the cavity, allowing the knockout pins to push the part off the force. Then the mold is closed again to repeat the cycle.

the connecting material is called a *shot*. It is possible to design the mold with *pinpoint gating* or *submarine gating* or as a *three-plate mold* for automatically separating the parts from the runner piece. Another technique is to use hot runner molds to avoid having to regrind the runner pieces, but these sophistications are justified only on very high-production tooling, and with certain plastics. (See Fig. 9.)

Design. In planning a part to be made by injection molding, consideration should be given to a number of points relating to the best operating conditions and a good appearance in the finished piece. Undercuts should be avoided, as they usually require cam-actuated parts in the mold, which add to the cost and limit the useful life of the mold.

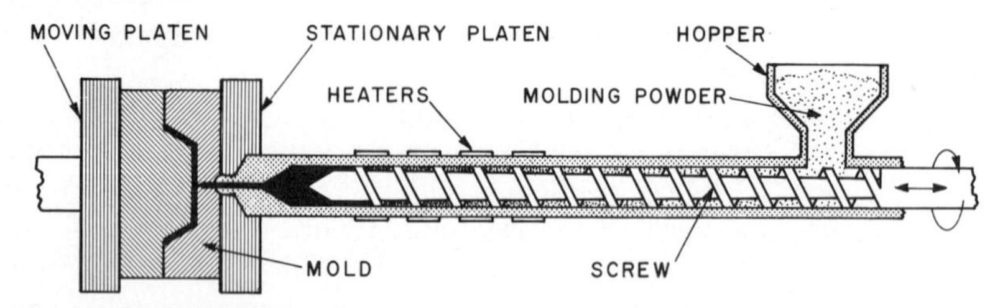

Fig. 9. Molding machine. Pellets of plastic from the hopper are carried by the screw toward the mold. As the screw turns, it moves toward the right, away from the mold. Note that the channel of the screw gets shallower as it approaches the mold. Kneading of the plastic by the action of the screw generates most of the heat needed to melt the plastic, and only a small amount is added by the heater bands. When the mold has closed, the screw stops rotating and moves forward, pushing the plastic into the mold. Then the screw starts to rotate and move back from the mold again, to prepare for the next shot.

A thick section next to a thin section may crack because the different shrinkage rates will set up strains at that point. Inside corners should not be sharp, as sharpness weakens the piece by "notch effect"; the minimum radius should be 0.020 in. (See Fig. 10.)

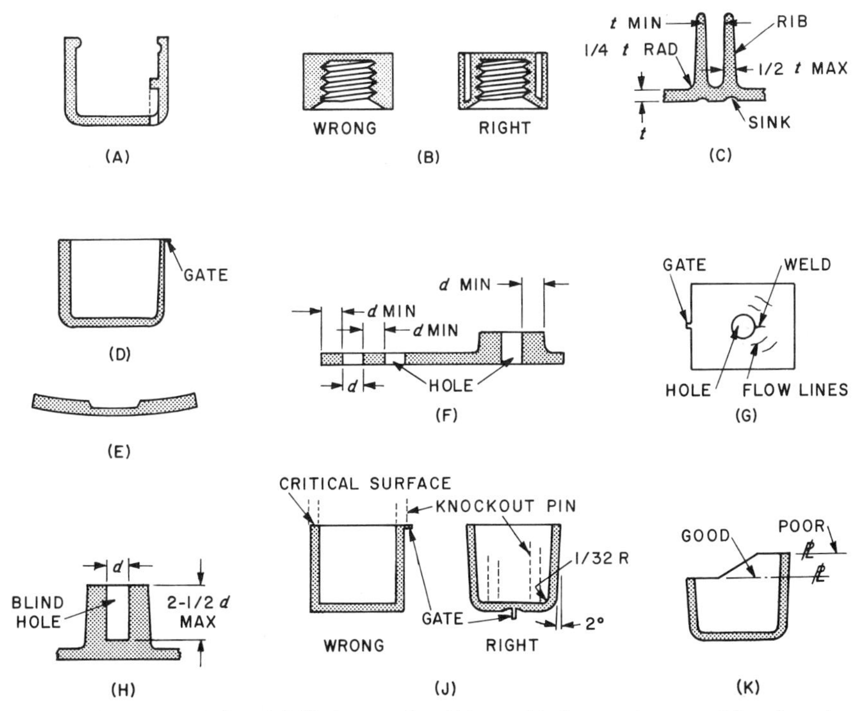

Fig. 10. Design principles. (*A*) Undercuts should be avoided as much as possible; otherwise a complicated mold may be required to release the part from the core of the mold. One trick to avoid this is to let the cavity project up through the part to form the underside of the undercut, as shown in the lower right. (*B*) Uniform thickness should be provided by coring out heavy sections. This will reduce cost and will avoid unsightly sink marks. (*C*) Ribs should not be more than half as thick as the adjoining wall; otherwise shrinkage of this large amount of plastic will cause a sink mark on the opposite side. For the same reason, ribs should not be closer than the wall thickness (*t*) apart. Avoid sharp inside corners; a radius of at least one-fourth the thickness of the wall is preferred. (*D*) The gate should be at the thinnest part of the piece to minimize the resistance to the flow of the plastic. The plastic entering the mold should impinge upon some part of the mold, to avoid laminations in the finished piece. (*E*) Changes in thickness will cause warpage due to differences in shrinkage from one section to another. (*F*) Holes should not be closer together or closer to an edge than the diameter of the hole. The wall of a boss should not be less than the diameter of the hole. (*G*) Holes should not be directly opposite a gate, where they may cause flow lines and a weak weld. (*H*) Unsupported core pins may bend from the injection pressure if they are longer than 2½ times the diameter. (*J*) Sharp corners should be avoided. Draft of at least 2° is preferred for easy removal of the part from the mold. The gate should be recessed if possible, to avoid a degating operation. Knockout pins that may leave marks on a critical surface should be relocated. (*K*) Parting line of the mold should be kept flat whenever possible, to minimize problems of fitting mold parts together.

Try to maintain a uniform thickness throughout the mold, because changes in cross section waste material, cause unequal shrinkage, and tend to cause *sink marks.* The sink marks occur opposite ribs or bosses and are caused by additional shrinkage of the larger mass of material. They can sometimes be camouflaged by decorative treatment. Bosses or other projections should be round so that the mold can be drilled or turned on a lathe. The height of a boss should not be more than twice its diameter. Raised lettering on a plastic piece is much cheaper than depressed lettering because it is easier to cut the letters into the mold than to cut around them.

Draft of 1° or 2° on each side is necessary to ensure ejection of the parts from the mold. The location of the gate is an important decision; it should be in an inconspicuous spot, but it should be placed so that the plastic entering the mold will impinge upon some part of the mold, to avoid laminations which would cause bad welds and surface blemishes. The gate should be at the thinnest part of the piece to minimize any resistance to the flow of the material as it fills up the mold.

Deep and narrow holes cause problems because the projections in the mold to make them are liable to get bent by the pressure of the plastic as it fills the mold. About 2½:1 is the maximum depth-to-diameter ratio for small holes. If variations in wall thickness are necessary, it is preferable to make the change gradually, as with a 45° slope, but if the change must be made abruptly, try to have a small radius in the corners. A radius of one-fourth to one-half the wall thickness is ideal. If the change in thickness is more than 50 percent, there should be ribs and generous fillets or radii. Ribs provide stiffness and minimize warpage. They should not be thicker at the base than one-half the wall thickness to which they are attached. Keep in mind that stiffness varies as the cube of the height. If ribs are spaced close together, they should be at least one wall thickness apart. Holes should not be closer together or closer to an edge than the diameter of the hole. The wall of a boss around a hole should not be less than the diameter of the hole, and should have a 1° taper for easy removal from the mold. To prevent a weak weld line, holes should not be directly opposite a gate.

Shrinkage will depend on the type of plastic used; it will vary from 1/10 to 4 percent. Polystyrene, for example, will shrink about 0.003 in. per in. Shrinkage can be minimized to some extent by increasing the temperature and pressure of the plastic entering the mold, using a longer cycle, and keeping the mold cool. Undercuts can be ejected from the mold without the necessity of cam-actuated molds if the amount of deformation does not exceed about 7 percent and the approach angle is at least 30°. Undercuts up to 0.040 in. can be tolerated in polypropylene, if the temperature at ejection is 200°F or higher. Hinges in polypropylene should be 0.010 to 0.015 in. thick and as short as possible. For very large parts the thickness may have to be slightly more than this.

Special types of gates include jump gates, used if a fast flow of material into the mold is desired, and tab gates, which have the effect of a restricted gate on heavy-sectioned parts, especially where the material cannot impinge on a part of the mold. Multiple gates are used where the flow distances are great. For ejecting parts from the mold, knockout pins are the easiest and simplest devices to use; but if these might distort the piece, it is better to use a blade knockout or a stripper plate which may completely encircle the piece and thus distribute the force all around the edge. Vacuum valves or poppet valves are sometimes used on deep-cored parts, or air ejection might be used to help in getting the parts out of the mold.

Cost of injection-molded pieces can be approximated by weighing a model, or by calculating the weight, and applying the current price of the material. To this is added a factor for machine time, usually equal to the material cost, or more if it is a small or intricate piece. Amortization of the molds also should be included, figured over a 2- or 3-year period. If the mold cost is not known, add about one-fourth of the material cost for a rough figure, and add a like amount for packing and shipping. As an example: a 1-qt container that is injection-molded weighs 4 oz and takes about 4 cents worth of material. Adding another 4 cents for machine time, 1 cent for mold amortization, and 1 cent for packing and shipping brings the total up to 10 cents.

Molds. An important factor in the design of injection-molded plastic packaging components is the specification of the tooling. Although this is frequently left for others to decide, it is well to give some thought to the various aspects of mold making while the package is still under development. If the expected quantity of finished packages is large, the cost of the molds and other equipment is insignificant, and the very best molds will be the cheapest in the long run. (See Fig. 11.) On the other hand, for short runs the cost of the molds can add considerably to the final cost of the parts, and it would be well to weigh the merits of the various kinds of tooling that might be used. With blow molding taken as an extreme example of the range of costs, a set of cavities for one type of machine can cost five times as much as for another, depending on whether the neck finish on the container is molded or blown. Some parts lend themselves to either injection molding or thermoforming, and the choice may be strictly one of economics. The tooling costs are a great deal less, but the piece cost may be higher in the end because of scrap loss and higher raw material prices, with thermoforming.

If the parts are to be molded outside one's own plant, it is well to choose a supplier as early as possible and to work closely with him in the development of the final design. An agreement should be worked out with the supplier in the beginning as to secrecy and patent assignment, ownership of tools, cost of development, and similar details. It is much easier to get such matters settled at the beginning than to try to reach an

Fig. 11. Mold parts. An injection mold is a complicated tool with many parts, as can be seen in this photograph of a disassembled mold. (*Newark Die Co.*)

agreement after considerable time and money have been expended. Following are some of the points that should be settled with the supplier at the outset:

1. Sample tools, estimated cost, ownership, removal surcharge (usually 30 percent), estimated life, payment terms
2. Sample parts, estimated cost, approvals in writing, delivery dates
3. Patent assignment, assignment costs and legal fees, infringement protection, secrecy agreement
4. Production molds, number of cavities, family type molds, estimated cost, payment terms, ownership, removal surcharge, estimated life, maintenance and repairs, cancellation terms, adaptability to other (competitor's) molding machines, exclusive use, liability and casualty insurance, tooling prints and part prints, replacement costs
5. Quality control, warpage, gates, dirt, tolerances, AQL (see Sec. 19, "Quality Control"), sorting costs, limit samples, color tolerances, test methods, gauge costs, time limit for claims, contingent losses (assembly costs)
6. Packing and shipping, overruns and underruns

Highly polished and chrome-plated molds will produce parts with a high gloss, but they may cause problems in ejecting from the mold because of the difficulty in breaking the vacuum. Vapor honing or sand blasting will provide a slight surface roughness to minimize this problem. Molds that are used with polyvinyl chloride have to be plated with chromium, nickel, or in some cases gold in order to resist the corro-

TABLE 10 Comparative Mold Costs for a Typical Small Part

Type of mold	Mold cost	Piece cost
Injection mold. . . .	$ 8,500	$0.020
Compression mold .	13,000	0.013
Transfer mold. . . .	14,000	0.016
Plunger mold	21,000	0.022

sive action of the degradation products of PVC. Venting of the mold consists of grooves about ½ in. wide and 0.0015 in. deep at the parting line, to allow the escape of air. They can also be put in by grinding flats on the knockout pins or by grooving the inserts in the mold.

Molds are usually made of tool steel, but sometimes experimental molds are made of beryllium-copper. These are cheaper to make, are easier to alter or repair with a heliarc welding unit, and conduct heat more rapidly and so can be cycled faster. Being softer than steel, beryllium-copper requires a heavier section around the side walls for support against the molding pressures. Because it does not hold up well under abrasive conditions, it cannot be used for compression molds. Meehanite and Kirksite are other materials that can be used for special types of molds. (See Table 10.)

Various techniques are used in making the molds. The simplest is direct machining of a solid block of metal. Where a number of shallow cavities are needed, a *hob,* which is a replica of the part in hardened steel, is forced into the die block under tremendous pressure. It takes as much as an hour for the hob to sink into the die in this way, but every cavity made with the same hob will be precisely the same. Electrical discharge machining (EDM) is used for at least a part of the work in most mold shops. It is useful in matching the mating faces of the mold, to reduce flash, especially with a complicated parting line. By bringing the faces together in the EDM machine, a perfect fit can be produced. This system is used also for roughing out the cavities, and fairly good accuracy is obtained with very little finishing and polishing required. A few mold shops have tape-controlled jig borers, which reduce cost on standardized parts and highly repetitive tooling.

Molding Machines. The earliest machines for injection molding were straight *plunger* types in which a piston moved back and forth in a cylinder to force the hot plastic into the mold. (See Fig. 12.) An improvement on this type in the early 1950s consisted of a two-stage plunger system. One cylinder was called the *preplasticizer,* and since it did not have to inject directly into the cylinder, it could be made with a larger

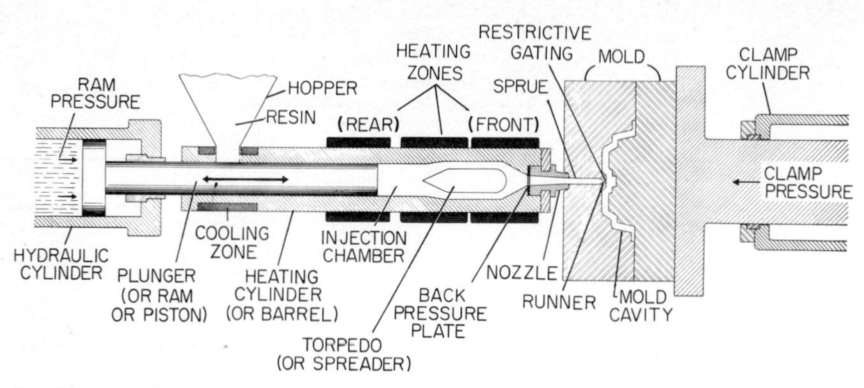

Fig. 12. Injection molding. The simplest type of molding machine is the plunger type. Resin from the hopper feeds into the barrel and is carried forward by the plunger. Heater bands melt the plastic, which is forced around the torpedo and through a screen into the mold. (*U.S.I. Chemicals*)

heat transfer surface. The injection cylinder, on the other hand, did not have to heat the plastic, and so it could be made with a larger diameter and stroke.

It was known that extrusion machines provided a more homogeneous melt than injection molding machines, and so it was logical to adapt the screw feeding principle to this purpose. The first screw machine, introduced in 1956, was the *piggyback* type. In this machine the screw section, which angled up from the mold area, heated the plastic and forced it into the horizontal plunger directly below. At the proper moment the plunger then moved forward to fill the mold. The newer "in line" machines use the screw to plasticize the material and also to inject it into the mold. They do this by moving the screw back away from the mold as it rotates so that enough material is pushed ahead for a shot. Then when the mold is ready to receive the plastic, the screw moves forward like a plunger to fill the mold. These are called *reciprocating screw* machines. (See Fig. 9.) For rigid PVC, which is very heat-sensitive, a reciprocating screw is almost the only means of handling it.

The advantages of the screw are better mixing, which provides for more uniform coloring, less thermal degradation, reduced cycle times, better appearance of the moldings, less internal stress in the moldings, and the possibility of using materials with higher melt viscosities. In a plunger machine the material moves faster in the center than near the cylinder wall, where the temperature is highest. The screw machine avoids this unevenness by continuously sweeping the walls, and it does not build up the temperature in the outer layers. Also, a substantial part of the heat for melting the plastic is generated in the polymer itself by conversion of the mechanical energy from the screw through internal

Fig. 13. Injection molding machine. Power for the reciprocating screw feed is supplied by the motor mounted on the upper right. The hopper is above the screw itself, and heater bands can be seen to the left of the hopper. The mold platen is visible in the open position just below the nameplate. There is no mold in the machine. The platen is actuated by the hydraulic cylinder behind it, to open and close the mold. (*National Automatic Tool Co.*)

friction of the molecules of the plastic as it is kneaded, and only a small amount of heat must be added by convection from the walls of the barrel. The apparent heat conductivity of a screw machine is nearly 100 times that of a plunger machine. The cost of an automatic 2-oz molding machine is around $20,000, and the larger machines are proportionately more, up to $300,000 for the biggest. (See Fig. 13.)

COMPRESSION MOLDING

In 1907 Dr. Leo Hendrik Baekeland mixed phenol-formaldehyde resin with wood flour and put it into a mold that was used for making rubber parts. With heat and pressure he was able to produce the first organic plastic pieces. The same basic method is still being used today for most of the thermosetting parts that are made. In compression molding the molds are heated to about 300°F, and granular material is put into the mold with a slight excess. The excess is squeezed out as the two parts of the mold come together, and must be trimmed off in a subsequent operation. (See Figs. 14 and 15.) Pressures of about 2,000 psi are applied, and the material is allowed to *cure* for a minute or so, depending on the size of the piece, and it is then ejected from the mold. The flash is removed by tumbling in a revolving barrel type of machine made for this purpose.

The materials used in compression molding are called *thermosets,* and unlike the thermoplastic compounds, they will not soften again when heated, once they are cured. Phenol and formaldehyde, or urea and

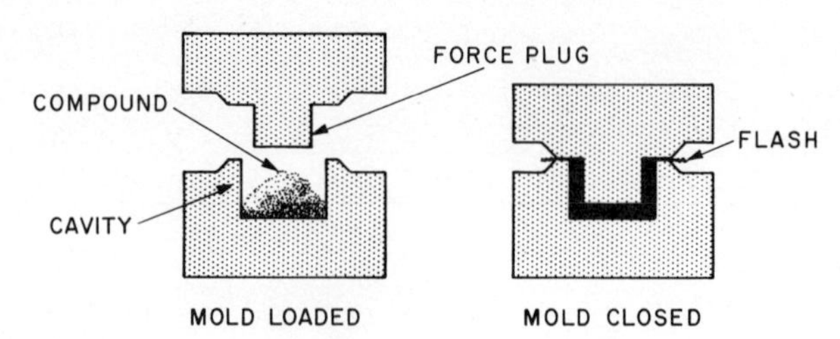

Fig. 14. Compression molding. A slight excess of molding powder is put into a heated mold. When the mold closes under high pressure, the excess plastic squeezes out as flash, and must be removed in a separate operation. The plastic cures in a minute or so, from the heat of the mold, after which the mold opens and the part is ejected.

formaldehyde, are combined to form resins which will soften when first heated, and can be pressed into a mold of the desired shape. Continued heating at a slightly higher temperature then causes the material to "set up" into a very hard and durable piece. Usually *fillers* are mixed with the resin to get better processing qualities, greater toughness, or lower cost. The fillers can be wood flour, asbestos, graphite, chopped canvas, mica, sisal, paper, synthetic fibers, glass, or similar materials. The colors

Fig. 15. Compression molding machine. This is a 30-station rotary-type press which makes small parts such as screw caps directly from powder, without preforms, at a rate of 30 parts per minute. (*New England Butt Co.*)

of phenolics are limited to the dark shades; ureas are available in any opaque color, but they are more expensive than phenolics. Some of the colors used in ureas, in the yellow and orange family, have a tendency to fade in sunlight; and pink and orchid shades are particularly bad in this respect.

The primary use of thermosets in packaging is for closures. They are more expensive than metal closures, but their appearance is better suited to pharmaceutical and cosmetic packaging. (See Sec. 9, "Closures, Applicators, Fasteners, and Adhesives," for more information on this subject.) In addition to compression molding, it is possible to use thermosetting materials in injection molds that are made for the purpose. This is called *transfer molding.* The mold costs are higher than for compression molding, and the molding machines are more expensive, but the piece price is significantly lower. (See Table 11.) The molds are similar to the injection molds previously described, with the same principles of design. The basic difference is that the mold is heated rather than cooled. A reciprocating screw machine is used, and the material is heated to about 200°F before it is injected into the mold. After curing

TABLE 11 Approximate Manufacturing Costs of Thermoset Screw Caps

Size of closure	Weight, g	Cost each*
15-400	1.4	$0.0011
20-400	2.5	0.0015
24-400	3.2	0.0018
28-400	4.2	0.0021
33-400	5.5	0.0028
38-400	6.9	0.0038
48-400	10.1	0.0055
63-400	16.3	0.0076
89-400	34.8	0.0141
120-400	52.7	0.0267

* The costs are made up of the following components: material 30 percent, machine time 50 percent, mold amortization 10 percent, packing and shipping 10 percent. If the liner were figured in, we would add another 10 percent to the above costs. As an example, a closure that takes 10.1 g such as the 48-400 size above, with phenolic at 21 cents per pound in an eight-cavity mold and a 30-sec cycle, the costs would be material 0.45 cent, heat and power 0.03 cent, labor 0.025 cent, and machine amortization 0.05 cent. Selling costs are not included.

for 15 to 30 sec, the mold opens and the parts are ejected by knockout pins, along with the runners and sprue. The parts are hot when they are ejected. The mold then closes and repeats the cycle. Costs are less than for compression molding because preforming, preheating, and finishing expenses are eliminated, and the operation is fully automatic.

Molded phenolic parts will continue to shrink for several hours after they are removed from the mold. Actually the shrinkage even continues in a very slight amount for some months afterward, but this long-term shrinkage is significant only for very closely fitting parts. Where this must be controlled, it is possible to "postbake" the pieces and virtually stop all further dimensional changes. Eight hours at 350°F will provide a shrinkage of about 0.005 in. per in. in a typical phenolic item, which is sufficient for most purposes.

BLOW MOLDING

The history of blow molding goes back before World War I, when baby rattles, dolls, and ping-pong balls were made of celluloid by this process. In fact, if we go back further, we find the ancient Egyptians blow-molding articles of amber. For practical purposes, however, the industry got its start when Enoch T. Ferngren and William Kopitke combined the extrusion process with blow molding and sold their idea to the Hartford Empire Company in 1937. The Plax Corporation was set up under James Bailey to develop the process. During World War II polyethylene was introduced and proved to be an ideal material for this technique. Other companies soon followed Plax, designing their own machines and molding "squeeze" bottles, but their machines were kept secret, and it was not until 1958 that such machines became commercially available.

One of the earliest retail packages to use the new squeeze bottle principle was Stopette deodorant, introduced around 1947, and it was an instant success. In 1957 the Ziegler low-pressure process for making high-density polyethylene caught the attention of the bottle manufacturers, and by 1959 five companies were making detergent bottles of this material. Today there are more than 350 manufacturers making nearly 3 billion blown plastic containers every year. (See Table 12.)

TABLE 12 Plastic Bottle Production per Year in the United States

Household chemicals	1,800,000,000
Toiletries and cosmetics	700,000,000
Medicinals	290,000,000
Foods and beverages	160,000,000
Industrial chemicals	120,000,000
Automotive and marine	28,000,000

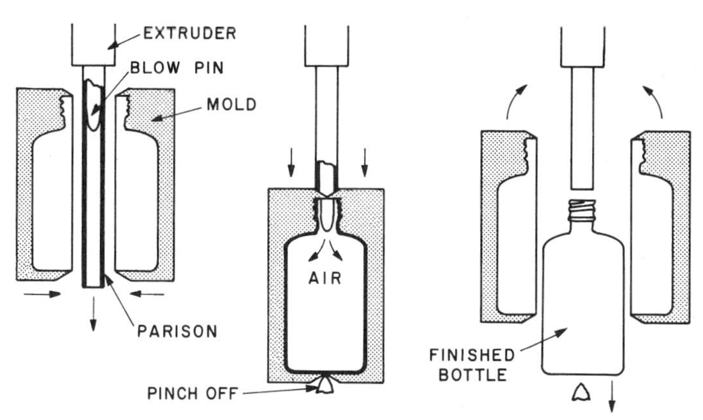

Fig. 16. Blow molding. A hollow tube of plastic is extruded down between the two halves of the mold. The mold closes and moves down with the parison as it continues to extrude. Air is injected to stretch the warm soft plastic to the shape of the mold. The mold opens to release the piece and moves up to meet the parison again.

The method most widely used is called *extrusion blow molding*. It consists of extruding a hollow tube, called a *parison*, downward between two halves of a mold. As the mold closes, it pinches off the bottom, but leaves the top open for the injection of air. The warm soft plastic stretches out under this pressure and takes the shape of the mold. (See Figs. 16 and 17.) In some processes a small tail remains attached to the bottom of the bottle at the pinch-off and must be removed in a secondary operation, and the sealing surface must be reamed. One of the disadvantages of this method is that chips sometimes get into the bottles and are difficult to remove. A variation of this method called *injection blow molding* consists of injection-molding the parison instead of extruding it. This permits greater accuracy in the neck finish, which is important when plugs or other fitments are to be used. The mold costs are much higher, and this process is used only for small containers, although some have been made as large as quart size experimentally.

It is possible to have more than one extruder feeding the die, and containers have been made with as many as five differently colored vertical bands in the same bottle. Another form of blow molding uses two extruded sheets which are trapped by the mold as it closes; at the same time the mold seals the edges together and pinches off the excess, which can then be reground and put back into the process. This method is most often used for very large containers. The appearance of small containers made by this method is not very attractive because of a fin type of seal which goes completely around the bottle. Screw-type closures are not practical, and any type of reclosure is a problem. It is probably the lowest-cost method, however, since it can operate at nearly three times the rate of regular blow molding and can use thinner walls for the containers.

Fig. 17. Blow molding machine. A four-cavity mold is shown in the open position, ready to receive the four parisons in the foreground, which are moving down from the extrusion die. The sharp edge which pinches off the bottom of the parison can be seen clearly at the bottom of each mold cavity. (*C. Tennant, Sons & Co.*)

The *cold tube* process was introduced in England in 1963 and is beginning to be used in this country. The tubes are extruded, cut to length, and stored for future use. Later in a subsequent operation they are reheated and blown to shape. The tubes can be made at temperatures lower than those used for blow molding, which is desirable with heat-sensitive materials such as polyvinyl chloride. It does not work well, however, with the olefins, which do not hold their shape when they are reheated.

Design. One characteristic that must be kept constantly in mind by the designer is shrinkage. Since the material is being stretched in a semi-molten state, stresses will be built into the piece, or in common parlance the "plastic memory" makes it want to return to the shape of the original parison. Most of the shrinkage takes place as the container cools after molding; for example, polyethylene shrinks about 3 percent, but days later a small amount of deformation will still be taking place. This is not likely to be a problem unless there are close-fitting parts that must move freely.

The easiest shape to blow is a cylinder. This yields a uniform wall thickness and consequently a very economical container. Filling equipment can be run twice as fast for cylinder shapes, and the equipment costs only half as much. However, if the product is apt to cause *bottle collapse,* the round shape will show this up more quickly. Detergents, which absorb oxygen from the headspace, exhibit this phenomenon; it starts to occur in 3 or 4 days, becoming severe usually within 2 weeks and developing a vacuum equivalent to about 5 in. of mercury. For this type of product it becomes necessary to use an oval bottle with flat panels to conceal the distortion.

Ovals are the next best shape, and rectangular or odd shapes are the least desirable. As the parison stretches out to fill the mold, it becomes thinner and the farthest part therefore has the thinnest wall. There is also a certain amount of *drawdown,* which results from the stretching of the parison by its own weight and causes the top portion to be thinner than the bottom. It is possible to compensate for these conditions by varying the opening in the extrusion die so that it is thicker where it has to stretch farthest. The wall can be increased all around to take care of the drawdown by moving a tapered mandrel upward in the die as the parison moves down. (See Fig. 18.)

Stress cracking sometimes causes failure of blown containers. This can be reduced in part by avoiding sharp corners and sudden changes in thickness. A generous radius at the bottom edge and at the shoulder, for example, will prevent thin spots which are prone to stress-crack, especially where they are adjacent to the heavy section of the pinch-off. Even more important is the right choice of material to suit the product to be packed.

Stability on the production line may be a problem because of the light weight of plastic bottles, unless a broad base is made part of the design. Sometimes vacuum can be used to control the bottles on a high-speed filling line. The size of the finish and the neck opening should be as large as possible to get the maximum efficiency on the filling line.

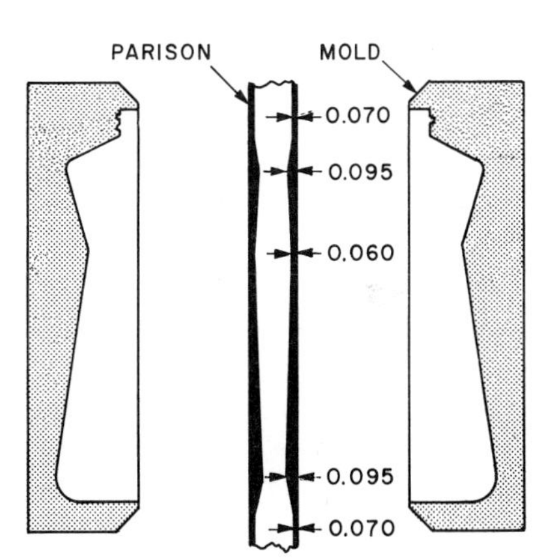

Fig. 18. Programmed parison. The wall thickness can be made heavier where it has to stretch farther to fill the mold, so that the final wall thickness will be more uniform. The neck also can be made heavier to withstand capping stresses.

Application of screw caps and stacking in the warehouse may put a strain on the shoulder section. A conical shape gives the greatest strength under such conditions, and flat shapes or horizontal ribs are the poorest. A gripping surface near the shoulder, for the capping machine, will resist the torquing forces better than one near the bottom. Screw threads of the buttress type are best, since they are not so likely to jump threads when they are torqued down as the conventional threads that are used on glass containers. Also the transfer bead below the thread can usually be omitted because it does not serve the same function in a plastic bottle as in a glass bottle. (See Table 13.)

The base of a container should be indented, or, to borrow a term from the glass industry, it should have a *push-up*. The flat area, however, should be at least 1/4 in. wide to avoid catching in gaps in conveyors or dead plates. *Control areas* should be provided for handling in pocketed turrets by having flat areas near the bottom and top of the container. If it is necessary to transfer from one turret to another, it may be feasible to provide two flat surfaces close together. The bottom control area should not be too close to the bottom; if it is, the turret plate or star wheel will not clear any closures that might fall onto the working surface. Clearance is required also for cleaning up spilled liquids. Cylindrical and rectangular shapes are the easiest to handle on conveyors. Ovals, triangles, and diamonds tend to jam between the guide rails. (See Figs. 10 and 11 in Sec. 6, "Glassware," page 6-18.) Rectangular shapes should have rounded corners so that they do not catch on the edges of rails and star wheels, and also to allow mechanical fingers to enter between them. Tapered containers are less desirable than straight-sided containers, for they can be troublesome if they are lifted under a filling head at the same time that they enter a star transfer wheel.

Threaded finishes should have double or triple lead threads if possible, and the number of turns for sealing should be kept to a minimum. A generous straight section before the start of a thread, known as the *S* (for start) dimension, helps to avoid cocked caps. If the thread is depressed slightly at the parting line, it will keep any flash from interfering with the fit of the closure. Plugs and fitments should be well tapered for easy assembly, but they should not taper all the way up; if they do, they may come loose with vibration. A slight undercut will help to keep them in place.

Molds. The two most important factors to be considered in making a mold for blow molding are heat transfer properties and resistance to wear. Copper with 1.7 percent beryllium has these qualities, and it can be sand-cast or pressure-cast to make a very serviceable mold. If it is to be machined rather than cast, an alloy with 2 percent beryllium is preferred. Minor repairs can be made by peening or drifting the metal

TABLE 13 Standard Finish Dimensions for Plastic Containers

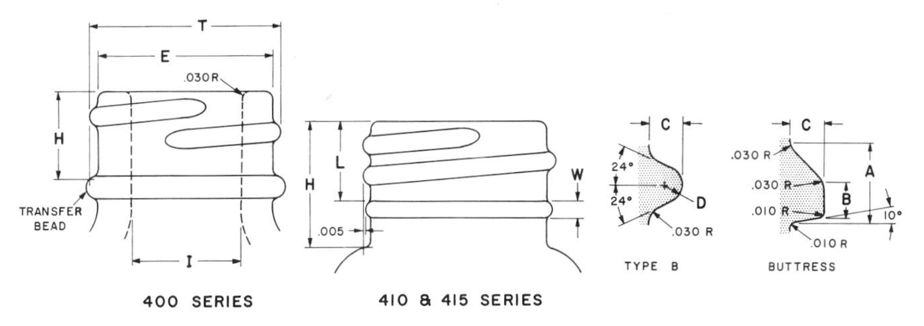

Type	Series	Millimeter size	T min.–max.	E min.–max.	H min.–max.	I min.
Shallow continuous thread	400	18	0.688–0.704	0.604–0.620	0.356–0.386	0.325
		20	0.767–0.783	0.683–0.699	0.356–0.386	0.404
		22	0.846–0.862	0.762–0.778	0.356–0.386	0.483
		24	0.924–0.940	0.840–0.856	0.385–0.415	0.516
		28	1.068–1.088	0.974–0.994	0.385–0.415	0.614
		30	1.107–1.127	1.013–1.033	0.388–0.418	0.653
		33	1.241–1.265	1.147–1.171	0.388–0.418	0.791
		38	1.457–1.476	1.358–1.382	0.388–0.418	0.987
		43	1.624–1.654	1.530–1.560	0.388–0.418	1.165
		48	1.840–1.870	1.746–1.776	0.388–0.418	1.381
		53	2.032–2.067	1.938–1.973	0.393–0.423	1.578
		58	2.189–2.224	2.095–2.130	0.393–0.423	1.735
		63	2.426–2.461	2.332–2.367	0.393–0.423	1.972
		66	2.544–2.579	2.450–2.485	0.393–0.423	2.090
		70	2.701–2.736	2.607–2.642	0.393-0.423	2.247
		83	3.233–3.268	3.113–3.148	0.472–0.502	2.753
		89	3.476–3.511	3.356–3.391	0.520–0.550	2.918
		120	4.689–4.724	4.569–4.604	0.670–0.700	4.131
Medium continuous thread	410	18	0.688–0.704	0.604–0.620	0.508–0.538	0.325
		20	0.767–0.783	0.683–0.699	0.539–0.569	0.404
		22	0.846–0.862	0.762–0.778	0.570–0.600	0.483
		24	0.924–0.940	0.840–0.856	0.631–0.661	0.516
		28	1.068–1.088	0.974–0.994	0.693–0.723	0.614
Tall continuous thread	415	13	0.502–0.514	0.442–0.454	0.437–0.467	0.218
		15	0.569–0.581	0.509–0.521	0.542–0.572	0.258
		18	0.688–0.704	0.604–0.620	0.602–0.632	0.325
		20	0.767–0.783	0.683–0.699	0.727–0.757	0.404
		22	0.846–0.862	0.762–0.778	0.822–0.852	0.483
		24	0.924–0.940	0.840–0.856	0.942–0.972	0.516
		28	1.068–1.088	0.974–0.994	1.067–1.097	0.614

to fill up nicks and dents. Major repairs are made with a gas torch. The cost of this material is around $1.50 a pound, which is expensive, but for high-production tools it is not a significant factor. Aluminum also makes a good mold, but it will not last as long as beryllium-copper. Zinc is easy to cast, it machines well, and it is cheap. It does not carry off the heat as well as the other two metals, but the resistance to wear is fairly good. Cast iron is the strongest and cheapest material, since it can be made with very thin walls and has a long service life. Maintenance can be a problem, however, because of the tendency of cast iron to rust. The average cost for a mold is about $500 for extrusion blow molding polyethylene, going up to $5,000 or more for injection blow molding polyvinyl chloride.

The neck and base portions of a mold are generally made as separate pieces so that they can be machined and finished more easily. Molds for blow molding are not usually polished but made rough so that air will escape more easily between the plastic and the mold surface. The two halves of the mold must line up perfectly, as any misalignment will result in thinning along the parting line. This is caused by the plastic not being in close contact with the mold at this point and staying hot. The hoop stresses resulting from the internal air pressure, as the part cools and shrinks, will stretch the warmest part.

In the area where the lands pinch the parison together, they must be relatively narrow, to cut and squeeze out the excess material, and yet wide enough to effect a good plastic-to-plastic seal. In other parts of the mold the lands can be any width, and are usually much wider. Proper cooling of the mold is necessary to get the shortest possible cycle time and also to get a uniform surface appearance. Water channels in each mold half can be drilled out, or in a cast mold they can be cored. Cast aluminum is often porous, making it difficult to provide cooling channels. Clamping pressures are relatively light, in comparison with injection or compression molds, since they need only to resist pressures of around 5 psi in the cavities. (See Fig. 19.)

Machines. Extruders of 2- to 4-in. size are generally used in blow molding machines; these figures refer to the inside diameter of the barrel. The length-to-diameter ratio is preferably around 24:1, and the screw is a constant pitch with varying flight depth. The end where the pellets come in from the hopper is where the flight is deepest. As the pellets move along and start to soften and melt, they take up less space, and so the flight is made gradually shallower. (See Fig. 20.) The difference between the channel depths at the feed end and at the delivery end, is called the compression ratio of the screw. This will be in the range of 1½:1 up to 4:1, depending upon the plastic used. In the third portion, called the metering section, which is about 25 percent of the total length, the flight maintains a constant

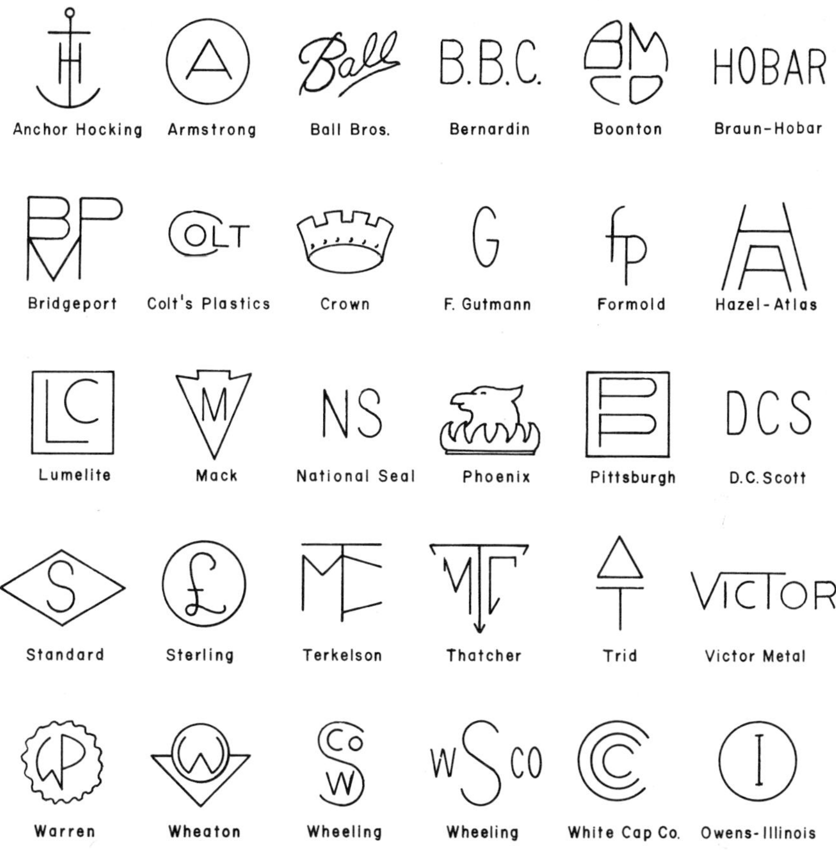

Fig. 19. Molder's marks. A manufacturer's identifying symbol can sometimes be found in a cap under the liner, or on the bottom of a plastic bottle. The chart may be helpful in locating the source of a molded piece from the mark on a specimen.

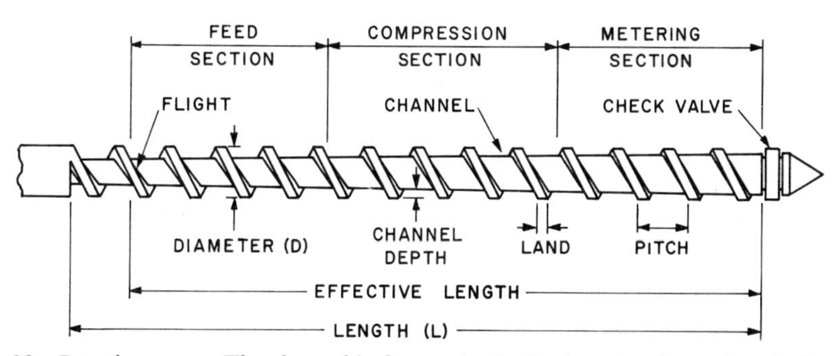

Fig. 20. Extruder screw. The channel is deepest in the feed section, becoming shallower where the plastic softens and runs together in the compression section, and still shallower in the metering section, where the maximum heat is generated by mastication of the plastic under pressure.

depth. The screw turns at about 150 rpm, and most of the heat for melting the plastic comes from the friction of mixing with the screw. A small amount of heat is added by convection from the barrel. Pressures of about 2,000 psi and temperatures starting at 300°F at the feeding end and going up to 340°F in the metering section are common. (See Fig. 21.)

The ratio of parison size to finished bottle size, or the *blow ratio* as it is termed, is best kept in the range of 2:1 to 3:1. The size of the parison for this purpose is measured after the *swell* has taken place. The swell is usually about 2 to 3 percent. With a larger ratio than this, there is risk of uneven wall thickness, but for practical purposes it is often necessary to go well beyond the hot melt elastic limit to fill out the mold. Most often the parison is made the same size as the neck finish;

Fig. 21. Blow molding machine. Plastic pellets are put into the hopper on top, where they feed down into the barrel of the extruder, and are carried by the screw through the heating process to the extrusion die shown at the top right. The parison is extruded down between the mold platens where it is blown to shape. No mold is shown in the illustration, but the two halves would be bolted to the platens, which are the flat plates with holes. (*Fischer Maschinenfabrik*)

this is called "inside the neck" blowing, in which there is no flash or pinch-off at the threads.

Sometimes there is a weld line where the extrudate passes around the mandrel and reknits on the other side as it changes from horizontal to vertical flow downward through the die. A long land on the mandrel, especially where the blow air passes through the mandrel, will help to avoid this condition. Blow molding machines cost around $5,000 for the simplest type and up to $20,000 for the larger, more sophisticated machines.

Processes. There are several techniques for blow molding, but for purposes of illustration we will describe the rising mold process. The soft plastic from the extruder comes through a screen and enters the crosshead, which directs it downward over a *torpedo,* and then out through the die to form the parison. The two halves of the die close over the parison and move downward at the same rate as the parison is extruded, while air is being injected into the mold. When it reaches the bottom of its stroke, the mold opens and the bottle is ejected by compressed air. By this time the parison has extruded far enough to make another bottle; the mold rises to the die and again closes around the parison, to repeat the blowing operation as it moves downward. (See Fig. 16.)

Materials. The most widely used material for bottles is *polyethylene.* The low-density variety is used if some transparency is desired or if a soft, flexible container is necessary. High-density polyethylene costs a little more, but because of its stiffness and low permeation rate, it can be made into bottles with thinner walls, and therefore the final cost is less. (See Tables 14 and 15.) Low-density materials are classified as those in

TABLE 14 Approximate Costs of Plastics for Blow Molding

Plastic	Cost per lb
Polystyrene	$0.15
Polyethylene, low-density	0.16
Polyethylene, high-density	0.18
Polypropylene	0.19
Polyvinyl chloride	0.21
Acrylic	0.46
Cellulose propionate	0.62
Acetal	0.65
Phenoxy	0.75
Nylon	0.86
Polycarbonate	0.90

TABLE 15 Costs of Bottles Made from Various Materials*

Capacity, oz	Low-density polyethylene	Polyvinyl chloride	Glass	High-density polyethylene
2	$ 29.78	$ 38.15	$22.55	$25.10
4	41.20	43.38	26.18	28.65
6	50.50	47.16	31.75	31.62
8	62.00	52.12	37.42	33.00
12	71.08	61.80	43.65	42.62
16	93.65	67.32	51.08	46.92
24	121.45	83.15	66.60	57.50
32	149.25	101.28	84.63	67.95

* The costs for plastic bottles can be broken down as 50 percent for material, 40 percent for machine time, 9 percent for packing and shipping, and 1 percent for mold amortization. Decoration will add about $7.00 per color per side. These costs are typical, but 64-oz milk bottles have been reported as low as $37.50 per thousand.

the range of 0.910 to 0.925 g per cc; high-density resins are those in the 0.940 to 0.965 g per cc range. The highest-density polyethylene is subject to stress cracking and is sometimes called *bleach grade* resin; a 0.950 material, which is more resistant to stress cracking, is called *detergent grade* resin. (See Tables 16 and 17.)

Polyethylene has a low moisture-vapor transmission (MVT) rate and so is a good material for containers to hold aqueous products. It is not recommended for oily products, since they may pass through the walls and make the outside sticky to the touch. Polyethylene is a poor barrier for gases and should not be used with products that are sensitive to oxygen. The permeability of oxygen through polyethylene is about 150 cc per 100 sq in. per 0.001 in. thickness per 24 hr. Polyethylene also has a strong tendency to develop a static charge that attracts dust, and this can be bothersome on a retail shelf. Some additives will be partly effec-

TABLE 16 Effect of Density on the Physical Properties of Polyethylene

Physical property	Density		
	0.940	0.955	0.960
Melt index	0.2	0.2	0.2
Stress-crack resistance	1,000	250	60
Tensile strength, psi .	2,100	3,800	4,400
Elongation, percent .	100	70	30

TABLE 17 Effect of Melt Index on the Physical Properties of Polyethylene

Physical property	Melt index				
	0.2	0.9	1.5	3.5	5.0
Density	0.960	0.960	0.960	0.960	0.960
Stress-crack resistance	60	14	10	2	1
Tensile strength, psi	4,400	4,100	4,100	4,100	4,100
Elongation, percent	30	25	20	15	12
Impact strength, ft-lb	14.0	4.0	2.5	1.5	1.2

tive against this condition, but a good answer to this problem has yet to be found. Usually the only additive used in the polyethylene for bottles is a small amount of dibutylparacresol as a mold release agent. Any opaque color can be used for bottle resins, but the metallic colors such as gold and silver do not have very good brilliance. Polyethylene cannot be made crystal clear, but in its natural form it is translucent.

The next most important material for blow molding is *polyvinyl chloride.* It has good impact strength, is relatively inexpensive (base resin is being sold for 8 cents per pound in some areas), and is available perfectly clear and colorless. PVC is compatible with oils and most of the organic solvents. It is not so good as polyethylene for retaining aqueous formulas, and the water loss can range from 2 to 20 percent per year. Polyvinyl chloride is difficult to handle in the molding machine because of its tendency to degrade when heated. For this reason extreme cleanliness must be observed, and the equipment must be designed with no pockets or crevices where the plastic can hang up. Between the critical degradation temperature of 460°F and the operating temperature of 400°F there is not much margin for fluctuations. One important disadvantage of PVC is that it scratches easily; this defect is true of most plastics, but the clarity of PVC makes it more apparent. It is recommended that vibration tests and drop tests be made on bottles after they have been standing for at least 24 hr with the actual product in them. (See Sec. 18, "Test Methods.") Also, because of an increasing brittleness at low temperatures, it is suggested that drop tests be made at a temperature of 0°F.

It took the industry a long time to learn to work with this material, but it is now possible to buy blow molding equipment specially designed for PVC, and good-quality bottles can be produced as readily as with polyethylene. The so-called rigid PVC material contains no plasticizer, or at the most it would not be over 5 percent and usually of the permanent type of plasticizer. (See "Plasticizers," page 8-41.) There may also be some acrylic polymers added as a processing aid, with some stabilizers like a tin mercaptide or one of the barium-cadmium compounds, and possibly an impact modifier such as a nitrile rubber or a chlorinated polyethylene. Internal and external lubricants are sometimes used. The internal lubricants decrease the friction between the molecules and thereby lower the melt viscosity. The external lubricants migrate outward and form a film on the outer surface. An excess will prevent adhesion of the plastic to the walls of the barrel and interfere with proper mixing by the screw. For products that come under FDA regulations, the choice of a stabilizer becomes a problem. The heavy metal compounds have the best processing characteristics and yield the best-looking containers, but most of them are not sanctioned for use with foods or internal medication. It then becomes necessary to resort to zinc or calcium soaps, which have the approval of FDA, but these tend to produce off-color bottles lacking in clarity and brilliance. (See Table 18.)

TABLE 18 Typical Additives Used in Polyvinyl Chloride

Plasticizers (should not exceed 35 percent):
Di-2-ethylhexyl phthalate
Dibutyl sebacate
Tri-*n*-butyl acetyl citrate
Diphenyl 2-ethylhexyl phosphate
Stabilizers (should not exceed 1 percent):
Di-*n*-octyl-tin-dilaurate
Phenylindole (high cost, poor light stability)
Aminocrotonic acid ester
Zinc stearate
Lubricants (should not exceed 4 percent):
Amides and esters of high fatty acids
Higher fatty alcohols (no haze)
Organopolysiloxanes
Polyethylene

The advantages of polyvinyl chloride over polyethylene are in the area of gas transmission; the oxygen rate, for example, is only about one-fifteenth as much. This permits PVC to be used for hair-waving formulas and shampoo products that are vulnerable to oxidation. The loss of carbon dioxide from beer in PVC bottles with a 60-mil wall is about 0.5 volume per month at 100°F. The oil resistance is also good, which makes this material a good choice for packaging household cleaning materials containing pine oil, and even cigarette lighter fluids are satisfactory in these containers. (See Table 19.) The rate of

TABLE 19 Compatibility of Polyvinyl Chloride

Compatible	Incompatible
Alcohols	Aldehydes
Aliphatic amides	Aromatic ethers
Aliphatic ethers	Aromatic hydrocarbons
Aliphatic hydrocarbons	Esters
Inorganic liquids	Halogenated hydrocarbons
Organic acids	Ketones
	Nitrogen compounds
	Phosphorus compounds
	Sulfur compounds

shrinkage on cooling is very low, but PVC has a lot of elastic memory and it will snap back if it is not thoroughly cooled before it is taken out of the mold.

A very useful plastic for blow molding is *polystyrene*. It is odorless, tasteless, and rigid and has good dimensional stability. The outstanding quality is its clarity and gloss, which give it a brilliance equal to glass, and it has the lowest cost of all the plastics used for blow molding. One important disadvantage is its brittleness. With good design and proper protection in transit, however, it can be used successfully, as attested by several packages now on the market The brittle nature of polystyrene precludes its use in the extrusion type of blow molding equipment. Instead, the parison is injection-molded, and in a subsequent stage air is introduced to blow out the body to fill the mold. Thus, there is no tailpiece or other trim from pinching off the parison, as there would be in extrusion blow molding. The neck portion can be held to close tolerances because it is not altered in the blowing stage.

Polystyrene is not suitable for packaging products with a large amount of perfume or flavoring oils, as the esters and ketones in the essential oils will dissolve the polystyrene. It is a fairly good barrier for moisture, and weight losses of about 1 percent per year can be expected with aqueous products. Styrene has a low gas transmission rate, and so it should be suitable for oxygen-sensitive materials. As with any transparent plastic, scratches and other defects are very noticeable, and shipping tests are strongly recommended before adopting this type of package. Polystyrene is generally used in pure form, although a small amount of acrylic will improve its chemical resistance for certain applications. Some formulas have FDA approval for internal medicinals and food products.

Although they have not been used to any great extent, *cellulose acetate* and *cellulose propionate* offer some very desirable properties for certain applications. They have good transparency and gloss, excellent impact strength, and good dimensional stability. Cellulose propionate can be formulated to include only acceptable materials if it is to be used with medicines and foodstuffs. Products containing strong acids and alkalies are not recommended for these containers, although dilute solutions should be compatible. Propionate is better than acetate in this respect. Organic solvents sometimes attack cellulosics, and they should be tested thoroughly before being used in these containers. Cellulosics must be thoroughly dry before processing, or trapped moisture will cause blemishes at forming temperatures. It is also necessary to control the temperature of the melt closely, as the cellulosics exhibit large changes in melt viscosity with small changes in temperature. Overheating results in a "runaway" parison that cannot be blow-molded.

Some bottles have been blow-molded of *polypropylene,* but their size has been limited by the melt characteristics of this material. Very large parisons tend to sag under their own weight and cause uneven wall thickness. Resins of higher molecular weight provide a better melt flow rate for this purpose than the low-weight material. Large flat surfaces should be avoided because warpage is very hard to control. The impact strength is fairly good at room temperature, but it falls off badly at sub-zero temperatures. This can be helped to some extent with blends or copolymers of polyethylene with polypropylene. The copolymers have less clarity than the pure polypropylene, and they are lower in stiffness and heat resistance. Polypropylene has more resistance to oil, grease, and heat than polyethylene, and it is less likely to stress-crack with products that promote this type of failure. An improved method of blow-molding polypropylene has been developed by Phillips Petroleum Company; known as the Orbet process, the plastic is oriented biaxially as it is blown, to provide exceptional clarity, greatly improved tensile and impact strength, and better barrier properties than unoriented polypropylene. Barex 210 is a promising new copolymer made by Vistron Corporation; it is a nitrile rubber-modified acrylonitrile/methyl/acrylate, and was used to make Pepsi-Cola bottles for testmarketing.

Other materials which are less important commercially, but which may have special applications, include *polycarbonate,* which has amazing strength characteristics, excellent clarity, and good resistance to oils; it is stain-resistant, has no odor or taste, and can be steam-sterilized; some grades are approved by FDA for food products. The biggest drawback is that it costs several times as much as polyethylene or polyvinyl chloride. The resin must be perfectly dry when it is processed, or there will be degradation of the plastic, and the moisture will cause silvery streaks and "chicken tracks" that spoil its appearance. The temperature must be controlled very carefully, as the viscosity changes rapidly with small changes in temperature. Large parts can be a problem because of parison sag. Some of the difficulties with degradation mentioned previously in connection with PVC will also apply to polycarbonate. *Phenoxy* is being used to a limited extent, but it has a tendency to stress-crack under certain conditions and its cost is quite high. Another transparent material being tested for this purpose is *methacrylate*. One of the newer materials that has been used for shampoo bottles is acrylic copolymer. It is medium high in cost, has good clarity, and can be formulated to meet FDA requirements. The major drawback of this material is its tendency to "blush" or show opaque white spots when it is bruised. These disappear slowly at room temperature and very quickly when heated. It is a good barrier for oxygen, is resistant to oils and dilute alcohols, has good impact resistance, is four times as stiff as regular polyethylene, can be used for hot filling, and is resistant to static collection of dust; flavors and odors will hold up well, but loss of weight with aqueous products is high—around 4 percent per year.

Coatings. To improve the permeation rate, or for a better appearance, or to resist scuffing and abrasion, it is possible to coat the outside or the inside of plastic bottles. An exterior coating can be put on much heavier, but since it is usually rolled on, it is not possible to cover the entire outside surface. For polyethylene bottles, outside coating is the only practical way, inasmuch as it is necessary to treat the surface in order to get a good bond.

Interior coatings give complete coverage, but they are more expensive to apply. It is suggested that the cheaper method of outside coating be tested thoroughly before ruling it out. The most commonly used materials for coatings are the epoxies, although vinylidene chloride (saran) also can be used for this purpose.

Testing. There is a misconception that plastics are unbreakable. Although the amount of abuse which plastic containers can withstand is considerable, they are far from indestructible. When plastic is substituted for glass, it is important to test the new containers adequately so that the proper protective packing can be provided. It will be found that plastic containers, when stacked in the warehouse, cannot support the same amount of top loading as glass containers; they may require additional vertical support pieces.

Impact resistance is generally tested by dropping individual filled bottles on a hard surface. A typical bottle will withstand drops of 6 to 8 ft without failure, and rupture usually occurs at the pinch-off. Bottles should be filled with the actual product for reliable results, and they should be filled 24 hr before testing. It has been found that it takes a certain length of time for the inside surface to reach a state of equilibrium with the contents, and so freshly filled bottles should not be used for testing.

ROTATIONAL MOLDING

For small quantities of large items for which mold costs are an important factor, or in the development of a new item in which changes must be made quickly and inexpensively and various wall thicknesses must be tried, rotational molding might be the method of choice. This technique produces parts of uniform thickness and without the internal stresses that thermoforming might produce. It is generally used for large containers such as drums and carboys. The minimum wall thickness that can be produced is about 0.030 in. The molds are less expensive than injection molds, but a little higher in cost than thermoforming molds. A typical mold would cost around $2,000 to make a part 1 cu ft in size. Material costs are higher for the powder than for the beads that are used for injection molding, and the production rate is low, depending on the number of molds in operation.

Vinyl plastisols were the first materials used for this process, followed by urethanes. Now polyethylene, polypropylene, cellulosics, acrylic, and polycarbonate parts are being produced by this method. A weighed amount of very fine powder is put into the mold by hand. The mold is mounted on a spindle which carries it into an oven. While it is in the oven, the mold is heated to about 500°F by hot air, or sometimes by a hot spray, while the spindle rotates the mold so that the inside surface becomes uniformly coated by the plastic as it melts. (See Fig. 22.)

The mold is mounted as close to the axis of rotation as possible, so that the centrifugal force will not cause too much variation in wall thickness. The rate of rotation is about 12 rpm around the major axis and about 3 rpm around the minor axis. Higher speeds are sometimes used to force the plastic into very narrow recesses in the mold by centrifugal force. It

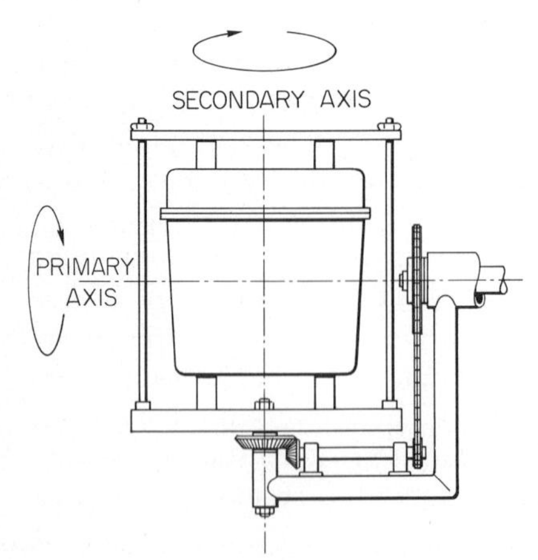

Fig. 22. Rotational molding. Powdered plastic is put into a cold mold, and heat is applied to the outside of the mold while it rotates in two directions, so that the inside surface becomes uniformly coated by the plastic as it melts. The mold is then cooled, and the piece is removed, ready for the next cycle. (*U.S. Industrial Chemicals Co.*)

may take as long as 10 min to heat the plastic enough to coat the inside of the mold uniformly; the spindle then carries the mold into the cooling chamber. Here the mold is sprayed with cold water to solidify the material. Cold air also is used for this purpose, either with or without the water spray. Some machines, as shown in Fig. 23, are self-contained; that is, a circulating system brings the heating and cooling oil to the jacketed mold, so that no oven is required.

Fig. 23. Rotational molding machine. This type of machine does not require an oven, but is heated and cooled by circulating oil through the double-walled mold. Large parts can be made from such materials as polyethylene, polystyrene, ABS, or nylon, or from two or more layers of different plastics. (*Krauss-Maffei Corp.*)

The molds are generally made of steel, although aluminum or electroformed copper-nickel is sometimes used for small molds. The walls of aluminum molds are about 3/8 in. thick, and steel molds are generally about 1/8 in. thick. The molds are vented to atmosphere so that pressure does not build up during the molding cycle, although a small amount of pressure is sometimes desirable to keep the part from shrinking away from the mold, which would slow down the cooling. No draft or taper is usually necessary in the design of the parts. Flat panels should be avoided, for they will shrink away from the mold as they cool, especially if they are cooled too rapidly, with consequent warpage and brittleness.

SLUSH CASTING

In this process resins are mixed at room temperature with nonsolvent types of plasticizers along with any fillers, stabilizers, or colorants that may be desired, to form pastes or viscous fluids called plastisols. These are poured into a warm mold (250 to 300°F) until a gel has formed, and the excess is poured out, leaving the coating or shell in the mold. The thickness can be controlled by the temperature of the mold and the viscosity of the plastisol. The temperature is then raised rapidly to 300 to 350°F and causes the mixture to set up by fusion of the resin with plasticizer.

COLLAPSIBLE TUBES

The plastic collapsible tube was introduced into this country by Bradley Dewey when he acquired the rights to a European process in the early 1950s. This is an extrusion molding process in which the body is extruded while the shoulder and neck are injection-molded in the same machine. It is thus possible to have different colors for the body and the shoulder. The body is extruded directly into the cavity of an injection (heading) mold. Then, while the plastic is still hot, the shoulder piece is injected around it and the two parts fuse together at the corner of the shoulder. Another method allows the body to cool before it is transferred to the injection mold. The shoulder and neck are then molded around the body so that the weld line is halfway up the shoulder. This process does not provide as strong a joint as the first method.

These tubes are generally made of low-density polyethylene, although other materials can be used, such as high-density polyethylene, polypropylene, and polyvinyl chloride. They are decorated by offset printing in three or four colors, and can be given a coating of clear epoxy resin for gloss or for extra protection of the product. Originally the standard dimensions and tools were those of the European system. More recently, however, the threaded finish and the body sizes have been changed to conform to the standards used by the metal tube industry, as far as possible. (See Table 20.) The same closures and fitments, as well as the handling equipment, can be used, although the method of sealing the ends of the tubes is quite different.

Sealing filled plastic tubes usually consists of flattening the end between cold jaws and passing the protruding portion (about 1/8 in.) between radiant heaters until the plastic melts and forms a bead. If any of the product splashes on the inside sealing surface, it may interfere with a good seal. Frequent checks should be made during a production run if this is known to be a problem. Equipment is being developed for ultrasonic sealing of polyethylene tubes to give better seals with contaminated surfaces. This method uses a cold hammer and anvil actuated by high-frequency crystals so that the very rapid pounding causes sufficient friction at the interface to raise the temperature above the melting point.

An important difference between plastic tubes and metal tubes is the "suck-back" characteristic of the plastic tubes. This is both an advantage and a disadvantage. From an appearance standpoint it is an advantage.

TABLE 20 Standard Dimensions of Plastic Tubes

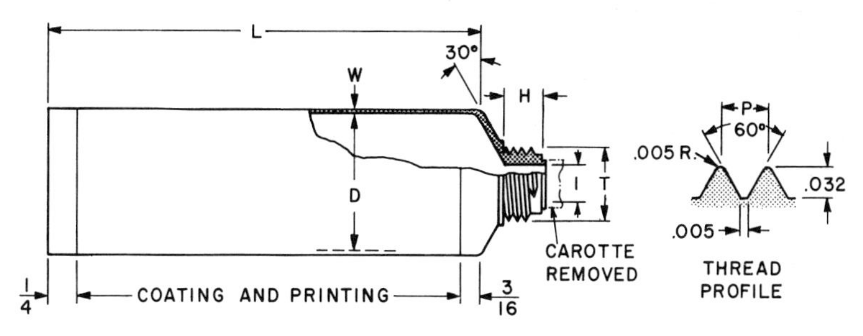

Neck size*	I Max. orifice ±0.010, in.	T Thread OD ±0.005, in.	P Thread pitch ±0.003, in.	H Neck height ±0.010, in.
M4	0.060	0.250	0.038	0.235
12	0.125	0.312	0.050	0.187
16, S16	0.187	0.375	0.050	0.235
M5	0.187	0.420	0.059	0.235
20, S20	0.250	0.438	0.050	0.235
M8	0.320	0.563	0.059	0.235
28	0.375	0.563	0.050	0.235

D Inside diameter ±0.015, in.	W Standard wall ±0.003, in.	L Max. length ±0.031, in.	Capacity at max. length, fl oz
0.595	0.015	4 1/16	1/2
0.715	0.015	4 3/16	3/4
0.835	0.015	5 3/4	1 1/2
0.950	0.018	6	2
1.150	0.018	6 5/16	3
1.345	0.018	7 9/16	5
1.505	0.018	7 11/16	6 1/2
1.885	0.018	8 1/8	10 1/2

* M = metric series; S = U.S. standard series. Numbers without letters originally indicated the diameter of the orifice in sixty-fourths of an inch. Bottle threads are sometimes used on tubes, particularly the 22-400 size with H increased by 0.029 in. (See Table 13.)

The tube always looks full and is never wrinkled or misshapen. When it is partly empty, however, it is a nuisance because the air must be expelled each time before the product can be dispensed. This can be avoided by using a stand-up type of cap, which is as large in diameter as the body of the tube. It can be mushroom style, but more often it is straight-sided with a flat top. Printed copy on the tube is usually inverted to call attention to the stand-up feature and to make it readable when it is put on the shelf with the cap down.

The incompatibility of plastic tubes with some products may cause problems similar to those with blow-molded bottles made from the same materials. See "Materials," page 8-65, and also the chemical properties which are listed under the individual plastic materials. When tubes are molded, a piece of the sprue, called a *carotte,* is left on the neck. This is usually trimmed off in the finishing operation, but it could be left on for a snip-off tip if desired. Various styles of threaded finishes and closures are available; they are generally the same as those used for metal tubes. (See Sec. 7, "Metal Containers," page 7-29.)

THERMOFORMING

The process of heating and forming sheet plastics originated in 1936, when the use of shrinkable film on meats was introduced in France. During World War II this method was used for making contour maps, but it was not until the early 1950s that it came into general use. Today there are more than 500 companies producing 5 billion thermoformed packages worth over $150 million each year.

The various types of thermoformed packages include skin packaging, blister packaging, and formed primary containers and closures. The common characteristic of all these forms of packaging is that they start with flat film or sheet plastic. This material is heated until it is soft and pliable, and then it is shaped with vacuum or pressure, or with forming dies, to the desired shape.

Much thinner walls can be made by this process than by any other method of fabrication. The cost of the material is therefore lower but this is offset to some degree by high scrap losses. Any of the thermoplastic resins that can be made into sheets can be used for thermoforming. They may be transparent or opaque, and they can be made in a variety of colors and patterns. The materials are described more fully

TABLE 21 Typical Mold Costs for Thermoforming

Epoxy resin	$100 per sq ft
Cast aluminum	$200 per sq ft
Sprayed metal	$250 per sq ft

in Sec. 3, "Films and Foils." The molds for thermoforming are much less expensive than for other plastic processes, ranging from $500 to $1,000 for the simplest type of mold. (See Table 21.) Tooling-up time is only a matter of days instead of the weeks that are necessary for blow molding, or the months for injection molding.

Design. A mold for thermoforming can be either male or female, depending on the requirements of the finished piece. A female mold will give better detail to the outside surface, which is in contact with the mold. The greatest thinning will take place at the top of the dome, especially near the corners, and the heaviest sections will be near the flange. In the case of a male mold the opposite is true. A male mold must have at least 3° draft or taper, whereas no taper is needed in a female mold, since the piece pulls away from the mold as it shrinks. More than 3° may be required on very wide pieces or if the shrinkage rate of the material is high. The depth of draw should generally be less than the width of the cavity, to avoid excess thinning and webbing. Some special techniques for making very deep draws are described under "Processes," page 8-79. (See Fig. 24.)

Use generous radii wherever possible, especially where thinning is likely to occur. A radius should be at least twice the thickness of the starting material. A flange is nearly always necessary around the outside edge; try to use it to advantage for stiffening or fastening to a card. Ribbing will help to strengthen flat panels and wide spans. For vacuum forming, the mold has to resist stresses up to only 14.7 psi, which is normal atmospheric pressure. Pressure forming, on the other hand, may develop forces up to 300 psi. Large flat sections should be domed or crowned, if possible, to allow the air to escape as the sheet is pulled down over the mold. Avoid square holes or notches of any kind because of the stress concentration in the corners.

Molds. The materials for making molds are many and varied. Wooden molds can be used for developmental work, but they will soon get scorched from the heat of the plastic. Plaster of paris is often used for short-run molds, but it is not very durable, being inclined to chip and crack with prolonged use. Epoxy resins are sometimes used for forming light-gauge material up to 0.030 in., but their poor thermal conductivity makes them unsuited for heavyweight formings. The life of an epoxy mold is not much over 10,000 formings. Sprayed metal molds are costly, but they give perfect detail with no shrinkage. In this process zinc-aluminum wire is fed to a special spray gun, melted, and blown out as a very fine spray. It freezes instantly on the pattern with no significant amount of heating of the sprayed surface. In this way coatings up to 1/2 in. thick can be built up.

Cast aluminum or beryllium-copper is best for large-scale production,

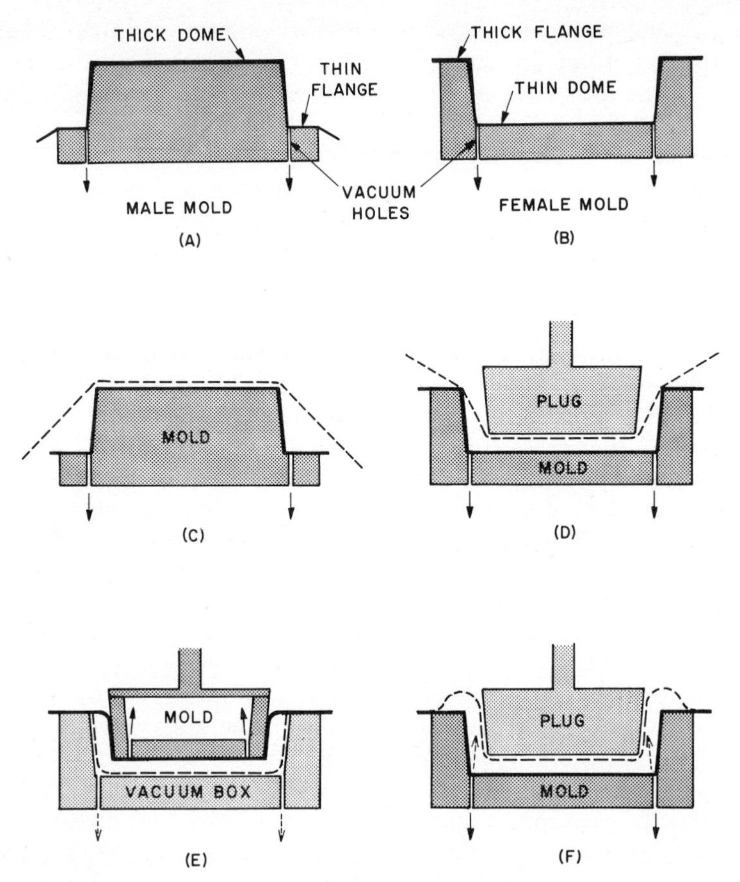

Fig. 24. Thermoforming. (*A*) Male mold will make parts with dome section nearly as thick as the original sheet, but thinner where it is drawn down to make the flange. (*B*) Female mold will provide a heavy flange and will be thinnest in the dome, especially at the corners. (*C*) Draping the warm sheet down over a male mold before the vacuum is applied will make the wall more uniform in thickness. (*D*) A plug assist is sometimes used to stretch the softened plastic, as shown by the dotted line, to minimize thin spots in the corners. (*E*) Snapback forming consists of pulling the heated plastic into a vacuum box first, and then forcing it back up against a male mold. (*F*) Billow forming is another variation in which compressed air stretches the sheet upward, after which a plug carries it down into the mold; vacuum then takes over to pull the sheet into place.

and the walls are usually made about ½ in. thick. Either of these metals will carry off the heat rapidly and will give a high rate of production. Holes must be provided in sufficient number and in the right places to draw the plastic down into every crack and crevice in the design. The holes should be about 1/64 in. in diameter for thin materials, and about 1/32 in. for the heavier gauges. Smaller holes may be required for polyethylene because hole marks show up more in this material. These holes are placed where the sheet will be formed last, usually in a corner. They are usually back-drilled to a larger size, as close to the surface as possible. Deep sections in the mold should have more holes than shallow parts, especially when the deep and shallow parts are adjacent. They can be as close as 1/2 in. apart if necessary.

Cooling is provided by cored or drilled holes in the mold. These must be placed so that they do not interfere with the air vent holes, and of course they must be waterproof. If copper or stainless steel tubing is used for cooling, it should be at least 1/2 in. in diameter and spaced 2 1/2 in. apart, well soldered to the back of the mold. Some sections of the mold may require more cooling than others, but this can be determined only with experience. For some materials, such as impact styrene, the mold should be kept fairly warm, around 150°F, to avoid stresses due to rapid cooling that may cause warpage.

If undercuts are necessary, they can be handled in several ways. Split sections can be removed from the mold to release the part. Hinged sections can fold out as the part is removed. Knockout pins can be incorporated in the mold for shallow undercuts, or cam-actuated sections can be pulled out of the way for releasing the piece. A loose piece called an *orphan insert*, usually of the same plastic material although it may be boxboard or other inexpensive stock, is inserted in the mold and remains with the finished part when it is taken out of the mold. Compressed air will often release a formed part if the undercuts are not too deep. The surface of the mold should be sand-blasted to allow passage of air between the plastic and the mold, during the forming operation, as well as for removal of the piece from the mold.

Processes. The basic technique in thermoforming is to suspend a sheet of plastic in a frame that grips it all around the edges. The sheet is heated until it softens, and then it is sucked down over a mold by vacuum. When it is cool, it is stripped from the mold and trimmed. The simplest machines are used for development work, and cellulose acetate is often chosen for this purpose because the temperatures need not be controlled very accurately and the strength of the material in the softened state is good. Heat can be applied from one side or both sides, and in a manually operated process it will be observed that the sheet sags as it softens. Then with further heating the sheet tightens up until it is perfectly flat. At this point it is nearly ready for forming, but it is usually better to allow another 15 or 20 sec before applying the vacuum. Some smoking of the plastic will be noted in the case of acetate, caused by the evaporation of some of the monomer that was not polymerized during manufacture of the plastic. The sheet is then sucked down into the mold. Milky white spots that may appear are caused by stresses due to insufficient heat. Webs where the sheet folds in against itself are caused by deep and shallow sections being close together, and they can be eliminated by *assist* plugs or rings, which are described below. Thin spots also will be helped by these assists.

The equipment for large-volume production is much more sophisticated than the laboratory-type machine just described. It is possible to extrude the sheet and form it in a continuous operation. This is not very practical, however, because the extruder operates at a much higher rate than the forming machine. It is therefore customary to extrude the sheet material and roll it up for storage until it is convenient to process it further. The rolled stock is fed into the forming machine by *tenter* chains, which grip the edges and carry it through the process. Heat is applied to both sides of the sheet, and this can be varied in different *zones* so that distribution is controlled and thin spots in the finished piece are avoided. The heated material is then carried over a mold, and with vacuum and pressure it is forced to conform to the shape of the mold.

To keep the thickness as uniform as possible throughout the finished part, various techniques can be used. Pushing the mold up into the sheet so that the plastic pulls down over it like a tent is called *draping.* A plug or ring of the proper shape, called an *assist,* can be used to stretch the material into the deepest parts of the mold before the vacuum or pressure is applied. Assists can be made of wood or wire, but in high-production equipment they are made of cast aluminum with water circulating through them, or with electrical heaters to maintain a temperature around 180°F so that they will not chill the sheet. The *plug* size is about 80 percent of the volume of the cavity, but not closer than ½ in. from the mold, with generous radii on all corners. It should conform to the general shape of the cavity, but with none of the details. The speed with which the plug stretches the sheet is important to the quality of the finished piece.

Sometimes air is applied to the plastic at about 3 psi so that it balloons out before the assists come into play. This is called *billow* forming; it is used to thin out the perimeter and put more material in the center. There are several variations of this method. In *air slip* forming, the sheet is ballooned out and then pulled in over a male mold. In *snap-back* forming, which is an upside-down version of the air slip technique, the sheet is drawn down while the mold enters from the top, and then is forced up against the mold. The *billow snap-back* method is a combination of two of the systems just mentioned, as the name implies. Plug assists also can be incorporated in any of these methods.

Some molders allow a controlled amount of slippage where the sheet is clamped at the edges, to get a little more material into the part in that area. Some materials such as the expanded or foamed plastics require *matched molds,* as the sheet wants to continue expanding when it is heated, and it is necessary to confine it between the male and female parts of the mold. There are materials which can be heated by pressing against a porous hot plate with air pressure, while the edge of the mold is pressing against the plastic. This process is called *trapped sheet* forming. When the plastic has softened, the air pressure is reversed so that the sheet is forced down into the mold. Both air pressure and vacuum are used, either separately or together, in all these methods. Pressure forming is much faster than vacuum forming, as the latter is limited to 15 lb of atmospheric pressure to do the work, whereas compressed air can be several times this amount. Nearly all the machines used for high-production work are pressure formers. (See Fig. 25.)

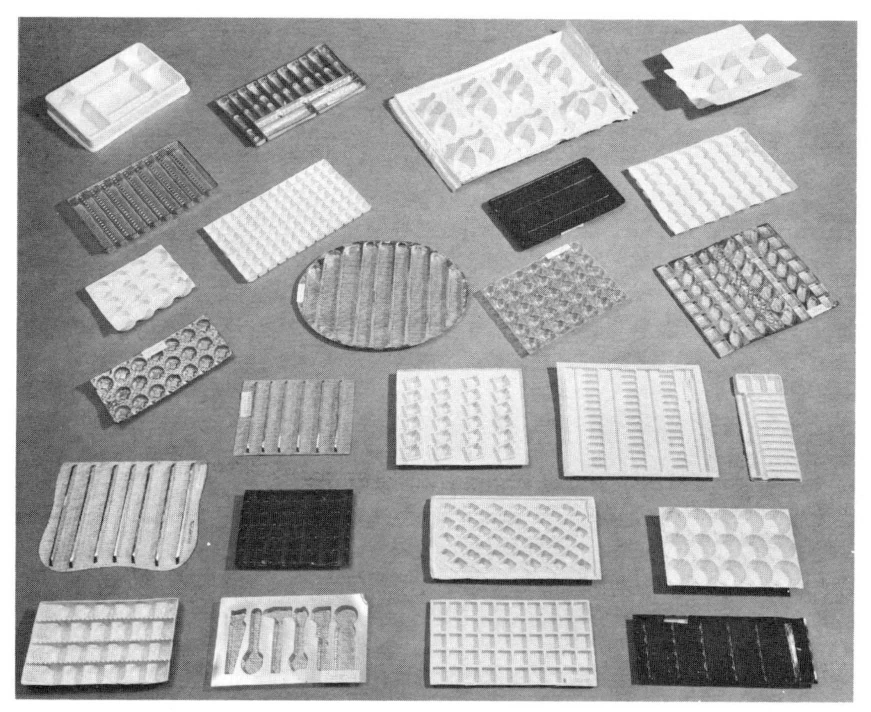

Fig. 25. Thermoformed pieces. Typical trays and inserts made from various colored plastics by thermoforming. (*The Pantasote Co.*)

After the piece has been formed over the mold, cold air may be blown over it to speed up cooling. Some formers find that a vapor spray is more efficient than air for cooling. The part is then separated from the mold by blowing it off or with knockout pins or other mechanical devices. The web is still supported between the tenter chains, and it is carried to the next station for trimming. A steel rule die or a clicker die is used to crush-cut the parts out of the web. The size and spacing of this cutting die must allow for subsequent shrinkage, as the sheet is still warm at this point. The skeleton of the web is then wound up on a reel, while the parts are conveyed to the packing station or to the next process. (See Fig. 26.)

An offshoot of blister packaging is *skin packaging*. In this process the sheet is heated and drawn down over the product, instead of preforming a blister to contain the item. It is used for small toys and hardware, which are placed on cards that have been printed, coated with a heat-activated material, and pierced with needlepoints so that a vacuum can be pulled through the card stock. As the film is sucked down around the objects, it adheres to the card, holding the pieces firmly in place. Thinner films are used—around 5 mils in thickness—so that it is basically cheaper than blister packaging, unless the card is much larger than the item. The film must cover the entire card in skin packaging, whereas a blister needs to be only slightly larger than the object it contains. There is a tendency for the cards to warp, owing to the shrinkage

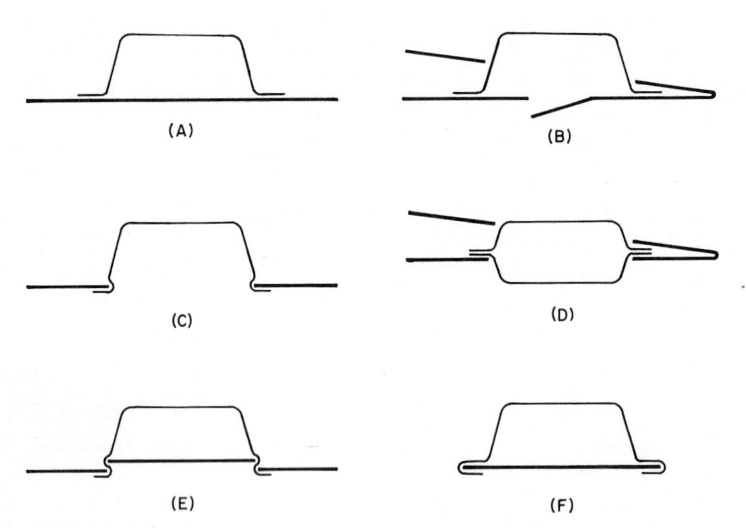

Fig. 26. Carding methods. Various techniques are used for attaching transparent blisters to printed cards. (*A*) Flange may be heat-sealed to a card which has a suitable coating. (*B*) Folded card with die-cut hole for blister makes a strong unit. Self-stick adhesive is generally used to fasten card together. Perforated trapdoor is indicated for removing parts from the back of the card. (*C*) A product can be snapped into a blister, and the blister snapped into a display card, for an easily assembled combination. (*D*) Double blister in folded card makes product appear to float in space. (*E*) Double undercut uses cutout from card as filler piece. (*F*) Folded edges of blister make tracks for sliding card. For inspection by customer, or dispensing of contents, backing can be retracted and subsequently reclosed.

of the film as it cools. Not much can be done to eliminate this, although it can be minimized by zone heating and by using a heavy weight of paperboard for the card.

Vinyl films are generally used for skin packaging, but polyethylene can be used with uncoated paperboard if the film is treated by flame or corona discharge, to oxidize the surface so that it will adhere to the board. It is usually not necessary to perforate the board when it is uncoated, as there is generally enough porosity for pulling the vacuum through the board. Surlyn A film also is being used for skin packaging if toughness is required, but it is considerably more expensive than the other films just mentioned. Cycle time for skin packaging is about 40 sec, and so machine time is worth about 12 cents a sheet. Film and paperboard cost an additional 4 cents. The cost of the individual cards of course depends on the size, but if the sheet is cut into 5- by 9-in. units, the cost is about 2 cents per card.

Materials. For thermoforming it is possible to use any sheet thermoplastic material. In practice, however, certain materials have been found to satisfy the requirements of packaging better than others because of cost, transparency, odor, or processing characteristics. (See Table 22.) The most widely used ones are impact-type *styrenes,* with differing proportions of synthetic rubber to suit the stress conditions that are anticipated. These are completely opaque. If transparency is desired, *cellulose acetate* or *cellulose propionate* will meet the usual requirements of low cost, high strength, and good processing properties. Extruded expanded polystyrene (EEP) is finding some applications in packaging, but it is difficult to form and it cannot be drawn much deeper than about half the width of the piece, even with matched molds, which are usually required.

TABLE 22 Relative Costs of a 3- by 5- by 1-in. Thermoformed Container in Large Quantities

Material	Thickness, in.	Sheet cost per lb	Cost per thousand containers
Oriented styrene	0.010	$0.45	$12.75
Hi-D polyethylene	0.015	0.55	13.00
Rigid polyvinyl chloride	0.010	0.55	13.75
Impact styrene	0.015	0.50	14.00
Cellulose acetate	0.010	0.85	14.50
Cellulose propionate	0.010	0.95	16.00
Extruded expanded styrene	0.060	0.85	17.00

Although *polyethylene* is somewhat difficult to form, its low price makes it a useful material for packaging. The low-density type lacks stiffness and is only fairly good as far as transparency goes. The high-density material is much stiffer and is more translucent than transparent. It is often used for food containers because it is chemically inert and does not usually affect flavors, but it should be thoroughly tested before being used on a large scale. For some types of food containers *polypropylene* is preferred, in spite of its higher cost and its brittleness at low temperatures. It has good machinability, especially for deep drawing.

Other materials which are not as suitable for packaging but deserve mention include *acrylonitrile-butadiene-styrene* (ABS), which has exceptional toughness for tote bins and luggage cases. The most important application in the packaging field has been for margarine tubs. For lighting panels, signs, and skylights, where transparency and weatherproofness are essential, the *acrylics* can be thermoformed nicely. For skylights they are sometimes blown to a dome shape without molds, depending on the air pressure to give it the right shape. The toughest material that can be thermoformed is probably *polycarbonate.* It has good transparency, and it is sometimes laminated with polyethylene for food packaging, such as the heat-and-serve containers, because of its high melting point. Another material for heat-and-serve containers is *polyester* with a polyethylene coating, but it is suitable only for shallow containers and cannot be deep-drawn. Rigid *polyvinyl chloride/polyvinyl acetate* copolymer is used to some extent in blister packaging, but it costs more than the cellulosics.

FOAMED PLASTICS

Foamed plastic can be defined as an expanded resinous material with a cellular spongelike structure, usually made by the introduction and dispersion of a gas in the liquid resin, and the subsequent setting or curing of the expanded mass. The resulting structure can be open-cell or closed-cell; thus it can be buoyant or absorbent, depending on the tensile strength of the resin in the liquid state and the postcuring treatment which may be used to crush the foam and rupture the cells.

Foamed plastic was first produced on a commercial scale by the Dow Chemical Company in 1943, and the total output has reached a rate of 1 billion lb per year in this country. Flexible urethane accounts for one-third of this amount, and rigid urethane and polystyrene for about one-fourth each. Over 700 firms in this country are engaged in the production of plastic foam. These include oil companies, rubber manufacturers, textile mills, and chemical companies.

The advantages of plastic foams are resistance to mold and bacteria growth, good thermal insulating properties (*k* factor 0.24 Btu/in.-sq ft-hr-°F at 40°F 2 lb/cu ft molded expanded polystyrene), light weight, good strength-to-weight ratio, excellent cushioning and blocking characteristics, nonabrasiveness, and easy moldability.

Costs will vary, depending on a number of factors. With general-purpose polystyrene selling at 14 cents per pound, the expandable beads for molding are 24 cents per pound, and the extrusion grade is 27 cents per pound. A molded box for shipping a single pint-size bottle costs about 15 cents, and 10-mil sheet material is around $5 per thousand square feet. Meat trays 5½ by 8 in. have cost about $7.50 per thousand, which figures out to about 60 cents per pound, but new techniques have brought the price down to about half this figure. Egg cartons cost about $25 per thousand at the present time.

Suppliers are Sinclair-Koppers Company (Dylite closed-cell molded expanded polystyrene); Dow Chemical Company (Styrofoam open-cell slab polystyrene); Monsanto Chemical Company (Santofome extruded expanded polystyrene sheet); Armstrong Cork Company (Resilo-Pak crushed expanded polystyrene); Scott Paper Company, and most of the rubber companies (urethane foam); Serex, Inc. (polyvinyl chloride); Dow Chemical Company (polyethylene); Marbon Division, Borg-Warner Corporation (ABS); Hercules, Inc. (polypropylene); Gilman Brothers Company (ionomer); Dow Corning Corporation (silicone); E. I. du Pont de Nemours & Company (urea); and Emerson & Cuming, Inc. (epoxy).

Polystyrene Foam. There are two types of polystyrene foam: an open-cell material that is available in the form of slabs and logs, and a closed-cell foam that can be molded to any desired shape. The open-cell material is made by injecting a volatile liquid such as methyl chloride into molten polystyrene which is kept under a pressure of 500 to 1,000 psi. This can then be extruded as logs or it can be cast into slabs.

Closed-cell polystyrene foam is made from beads of polystyrene containing petroleum ether or pentane. These beads are preexpanded in a large drum, by the introduction of steam until they are about 25 times their original size. They must be stirred while they are expanding to keep them from sticking together. The preexpanded beads are held for several hours until they have stabilized, and then are transferred to the mold, where steam is injected directly into the cavity through a series of tiny ports in the mold shell. This causes the beads to soften and expand still further, packing them tightly in the mold so that they stick together in a solid mass. The mold is then cooled with water circulating through the jacket. Since the foam is such a good insulator, it takes some time to bring the temperature down to the point at which it can be removed from the mold without continuing to expand. A ¼-in. wall takes about

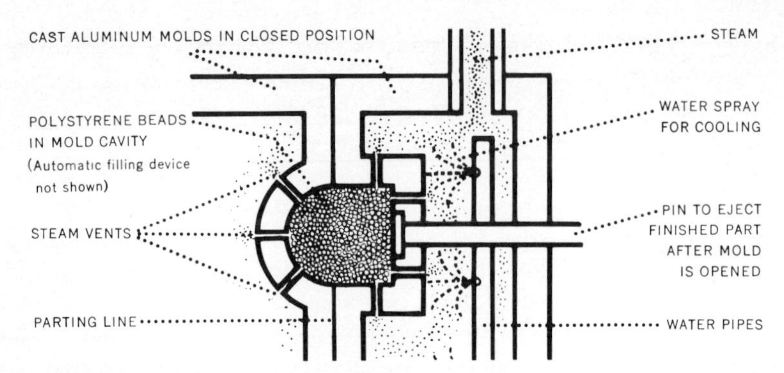

Fig. 27. Molding process for foam. Schematic diagram showing the various parts of a mold for making expanded polystyrene. Not shown is the filling device for packing the preexpanded beads into the cavity. Steam first enters the jacket through the pipe at the top right of the drawing, and passes through the vent holes into the mold cavity, causing the beads to expand and stick together. The steam is turned off, and cold water enters through the pipe coming up from the lower right, and is sprayed on the aluminum mold to cool the plastic foam in the cavity and stop the expansion process. The mold then opens, and the piece is pushed out by the ejector pins, after which the mold closes and is ready for the next cycle. (*S. Curtis & Son, Inc.*)

2 min to cool, and heavier sections take proportionately longer. (See Figs. 27, 28, and 29.)

A method of making extruded expanded polystyrene (EEP), in contrast to molded expanded polystyrene (MEP) just described, was developed in Japan in the mid-1950s as a substitute for paper, which is scarce in Japan. Extruded expanded polystyrene is produced in gauges of 8 to 100 mils, and in densities of 1.6 to 5.5 lb per cu ft. Various pastel colors can be made in addition to white, and the satin finish which is characteristic of this material makes it very attractive for packaging. It can be thermoformed, but the maximum depth-to-width ratio is only about 1:2, and because of the critical softening temperature it is necessary to use matched molds.

The closed-cell type of polystyrene foam was developed by Badische Anilin- & Soda-Fabrik AG in 1950. It is the most widely used type of foam for packaging, and it is finding applications in such diverse uses as disposable pallets, laminated shipping containers, coffee cups, and counter displays. A flexible type of foam is made by passing slabs of expanded polystyrene between rollers, which compress it to about half its original thickness, and then allowing it to recover. This ruptures some of the closed cells and increases the flexibility and compressibility of the foam.

Urethane Foam. When a polyol is mixed with an isocyanate, there is an exothermic reaction. A certain amount of water is included in the formula to cause foaming. Densities of from 1 to 40 lb per cu ft can be

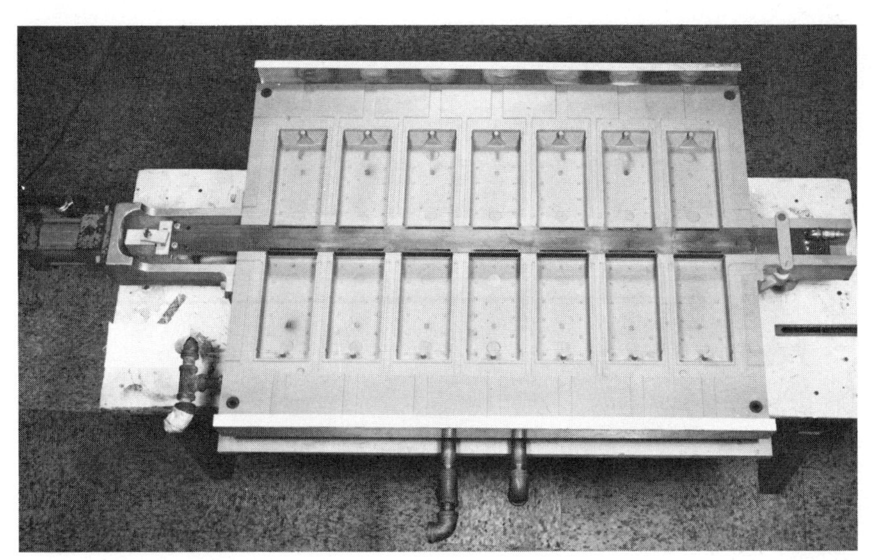

Fig. 28. Mold cavities for foam. Preexpanded beads are forced through the manifold from right to left between the two rows of cavities until the cavities are filled. A slide cutoff is operated by an air cylinder on the left. Two rows of holes on each side of the cavities are steam ports, and the two large circles are ejector pins. An air eject vent is in the center of each cavity. (*S. Curtis & Son, Inc.*)

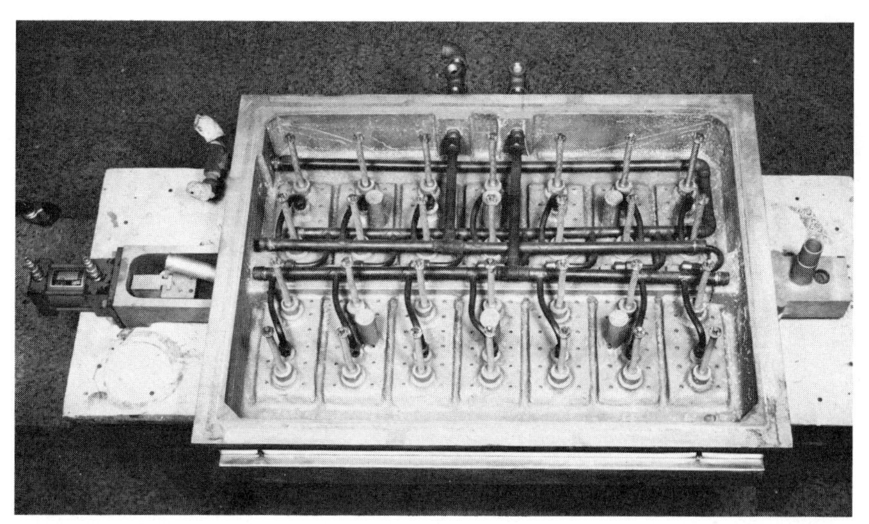

Fig. 29. Back of mold. Bead entry is on the right, and slide cutoff cylinder is on the left, with bead return pipe angled up between the mold and the cylinder. Toward the front are spring-loaded ejector pins, with air ejector tubes directly behind, and the air inlets projecting through the back. Boost air comes in on the left rear and goes into the center bead channel between pairs of cavities. Steam and water connections are in the cover plate not shown. (*S. Curtis & Son, Inc.*)

made, depending on the amount of water used. The water reacts with some of the isocyanate to form carbon dioxide, which expands from the heat of reaction, creating the cellular structure of the foam. Because toxic vapors are produced during the foaming process, the work space must be well ventilated.

TABLE 23 Properties of Urethane Foams

Property	Polyester	Polyether
Density, lb/cu ft	1.5–6.0	1.0–3.0
Resilience (falling ball), percent	20–35	35–50
Compression deflection at 25%, psi	0.55–1.20	0.15–0.70
Tensile strength, psi	20–40	15–25
Elongation, percent	250–500	200–300
Thermal conductivity, Btu/in.-sq ft-hr-°F	0.2–0.25	0.2–0.25
Dielectric constant, 10^6 cycles per sec	1.0–1.3	1.0–1.3

The polyols may take the form of polyethers or polyesters; the polyethers are cheaper and have a livelier recovery when used for cushioning, but polyesters have better heat stability and flame resistance. (See Table 23.) A typical mixture would be tolylene diisocyanate as an 80:20 ratio of the 2,4 and 2,6 isomers, mixed with diethylene glycol and a small amount of water to make the ester; substituting methyl glucoside for the glycol would produce a polyether. The stiffness or flexibility will depend to a large extent on the polyol that is used. If it is highly cross-linked, it will produce a more rigid foam than a simpler straight-chain molecule. More information is given under "Polyurethane," page 8-30.

Section 9

Closures, Applicators, Fasteners, and Adhesives

Caps 9-2
Screw Caps 9-2
Lug Caps 9-8
Crown Caps 9-9
Specialty Caps 9-9
Seals 9-10
Shrink Bands 9-10
Quality Control 9-11
Liners 9-12
Backings 9-12
Pulpboard 9-13
Newsboard 9-13
Gray Feltboard 9-13
White Feltboard 9-13
Composition Cork 9-13
Fine-grain Composition Cork 9-13
Rubber 9-14
Wax Treatments 9-14
Quality White Waxed Paper 9-15
Facings 9-15
Kraft Oil (Yellow Oil) 9-15
Black Alkali-resistant 9-15
Casein-coated Varnished Papers 9-15
Vinylite 9-15
Solvent-resistant Papers 9-17
Polyethylene 9-17
Saran 9-17
Pliofilm 9-17
Mylar 9-17
Cellophane 9-18
Tinfoil 9-18
Aluminum Foil 9-18
Lead Foil 9-18
Vinyl-coated Foil 9-18
Saran-coated Foil 9-18
Glassine Inner Seal 9-19
Vented Liners 9-19
Stoppers 9-19
Tapes 9-20
Adhesives 9-25
Principles of Adhesion 9-25
Theory of Adhesion 9-26
Types of Adhesives 9-27

The smallest part of a package, and often the most critical, is usually the closure. The security of the whole assembly and the integrity of the contents are dependent upon the cap or tie or whatever is used to complete the package. Not only must the closure remain intact throughout all the hazards of assembly, storage, handling, and shipping, but it must be easy to open and reclose when it reaches the consumer. There is often a very narrow line between a closure that is easy to open and strong enough to travel, and one that fails in shipment or is next to impossible for the average person to open.

In addition to its functional qualities, a closure can sometimes enhance the appearance of the package. A ribbon bow, an embossed metalized seal, or a decorative cap can provide the extra touch that spells the difference between success and failure in the marketplace. A message or brand identification can be included in the design of the closure, on which it is most likely to be noticed. A closure is the finishing touch to the package, and it deserves special attention if it is to make the most of its prominent position.

CAPS

Screw Caps. The origin of the screw cap is lost in history. Bottles with threaded finishes are known to have been used, although rarely because of the difficulty of making them, in the early nineteenth century. The Espy patent issued in 1856 disclosed a screw cap with a disk of cork that, when screwed down on the neck of the jar, brought "the said cork in compressing contact around the upper edge or mouth . . . the cork shall entirely cover the said mouth." The mason jar was patented in 1858 by John L. Mason. His chief contribution was an improvement in the design of the thread on the glass container. He started his thread a little below the top surface and made it fade away before reaching the shoulder. Previously the top edge tended to break away and the bottom would jam the cap thread before the cap was all the way down.

The crown cap was invented by William Painter, of Baltimore, who was granted a patent in 1892. The invention of the Phoenix cap, a flat disk with a separate split ring to draw it down tight, is interesting because its various sizes were designated in millimeters. In 1892 Achille Weissenthanner developed this cap in Paris, using the French system to specify the dimensions. The caps were shown at the Columbian Exposition in Chicago the following year in four sizes: 48, 53, 58, and 63 mm. A method of putting an unthreaded aluminum cap over a threaded glass bottle, and pressing the metal into the threads with rollers, was introduced in 1924.

Some plastic closures were used in the early twenties for toothpaste tubes, but it was not until Bakelite phenolic became available at a reasonable price, around 1927, that they came into general use. More expensive than metal caps, the plastic closures are used mostly for luxury items. The advantages of freedom of design, beautiful colors and textures, and heat resistance make them very desirable for premium products. They are not as strong as metal caps, and some breakage is to be

TABLE 1 Relative Numbers of Caps Used, by Type

Crowns	27 percent (50 billion in 1959)
Molded plastic	19 percent thermoplastic 6 percent thermoset
Vacuum screw and lug	20 percent
Metal screw and lug	14 percent
All other	14 percent

expected, but this has not been a serious problem. Nearly 10 billion plastic closures are used each year in this country. (See Table 1.)

The closure industry has not standardized on dimensions to the extent that the glass industry has. The attitude has been that they will make caps to suit the containers, whatever they may be. This is due in part to the variety of materials that are used; threads that are rolled into a metal shell cannot be the same as those that are compression-molded of a thermosetting plastic, or of a more flexible material like polyethylene. Furthermore, the caps from one supplier cannot always be interchanged with those from another source; this can be a bit of a nuisance, and for this reason there is some advantage in buying both containers and closures from the same supplier, when this is possible. Caps for narrow-mouth containers should be made about 0.016 in. larger than the mean dimensions of the containers, with tolerances of ± 0.005 in. for a good fit with a minimum of trouble. Tables 2 and 3 correlate the standard dimensions of the most popular sizes of continuous-thread (CT) closures.

The *size* and type of thread are designated by the diameter in millimeters, coupled with a number which signifies the style, such as deep, shallow, or interrupted thread. Thus 22-400, or, as it is sometimes written, 400-22, means 22 mm in diameter and a shallow continuous thread. (See Table 7 in Sec. 6, "Glassware," page 6-20.) All dimensions and tolerances are given in inches, and the only reason for the millimeter designation is that one of the very early metal closures was developed in France, and the sizes that were applied to them have continued to the present day. The outside diameter of a metal screw cap measured near the top is approximately the nominal size.

There is very little similarity between a bottle thread and a bolt thread. Since it is difficult to press molten glass into sharp corners, and a sharp change in profile causes stress concentrations that weaken the container, there has evolved a thread contour that consists of very large radii that blend into one another. This may not be ideal from a purely mechanical standpoint, but it permits a high rate of production of glassware and is generally adequate for the purpose. There have been some recent attempts to improve on this, because some of the newer, softer plastics

TABLE 2 Standard Dimensions for Screw Closures

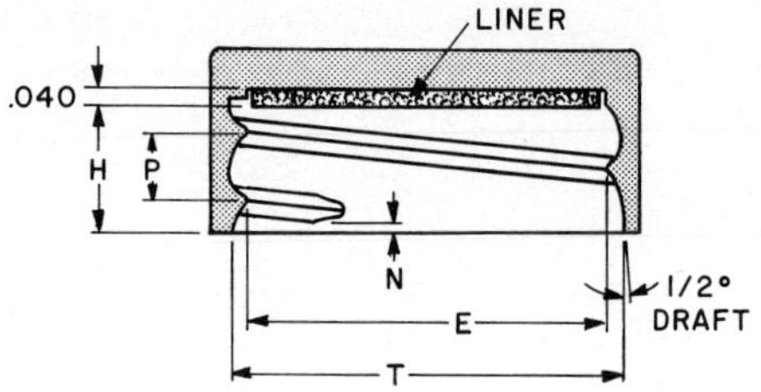

Type	Series	Millimeter size	T min.–max.	E min.–max.	H max.	P pitch	N max.
Shallow continuous thread	425	8	0.360–0.363	0.310–0.313	0.247	0.0714	1/32
		10	0.415–0.418	0.355–0.358	0.255	0.0714	1/32
		13	0.516–0.520	0.456–0.460	0.280	0.0833	1/32
		15	0.583–0.587	0.523–0.527	0.280	0.0833	1/32
	400	18	0.706–0.712	0.622–0.628	0.344	0.1250	1/32
		20	0.785–0.791	0.701–0.707	0.344	0.1250	1/32
		22	0.864–0.870	0.780–0.786	0.344	0.1250	1/32
		24	0.942–0.948	0.858–0.864	0.373	0.1250	1/32
		28	1.090–1.096	0.996–1.002	0.373	0.1666	1/32
		30	1.129–1.135	1.035–1.041	0.373	0.1666	1/32
		33	1.267–1.273	1.173–1.179	0.373	0.1666	1/32
		38	1.478–1.484	1.374–1.380	0.373	0.1666	1/32
		43	1.656–1.662	1.562–1.568	0.373	0.1666	1/32
		48	1.872–1.878	1.778–1.784	0.373	0.1666	1/32
		53	2.069–2.077	1.975–1.983	0.373	0.1666	1/32
		58	2.226–2.234	2.132–2.140	0.373	0.1666	1/32
		63	2.463–2.471	2.369–2.377	0.373	0.1666	1/32
		70	2.738–2.746	2.644–2.652	0.373	0.1666	1/32
		83	3.270–3.280	3.150–3.160	0.452	0.2000	1/32
		89	3.513–3.525	3.393–3.405	0.500	0.2000	1/32
		120	4.726–4.740	4.606–4.620	0.650	0.2000	1/32
Medium continuous thread	410	18	0.706–0.712	0.622–0.628	0.496	0.1250	9/64
		20	0.785–0.791	0.701–0.707	0.527	0.1250	11/64
		22	0.864–0.870	0.780–0.786	0.558	0.1250	3/16
		24	0.942–0.948	0.858–0.864	0.619	0.1250	9/32
		28	1.090–1.096	0.996–1.002	0.681	0.1666	17/64
Tall continuous thread	415	13	0.516–0.520	0.456–0.460	0.425	0.0833	7/64
		15	0.583–0.587	0.523–0.527	0.530	0.0833	7/32
		18	0.706–0.712	0.622–0.628	0.590	0.1250	11/64
		20	0.785–0.791	0.701–0.707	0.715	0.1250	9/32
		22	0.864–0.870	0.780–0.786	0.810	0.1250	21/64
		24	0.942–0.948	0.858–0.864	0.930	0.1250	7/16
		28	1.090–1.096	0.996–1.002	1.055	0.1666	15/32

TABLE 3 Standard Dimensions for Collapsible Tube Caps

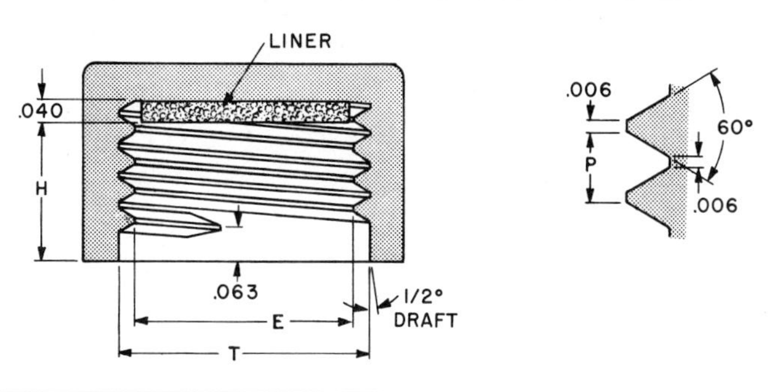

Neck size	Hard plastic			Polyethylene			P pitch
	T min.–max.	E min.–max.	H min.–max.	T min.–max.	E min.–max.	H min.–max.	
M4	0.262–0.275	0.198–0.211	0.177–0.187	0.257–0.270	0.193–0.206	0.146–0.156	0.038
12	0.324–0.337	0.260–0.273	0.146–0.156	0.319–0.332	0.255–0.268	0.115–0.125	0.050
16, S16	0.387–0.400	0.323–0.336	0.177–0.187	0.382–0.395	0.318–0.331	0.146–0.156	0.050
M5	0.432–0.445	0.368–0.381	0.177–0.187	0.427–0.440	0.363–0.376	0.146–0.156	0.059
20, S20	0.450–0.463	0.386–0.399	0.177–0.187	0.445–0.458	0.381–0.394	0.146–0.156	0.050
M8	0.575–0.588	0.511–0.524	0.177–0.187	0.570–0.583	0.506–0.519	0.146–0.156	0.059
28	0.575–0.588	0.511–0.524	0.177–0.187	0.570–0.583	0.506–0.519	0.146–0.156	0.050

tend to jump threads, or they lose their holding power in storage and shipment. The modified buttress thread is coming into general use in the smaller sizes and appears to offer some advantages. The *A style thread* is another alternative that is usually avoided by the glass manufacturers, but it has been found superior for large sizes of shallow threads. It provides a more complete wraparound and consequently more uniform pressure on the sealing surface. (See Table 9 in Sec. 6, "Glassware," page 6-24.)

The metal cap manufacturer finds it easier to make a thread profile which has generous curves, so as not to fracture the metal as it is stretched and rolled into shape. Because of the wide tolerance in the dimensions of the bottles, and the variations in the caps as well, there is a very sloppy fit between the two parts, in terms of machine shop practice. The area of contact and the angle of forces is far from ideal, but for a hard cap on a rigid bottle it is satisfactory. As we get into the softer plastics, however, the inadequacies of the system become apparent. It is for this reason that other thread shapes, such as the modified buttress, become necessary.

The security of the closure depends upon a number of things, such as the resiliency of the liner and the flatness of the sealing surface on the container, but more than anything upon the tightness or *torque* with which it is applied. The capping machine should be adjusted so that the chuck releases at the proper time, and a torque wrench can be used to set the tension on the clutch. The actual application force will be a little bit higher than this, owing to the inertia of the machine working at high speed. (See Table 4.) The only practical way to check the tightness is to measure the removal torque. This should be done a specified time after application, usually 5 min. Less than this will not give a true reading, and a longer time would allow more unusable material to be produced while the machine might still be out of adjustment. It would be well to check again after several days, to see whether there is any buildup or drop-off of torque. (See Table 5.)

Some liner materials, notably vinyls, seize upon the surface of the container after standing for a period of time and give a higher reading. In other cases, polyolefins will *cold-flow* so that there is a relaxing of tension on the sealing surface. Adjustments should be made in the application torque accordingly, since too high a figure will make it difficult for the consumer to remove the cap, and at the other extreme, there is the risk of leakage developing during shipment. Elevated *temperatures* also cause a loss of torque due to the expansion of the cap. This varies with the thermal properties of the material, but it can be as much as one-third less at 120°F for polyolefin caps. It is also worth noting that plastics which absorb *moisture,* such as the filled thermosetting resins, may lose as much as half of their holding power if they are stored under humid conditions after being put on the bottles. Every effort should be made to keep the product off the threads when filling, as this may act as a

TABLE 4 Recommended Tightness of Screw Caps

Cap size, mm	Rigid caps on plastic containers		Rigid caps on glass containers		Polyolefin caps on glass containers	
	Application torque, in.-lb	Minimum removal torque	Application torque, in.-lb	Minimum removal torque	Application torque, in.-lb	Minimum removal torque
15	6	3	8	4	12	7
20	8	4	10	5	15	9
24	10	5	12	6	18	11
28	12	6	14	7	21	12
33	15	7	18	9	24	14
43	18	9	22	11	33	20
48	20	10	24	12	36	22
58	24	12	28	14	44	26
70	28	14	35	18	52	32
89	36	18	45	22	65	40
120	48	24	60	30	90	55

TABLE 5 Retention of Torque for 28-mm Caps Applied with 20 in.-lb of Force

Type of cap*	Removal torque, in.-lb			
	5 min	1 day	1 week	1 month
Metal, P/TF liner	11	12	9	8
Metal, P/V liner	11	12	13	14
Phenolic, P/V liner	10	11	12	13
Urea, P/V liner	14	18	22	25
Styrene, P/V liner	14	13	13	13
Polypropylene, P/V liner	14	16	20	18
Polypropylene, linerless	16	15	15	14
Polyethylene, HD, linerless	13	13	12	9
Polyethylene, LD, linerless	14	12	12	12

* P = pulp; TF= tinfoil; V = vinylite; HD = high-density; LD = low-density.

lubricant, resulting in complete loss of torque during shipment in some cases.

In the *design* of caps, depressed surfaces on the top should be avoided because they will catch dust. A raised section is sometimes provided in closures for wide-mouth jars, to fit up into the "push-up" in the bottom of a jar that is *stacked* on top of it, to stabilize the stack on a dealer's shelf. The threads show on the outside of most metal caps, but there are some patented designs in which the metal is turned up inside to form the thread, leaving the outside straight and smooth. Plastic caps do not reveal the threads on the outside if they are made of thermosetting resins. The thermoplastic materials, however, will sometimes show the thread as a sink mark, caused by the shrinkage of the plastic in the mold. To hide this, it is common practice to put a fluted or knurled design around the sides of the cap. A slight doming of the top usually improves the appearance of a cap. The bottom edge of the cap should come as close to the surface of the glass as tolerances will allow, for the best cosmetic appearance. A generous radius on the inside of the bottom edge will more nearly conform to the shape of the container and avoid touching at that point. The *liner* should preferably be snapped in place instead of glued, so that it can rotate inside the cap, thereby reducing friction when the cap is being tightened or loosened. This is easily done in a molded closure, but it is more difficult with a metal cap. (See Fig. 1.)

In selecting materials for caps, the effect upon flavor or odor of the product may be a factor. Thermoplastics are generally better than thermosets in this case. Compatibility with the product is another important consideration, and to test it, simply immersing the closure for a period of

Fig. 1. Cap mold. An unscrewing type of mold for plastic caps is an intricate tool made to very precise dimensions. The mold maker is resting his hand on an air cylinder that draws the gear racks shown extending above the mold; these turn the pinions which unscrew the cores out of the caps. See also Fig. 11 in Sec. 8, "Plastics," page 8-50, which shows a mold disassembled. (*Newark Die Co.*)

time is not enough. When a cap is in place on a container, it is under stress, and under that condition the product may cause cracking, which would not show up in a simple immersion test. (See Table 6.)

Lug Caps. The quick-acting lug style of cap is used primarily in the food field, where it is well known for its use on wide-mouth jars for mayonnaise, mustard, and similar items. It is starting to come into more general use on small-mouth bottles for catsup and sauces. It has several advantages over the continuous-thread style of cap. For one thing, it is simple for the consumer to remove with a quarter turn. Another advantage is that the pressure of the cap is more uniform around the sealing surface of the container. Particularly with the larger sizes, a continuous-thread cap is only pulling down about three-quarters of the way around, and there is no thread contact the rest of the way.

In the past there has been some difficulty in getting a good match between the cap and the glass finish. As the molds would wear and the

TABLE 6 Approximate Cost of Screw Caps, per Thousand*

Size	Tinplate	Styrene	Phenolic	Urea	Aluminum
15	$ 2.14	$ 4.17	$ 3.42	$ 4.38	
18	2.26	4.34	4.31	5.67	$12.07
20	2.38	4.51	4.63	6.51	10.43
22	2.51	4.60	5.06	7.45	10.72
24	2.83	4.72	5.56	8.01	11.88
28	3.07	5.14	6.23	8.93	7.46
33	3.71	7.24	8.48	11.50	8.23
38	4.23	8.22	11.35	14.88	14.07
43	5.27	10.98	12.84	17.64	17.95
48	5.87	13.26	15.04	20.69	18.79
53	6.33	....	16.72	23.78	19.63
58	8.50	....	18.87	27.79	
70	10.05	....	27.33	37.91	
89	17.42	....	42.24	60.13	
120	33.29	....	80.23	98.54	

* Prices are for 400 series caps, standard colors, unlined, in large quantities. Add 10 percent for liner and 3 percent for lithography on metal caps.

glass would grow, the fit became tighter. This objection has been overcome to a large extent by a change in the design of the quarter-turn thread. Whereas previously the helix angle changed to a flat section at the bottom, and the distance from the sealing surface to this flat portion was very critical, now the helix angle is carried all the way down. This makes it self-adjusting, and the seal depends on the amount of torque applied and not on the L dimension to the flat.

Crown Caps. This style of cap, which is so widely used as a closure for beverage bottles, owes its development to William Painter, whose chief occupation was the manufacture of patent leather. He worked on it from 1884 until 1892, when he patented his final design, which was very much as we see it today. The Crown Cork & Seal Company was the first to make these caps, and they were followed by others until today their combined output is around 50 billion annually.

Originally the liners were cut from solid cork, but in 1912 composition cork was introduced, and it was quickly adopted because of its more uniform quality. *Spot crowns* consisting of disks of paper, foil, or film centered on the cork liner were introduced in the late 1920s to better preserve the sensitive flavors of ginger ale, orange soda, and beer. Not the least of its desirable qualities is the low cost of the crown cap. Large-volume users pay about $1.75 per thousand.

Specialty Caps. There are many types of applicator closures, and the types and designs vary with the different manufacturers. Caps with a boss on the inside to hold a brush or a glass rod, or sometimes a sponge, are used with cleaners, polishes, and adhesives. Polyethylene rods with the cap liner molded as an integral part are made to be used with standard caps. Shaker closures for sauces and hair tonics are fitted with solid liners that seal the bottle but can be removed by the customer for dispensing. Sifter tops are made for powders and granular products, such as ground spices. Snap caps of polyethylene are used on vials for tablets and capsules. Spout caps for lighter fluid and lubricating oil are well known. Caps with a center hole are available for use with medicine droppers or pump dispensers.

SEALS

Shrink Bands. Supplementary seals are often applied to luxury products. These include lead foil capsules, fish skin membranes, vinyl stretch bands, and vinyl shrink bands. The last is a tube of oriented (stretched) polyvinyl chloride 0.002 to 0.004 in. thick, made to slip easily over the cap and neck of the bottle. It can be put on by hand or machine. When the bottles are passed through a tunnel heated at around 285°F, the band shrinks in about 3 sec to form a secure tamperproof seal. The stretch-type bands, on the other hand, are made about 20 percent smaller than the package, and they must be stretched mechanically just before they are put in place. They return to their original size in a few minutes, to make a tight fit on the package. (See Table 7.)

Probably the most popular of these tamperproof seals are the viscose shrink-on bands and caps. They are available in a variety of stock colors and designs, and special messages and trademark designs can be supplied if the quantity is sufficient. They are made from regenerated cellulose, in somewhat the same manner as cellophane. Wood pulp is dissolved in caustic soda and carbon bisulfide, along with plasticizers, humectants, and colorants. The viscose that is thus produced is

TABLE 7 Approximate Costs of Vinyl Stretch Bands, per Thousand*

Width, in.	Circumference					
	12 in.	16 in.	20 in.	24 in.	30 in.	36 in.
3/16	$ 2.84	$ 3.24	$ 3.64	$ 4.04	$ 4.85	$ 5.22
3/8	6.60	7.40	8.20	9.00	10.24	11.40
5/8	10.24	11.62	13.00	14.32	16.40	18.32
1	12.65	14.60	16.55	18.68	21.65	24.58

* Add about $15 per thousand for imprinting, and a plate charge of $75. Minimum quantity for special imprints is about 5,000 bands.

TABLE 8 Approximate costs of Viscose Shrink Bands and Caps, per Thousand*

Diameter × length, mm	Caps	Bands
20 × 40	$ 6.10	$ 2.34
28 × 50	8.65	3.14
38 × 60	13.50	4.28
48 × 65	17.50	5.79
58 × 70	22.80	7.43
70 × 75	28.00	10.61

* Minimum quantities are around 25,000. Printing adds 25 to 50 percent to the cost. Special color adds 10 to 25 percent to the cost. Special die cuts add from 15 cents to $2.20 per thousand.

extruded as tubing from a die of the proper size for a particular cap size, into an acid bath which causes the viscose to coagulate. Special colors are possible, but because they are difficult to control in going from an alkaline medium to an acid medium in the process, opaque white is safest in this respect. The bands can be printed with instructions or a sales message in any color, except that the background color will affect the print color; for example, blue ink on a yellow background becomes green. Caps may be used in place of bands, but they are so much higher in cost that they are seldom used. Viscose caps can be used as primary closures for dry products, for example, to cover the sifter top on a powder container. The material is a poor barrier for moisture, however, and its chemical resistance is similar to that of uncoated cellophane. (See Sec. 3, "Films and Foils.") The difficulty of removal by the consumer can be overcome by perforating across the tube or die-cutting a tab in the middle. (See Table 8.)

Viscose bands are stored in water with some glycerine and mold inhibitors added. They are applied to the package while wet, and as they dry out, they shrink, just as any other cellulosic material such as wood or paper would. Specifications for size are given in millimeters, diameter first and then the length. In the case of caps, length is given from the bottom edge to the midpoint of the top surface. This is for the desired finished size and not the size of the wet band or cap.

QUALITY CONTROL

There are numerous defects in closures that must be controlled: the fit of the cap on the container, appearance and color match, integrity of

TABLE 9 Quality Control Criteria for Screw Caps

Critical defects, AQL 0.6 percent	Major defects, AQL 1.5 percent	Minor defects, AQL 4 percent
Incorrect liner facing	Cocked liner	Incomplete decoration
Incorrect liner backing	Incomplete threads	Smeared decoration
Liner thickness	Liners improperly glued in cap	Poor registration of decoration
Notched liner	Scratched or chipped surface	Dirt
Short shots (plastic)	Cracked or broken caps	Soft finish (coating)
No liner	Dents in metal caps	Glue on threads
Wrong color or copy	Mold flash, if sharp	
Sharp points or edges	Incomplete cure of plastic	
Wrinkled facing (delaminated)	Breakage when applied to bottle	

the sealing surface, and the usual accumulation of dirt and mechanical damage that any packaging material is subject to. However, because of the vital part the closure plays in the total package performance, it is more important to do a thorough job of evaluation here than for any other component. The list of defects and the suggested acceptance quality levels (AQL) shown in Table 9 may be helpful.

LINERS

The smallest component part of the package, but oftentimes the most important, is the cap liner. It must provide a tight seal, be compatible with the product, and not be a source of contamination. In many cases not enough attention has been paid to this essential element, and spoilage amounting to many thousands of dollars has resulted.

Liners are usually made with a resilient backing, plus a facing material. Sometimes a lubricant is put on the facing. The backing must be soft enough to take up any irregularities in the sealing surface of the container. It must be elastic enough to recover some of its original shape when it is removed and replaced. Finally, it must not be affected by the product or have any deleterious effect on the product. It is usually glued into the cap with a casein-latex type of adhesive, although polyolefin caps can be made with an undercut so that the liner is merely snapped into place and is free to rotate, thereby providing a more effective seal.

The facing must be carefully selected to give the barrier properties required, must withstand abrasion from the sealing surface when it is applied or removed, and must be compatible with the product.

BACKINGS

The resilient material which makes up the bulk of the cap liner can be various types of paperboard, cork, rubber, or plastic. These are sometimes used alone, or the paperboard can be coated or saturated with wax.

Pulpboard. The most widely used backing for pharmaceuticals, toiletries, and cosmetics is white pulpboard. This is made from groundwood pulp, with a small amount of sulfite pulp added for strength. It is supplied in thicknesses of 0.020, 0.030, 0.035, 0.040, 0.045, 0.050, and 0.055 in. Most common is 0.035 in. when used with a facing, or 0.040 in. when used alone. Pulpboard compresses 4 to 5 percent with about 80 percent restitution.

Newsboard. This is made of reprocessed newspapers. It is furnished in thicknesses of 0.020, 0.035, 0.040, and 0.045 in. Newsboard compresses about 6 percent with 80 percent restitution.

Gray Feltboard. Made from reprocessed wool rags, it is softer and more compressible than pulpboard or newsboard. It is available in thicknesses of 0.040, 0.045, 0.060 0.068, and 0.072 in. The two heavier weights are often used with roll-on (RO) aluminum caps.

White Feltboard. This is made from soda pulp combined with sulfite pulp and mixed with cotton linters and processed rag stock. It is used where a clean white material is needed. It has greater compressibility than either pulpboard or newsboard. It is often used with a foil facing.

Composition Cork. This is made from ground cork with a plasticized phenolic resin binder in a ratio of about 4:1. Animal glue, blood albumin, or gelatin is sometimes used as a binder, with aqueous glycerin as a solvent-humectant-plasticizer. (See Table 10.) Cork is composed of tiny cells, so small they can be seen only with a microscope. There are about 200 million of these in a cubic inch of cork. More than 75 percent of the volume is air. Cork has considerably more compression than the paperboard materials mentioned above, and recovery is also quite good. It can be coated, but it cannot be wax-saturated. Cork comes in thicknesses of 3/64, 1/16, 1/12, 3/32, and 1/8 in. Fragrances and flavors may be affected if a facing is not used.

Fine-grain Composition Cork. This is made from finely ground cork with a nontoxic, mold-resistant, plasticized resin binder. It cuts out more cleanly than regular composition cork, and it has better working qualities, especially for small liners. Cork has more compressibility and

TABLE 10 Properties of Cork

Temperature use range	−300 to 200°F
Density	6.7 lb/cu ft
Deflection at 10 psi	3.5 percent
Recovery	95 percent
Water absorption, by weight at 70°F	2.7 percent
Dimensional change vs. moisture content	1 percent
Capillarity	None
Air permeability	395 cu ft/hr/sq ft/in.
Vapor permeability	3 g/hr/sq ft/in. Hg

better restitution than pulp or felt. It is available wax-coated, but cannot be wax-saturated. It is supplied in thicknesses of 3/64, 1/16, 1/12, 3/32, and 1/8 in. If flavors or fragrances are involved, a facing may be required. The cork is ground to a uniform size, around 10 mesh, cleaned, and graded. For liners it must be composed of discrete, resilient particles, free from foreign matter, fines, or hardback. Cork will *compress* about 30 percent, and restitution after 5 min is around 90 percent. In extremely dry climates cork is less effective in producing a tight seal. For best results, cork liners should be stored where the air is not too dry. The liners for crown caps, being of uniform size, are more economically made by extruding the cork as rod stock and then slicing the liners from the cork rods. For screw caps the liners are generally punched from sheet stock. In damp, hot climates, where beverage use is high, mold growth in the cork liners for crown caps has always been a problem. *Mold inhibitors* should be added to the binder if this condition is likely to be encountered.

Rubber. In the form of cut or molded disks, or rings, rubber is used for serum vials and similar containers for biological products. It is used also for food containers, but usually as a *flowed-in* rubber gasket. Resistance to acids is good, but hydrocarbons and oils cause swelling, and eventually they will dissolve the rubber, as will esters and ketones. The water-vapor transmission rate (WVTR) is very low, being less than 0.5g/100 sq in./24 hr for natural rubber and even less for butyl rubber.

Wax Treatments. Wax has a twofold purpose. It provides an excellent barrier for water vapor, and it serves as a lubricant on the mouth of the container. It has good resistance to acids and alkalies, but it should not be used with organic solvents. Pulpboard can be saturated with refined paraffin (melting point 120°F) and then coated with microcrystalline wax (melting point above 140°F) about 0.007 in. thick. The water-vapor transmission rate is around 0.01 g/100 sq in./24 hr.

A thin wax *coating* is often used on liner facings made of metal foil or plastic film. It improves the barrier properties against moisture loss, lubricates the sealing surface, and fills up any pinholes that might be present in the facing material. (See Table 11.) If the coating is too thick, it may flake off under certain conditions.

TABLE 11 Thickness of Wax Coatings

Type	Minimum, in.	Maximum, in.
Special waxed .	0.0001	0.00025
Flash-waxed . .	0.00025	0.0005
Lightly waxed .	0.0005	0.001
Waxed	0.001	0.002
Heavily waxed .	0.002	0.003

Quality White Waxed Paper. Clay-coated sulfite paper is coated on both sides with 0.002 in. of refined paraffin wax (melting point 135°F). It is generally used with dry products. The water-vapor transmission rate is 0.01 g/100 sq in./24 hr. It has fair resistance to weak acids, weak alkalies, and water, but is not recommended for organic solvents.

FACINGS

The most critical part of the closure, and for that matter the whole package, is the facing on the liner. This disk is the key to the whole package system. In the early days there were not many facings to choose from, but today we have materials to suit almost any situation. In the late 1920s the list of facings available was limited to tinfoil, kraft oil, black oil, quality white waxed paper, and waxed pulpboard. The mainstay in the food line was kraft oil, either plain or waxed. The biggest problems were in the cosmetic, drug, and chemical products. Vinylite and Pliofilm were introduced in the late 1930s and filled a long-felt need. These were followed by saran and then by polyethylene, so that we now have a fairly good choice of materials to use, either alone or in combination. (See Tables 12 and 13.)

Kraft Oil (Yellow Oil). Supercalandered 0.004-in. kraft paper is coated with about 0.0001 in. of oleoresinous varnish, on one side or both sides. Sometimes a bleached kraft or an 80-lb (0.005-in.) drab express paper is used. Paper may be printed to look like drab express and is called simply printed drab express. When the varnish is mahogany-colored, it is called dark brown acid-resisting.

Black Alkali-resistant. This is a special alkali-resistant varnish that is colored black. It can be put on any of the papers that are used for kraft oil, but is usually made with supercalendered 0.004-in. kraft paper. Varnished papers have good *chemical resistance* to water, oils and greases, weak acids, and weak alkalies. They are not resistant to alcohols, hydrocarbons, chlorinated hydrocarbons, ketones, or ethers. The water-vapor transmission rate is about 0.4 g/100 sq in./24 hrs.

Casein-coated Varnish Papers. All the previously varnished papers can be casein-coated to provide additional resistance to organic solvents or to products which contain these solvents. They have excellent resistance to hydrocarbons, chlorinated hydrocarbons, ketones, esters, ethers, cellosolves, and absolute alcohol, but not 95 percent alcohol. They should not be used for products containing water. The water-vapor transmission rate is about 8 to 9 g/100 sq in./24 hr.

Vinylite. This is a white-pigmented vinyl chloride–vinyl acetate copolymer, which is usually calendered onto a white sulfite paper in a 0.002-in. coating. It has good resistance to oils and greases, water, alcohol, weak acids, and weak alkalies. It is not resistant to hydrocarbons, chlorinated hydrocarbons, ketones, or ethers. The

TABLE 12 Identification Tests for Liner Materials

Facing	Solvent test*	Flame test†	Hot-wire test‡	Acid test§
Kraft oil	Dissolves slowly	Does not melt, turns dark, oily odor	....	Darkens slightly
Casein varnish	Resistant	Does not melt, turns brown, protein odor	....	Attacked rapidly
Vinylite	Dissolves	Melts, hydrochloric odor	Green color	Turns brown at once, later becomes pink
Urea-formaldehyde	Resistant	Does not melt, turns brown and brittle	....	Darkens and dissolves at once
Polyethylene	Resistant	Melts quickly, waxy odor	....	Resistant
Saran	Dissolves	Melts, leaving black spot with brown edges	Green color	Turns yellow slowly
Pliofilm	Dull and mottled when rubbed	Melts, turns brown and sticky, rubber odor	Green color	Resistant
Mylar	Resistant	Melts slowly, sharp odor	....	Resistant
PT cellophane	Resistant	Bubbles, then burns like paper	....	Darkens slightly
Tinfoil	Resistant	Melts	....	Darkens at once
Aluminum	Resistant	Resistant	....	Resistant
Lead foil	Resistant	Melts	....	Resistant

* Solvent test: Dip rag in methyl ethyl ketone and rub.
† Flame test: Hold lighted match close to surface for a few seconds.
‡ Hot-wire test: Heat clean copper wire and touch surface, then hold in flame and note color.
§ Acid test: Put drop of 95 percent sulfuric acid on surface for a few minutes, then wash off.

TABLE 13 Permeation Rates of Liner Facings*

Facing	Water	Alcohol
Polyethylene, 2-mil	0.07	0.06
Saran, 75-gauge	0.07	0.08
Aluminum foil, 1-mil	0.04	0.12
Tinfoil, 1-mil	0.09	0.20
Polyester, 50-gauge	0.12	0.10
Vinylite	0.20	0.03
Solvent-resistant	0.23	0.61
Yellow oil	0.28	0.85

* Loss in grams from a 2-oz bottle with 22-mm cap with cork-backed liner, 13 weeks at room temperature.

water-vapor transmission rate is about 2 g/100 sq in./24 hrs. A *flash-waxed* coating will reduce this to about 0.5 g/100 sq in./24 hr. Foamed vinyl plastisols are being used for flowed-in gaskets and also as a substitute for cork in crown caps. They provide an excellent seal for carbonated beverages, but moisture loss and chemical compatibility of foamed vinyl are about the same as for Vinylite.

Solvent-resistant Papers. A coating of urea-formaldehyde-melamine resin, either clear or white-colored, is applied to white, kraft, or drab express paper. It has excellent resistance to oils and greases, weak alkalies, alcohols, hydrocarbons, and ketones. It is not recommended for acids or chlorinated hydrocarbons. Heat resistance is excellent, but it is not as flexible as some other facing materials. The water-vapor transmission rate is about 0.4 g/100 sq in./24 hr.

Polyethylene. White sulfite paper is coated with 0.002 in. of polyethylene, either high-density or low-density; or 0.002- to 0.010-in. film can be laminated to the paper. Most solvents will not attack polyethylene, and it is unaffected by strong acids and alkalies. Oil and grease resistance is generally good, but there are some exceptions, and tests should be conducted with the particular product to be packaged. Polyethylene is not a good gas barrier and should not be used for oxygen-sensitive products. Moisture protection is good, and the water-vapor transmission rate is about 0.4 g/100 sq in./24 hr for 0.002-in. thickness. Heavier films give lower figures.

Saran. White sulfite paper with a 0.001-in. coating of saran, or 75-gauge (0.00075-in.) film, either clear or white-pigmented, is coming into wide use because of its excellent barrier properties. The water-vapor transmission rate is 0.4 g/100 sq in./24 hr and the gas transmission rates are likewise very low. It is resistant to weak acids, weak alkalies, oils, and greases. It should not be used with hydrocarbons, ketones, or ammonium compounds. It is suitable for hot packing of foods.

Pliofilm. Rubber hydrochloride film 0.0012 in. thick is laminated to white sulfite paper. It has good resistance to acids, alkalies, alcohols, water, oils, and greases. It is not recommended for hydrocarbons, ketones, esters, or essential oils, and it should not be used for hot-packed foods. The film becomes brittle with age. Water-vapor transmission rate is 0.5 g/100 sq in./24 hr.

Mylar. This polyester film in various thicknesses is laminated to a white sulfate paper. It is available in 35-gauge (0.00035-in.), 50-gauge (0.0005-in.), or 100-gauge (0.001-in). It has good resistance to alkalies, alcohols, hydrocarbons, chlorinated hydrocarbons, ketones, esters, cellosolves, oils, and greases. It should not be used with acids, phenols, cresols, or benzyl alcohol. The water-vapor transmission rate for 50-gauge film is 2½ g/100 sq in./24 hr.

Cellophane. Various types of cellophane are used as liner materials and also as tamperproof seals that are adhered to the mouth of the container. *Plain transparent* (PT) cellophane is made of regenerated cellulose. When used alone, it has poor moisture and water barrier properties. It does, however, have good resistance to absolute alcohol, hydrocarbons, chlorinated hydrocarbons, ketones, esters, essential oils, mineral oils, and greases. It is sometimes used as a double or triple laminate, with itself or with coated cellophanes. Single films come in thicknesses of 0.0008 in. (yield code 250), 0.0009 in. (yield code 210), 0.0010 in. (yield code 195), 0.0011 in. (yield code 182), and 0.0014 in. (yield code 140).

More often the *moistureproof sealable transparent* (MST) types of cellophane are used; their properties depend on the coatings combined with them. The standard coating is nitrocellulose, but other coatings are available, notably vinylidene chloride, which has good gas and moisture barrier properties. See Sec. 3, "Films and Foils," for more information on this material.

Tinfoil. A lamination of 0.0015- or 0.002-in. tinfoil with paper, or directly onto the backing, will provide excellent moisture protection. The moisture-vapor transmission rate is generally regarded as zero. It has good resistance to hydrocarbons, chlorinated hydrocarbons, alcohol, ketones, esters, oils, greases, and water. It should not be used with acids or alkalies.

Aluminum Foil. Available in thicknesses of 0.0015 and 0.002 in., it can be laminated to paper or to the backing material. Because aluminum is fairly stiff and hard, it is better to use one of the softer backing materials, such as feltboard or cork, rather than a pulpboard. It can be used with hydrocarbons, chlorinated hydrocarbons, oils, and greases. It is not recommended for aqueous products, acids, or alkalies.

Lead Foil. Two types of lead foil are used for cap liners. *Chemically pure* (CP) foil is 100 percent lead, and *composition foil* has a coating of tin which is ½ to 1 percent of the total thickness. Lead foil is supplied in weights of 0.002 to 0.010 in. It is laminated to paper or directly to the backing material. It has good resistance to solvents, but is affected by acids and alkalies. It should not be used with foods, internal medication, or topical drug products.

Vinyl-coated Foil. A lamination of 0.00035-in. aluminum foil on kraft paper, with a thin vinyl coating, and usually lightly waxed, is widely used for toiletries. It is satisfactory also for mayonnaise and similar food products. Its water-vapor transmission rate is extremely low, and it is resistant to weak acids, weak alkalies, alcohol, oils, and water. It is not recommended for hydrocarbons, ketones, ethers, or essential oils.

Saran-coated Foil. A thin coating of saran, usually waxed, backed up with 0.00035-in. aluminum foil on white sulfate paper is a good choice for cosmetics. The moisture loss is near zero, and resistance to oils is good. It should not be used with hydrocarbons, ketones, or ammonium compounds.

Glassine Inner Seal. A lamination of two or three plies of 28-lb glassine, joined with a resin adhesive, is used as a loose component of a cap liner. When adhesive is applied to the mouth of a bottle or jar after filling, and the cap put in place, the liner adheres to the container, making a tamperproof seal. The backing, which stays with the cap when it is removed, must have a suitable facing for resealing the container when the inner seal is removed. Pressure-sensitive coatings can be used to adhere the inner seal to the mouth of the jar, or special heat-activated coatings are induction-heated by passing the finished packages through a high-energy electrical field. With induction heating the caps must be plastic, not metal, and an aluminum facing is used to create interference within the high-frequency magnetic induction field. The heating coil is made of copper tubing, with water circulating through it for cooling. The coil is just large enough to fit over the cap without touching it. Since glassine has poor resistance to any kind of liquid, it is used only with dry powders or granular products. Water-vapor transmission rate is about 2 g/100 sq in./24 hr.

Advantages of an inner seal are that it gives extra protection from leakage if the caps get loose; provides a tamperproof seal for the consumer; increases protection of the contents by reducing moisture and gas transmission rates; can be hot-packed to provide partial vacuum in the package; improves the implication of quality in the product; can be printed with special instructions or sales message; can be used as a return coupon for promotions.

Vented Liners. Some products such as laundry bleaches require special closures that will relieve any buildup of pressure in the bottle. There are various designs for these caps, but most of them consist of two or three notches in the edge of the liner backing, extending across and just beyond the sealing surface of the bottle. The liner facing will have several pinholes near the center, placed so that they will not coincide with the notches in the liner. The cap itself may have some small holes near the center.

STOPPERS

The use of cork as a stopper material dates back to the middle 1500s and even earlier. The stoppers were tapered corks, and when champagne was invented in the 1660s, they were tied in place with string. The bark of the cork oak (*Quercus suber*) from Spain and Portugal is still used for wine bottles, although plastics are gaining in popularity. When the cork tree is at least 12 years old, the first stripping of the bark takes place. The tree is not stripped again until 9 years later; in Portugal this is by decree of the Junta Nacional de Cortica. Cork suitable for wine stoppers is not obtained until the third harvest. After air drying, the cork is taken to the factory to be boiled, to destroy insects and molds, and to shrink the pores. After rinsing and bleaching, the cork is sorted for quality. Some cork stoppers are decorated with embossed wood tops made on the same equipment that is used for making checkers. Wine corks are generally divided into three grades: *second quality* cork for wines that are aged a relatively short time; *first quality* for longer periods of aging; and *extra select quality* for premium wines.

The wine and liquor industry in this country uses three types of stoppers: straight wine corks, chamfered whiskey corks, and champagne corks. Straight wine corks are either 1½ in. long for tenth bottles or 1¾ in. long for fifth bottles. A champagne cork is 48 mm long by 31 mm in diameter. The lower half is squeezed down to half its diameter and inserted into the champagne bottle, and the upper half is then mushroomed and a wirehood, or muselet, is applied to hold the cork in place.

Cork is favored for wine bottles because of tradition and also because of the satisfying "pop" when it is drawn from the bottle. Actually a screw cap with a liner having a foil or film facing will better preserve the sensitive flavor of a fine wine, but it is difficult to change the long-standing customs in the wine industry.

The largest use of rubber stoppers is in chemicals and biologicals. Sleeve stoppers have a skirt which is molded upward from the solid plug portion. After insertion in a serum bottle, the skirt is pulled down around the mouth of the bottle. Other types of rubber plugs have a thin section in the center for insertion of a hypodermic needle. These plugs are held in place with aluminum seals that are crimped under the lip of the bottle, and have tear-off tabs.

Suppliers of rubber stoppers usually also make *rubber bulbs* for medicine droppers. These are sometimes made with a flange on the bottom, and when inserted in a cap with a suitable hole, and with a glass or plastic pipette, they form a closure for dispensing drop dosages.

TAPES

A great variety of tapes are used as a means of closure in packaging. Probably the best known is the *gummed tape* that is moistened and applied to bundles and boxes. For corrugated shipping containers the width and thickness of tape are specified in the carrier regulations. (See Sec. 16, "Laws and Regulations.") Basically it is made from 60-lb basis weight kraft paper 2 in. wide. If the tape is made from two 30-lb sheets with reinforcing fibres in between, such as sisal, glass, or rayon, less tape is required to meet the regulations. These tapes are available in bleached white or skim white, as well as kraft, to match the color of the shipping container. They can be vat-dyed a solid color, or they can be printed with a sales message or brand identification. (See Tables 14, 15, and 16.)

There are also *pressure-sensitive* tapes that can be used for sealing boxes, but the cost is so much higher than the gummed tapes that they are seldom used. They are not as secure, since they will come loose

TABLE 14 Cost Comparison of Closing Methods for Corrugated Boxes*

	Glued flaps (50 percent coverage) at 0.25¢/sq ft	Reinforced tape (2 pieces per box) at 0.5¢/ft	Dustproof seal (6 pieces per box) at 0.1¢/ft	Stapled flaps (preformed staples) at 0.1¢ each
Material .	0.5¢	2.2¢	1.3¢	3.8¢
Labor . . .	1.2	0.9	3.0	1.6
Total. .	1.7¢	3.1¢	4.3¢	5.4¢

* Costs based on a box 24 by 18 by 18 in. hand-sealed.

TABLE 15 Approximate Costs of Gummed Paper Tapes, Cost per 100-yd Roll*

Width, in.	60-lb kraft	Asphaltic reinforced one way	Asphaltic reinforced two ways	Nonasphaltic reinforced one way	Nonasphaltic reinforced two ways
1	$0.12	$0.41	$0.44	$0.52	$0.56
2	0.24	0.82	0.88	1.04	1.12
3	0.36	1.23	1.32	1.56	1.68

* For special colors add 10 percent.

TABLE 16 Comparison of White and Brown Paper Tape Costs per 100-yd Roll

Width, in.	White tape				Brown tape			
	35-lb	60-lb	90-lb	Reinforced asphaltic	35-lb	60-lb	90-lb	Reinforced asphaltic
1	$0.15	$0.16	$0.25	$0.45	$0.13	$0.19	$0.22	$0.40
2	0.30	0.32	0.50	0.90	0.26	0.38	0.44	0.80
3	0.45	0.48	0.75	1.35	0.39	0.57	0.66	1.20

under prolonged stress, especially if the force is in the direction of peeling rather than sliding. The adhesion of pressure-sensitive tapes improves with time for the first couple of days. The adhesive is formulated to suit the application; thus masking tape is less tacky because it must be easily removable without affecting the surface on which it is placed. For tape that must resist stress as soon as it is applied, quick tack can be built into the formula. (See Tables 17, 18, and 19.)

Backing materials are various grades of paper, cloth, or film such as cellophane or polyester, sometimes reinforced with rayon fibers. Tear tapes for corrugated boxes are a combination of rayon and film, with high strength in the machine direction (MD) but very weak in the cross direction (CD). Similar tapes are used for strapping full pallet loads together, in place of steel straps, and for banding metal rods and lumber into bundles. Fiber drums with telescoping covers are usually sealed with a strip of pressure-sensitive tape over the joint, and multiwall bags can be made siftproof by covering the closure with tape known as a tape-over-sewn (TOS) closure.

Testing of pressure-sensitive tapes should cover the backing as well as the adhesive. *Peel adhesion* is measured in ounces per inch of width, when peeled from a finely ground stainless steel surface at an angle of 180° and at a speed of 12 in. per min, 15 min after application. *Sheer adhesion* is a measure of the time in minutes for 1-in.-wide tape, with 1/2 in. of its length attached to a finely ground stainless steel plate in a vertical position, and with a 500-g weight fastened to its lower end, to separate from the panel. If tape is to be subjected to low temperatures, the peel adhesion and sheer adhesion tests should be repeated at refrigerator or freezer temperatures. Check also for adhesive transfer as a result of the peel adhesion test. *Unwind adhesion* is a measure of the force in ounces per inch of width to pull tape from the roll at 12 in. per min. *Tensile strength* is measured in pounds per inch of width when a 4-in. length is stressed at a rate of 12 in. per min. *Elongation* is determined under the same conditions as tensile strength above (usually done at the same time) and reported as a percentage of length. *Moisture-vapor transmission rate* is measured in grams per 100 sq in. per 24 hr, when used to cover a test dish containing calcium chloride and exposed to 95 percent relative humidity at 100°F. *Quick stick* is measured as the ounces required to peel tape at an angle of 90°, from a finely ground stainless steel plate, at a speed of 12 in. per min, within 1 min of laying the tape horizontally on the plate with only the pressure of its own weight. If tape is to be applied to a painted surface, it should be tested for staining by applying and exposing to ultraviolet light for 4 hr, and such other conditions as may be expected.

TABLE 17 Typical Properties of Pressure-sensitive Tapes

Property	Sealing	Print cloth	Polyethylene-impregnated cloth	Vinyl-coated broadcloth	Vinyl film	Aluminum foil, 3-mil	Cellophane	Cellulose acetate, 2-mil
Peel adhesion, oz/in. width	16	28	50	32	35	32	30	30
Tensile strength lb/in. width	11.5	38	34	65	21	27	30	25
Thickness, mils	8	12	9.5	14	6	5	2.7	2.8
Elongation, percent	10.5	. . . . .	. . . .	. . .	150	5	12	12
Color	Natural	Natural	Silver	Any	Any	Natural	Clear	Clear

TABLE 18 Pressure-sensitive Tapes

Description	Feature	Application	Government specification			Cost per yd, 1 in. wide	Supplier's product number		
			Specification	Type	Class		Permacel	3M	Technical
Clear cellophane	Transparency	Light packaging	LT-90C	I	A	$0.014	44	600	200
Transparent acetate	Dimensional stability	Label protection	LT-90C	II	B	0.030	90	800	
Colored acetate fiber.	High holding power	Can sealing	LT-101 PPP-T-60 Jan-P-127	III III III	 3 B	0.032	98	710	
Clear acetate fiber	High holding power	Government packaging	PPP-T-60 Jan-P-127	III	2 B,C	0.030	991	711	
Waterproof paper	High holding power	Carton sealing	PPP-T-76			0.012	17	260	
Rope fiber flatback paper . .	Stain-resistant, high tensile strength	Holding	. . . .	. . .	. . . . .	0.019	72	251	162
Extra strength flatback paper	Extra high tensile strength	Bundling	. . . .	. . .	. . . . .	0.028	729	280	
Crepe paper	All-purpose	Masking	UU-T-106C	I		0.013	703	271	130
Crepe paper	Stain-resistant	Masking		. . .	. . . . .	0.013	70	215	122
Double-faced paper	High holding power	Holding	UU-T-91A	. . .	. . . . .	0.053	02	400	405
Aluminum foil	Waterproof	Protective wrapping	MIL-T-11291	. . .	. . . . .	0.054	11	430	
Reinforced strapping	Low Cost	Heavy packaging	PPP-T-97A,B,C	II	Opaque	0.036	15	325	1500
300-lb/in. strapping	High tensile strength	Heavy packaging	PPP-T-97A,B,C	II	Trans.	0.046	162	898	
Hi-tack strapping	High impact strength, colors	Heavy packaging, bundling	PPP-T-97A,B,C	I	. . . . .	0.039	16	880	188
500-lb/in. strapping	Very high tensile strength	Palletizing	PPP-T-97A,B,C	III	. . . . .	0.046	161	890	189
500-lb/in. suspension	Waterproof, noncurl	Suspension packaging	PPP-T-97A,B,C	IV	. . . . .	0.048	164	870	
Cotton cloth	Low cost	Utility packaging	PPP-T-60 Jan-P-127	II I	. . . .	0.020	64	380	60
Waterproof cotton cloth . . .	Waterproof, colors	Packaging	PPP-T-60	III	I	0.030	691	. . .	90W–99

TABLE 19 Approximate Cost of Pressure-sensitive Tapes, per 60-yd Roll

Width, in.	Masking tape	Strapping tape	Cellophane tape
1/4	$0.22	$0.60	$0.29
3/8	0.28	0.65	0.39
1/2	0.33	0.84	0.53
5/8	0.38	1.00	0.63
3/4	0.42	1.25	0.76
1	0.55	1.55	0.83

ADHESIVES

The amount of adhesive that is used in a package is a relatively small part of the total, in terms of the quantity used or as a part of the final cost. But just as a chain is as strong as its weakest link, so the package may be only as good as the adhesive that is used. The importance of the adhesives industry in the total packaging field is indicated by the production output in this country of over a billion pounds a year, worth somewhere around $125 million. About half of this, on a pound basis, is in vegetable-type adhesives, with resins nearly the same amount, whereas animal glues account for only about 10 percent of the total. Mineral-base materials (silicates) used in making corrugated board represent a still smaller segment of this business. Over 75 companies make adhesives, but the five giants do 65 percent of the total business (excluding corn products).

Principles of Adhesion. The objective in joining materials by means of an adhesive is not to get the substrates as close together as possible, but rather to get a sandwich of substrate-adhesive-substrate. Although it may be desirable to keep the adhesive to a minimum, it is poor practice to get it too thin, either with pressure or by dilution. In fact a thicker film of adhesive is nearly always stronger than a thin film.

Rough or porous surfaces that have a "tooth" can be joined with materials that harden or crystallize. Very smooth surfaces on glass or plastic containers, on the other hand, present a very different problem. For these it is often better to use an adhesive that remains flexible and a little tacky. This provides what is sometimes called a suction bond. It may be necessary to *treat* some plastic surfaces, particularly the olefins, with a flame or corona discharge to oxidize the surface. The use of *primers* to improve the bonding characteristics of certain films and foils is also important, particularly in the area of laminations, where extrusion techniques provide economical combinations of flexible materials. Do not overlook the importance of plasticizers or other *additives* in plastic films which may have an adverse effect on adhesion. A typical dextrin adhesive will cost about $1.20 per gallon. One gallon will cover about 50,000 sq in. if applied by hand, and 100,000 sq in. if applied by machine.

Theory of Adhesion. The mechanism of adhesion is not clearly understood, but it appears to be a combination of physical and chemical forces. The following are some highlights on a very complex subject which may help to provide a better understanding of adhesive phenomena. In the first place, *mechanical bonds* will result from the penetration of the adhesives into porous materials. When the sealant soaks into the fibers and hardens, there is an interlinking which gives a bond that depends for its strength upon the cohesive forces within the adhesive or the substrate, whichever is weaker. The viscosity of the adhesive and the porosity of the material will, of course, determine the amount of penetration into the substrate.

If the surfaces to be joined are not porous, we must depend upon molecular forces instead of the mechanical interlocking to form the bond. In this case the *roughness* or smoothness of the surface must be considered, for it is an important factor in adhesion. There is an ideal angle away from the parallel, for the slopes of the hills and valleys that make up the roughness of the surface, that will give the maximum holding power for the free energies of the particular materials involved. For paper or any other cellulose material, for example, this angle has been calculated to be 43°. The maximum work of adhesion in this case is between 40 and 60 dynes/cm, and so the adhesives should have a surface tension within this range. Shrinkage of the adhesive as it sets up will reduce the effectiveness of these forces, but it may be minimized to some extent by working with the adhesive at a higher temperature and chilling it as quickly as possible after it is applied. On perfectly smooth surfaces, however, the reverse may be true; that is, a low temperature and slow cooling will be less likely to set up stresses in the adhesive which tend to break up the intimate contact at the interface. A preliminary indication of this can be found by coating only one side of strips of the substrate, exposing them to various conditions of temperature, humidity, and aging, and then checking the condition of the adhesive.

As the surface becomes *smoother,* the molecular attraction between the adhesive and the substrate takes on a greater significance. There are several forces at work in these infinitely small spaces between the atoms and molecules, and the effectiveness of these forces is dependent in large measure on the distances involved. The "wetting" effect of the adhesive is very important in bringing the materials into intimate contact. Some adhesives have reactive groups in the molecule which can form primary bonds with the substrate. Examples of these reactive groups are oxirane, isocyanate, and cyano. They work better at the lower molecular weights, spreading out and giving a good wetting before they cure. When silicones are present as mold release agents, there is interference with the chemical bonding, due to a preferential orientation of electron-induced fields toward the oxygen atom next to a dimethyl silane group, which has room for additional electrons in the *p* subshell of the oxygen.

The *molecular diffusion* of one material into the other, which is on a much smaller scale than the "penetration" previously mentioned, is indirectly proportional to the ⅔ power of the molecular weight, and increases with temperature and time. This type of diffusion is more pronounced with polar compounds, provided the polar groups do not drastically interfere with the mobility of the molecular chains. Branchiness of the molecular chains also favors diffusion, and is proportional to the ⅔ power of the number of effective branches.

Dispersion forces, which are on a still smaller scale than "diffusion" phenomena, result from the changing position or "dispersion" of the electrons surrounding atoms which are in close proximity. This decreases with the inverse seventh power of the distance and becomes insignificant beyond a distance of 4½ Å. For comparison, carbon atoms in a chain are 1½ Å apart (six-billionths of an inch).

Electrostatic forces of polar groups are in addition to the dispersion forces, and they are three or four times as great. The amount of such energy can be demonstrated by the potentials which are developed in peeling polyvinyl chloride or the cellulosics from a glass surface. The discharge potentials build up to several thousand volts. One other factor that should be mentioned, although it is less significant than those already discussed, is *van der Waals* forces. These forces arise from the attractions that come into existence because of dipoles present in polar molecules. They are less than a tenth the power of a covalent bond such as a chlorine-chlorine bond.

TYPES OF ADHESIVES

The term *glue* usually refers to adhesive from animal sources, although vegetable materials are sometimes included in this category. *Paste* and *mucilage* are nearly always of vegetable origin, and they are used exclusively with paper products. *Cement* is a solvent type of adhesive for bonding such materials as leather, rubber, glass, etc. The word *adhesive* is generally taken to include the whole field, although in its narrowest sense it refers to synthetic resin emulsions. (See Table 20.)

Animal glue is made from the hides and bones of animals, chiefly beef steers. It is available in flake or granulated form and is dissolved in water at around 140°F. Bone glue and hide glue are usually blended to

TABLE 20 Properties of Adhesives

Adhesive	Resistance*					Effectiveness*						
	Cold	Fungus	Heat	Solvents	Water	Glass	Leather	Metal	Paper	Rubber	Textiles	Wood
Cellulose nitrate	F	E	P	P	F	M	E	F	E	P	E	G
Ethyl cellulose	G	E	M	P	P	P	M	P	E	M	G	M
Polyvinyl alcohol	M	E	G	E	F	P	G	P	E	P	G	M
Polyvinyl acetate	M	E	M	M	M	E	E	E	E	P	E	E
Asphalt	G	E	P	P	G	G	M	G	G	G	G	M
Wax	G	E	P	P	E	P	G	P	E	P	E	M
Phenol-formaldehyde	G	E	G	E	G	P	M	P	M	E	M	M
Urea-formaldehyde	E	E	M	E	M	P	G	P	E	P	E	E
Epoxy	E	E	E	E	E	E	E	E	E	E	E	E
Dextrin	F	P	F	M	P	P	M	P	M	P	M	F
Mucilage (gum arabic)	F	P	F	M	P	P	M	P	E	P	M	F
Silicate	G	E	E	E	P	E	F	G	E	P	G	G
Rubber cement	G	E	F	F	G	F	G	M	G	G	G	F
Casein	M	F	M	G	F	P	F	P	G	P	M	M
Animal glue	G	F	M	E	P	M	E	P	E	P	E	E
Soybean glue	P	F	F	M	P	F	F	P	G	P	M	F
Zein	G	G	G	G	M	F	E	F	E	P	E	E
Blood albumin	F	F	F	G	P	P	G	P	G	P	G	F

* E = excellent; G = good; F = fair; P = poor; M = no data.

make the standard product. It is generally medium brown in color, slightly on the acid side, and is fairly fluid in consistency. It has good tack and fairly fast setting and drying rates. Animal glue is used to a limited extent as an iceproof adhesive in the brewing and soft drink industries. It makes a strong joint with paper and wood products, but it has poor resistance to the effects of moisture, mold, fungi, and insects. It has good nonwarping qualities for setup boxes and tight-wrapping when used hot (140°F), with a minimum amount of water. *Fish glue,* made from fish offal, has limited application in packaging because of the odor.

Casein glue is a product of the dried curds of milk, sometimes combined with lime and other ingredients. It is prepared for use by mixing with cold water and is used as an iceproof glue for labeling beverage bottles. Casein glue is amber to dark brown in the unpigmented form, slightly alkaline, and fairly fluid in consistency. For fast setting, it must be applied in a very thin film. Dilution with water is difficult unless ammonia is used, and for greater water resistance alcohol is sometimes used as the diluent. It is mixed with clay to make clay-coated folding boxboard. It has poor tack, but very good adhesion when dry. *Soybean glue* is made from soybean meal after the oil has been extracted. It is combined with caustic soda, lime, and silicate of soda. *Zein glue* is made from a soft yellow powder derived from corn. It is soluble in alcohol and is pure, nontoxic, and practically odorless. It is used for laminating paper to paper. *Albumin glue* is prepared from soluble dried blood, with minor additives. It cures at around 170°F and makes a strong bond. The last four are widely used in the plywood industry for waterproof laminations. In packaging they have limited application, although casein is sometimes used as a label adhesive, and albumin is one of the binders used for making cork compound. They do not age well, and they are attacked by mold and fungi.

Resin adhesives that are used in packaging generally are polyvinyl acetate emulsions. They dry quickly and make a strong bond with paper and many films, coatings, and similar nonporous surfaces. They cost two to three times as much as dextrin glues, being priced around 35 cents per pound for 33 percent solids (f.o.b. factory). Some formulations can be reactivated with heat if a heat-sensitive coating is needed. Other types will form an easy-release bond. They are frequently used with plastic films, as for making windows in folding boxes and in envelopes. They are also used for the glued lap in corrugated cases.

Lacquers are similar to resin adhesives except that instead of being in a water-base emulsion, they are in a solvent solution. They are higher in cost and more difficult to use because of the solvent fumes. Their chief value is where the solvents are needed to "bite" into a smooth surface, as with plastic films. They are made from such materials as cellulosics, vinyls, gums, and resins, dissolved in alcohol, naphtha, benzol, methyl ethyl ketone, or other organic solvents.

Starch pastes are made from corn, tapioca, potato, wheat, rye, and sago. Starches in their natural form are used only for paper and paperboard packages, as in bag pastes and label pastes. More often the starches are converted by dry heat, acids, or enzymes into white or yellow-colored dextrins, which have a quicker tack and a higher solids content. Dextrin labeling glues are usually brown in color, relatively fluid in consistency, and generally on the acid side. They have good transfer and stenciling properties, and they dry fairly fast. However, they are not waterproof or iceproof. An identifying test is made with a dilute iodine solution; starch turns the test solution blue, whereas dextrin gives a red or brown color. *Borax* may be added for quicker setting and better adhesion. The borated dextrins do not work as well in the machines, however, because they lack shortness and tend to "feather" or string out. They should be used within 3 months of manufacture, although under ideal storage conditions they may keep for a year or more. Dextrins are the least expensive and have the best machining qualities of all adhesives, and equipment is easily cleaned with warm water. Tapioca dextrin is preferred for high-speed operations. Corn dextrin is lower in cost, but it does not set up as quickly. Starch ethers are a new class of compounds that are used in making bags, and also as a remoistening adhesive. Special types of dextrin can be combined with polyvinyl acetate emulsion to improve lie-flatness in making envelopes.

Around 1935 a new method of using starch adhesives for corrugated board was introduced. This is generally known as the Stein-Hall process, which uses raw rather than cooked starch, and by heating at the point of application, causes it to gel and become adhesive. Less water is required with this method, making it more practical for corrugated production. Cornstarch is generally used for reasons of cost and availability, although tapioca starch makes a better adhesive; a mixture of the two is sometimes used. Untreated starch makes up the bulk of the mixture, for the sake of economy, with small percentages of modified starches occasionally included in the formulation.

When raw starch is mixed with water and heated, the tiny capsules burst, releasing the soluble starch to mix with the water. The temperature at which this occurs is called the gel point; for cornstarch it is around 160°F. The gel temperature can be lowered to about 140°F by adding caustic soda, which permits faster machine speeds in making corrugated board. With enough caustic soda the starch will gel at room temperature, but this is not desirable. If it gels much below 140°F, there will not be enough penetration into the paperboard.

The viscosity of the solution is important in getting a good adhesive bond. The standard method of measuring the viscosity of a starch mixture is to determine the time that it takes for 100 cc at 100°F to flow through a 3/32-in. hole in a metal plate with the surface of the liquid about 3 in. above the hole. Cornstarch works best in the range of 29 to 34 sec. A thinner viscosity allows more penetration into the paper, but if penetration is too rapid, the starch will be starved of enough water to gel properly.

Borax has no effect on raw starch, but after it has gelled, the starch is converted from a thin stringy solution to a heavy short-bodied, tacky mixture by the addition of the borax. Too much borax will buffer the caustic soda, causing the gel point to rise.

Jelly gums are colloidal dispersions of modified starches that are used for labeling. They have fairly good resistance to high humidity and moisture, and they are generally considered semi-iceproof. Jelly gums are rubbery and cohesive, white to reddish brown in color, and usually alkaline. They are effective on hot, cold, wet, dry, greasy, or oily bottles made of flint, opal, or borosilicate glass, and on most of the coatings used to resist scratching, except the silicones. Jelly gums are widely used in the food, beverage, pharmaceutical, and cosmetic industries. Animal jelly glues are formulated to be liquid at room temperature and to gel in cold water; they are known as *iceproof gums* for beverage bottles.

Mucilage is used in the office and in the home, but it has very few industrial applications. It is made from a water solution of gum arabic, tragacanth, cerasin, or senegal. It is sometimes used for envelopes because of its nonblocking characteristics under humid conditions.

Silicate adhesive is often used to join corrugating medium to linerboard in the manufacture of corrugated cartons. It is a water solution of sodium silicate, about 60 percent water, with a viscosity around 40°Bé, having a Na_2O to SiO_2 ratio of 1:3.38; it is the lowest-cost adhesive available. A higher percentage of the alkaline Na_2O will make it set more slowly and may affect the color of the paper or the inks. Cleaning the machines is difficult after it has hardened. Silicate tends to stay on the surface of porous materials such as paper, and so it can be used in smaller amounts than most adhesives. When it dries, it makes a very brittle film, and for this reason it can be used only on rigid package structures.

The pH of linerboard is mildly acid, and this sometimes has an unfavorable effect on alkaline adhesives. Silicate adhesives, unlike most other adhesives, may have a caustic reaction under conditions of moisture and heat. This can cause dark streaks in the board, or it may attack the inks that are used in printing the box. It has even been known to etch glassware that came in contact with the linerboard. Since the silicate solution needs to lose only about 14 percent water, very little heat is necessary to drive off this small amount to make the silicate set. The proportion of water to solid is measured by the specific gravity of the solution on the Baumé scale. The higher the Baumé reading, the faster the solution will set. More important, however, is the ratio of soda ash to silica. The lower the soda ash, the faster the set.

Lower Baumé solutions will penetrate deeper into the paperboard and take longer to set These thinner grades should be used with dense stocks, and with high heats or slow speeds. The heavier mixtures are better suited to porous stock and high-speed operations. Sometimes silicate is combined with starch adhesive and with clay for special purposes. This should be done only by an expert, as silicate has a tendency to gel the raw starch, and there are also some patents covering these mixtures.

Cohesives or self-seal adhesives are made from blends of waxes and latexes. They have the characteristic of sticking to themselves and not to any other material. They must be kept clean because dust will interfere with their adhesion. The aging characteristics are poor, and they should be used within 6 months of being coated. Envelopes and folding boxes with cohesive material on their flaps are quickly and easily closed on the filling line. The gums are high in cost, but in some operations this is offset by the more efficient use of labor.

Latex adhesives are available as solvent solutions or as water emulsions. Rubber latex is a high-cost material, and these adhesives are not cheap. Natural crepe rubber dissolved in naphtha or benzol is used as a temporary or a permanent adhesive for paper, rubber, or cloth. Cost is around 20 cents per pound. Any trace of copper from the equipment will reduce the aging qualities of rubber adhesives.

Hotmelts made from thermoplastic resins such as polyethylene are generally used to tack things in place, rather than as the sole bonding medium, because of their high cost and also the difficulty of spreading in the short interval before the adhesive chills. One method is to use a flexible rope of adhesive, which is fed into a heated pump and applied immediately to the package. In another system, the adhesive is heated in a separate tank and is then circulated to the point of application through insulated tubes. Can labeling machines use hotmelts that have instantaneous high tack for label pickup. For bag and carton seaming and sealing, especially with nonporous surfaces, hotmelts permit high-speed operations. The definition of the term "hotmelt" excludes low-temperature waxes as well as high-melting resins, but spans an area somewhere between the two. Probably the temperature range of 300 to 400°F and viscosities between 500 and 15,000 centipoises would include most of the materials known as hotmelts. Cost of hotmelts is around 45 cents per pound.

Thermoplastic coatings that are applied to paper and paperboard can be reactivated by heat in a subsequent operation. This has permitted higher speeds in labeling, enveloping, and blister packaging. By printing the coatings in a special pattern, not only is the cost kept to a minimum, but easy-opening features can be built into the package.

Section 10

Bags and Envelopes

History 10-1
Advantages and Disadvantages 10-2
Nomenclature 10-2
Design Considerations 10-5
Processes 10-11
Testing 10-11

One of the oldest forms of packaging, and still one of the most popular, is the paper bag. It performs all the basic functions of packaging, that is, containment, protection, and communication, at the lowest possible cost. With a wide range of sizes and kinds of materials from which to choose, the bag offers a versatility that can scarcely be matched by any other type of package.

HISTORY

It would be difficult to say how the bag got its start and when it replaced the animal skins used by nomads for carrying water, wine, cheese, and other subsistence items. In more recent times, when hand operations were superseded by machines, bags became so generally available that now they are a staple household item. At the present time about 1 million tons of large shipping sacks are produced each year, and 1½ million tons of smaller bags, 90 percent of which are made of unbleached kraft, of the familiar type used in the grocery store.

The number of large shipping sacks produced in this country per year is over 3 billion, of which about 30 million are all plastic and the rest are multiwall paper sacks. In the high-barrier category of large sacks costing over 20 cents each, it is reported that nearly 100 million are produced, and for the medium-barrier sacks priced at 15 to 20 cents the figure is near 450 million. The remainder is assumed to be, from the way it is reported, the minimum-protection type of sack selling in the range of 7 to 15 cents each.

ADVANTAGES AND DISADVANTAGES

Of all the various package forms the lowest in unit cost is undoubtedly the paper bag, if we exclude certain sleeves and bands which are not really complete packages. Bags also keep shipping costs to a minimum since they have the lowest tare weight ratio, that is, the weight of the container in relation to the weight of the contents. Being securely closed on all sides, they are essentially dust-tight and thus provide protection of their contents from outside contamination. Bags can be tailored to fit snugly around the products they contain, and beyond this they will adjust to any shift in the shape of the contents. A fluffy product which tends to settle on standing, for example, will take up less space in storage because the bag settles with the product. Bags take up a minimum of space in storage and shipment, both before and after filling. Sizes can be made to suit almost any conceivable product, from the tiniest seed packet to huge wrappers for lumber.

On the negative side is the nonsupporting character of a paper bag. It may not stand as neatly on the dealer's shelf as some of the more rigid types of packaging, and the wrinkles and folds may be unattractive for certain purposes. Stacking in the warehouse or in a retail display may also present some problems. Durability is usually borderline, and in some instances is deliberately so. A bulky, low-cost product is often put into a minimum of packaging for economic reasons, and a breakage factor of ½ to 1 percent is built into the design. On the other hand, it is possible to strengthen a bag to almost any degree through the use of scrim and similar reinforcing materials laminated to the base sheet. This type of reinforcement adds considerably to the cost, but for export or armed services it is sometimes necessary. More often, a rigid form of packaging would be chosen in place of a bag under those circumstances, but there are times when the reduced tare weight and the savings in cubage will make the bag the best choice.

NOMENCLATURE

A flexible container which is open at one end is broadly called a *bag*. Although in any size it is also called a *sack*, this term is usually reserved for very large bags holding 50 lb or more. An *envelope* is usually smaller than a bag, but not necessarily so. Envelopes are die-cut and are folded

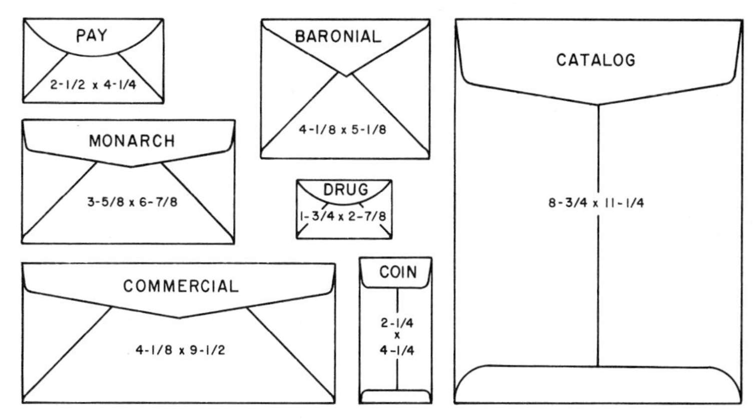

Fig. 1. Envelopes. Of the many sizes and styles of envelopes, a few are illustrated here. Each of these comes in several sizes, and the dimensions shown are typical although they are not the only dimensions that are being used.

differently from bags, as shown in Fig. 1, and they are made on a completely different type of machine. There are four general styles of bag construction: (1) *automatic bottom* or self-opening style (SOS); (2) *square bottom,* also known as pinch bottom; (3) *flat* bag; and (4) *satchel* bottom. (See Figs. 2 and 3.) Some other terms that apply more to the way in which a bag is used, or to materials from which it is made, should be mentioned here: A *baler* bag is used to hold a number of smaller bags; it may be satchel bottom or self-opening style. A *multiwall* sack is generally required for packages of chemicals, it is made from three to six plies of paper, or more, depending on weight, value, and type of export or domestic service required. A two-ply bag is more often called a *duplex* construction. The various *plies* are always described in the proper

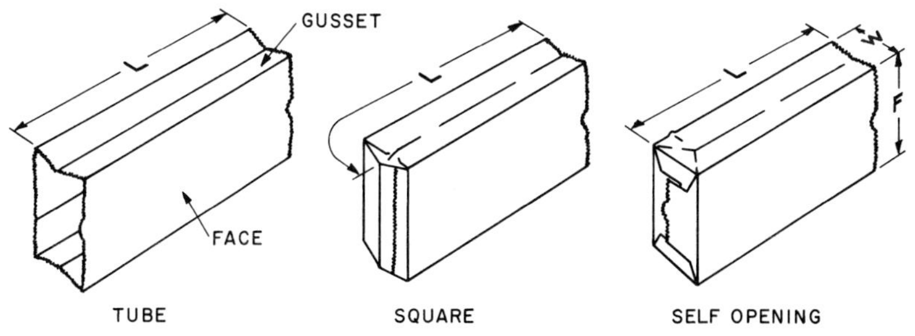

Fig. 2. Terminology. Dimensions should be specified as face, width (or gusset), and length. Note that the length (*L*) is measured differently for a square (pinch bottom) bag and for a self-opening (automatic) bag. The length of the tube before it is formed into a bag is sometimes given, and this should not be confused with the finished length.

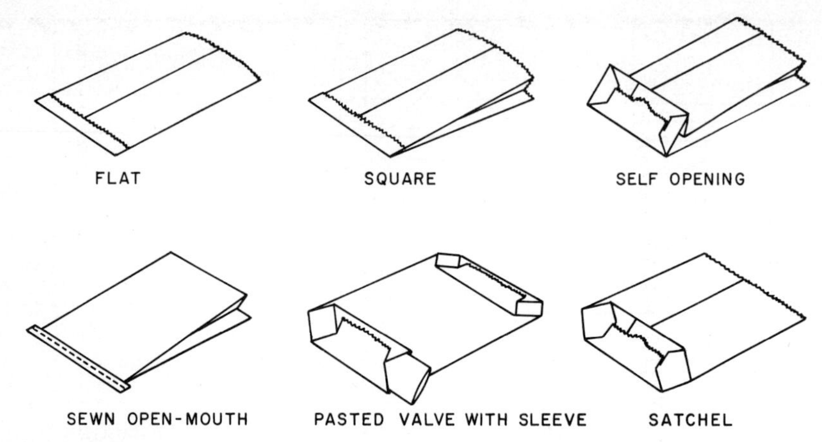

Fig. 3. Styles of bags. A continuous web of paper is formed into a tube and glued along the overlap. This is torn to length against a serrated bar, which gives the characteristic saw-tooth edge. The end is then folded over and glued to make the bottom.

sequence, which is starting from the inside and working outward. For example, a sack that is specified as 1/50, 1/90AL, 1/70 would indicate a three-ply construction consisting of a 50-lb basis (24 × 36—500) kraft sheet on the inside, an asphalt lamination in the middle, probably made of two 30-lb sheets with 30 lb of asphalt between them, and a 70-lb kraft sheet on the outside. The basis weight for kraft paper, as indicated, is the weight of 500 sheets 24 by 36 in. (3,000 sq ft).

Multiwall sacks are either *sewn* across the top and bottom, or they are of *pasted* construction. The side seams in either case are glued. Usually a starch or dextrin glue made from cornstarch is used for this purpose. If only one of the ends is closed by the manufacturer, it is called an *open-mouth* sack. In other cases both ends are closed except for a small *valve* in one corner, which may have an extended *sleeve* that is folded in after filling, or may depend on the check valve action of an internal sleeve for a tight closure. The folded sleeve in a pasted bag will give the least amount of sifting. (See Fig. 4.)

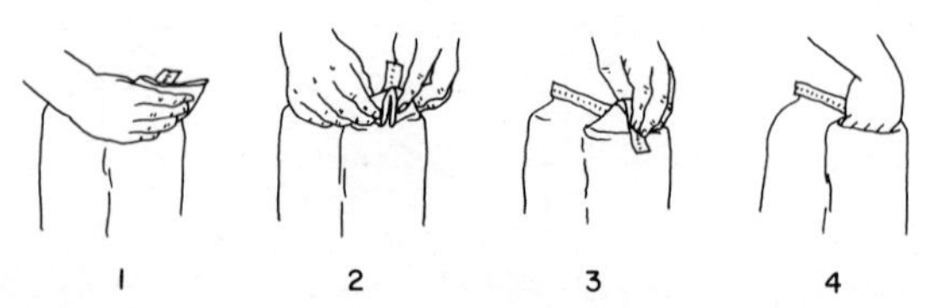

Fig. 4. Folding a sleeve. The proper method of folding the sleeve of a valve-type sack is shown. (1) Shake loose material into bag and flatten the end of the sleeve. (2) Fold corners down at a 45° angle so that the edges meet. (3) Fold point down, with any projecting tape. (4) Slide the folded point as far as possible into the pocket, and sharply crease the last fold.

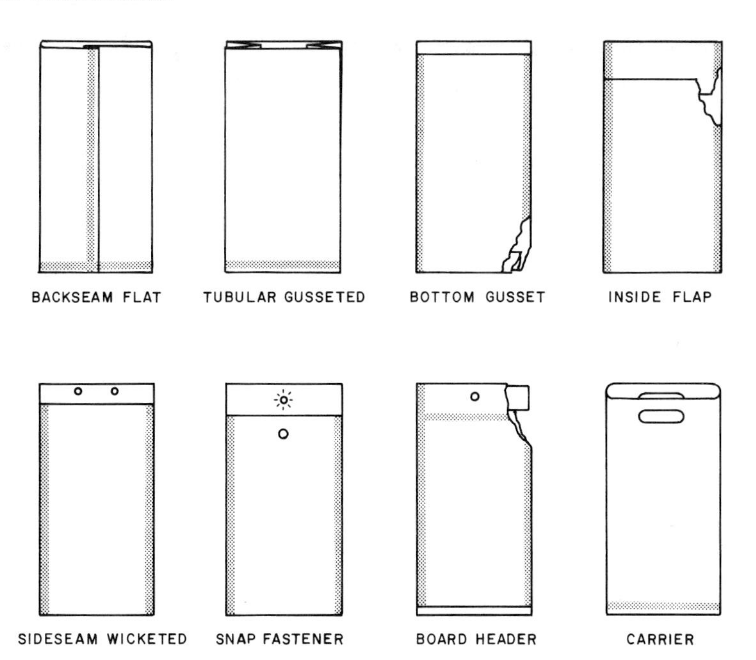

Fig. 5. Plastic bags. Tubing requires only a heat-seal across the bottom to complete the bag. The top is usually flush-cut and may therefore be difficult to open. Side-seam bags can be made with a lip at the top for ease in opening. Various attachments can be used for special purposes. See also Fig. 10 in Sec. 3, "Films and Foils," page 3-30, for forms of film in rolls.

Plastic bags and paper bags require completely different types of equipment, both for manufacturing and for sealing. The former may be made from plastic tubing or from a flat web that is folded and joined in a *back-seam* construction. (See Fig. 5.) Either of these can be *flat* or they can be *gusseted;* the ends are generally heat-sealed to complete the closure. In some cases a web of film is folded and heat-sealed to give a *side-seam* bag. The folded edge forms the bottom and can be accordion-folded if a bottom gusset is desired. The top edge usually has a lip for easy opening when filling, which is one advantage of a *side-seam* bag over the other types, which must be flush-cut.

DESIGN CONSIDERATIONS

Small paper bags can be made in several styles. For thin products a flat bag is the most economical. If the contents is bulky, it is better to have side gussets, which are accordion folds to permit the bag to "square up" with less bulging. (See Fig. 6.) The square-bottom style runs faster on the bag machine and is more economical by about 25 percent than the automatic bottom, although the latter is much easier to open for

Fig. 6. Bag machine. SOS (self-opening style) grocery bags are fabricated and printed in one operation on this machine. The paper feeds from the roll on the right, is glued along one edge by the small wheel at top right, and travels down through the former, which folds it into a continuous tube. The tube is cut to length and is carried around a ferris wheel toward the left, which glues and folds the bottom. The finished bags are stacked on the conveyor at the far left. (*H. G. Weber & Co., Inc.*)

filling, by grasping the lip and snapping it through the air so that the bottom pops out.

Of the larger sacks, the sewn paper sack is the strongest and least expensive, but it does not offer as much in the way of moisture protection or siftproofness as the pasted sack. The thread which closes the ends is usually chain-stitched, and instructions showing which end to pull for opening should be included in the printed copy. Where alkali is present, oil-dipped cotton thread should be specified for the stitching. In the case of strong acids, Dacron thread is recommended. With all highly reactive materials, the open-mouth sack must be bar-pasted at the top.

Some improvement in the siftproofness of a sewn sack is afforded by adding a thick filler of soft cotton with the sewing thread. This helps to plug up the needle holes. A 90-lb creped kraft paper tape folded over the end before sewing keeps the thread from tearing out of the bag. For better moisture protection the tape-over-sewing (TOS) is superior to the tape-under-sewing (TUS). In this case the tape must be glued or heat-sealed in place, or adhered with a pressure-sensitive coating. The kraft paper tape is available in several standard colors so that different products can be color-coded if desired.

The valve sack requires no equipment for closing, as both ends are closed by the sack manufacturer. It does, however, take a special filling machine with a nozzle to fit through the valve opening. One operator can service two or more filling machines, since it requires only sliding the bag on and off the nozzle to complete the operation, while the bottom of the sack rests on the table. The pasted sack is neater-looking than the sewn sack, because it fills out to a squarer unit, and product identification can be printed on the top and bottom more easily. It is not quite as strong, however, but with stepped-end construction, in which the ends of each individual ply are glued to themselves, it is quite serviceable.

Export multiwall sacks are usually made with five or six plies of paper having a total basis weight between 270 and 350 lb, with a moisture barrier and a wet-strength outer ply, even if the product is not hygroscopic. Domestic sacks, on the other hand, usually have three to five plies with a total basis weight of 140 to 280 lb. Two-ply bags made from 60- or 70-lb paper are now beginning to replace some of the three-ply bags made from the 40-lb kraft, which is becoming difficult to obtain. (See Table 1.) Length and width of a sack should have a ratio close to 2:1 so as to interlock on a pallet when each layer is turned 90° and thus make a neat, stable load without wasting space. Some packaging engineers prefer the shorter bag, with a ratio of 1½:1, because it is easier for handling. The gusset can be any convenient size, but usually is around one-fifth the width or between 3 and 5 in. For widths over 24 in. or lengths over 34 in., check with the supplier to be sure the size is not beyond the range of his machines. The size of a sack for a particular

TABLE 1 Moistureproof Paper for Multiwall Sacks

Barrier material, lb/3,000 sq ft*	MVT (creased paper), g/100 sq in./24 hr at 100°F, 90% RH	Cents per 1,000 sq in.
1/90 AL	4.00	2.07
6-lb Lo D PE (½-mil)	3.00	1.83
½-mil PE free film	3.00	1.34†
2/90 AL	2.00	4.14
10-lb Lo D PE	2.00	2.32
6-lb Hi D PE	2.00	3.27
7½-lb Hi D PE (½-mil)	1.50	3.51
15-lb Lo D PE (1-mil)	1.25	2.83
1-mil PE free film	1.25	1.85†
10-lb Hi D PE	1.00	3.96
K type cellophane	1.00	5.84
20-lb Lo D PE	0.75	3.28
15-lb Hi D PE	0.60	4.82
20-lb Hi D PE	0.50	5.62

*AL = asphalt-laminated; Lo D = low density; Hi D = high density; PE = polyethylene.

† Includes cost of 1/40 kraft paper for comparison purposes.

product will depend on its density, trapped air, and free-flowing qualities, and the only satisfactory way to determine the correct dimensions is by trial and error, making up a sample, fitting it, and adjusting the dimensions accordingly. The weight of the product in a sack, for convenient handling, should be from 50 to 80 lb. Although 100-lb sacks are fairly common, they are too heavy for one man to pick up easily and should be avoided.

Protection from damage or from the atmosphere is provided by the different plies of paper or film. Two thin sheets are more serviceable than one thick sheet of equivalent weight, as a general rule. Not only are they more flexible, but the forces seem to be more evenly distributed and less concentrated with the multiple construction. The outer ply, however, should not be less than 60-lb basis to resist snagging. Kraft paper is the strongest and cheapest type of paper and will usually make up the bulk of the plies. Semibleached or full bleached kraft on the outside improves the appearance of the finished sack at a slight additional cost.

For greaseproofness, one ply of glassine paper could be included, but for moisture protection a layer of asphalt between two layers of paper is generally used. It is the least expensive moistureproofing, but in cold weather it becomes stiff; it also tends to gum up the sewing needles, and with products that are packed hot, the asphalt may bleed through. A ply of plain paper is nearly always put on both sides of the asphalt lamination because of this bleeding problem. Polyethylene as a coating or as a free film is a much better moisture barrier than asphalt, especially the high-density type of polyethylene, but it is more costly. Other choices are wax paper, which gives excellent odor and moisture protection, and 0.00035-in. aluminum foil laminated to 40-lb kraft paper, which is considerably more expensive but is a nearly perfect barrier for water vapor and gases. The barrier material is generally put on the inside, as close to the product as possible, to avoid puncture from the outside, and when it is put directly against the product, it often helps the product to slide out of the bag easily.

Other special papers for sacks include wet-strength kraft, which has a melamine resin added during manufacture; antiskid coatings of colloidal silica or other treatments to minimize the shifting of loads in the warehouse and during shipment; fiber-free plies consisting of either a highly calendered or a coated paper to maintain the purity of critical products; and extensible paper (described more fully under paper manufacturing in Sec. 4, "Paper and Paperboard"), which has 10 to 15 percent greater impact strength than regular kraft paper.

A certain amount of damage is bound to occur in shipment, and the extent to which you should overdesign will depend on the value of the

TABLE 2 Approximate Costs of Multiwall Sacks, 19 by 4 by 34 In., per Thousand*

Ply construction†	Sewn open mouth	Pasted open mouth	Sewn valve	Pasted valve
3/50	$ 87.00	$ 92.00	$110.00	$105.00
1/50, 2/60	95.00	100.00	119.00	115.00
2/50, 2/60	115.00	122.00	143.00	139.00
2/50, 1/60, 1/60 WS	122.00	129.00	150.00	145.00
3/50, 2/60	136.00	144.00	168.00	164.00
0.0015 PE, 1/50, 2/60	134.00	145.00	165.00	161.00
1/10 PE 50, 1/50, 2/60	158.00	167.00	190.00	186.00
3/50, 2/60, 1/70	160.00	NA‡	203.00	200.00

* Add about 8 percent to the above figures for extensible paper.

† WS = wet strength; PE = polyethylene. 1/50, 2/60 indicates one ply of 50-lb kraft and two plies of 60-lb kraft, starting from the inside. Free film polyethylene is indicated as 0.0015 in. thick. Coated polyethylene on paper is shown as 1/10 PE (10 lb or about 0.00075 in. thick) on 50-lb kraft.

‡ NA = not available.

product; products that are worth more than $1 per pound might better be put into a fibre drum. Some engineers use 50 cents per pound as the breakpoint between bags and drums. For low-cost items, such as building materials, a ½ or 1 percent breakage is planned, and a few empty sacks are included with each carload or truckload to repackage the material from the broken sacks. (See Table 2.)

Polyethylene sacks for industrial products have come into limited use, but the high cost and difficulty of sealing have discouraged their widespread adoption. The advantages of moistureproofness, transparency, and chemical resistance will sometimes make these the best choice for fertilizers and certain chemical products. The heat-seal area has only about two-thirds the strength of the rest of the sack unless it is reinforced with another strip of polyethylene and sealed through this extra layer. (See Table 3.)

A 5-mil polyethylene sack to hold 50 lb costs around 7 cents, and a 10-mil bag for 100 lb is about 12 cents. They can be printed with up to

TABLE 3 Size Limitations of Polyethylene Industrial Sacks

Face	16 to 20 in.
Length	18 to 38 in.
Width	4½ to 5½ in.

four colors on both sides, with nonslip inks. Both valve and open-mouth styles are available in polyethylene. The best polyethylene resin for large bags has a density of about 0.915 and a melt index around 0.5. A few pinholes can be added to allow trapped air to escape without seriously affecting the moisture protection.

Small plastic bags made from tubing are most economical as they require only a cross seal and cutoff, and so can be run at high speed with a minimum of problems. (See Table 4.) If exceptional clarity is required and the film is not available in tube form, the back seam or side seam becomes necessary. If side gussets are not required, there is an advantage in the side-seam construction, since a lip can be made which will facilitate opening the bag for filling. Otherwise the open end must be flush-cut, which tends to stick the edges together and make it difficult to get the end of the bag open.

Polypropylene bags have exceptional clarity and good printability and are often used for fine hosiery and other luxury items. Cost is about double that of polyethylene bags of equivalent dimensions. They are more difficult to fabricate, and gusseted bags are especially hard to seal, so that the flat style of bag is generally used. Side-seam bags tend to string out as they are cut off with a hot knife against a Teflon-coated anvil, causing rough edges and "angel hair." The best sealing conditions are around 10 psi for ¾ sec at 375°F with a knife-edge radius of 0.010 in. or less.

Cushioned mailing bags are made of two plies of kraft with a cushioning material such as shredded or flaked paper evenly distributed between them. They are used for mailing books and small fragile objects. Cotton mailing bags are made with a drawstring at one end and an envelope or address tag sewn into the other end.

TABLE 4 Approximate Costs of Polyethylene Bags Unprinted, per Thousand

Size W × L, in.*	Tubular style, in.				Side-weld, in.	
	0.001	0.00125	0.0015	0.002	0.00125	0.002
3 × 6	$1.70	$1.90	$2.25	$2.50	$1.65	$1.95
3 × 10	2.20	2.75	3.05	3.70	1.90	2.60
4 × 6	1.95	2.20	2.50	3.00	1.85	2.35
4 × 12	3.05	3.60	3.90	4.75	2.45	3.70
6 × 8	2.85	3.20	3.60	4.40	2.60	3.85
6 × 14	3.80	4.50	5.00	6.35	3.65	5.35
8 × 10	3.65	4.15	4.90	6.15	3.60	5.20
8 × 14	4.75	5.50	6.40	8.10	4.50	6.85
10 × 12	5.15	6.00	6.95	9.70	5.00	7.30
10 × 14	5.75	6.85	7.95	10.50	5.60	8.40

* W = width; L = length.

Burlap is not used as much as it used to be, but for special purposes it can be laminated to kraft paper with asphalt as the laminant, or to polyethylene film with a special adhesive. Plain burlap sacks are available in 7½-, 10-, or 12-oz material, in limited widths, in sewn open-mouth or valve types, with sewn or cemented side seams. Woven mesh bags of polyethylene and polypropylene fibers also are available and are beginning to replace the woven fabrics made from natural fibers.

Various kinds of supplementary devices can be applied to small paper or plastic bags on the bag-making machine with little additional cost. Holes can be punched for hanging, for ventilation, or to relieve pressure or vacuum for faster packing. These are usually ¼ in. in diameter and are most conveniently spaced in even inches to avoid special tooling. A half-circle cut is sometimes used as a flutter valve instead of a hole, for better appearance.

Other additions include windows consisting of a large hole covered with a transparent film or a mesh fabric; handles cut into the bag and reinforced or added on; and paperboard saddles to be used for display purposes. Special equipment has been developed for these operations, and with a little ingenuity the bag manufacturer can produce a great variety of opening and reclosing features, carrying and display devices, and many other special constructions.

PROCESSES

Large sacks can be filled with granular material on automatic equipment at speeds above 20 per minute with a weight variation of less than ¼ lb either way. With semiautomatic equipment the filling rate will be about 2 per minute. For best results the moisture content of the paper should be between 6 and 8 percent. Printing of most bags is by flexography, although some rotogravure is used for long runs and fussy jobs.

TESTING

The standard drop test for sacks of 50 lb or more is to drop the filled and sealed sack from a height of 3½ ft alternately on its face and back until it breaks. The results are reported as the average number of falls to failure. Some laboratories prefer an edge drop from a height of 2 ft. Bags under 50 lb should be tested by dropping from a height of 2 ft on their bottoms until they break. In each case a bag should survive an average of two to four drops before spilling its contents.

Section 11

Aerosols

History 11-2
Principles of Operation 11-4
Advantages and Disadvantages 11-5
Selection Criteria 11-7
Propellants 11-8
Fluorocarbons 11-10
Hydrocarbons 11-13
Diluents 11-15
Gas Laws 11-15
Costs 11-16
Trade Names 11-16
Valves 11-16
Costs 11-22
Suppliers 11-22
Containers 11-22
Labels 11-26
Overcaps 11-26
Processes 11-26
Specifications 11-27
Testing 11-28

Pressurized containers offer many advantages for certain types of products. They permit controlled dispensing at the touch of a button. They protect the contents from contamination, and from oxidation by the atmosphere, down to the last drop. The product is converted to a better form—spray, stream, drops, or foam—without effort on the part of the user. There are even a few products that owe their very existence to the spray can, among which are the hair sprays, shaving creams, windshield deicers, and artificial snow. There are other products that are considerably more convenient for the consumer in a pressurized container, for example, insecticides, room deodorants, perfumes, and touch-up paint. The size of the spray particles and the pattern of the spray can be varied to suit a particular purpose, and the usefulness and effectiveness of many products are enhanced when they are in an aerosol form.

The precise meaning of the word "aerosol" is a "production of minute solid or liquid particles so fine that they remain suspended in the air for long periods of time." The more commonly accepted definition is a "container of liquid under pressure that is released through a push-button valve."

Various sizes and types of containers are available for aerosols; for example, purse-size aerosols holding a few grams are used for perfumes and pharmaceuticals, and giant 24-oz cans can be filled with cleaning and sanitizing liquids. Between these extremes is a wide choice of two-piece and three-piece tinplate cans, aluminum and stainless steel cans without seams, polypropylene and acetal plastic containers molded in attractive shapes, and glass bottles with and without plastisol coatings. Color and decoration can be applied in several ways, and the creative designer has ample opportunity to make these packages very attractive and appealing.

HISTORY

In the early days of World War II there were more casualties in the tropical areas from diseases such as malaria than there were from military action. To combat the insects that carried these diseases, the Department of Agriculture was asked to develop a better method of insect control. After several unsuccessful attempts by Lyle D. Goodhue and William N. Sullivan, working as a team in the USDA laboratories at Beltsville, Maryland, to vaporize insecticides by heat, the aerosol was finally born on Easter Sunday in 1942. Dr. Goodhue, working alone in his laboratory, mixed refrigerating gas (propellant 12) with pyrethrum and sesame oil in a metal cylinder, and attached an oil burner nozzle. When he released some of his mixture above a cage containing roaches that were being used for other tests, all the roaches died within a very few minutes. After further development by Goodhue and Sullivan this system was adopted by the Army and became standard equipment in the South Pacific islands. A patent was issued in their names in 1943 and assigned to the Secretary of Agriculture.

It was soon learned that a mixture of propellants 11 and 12 was just as effective as the straight 12, and the reduced pressure made possible the substitution of less costly containers for the welded steel units that were being used by the soldiers. A modification of the beer can was developed for this purpose, and in November, 1946, the first commercial carload shipment of these containers was made by Crown Can Company. In 1947 the Continental Can Company introduced a three-piece can with a concave top and bottom, and with a valve soldered in place. The first hydrocarbon propellant was used in 1953, when Dr. Daniel Terry developed the Bon Ami Window Cleaner, using isobutane as the propellant.

About 5 million "bug bombs" were sold in 1947, and this market has been expanding ever since, until today there are about 2 billion units of all types consumed in this country. Hair lacquers, deodorants, perfumes, and other personal products account for about half of this volume. Room fresheners, cleaners, and household aerosols make up the next largest category, followed by paints, insecticides, industrial products and foods.

There is still a large untapped market in foods, for this industry has been very slow to adopt pressurized containers for such things as mustard, ketchup, and salad dressings. There is also a completely new technique of administering drugs which is based on aerosols and has some very interesting possibilities for the future, such as inhalation therapy, which approaches intravenous therapy in its rapidity of action. Compared with an injection, this form of aerosol administration causes a great deal less discomfort for the patient, and it requires no sterilization techniques, nor does it take the services of any qualified medical personnel to administer the dose. Most drugs that are now given parenterally can be given more easily in the aerosol form. An aerosol throat spray, for example, will reach deep down into areas that a gargle does not touch. Furthermore, a topical application by means of an aerosol will permit direct contact of the medicine to the skin without any greasy "salve" carrier, resulting in faster absorption and the elimination of the irritation which often results from the manual application and rubbing in of the medication.

Although the aerosol has added a new dimension to packaging, it does not automatically bring success in the marketplace. Unless a packaging system satisfies a real need for the consumer, it is doomed to failure, and aerosols are no exception. Several examples of poorly conceived products can be cited: Chocolate syrup in a pressurized container was irresistible to children, and they could not wait to get home to try it; the result was messy displays and store shelves that infuriated the storekeepers. A coffee concentrate that required refrigeration and had an inferior flavor was a quick casualty. Toothpaste that was too thin to suit the popular taste and often failed to dispense, because of cavitation within the can, soon disappeared from the marketplace. A barbecue sauce was taken off the market when the cans failed at the side seam, and fruit-flavored beverage concentrate also had some technical shortcomings and had to be abandoned. There are many food products such as mustard and ketchup that would appear well suited for aerosol dispensing, but unless the package offers a real advantage to the consumer, the gimmick of push-button convenience is not enough to make it sell.

Another word of caution regarding the safety aspects of aerosols: Several years ago *Popular Mechanics* revealed that a 40-year-old woman in Willow Grove, Pennsylvania, threw an empty aerosol into a trash fire, and a scrap of flying metal cut her jugular vein, killing her. An attendant at an incinerator in Mamaroneck, New York, injured an eye in a similar accident. These things do not happen very often, but it is important to put the necessary warnings on the label and to avoid making potentially dangerous combinations such as oven cleaners with flammable propellants that may ignite from the pilot light. It should also be recognized that when aerosols are sprayed into a flame or on a hot surface, there is produced a very significant amount of phosgene, carbon monoxide, and hydrogen chloride which could be dangerous.

PRINCIPLES OF OPERATION

A typical aerosol package is made up of a liquid product plus a propellant partly liquid and partly gaseous, all of which are under pressure. In a filled container, the liquid portion will generally occupy more than three-quarters of the space, while the gaseous portion fills the space above. A dip tube usually goes from the valve down into the liquid. When the valve is opened by depressing the button, the gas inside the container, exerting pressure in all directions, forces the liquid up the dip tube and out through the valve. A small amount of liquid propellant will then evaporate inside the package to take the place of the liquid that was dispensed, increasing the gaseous portion by that amount. When a particle of liquid propellant changes to a gas, it will occupy as much as 250 times its original volume. Eventually there will be very little liquid left, and almost the entire space will be occupied by gas, but the pressure will always be about the same, as long as there is still some propellant in the liquid form.

There are several variations to this arrangement which will be discussed as we go along, but staying with our typical example, we will assume that the product and the liquid propellant have mixed together easily to make a homogeneous solution. When the mixture is dispensed into the air, the change in pressure causes the propellant to evaporate, breaking up the product into tiny droplets and forming the familiar spray mist. A 1-sec burst of a typical spray will produce over 100 million particles. If the propellant is emulsified with the product instead of dissolved in it, evaporation of the propellant produces a foam instead of a spray. When the product is suspended in the propellant, as in the case of paint, the solid particles are ejected onto the surface being covered, and allowed to dry. The important point to be noted here is that the *kind of mixture* of propellant with product is a determining factor in the end result.

In some cases the two will not mix at all, but separate into different layers. The pure product is then forced up the dip tube when the valve

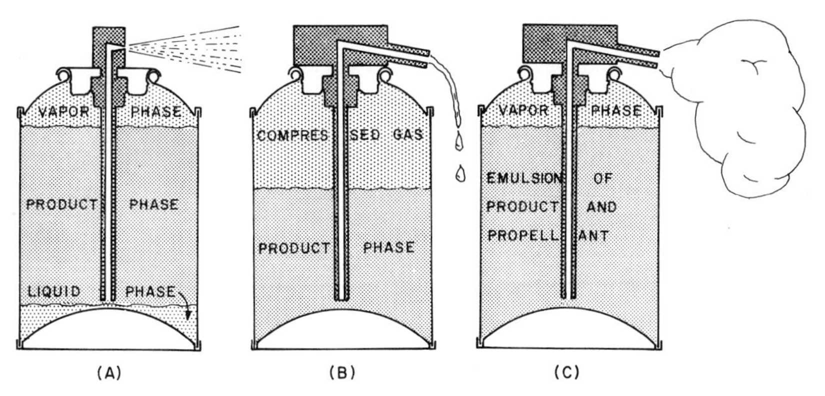

Fig. 1. Propellant systems. In a three-phase system at *A* above, the liquid propellant is in a layer separate from the product, with propellant vapor on top. With a compressed gas that does not liquefy, as in *B*, or if the propellant mixes with the product, as at *C*, it is a two-phase system. Different forms of product also are shown, with a spray being produced by a suitable actuator at *A*, a solid stream at *B* where the product and propellant do not mix, and a foam at *C* resulting from a mixture of product and propellant, and with a large orifice in the actuator.

is opened, and none of the propellant goes with it. In this case the velocity and the shearing action by the valve and actuator will break up the liquid into a spray, although not so effectively as with the dissolved gas. A *vapor tap* consisting of a small hole in the side of the valve body also can be used to help atomize the product. A small amount of gas escapes from the headspace through this vapor tap and goes into the product as it passes through the valve, to assist in breaking up the liquid.

When the product is in a layer separate from the propellant in this way, it is called a *three-phase system,* the three parts being gaseous propellant, liquid propellant, and liquid product. Usually the propellant goes to the bottom, with the product floating on the top, but with hydrocarbons the reverse is true. As the aerosol is used, the liquid propellant bubbles up through the product and fills the space left by the product that has been dispensed. (See Fig. 1.)

An aerosol that is to be used straight up will have a dip tube extending from the valve down into the product. If it is intended to be used upside down, as in the case of food toppings, the dip tube is omitted. It is possible to design an aerosol container that will work in either position, with a vapor tap in the side of the valve inside the can, but this usually means a compromise in the quality of the spray.

ADVANTAGES AND DISADVANTAGES

The convenience of dispensing materials from aerosols at the touch of a button is the outstanding feature of this type of package. There are some products that would not be used at all if it were not for their self-energized packages. The cost of all this convenience is rather high, when compared with the usual bottles or cans. The expense of the valve, actuator, container, and cap added together makes a costly package, but since the consumer seems quite willing to pay the premium for this kind of convenience, the price has not been very much of a detriment to sales in most cases.

The standardization of package forms in the aerosol industry has resulted in a sameness in appearance that is probably limiting the sales of aerosols in the marketplace. There are so many sprays in the standard $2^{11}/_{16}$-in.-diameter cans with white plastic overcaps that they all look very much alike. Only the label design distinguishes them. Rarely does the designer take advantage of the opportunities that are available to change the shape of the overcap, or to utilize the possibilities of the glass or plastic container to develop a really interesting package. Furthermore, the silly little buttons that are used on most aerosols are difficult to aim and make the user's finger sore if they are used for any length of time. Some better-designed actuator buttons are now becoming available, and these should be coming into more general use in the near future.

Visibility of the contents is not possible, except with glass aerosols, and this advantage is usually lost by covering the containers with an opaque coating. The consumer thus has no way of knowing whether the container is only partly filled when it is purchased. Too often there is a slow leak that cannot be detected in production; in storage or on the dealer's shelf over a long period of time the container may go completely empty. As a matter of fact, the majority of aerosols will have some loss before they are purchased, but it is usually less than 10 g per year and they can be overfilled by this amount to compensate for the loss. Normally the gasket in the valve swells, so that the rate of loss becomes less as time goes on. The aerosol industry considers 2 percent to be the maximum allowable loss from a container in 1 year, although there are some companies that operate with a 3 percent maximum. A method of measuring this type of leakage on the production line consists of an inverted column of water over the valve. Bubbles will accumulate at the top of the column over a period of time and can be used to determine the rate of loss.

One of the advantages of aerosols that is often overlooked is the protection of the contents from outside influences. Whereas most containers take in air as the contents are used up, pressurized packs never allow anything from the atmosphere to enter the container. The consumer can thus be assured that the product will always be as clean and pure as when it left the factory, and the manufacturer can be confident that there will not be any oxygen getting in to degrade the formulation.

SELECTION CRITERIA

The choice of the container, valve, and actuator depends on the type of product and also on the end results that are desired. Thus shaving soap that is to be dispensed as a foam requires a package different from that for a hair lacquer intended to come out as a fine spray. The best approach to this problem is to start with a consideration of the formula: is it miscible with the propellant, or will it form a separate layer in the container, or will it be somewhere in between, such as in a suspension or an emulsion? This then brings up the question of the propellant: will it be a liquefied propellant such as a fluorocarbon, or will it be a compressed gas, like nitrogen or carbon dioxide? The subject of propellants will be taken up in more detail a little further on, but at this point it should be established whether a constant pressure is necessary until the package is completely empty; this would dictate a liquefied type of propellant.

The type of formulation may also determine the method of filling. If the composition is nonaqueous, it can be cold-filled; that is, the product and the propellant can be chilled so that they are below the boiling point of the propellant. They can then be filled into the container like any other liquids, before the valve is put in place, at a good rate of speed. If it is a water-base product, on the other hand, it cannot be cooled or it would freeze solid; therefore it must be pressure-filled. This consists of putting the product in first, next securing the valve in place, and then adding the propellant, which must be filled through the valve: a relatively slow process. Food products and some pharmaceutical preparations require propellants that have been cleared by the FDA.

The character of the spray or foam will depend in large part on the combination of valve and actuator. For example, a wet spray made up of droplets larger than 50 μ in diameter, that will not blow away, is desirable for coating surfaces, whereas a fine dry mist is better suited for room deodorants. The kind of spray can be controlled by choosing the proper orifice sizes in the valve stem, the tailpiece, and the actuator button. Some actuators are designed to provide a mechanical breakup of the spray in addition to the dispersion resulting from evaporation of the propellant. These are described under "Valves," page 11-16. The compatibility of the valve with the formulation is another factor that should not be overlooked. When certain materials are in contact with some plastics, for example, in metering chambers, they may leach the antioxidants and plasticizers from the plastic. In other cases the ingredients in the formulation may be absorbed by the plastic.

Whether to use a container made of aluminum or glass or tinplate may depend entirely on marketing considerations, or in some situations it may be determined by the nature of the formulation. After these decisions have been reached, the compatibility and stability of the product should be checked very carefully with the actual containers that are to be used, and the tests should be for an extended period of time, under varying atmospheric conditions—at least for 3 months and preferably for 6 to 9 months. The things to look for in the tests are the settling out of solid particles, degradation of perfume, crystal formation, lumping together of particles, corrosion of the container, and leakage of any kind. There is often a delicate balance between the ingredients in the formulation, the propellant system, the materials and coatings in the container, and the various parts of the valve. Never fail to try them out in use, under all the possible conditions that can be anticipated, and when the results are satisfactory, do not change a thing without going through the complete testing procedure again.

Can manufacturers have made some progress in developing containers that will separate the propellant from the product in the container. A plastic bag is available that can be sealed into the seam of the can, allowing the propellant to be put into the space between the bag and the can by means of a hole in the bottom of the can, which is then sealed by a rubber plug. Another interesting new development is the Preval unit of the Precision Valve Corporation. This has the propellant in a separate metal tube that extends down into the product. The outer container can be of any material, since there is no pressure in this part of the system.

Summing up the various things to consider in selecting the components of an aerosol package, we have: (1) the product formulation: whether aqueous or nonaqueous; (2) method of application; as a fine or coarse mist, paste, or foam; (3) the type of container: glass, aluminum, tinplate, or plastic; (4) the valve to control the rate of application and to mix the product and propellant; and (5) the kind of propellant: compressed gas or liquefied gas.

PROPELLANTS

There are three broad classifications of propellants: fluorocarbons, hydrocarbons, and compressed gases. The fluorocarbons are more costly, but they can be formulated to produce any desired pressure in the range of 15 to 120 psig, and the pressure will remain constant until the product is all used up (if the temperature stays the same). This is true of the *hydrocarbons* also. On the other hand, *compressed gases* start out with a high pressure (about 100 psig), which drops off as the contents is dispensed. (See Fig. 2.) Temperature has a very significant effect on pressure, and some aerosols will not work outdoors in the winter because of the low temperature. An increase of 40 to 50°, for example, will

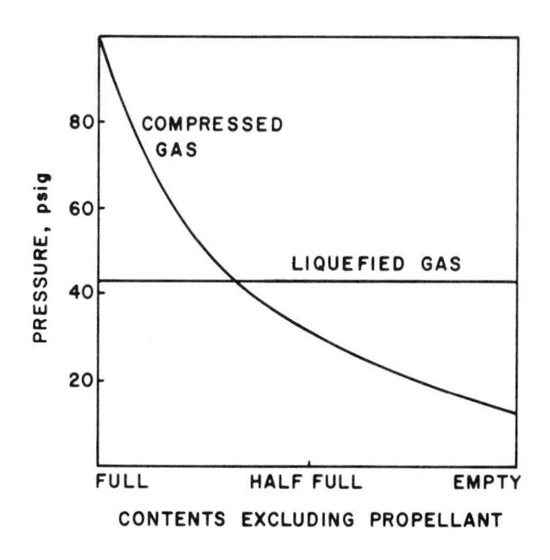

Fig. 2. Pressure drop-off. Liquefied gases such as the fluorocarbons and the hydrocarbons will maintain a constant pressure until the contents of the can is all used up, as long as there is any liquid propellant remaining. Compressed gases such as nitrogen and carbon dioxide will show a drop in pressure as the contents of the can is used up, due to expansion of the gas into the larger headspace. A typical curve is shown, but this will vary according to the gas used and the space occupied.

double the pressure of fluorocarbon-type propellants. (See Fig. 3.)

A recent decision by the courts has had a significant effect on the aerosol industry. On October 13, 1953, J. George Spitzer, along with Irving Reich and Norman Fine, was granted a patent for oil-in-water emulsions with fluorocarbon propellants. The original application had included hydrocarbons, but these were dropped from the claims in the course of prosecuting the patent through the Patent Office, although they remained in the specifications of the patent. Carter Products Company acquired an exclusive license under this patent and subsequently introduced Rise shaving cream. The Colgate-Palmolive Company challenged the validity of the patent, but in the end the courts forced Colgate to pay almost $5 million to Carter for patent infringement. As a result of this decision most of the manufacturers of shaving cream have switched to mixtures of propane and isobutane (usually in

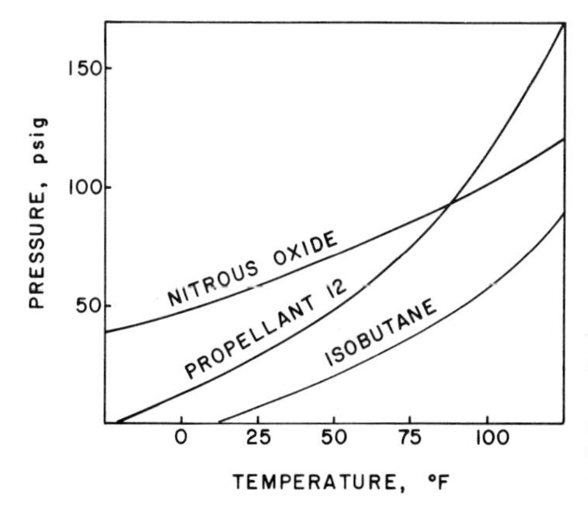

Fig. 3. Effect of temperature. An increase in temperature will have a marked effect on the pressure within an aerosol container. This is most significant with the fluorocarbons and least noticeable with the compressed gases.

proportions of 13:87) instead of the fluorocarbon propellants, to avoid paying a royalty to Carter.

Elsewhere in this book pressures are given as pounds per square inch (psi). It is customary in dealing with aerosols, however, to specify whether it is gauge pressure (psig) or absolute pressure (psia). The latter is used for calculations, but the former will be used for most of the discussion which follows. *Gauge pressure,* as the name indicates, is the pressure shown by a gauge, and is the amount above atmospheric pressure taken as zero. *Absolute pressure* is the intensity of pressure above a complete vacuum, and is higher than gauge pressure by about 14.7 lb at sea level. Absolute pressure is constant for a given set of conditions, whereas gauge pressure varies with changes in air pressure. For this reason the literature usually reports the results of research work as absolute rather than gauge pressure.

Fluorocarbons. Discovered in the 1880s, the fluorocarbons did not achieve any economic significance until they began to be used as refrigerants in place of ammonia and sulfur dioxide in the 1920s. The vapors are heavier than air, and they have a low toxicity, the most common ones being safe to breathe at a concentration of 1,000 ppm. However, large amounts can cause suffocation by displacing oxygen, since they are heavier than air. Fluorocarbons have a slight odor, somewhat like the chlorinated hydrocarbons. Fluorocarbons cannot be used in fire extinguishers that are to be used in enclosed areas, because of the formation of phosgene, hydrogen chloride, and hydrogen fluoride with heat. For the same reason aerosols containing fluorocarbons should not be sprayed on any hot surfaces.

A *numerical system* is used to designate the different fluorocarbon propellants, based on their molecular structure. The first digit indicates the number of carbon atoms minus 1. If this is zero (one carbon atom), it is dropped. The second digit shows the number of hydrogen atoms plus 1, and the third digit gives the number of fluorine atoms:

Propellant (0)11	$CFCl_3$
Propellant (0)12	CF_2Cl_2
Propellant 152	$C_2H_4F_2$

A convenient way to determine the number of different atoms is to add 90 to the propellant number, which will then give the actual number for each element:

$$\begin{array}{r} 152 \\ 90 \\ \hline 242 \\ ||| \\ \mathrm{CHF} \end{array}$$

Isomers are designated by subscript letters, as in the case of 142, 142_a and 142_b. If the molecular weight is equally balanced, or nearly so, the subscript letter is omitted, as in the case of CH_2—CHF_2, which is called simply propellant 142. The next most nearly balanced isomer is CH_2F—$CHClF$ with a molecular weight on the right half that is double that on the left; it is called propellant 142_a. CH_3—$CClF_2$ is the most unbalanced, with a molecular weight on the right side more than five times that on the left; it is designated 142_b. The letter C as a prefix indicates a cyclic molecule, as in the case of propellant C-318.

The standard notation for *mixtures* always indicates the highest pressure component first; thus 12/11(40:60) shows a mixture of 40 percent propellant 12, and 60 percent propellant 11. However, nonfluorocarbons are written last regardless of pressure; thus 12/11/iso(45:45:10) indicates that 10 percent of the mixture is isobutane. This mixture is called propellant A, which is often used for hair sprays.

When propellant 12 is used alone, it has a pressure of 70 psig at 70°F, but when it is mixed with equal parts of propellant 11, the mixture will have a pressure of about 38 psig. Increasing or decreasing the proportion of propellant 12 will increase or decrease this pressure, so that any pressure between 0 and 70 psig can be produced. (See Fig. 4.) Heavy formulations such as paints will require the higher pressures, while perfume in uncoated glass containers will usually be packed with a pressure down around 15 psig. (See Table 1.)

Propellant 11 is used mainly to reduce pressure by mixing with other higher-boiling gases. It has no pressure of its own at room temperature, since its boiling point is 74.7°F. It also increases solubility in certain products, but it has the disadvantage of being unstable in the presence of water, breaking down to form hydrochloric acid, which can play havoc

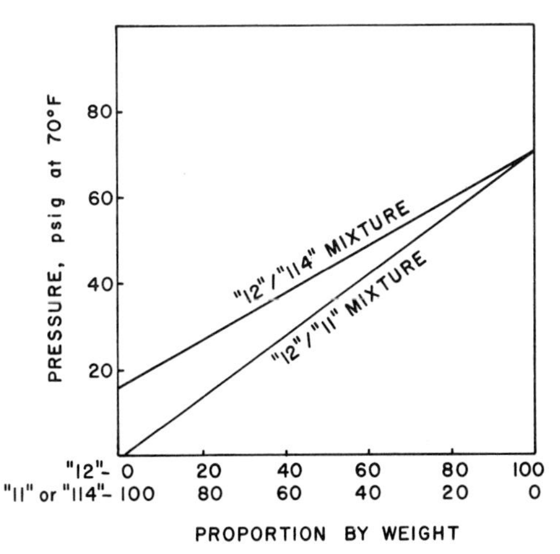

Fig. 4. Propellant blends. Various proportions of propellants in a mixture will provide different pressures. Thus any desired pressure over a wide range can be obtained by choosing the right amounts of each of the components.

TABLE 1 Properties of Fluorocarbons

11	Reduces pressure of other fluorocarbons. Increases solubility. Unstable with water. Has an off odor when used with fragrances.
12	High pressure, low cost. Most widely used. Low chilling effect with deodorants.
21	Low pressure. Very sensitive to water, more so than 11. Attacks gasketing material. Used for mold release spray.
22	High pressure and good solvent properties. Flammable and unstable with some formulations. Used as a refrigerant.
113	Selective solvent. Does not attack most plastics. Suitable for foam products and dry powders.
114	Widely used in place of 11, especially with formulas that contain water. Nonflammable. Low toxicity. Low odor.
115	Approved for foods. Used for dessert toppings.
142_b	Used for powder sprays and fragrance formulas. Low density, moderate cost. Better ratio of expansion and solubility in water than 114. Often blended with 114. Slight tendency to break down in the presence of water, especially alkaline solutions. Moderately flammable.
152_a	Low density. Good solubility. Flammable. Very low toxicity. Low taste and odor level.
C - 318	Approved for foods. Used for dessert toppings. High cost.

TABLE 2 Comparison of Propellants

Propellant	Formula	Pressure at 70°F, psig	Cost per lb
11	CCl_3F	0	$0.22
12	CCl_2F_2	70	0.23
22	$CHClF_2$	122	0.60
113	$CCl_2F—CClF_2$	19	0.59
114	$C_2Cl_2F_2$	14	0.36
142_b	$CClF_2$	29	0.55
152_a	CH_3CHF_2	62	0.55
Alcohol	C_2H_5OH	0	0.11
C-318	C_4F_8	25	2.50

with metal containers. Propellant 12 is the most commonly used material because of its high pressure and relatively low cost (23 cents per pound). (See Table 2.)

Hydrocarbons. Like the fluorocarbons, the hydrocarbons are liquids at room temperature when they are in a sealed container. Therefore the pressure will remain constant from beginning to end, as long as there is some liquefied gas remaining in the container. Butane, isobutane, and propane are very low in cost and are often used with water-base products. They are extremely flammable, however, and should be handled with suitable precautions. Label copy on the containers must conform to the Federal Hazardous Substances Labeling Act when hydrocarbon propellants are used. The flammability of any aerosol propellant is roughly proportional to the number of hydrogen atoms in the molecule, and hydrocarbons have the maximum amount. In mixtures it is best to keep the hydrocarbon portion below 12 percent to avoid flammability problems, although there is sometimes a head-and-tail effect, as in the case of 12/11/iso. That is, the propellant 12 evaporates first, and in a flammability test the isobutane is masked by large amounts of this gas. Later the propellant 11 comes into play to give the same effect near the end.

Vinyl chloride $CH_2{=}CHCl$ is sometimes used as a substitute for the hydrocarbons. It is less flammable than the hydrocarbons, and although it is more expensive (about 8 cents per pound), it can be used in amounts up to 27 percent without flammability problems. Vinyl chloride has an etherlike odor, and it may not be suitable for toiletry products, but it is widely used in paints and lacquers.

The hydrocarbons are identified by the letter A followed by a number which indicates the pressure. Thus, *n*-butane is designated propellant A-17 because it has a pressure of 17 psig at room temperature. Isobutane is called propellant A-31, and propane is known as propellant A-108. The pressures given may not agree with the textbook values, because aerosol-grade gases are not absolutely pure but contain small amounts of other gases. All the hydrocarbons have some odor; this is most noticeable when a beakerful has nearly all evaporated. Isobutane has a pressure of 31 psig at 70°F, and propane has 108 psig at that temperature. Mixing the two together in the right proportions will give any pressure in between. Butane has a pressure of 17 psig at 70°F. Since hydrocarbons weigh only one-third as much as the fluorocarbons, and the cost is only about 6 cents a pound compared with an average of 30 cents per pound for the fluorocarbons, they offer a great cost advantage. Since they are lighter than water, in a three-phase system they will float on top of the product, eliminating the problems of dip tube length associated with heavier propellants. Isobutane must be shaken in order to get a mixture of propellant and product.

Fluorocarbons sink to the bottom of the container, whereas the hydrocarbons tend to float on top of the product. By making a blend of fluorocarbons and hydrocarbons, it is possible to match the density of the product, thereby making it easier to mix when the container is shaken.

Compressed gases, as distinguished from the fluorocarbons, include carbon dioxide, nitrogen, and nitrous oxide. They do not maintain a constant pressure as liquefied propellants do, but the pressure drops at a steady rate as the container is emptied. When the headspace increases to twice the amount, the pressure that is exerted will be only half as much. The containers cannot be cold-filled but must be pressure-filled through the valve. Nitrous oxide and carbon dioxide are partially (3 to 5 percent) soluble in many products and so take on some of the properties of the liquefied propellants. A mixture of N_2O/CO_2(15:85) is sometimes used for food products because the sweet taste of the nitrous oxide offsets the acidic taste of the carbon dioxide.

Nitrogen is nearly insoluble and thus is useful for toothpaste and syrups to produce a solid stream of product. It is often used at a starting pressure of 100 psig at 70°F. The amount of pressure left when the product is all used up will depend on the amount of *headspace* at the start. With a 25 percent headspace, for example, an initial pressure of 100 psig will drop to 25 psig when the contents is depleted; also, with each 10°F rise in temperature, there is about 2 psig increase in pressure. Solubility of nitrogen in water is about 1½ percent, and in alcohol 14 percent. Pressure testing of filled containers at 130°F is not required by DOT regulations when nitrogen is used. Liquid propellants of the hydrocarbon type or the fluorocarbon type require about 7½ percent headspace for gas expansion. The government specifies only that there must be headspace remaining when the temperature of the contents is raised to 130°F. The compressed gases can be combined with fluorocarbons, and carbon dioxide is sometimes used with propellant 12 for insecticides. Carbon dioxide should not be used with soap products because of undesirable chemical reactions. Other combinations of propellants are used for special purposes, and some of these get very complicated. One such formulation for hair spray, for example, has water and alcohol in amounts sufficient to dissolve a limited amount of propellant, leaving the excess as a separate top layer. Since pure hydrocarbon would be too flammable, a mixture of hydrocarbon and fluorocarbon propellant is used, with enough hydrocarbon to cause the excess to float above the mixture of concentrate and the dissolved propellant.

Whipped toppings represent 90 percent of the food aerosols. For food products there are not very many suitable propellants. Nitrogen, nitrous oxide, carbon dioxide, and propellant C-318 are the ones most generally used, and blends of two or more of these are sometimes used. Butane and propane affect the taste, as do most of the fluorocarbons.

Diluents. A number of materials are used in propellant mixtures to increase solubility or to lower the pressure. Among these are methylene chloride, dimethyl ether, vinyl chloride, trichloroethane, mineral oil, and kerosene. Of these, dimethyl ether mixes best with water, but it is very flammable and it will attack gasket materials. The chlorinated diluents are used in low-price hair sprays, but they usually have an undesirable odor.

Gas Laws. Certain principles of physics cover compressed gases, and the following may help in understanding the behavior of aerosol propellants. Strictly speaking, these statements will apply only to a perfect gas at low pressure, and real gases will deviate slightly from these rules, but they are basically correct.

Boyle's Law. The pressure of a gas varies as the inverse of volume, and conversely the volume varies as the inverse of the pressure, provided the weight is constant and the temperature is constant:

$$\frac{P_{\text{initial}}}{P_{\text{final}}} = \frac{V_{\text{final}}}{V_{\text{initial}}}$$

Note that a mole of gas occupies 22.4 liters at 70°F and 14.7 psia.

If a cubic foot of gas at a barometric pressure of 29.92 in. Hg (14.7 psia) is compressed with a pressure of *two* atmospheres or 59.84 in. Hg, the volume will be the inverse of the change in pressure, or *one-half* cubic foot.

Charles' Law. The volume of a gas is directly proportional to the absolute temperature (°F + 459.2). For example, a cubic foot of gas taken at 70°F and heated to 90°F will increase to

$$\frac{90 + 459.2}{70 + 459.2} \text{ cu ft}$$

or 1.04 cu ft:

$$\frac{V_{\text{initial}}}{V_{\text{final}}} = \frac{T_{\text{initial}}}{T_{\text{final}}}$$

Note that if more than one gas law is involved, you should figure them successively, in any sequence.

Avogadro's Law. Equal volumes of all gases at the same temperature and pressure contain the same number of molecules. One molecular weight of a gas will occupy 22.4 liters at 70°F and 1 atm of pressure. To find the rate of expansion, first find the volume of 1 mole of the gas when it is in the liquid state:

$$\text{Volume} = \frac{\text{molecular weight}}{\text{density}}$$

For propellant 12 the density of the liquid is 1.325:

$$\text{Volume} = \frac{121}{1.325} = 91.5 \text{ cc} = 0.0915 \text{ liter}$$

Therefore the volume of propellant 12 in the gaseous state, 22.4 liters, is 245 times that in the liquid state, 0.0915 liter.

Raoult's Law. The vapor pressure of each component in a mixture of gases is proportional to its mole fraction. To review the definition of a mole, the atomic weight of hydrogen is 1 and that of oxygen is 16, as shown in the chemistry textbooks. The sum of atomic weights in a molecule of H_2O is 18. A gram-molecular weight, usually shortened to *mole,* is the same thing expressed in grams; thus a mole is 18 g of water.

In a mixture of 12/11(40:60) the molecular weight of propellant 12 is 121, and of propellant 11 it is 137. Therefore:

$$\frac{40\%}{121} = 0.331 \text{ mole} \qquad \frac{60\%}{137} = 0.438 \text{ mole}$$

To find the mole fraction, divide by the total:

$$\frac{331}{769} = 0.43 \qquad \frac{438}{769} = 0.57$$

Multiply by the absolute pressure of the pure gas in each case:

$$\begin{aligned} 0.43 \times 85 \text{ psia} &= 36.5 \\ 0.57 \times 13.4 \text{ psia} &= 7.6 \\ \text{total} &= 44.1 \text{ psia at } 70°\text{F} \end{aligned}$$

Costs. Several factors need to be considered in determining the cost of propellants. The cost per pound is only one element in the total analysis. Density also is important in the calculations; for example, hydrocarbons weigh only one-third as much as fluorocarbons. The filling rate also should be taken into account, and a propellant that can be cold-filled will have a lower packing cost than one that has to be filled through the valve. Also the cost of handling bulk propellants from the supplier, and special provisions for explosive gases, should be considered.

Trade Names. Air Reduction Co.; Commercial Solvents Corp.; Freon (E. I. du Pont de Nemours & Co.); Genetron (Allied Chemical Corp.); Humble Oil & Refining Co.; Isotron (Pennsalt Chemicals Corp.); Kaiser Aluminum & Chemical Corp.; Phillips Petroleum Co.; Stauffer Chemical Co.; Union Carbide Corp.

VALVES

There are many types of valves to choose from. One supplier claims 3,040 variations of a single basic construction, and considering the different orifice sizes in the stem, the body, and the vapor tap, plus the various dip tube diameters and actuator combinations, it is mathematically possible to have such a number. There are about 80 manufacturers throughout the world but only 25 are major producers, and 15 of them are in this country.

The primary purpose of the valve is to control the flow of the product out of the container: to keep it in when it is not needed, and to let it out when it is wanted. It also has an important effect on the character of the product that is dispensed. For example, a foam product will come out as a foam, a spray, or a stream with different valves and actuators. Although it is not so significant a factor as the formulation, the valve is nevertheless one of the variables that contribute to the quality of the final product.

Essentially the valve consists of a plastic stem which presses down on a rubber diaphragm contained in a hollow plastic body. When the valve is forced down, or in some cases when it is tilted, it stretches the rubber away from an opening in the body so that the pressurized contents escapes through the hole. A metal cup is used to attach the valve to the container, an actuator button may direct the spray off to the side, and a dip tube reaches down to the bottom of the container. (See Fig. 5.) With minor modifications by different manufacturers, they all work about the same way. Various sizes of orifices in the body, stem, and actuator from 0.013 to 0.040 in. can be obtained from all manufacturers,

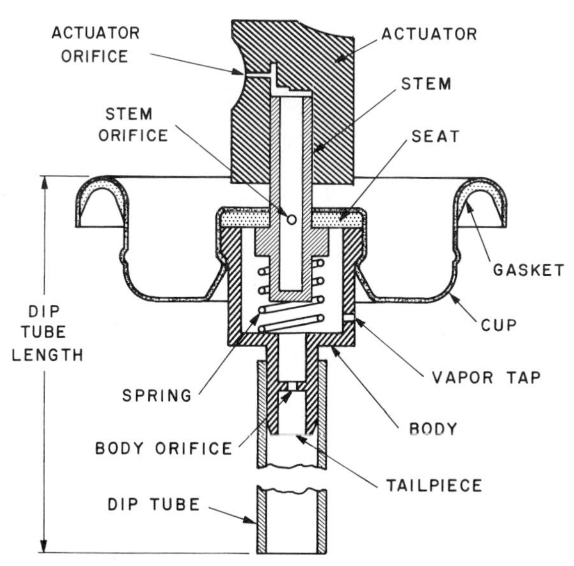

Fig. 5. Valve. Various parts of a typical aerosol valve are shown. Metal cup shown is for a standard 1-in. opening in a metal container. The four orifices shown vary in size to suit different products. Other types of actuators can be used interchangeably.

TABLE 3 Typical Orifice Sizes for Different Products

Product	Body orifice	Vapor tap	Stem orifice	Button orifice*	Delivery rate†
Hair spray	0.013	....	0.013	0.020 RT	0.6
Hair spray	0.062	0.013	0.013	0.018	0.8
Deodorant	0.018	0.018	0.018	0.015 RT	0.7
Room freshener	0.030	....	0.018	0.020	1.0
Room freshener (water base)	capillary	0.030	0.030	0.030	0.9
Oven cleaner	0.062	....	0.030	0.018 MBU	1.1
Insecticide	0.014	....	0.013	0.020 RT	0.5
Deicer	0.062	....	2 × 0.036	0.040	9.0

* RT = reverse taper; MBU = mechanical breakup.
† Grams per second at 70°F.

and these are the most critical factors. There will be differences in the shape of the mixing chambers which may affect the spray pattern, but each manufacturer of valves usually has several variations to offer. (See Table 3.)

Foam valves are quite different from spray valves. For food toppings they are usually of the tilt type, and they are designed to be used in the inverted position, and so no dip tube is used. Shave cream, however, is always used right side up, and it is made with a dip tube that goes down inside the can to the bottom. The actuator on a foam valve generally has a long stem and no restrictions so that it can act as an expansion chamber. The only orifice is the one at the valve seat, which is usually quite large.

Spray valves may have several orifices, with expansion chambers in between. The bottom of the valve body, known as the tailpiece, may be restricted to reduce the rate of discharge. The next orifice is in the valve stem, the third is in the actuator, and there may be a fourth if an insert is added to the actuator. These orifices usually become progressively larger, for example, 0.016, 0.018, 0.020, 0.022 in. The standard notation for specifying orifice sizes is to give the stem orifice first, followed by the vapor tap if there is one, and finally the body orifice in this manner: 0.018 × 0.020 × 0.080. An actuator with an insert will probably have a swirl chamber; it is called a mechanical breakup actuator. The swirling causes the product to issue as a wide hollow cone, breaking up as it spreads out in a fine mist. Without the mechanical breakup feature, the spray would come out as a narrow solid cone, and in the case of a three-phase system, in which the product does not mix with the propellant, it might be a coarse spray containing many large droplets. An insecticide spray which mixes readily with the propellant, on the other hand, would not need this extra breakup, and most of the particles would be under 30 μ (0.001 in.) in diameter just from the expansion of the propellant. Another variation in actuator orifices is the reverse-taper channel. The effect of this construction is to spread the spray over a wider angle.

A *vapor tap* is a hole in the side of the body of the valve which allows a small amount of propellant to escape into the valve and mix with the product as it is dispensed. This helps to break up the spray into a fine mist. It is particularly good with water-base products in which the propellant and concentrate are immiscible, but it requires a good balance between the size of the vapor tap and the amount of propellant.

A spray which is intended to coat the surface rather than remain suspended in the air will need larger droplets, above 50 μ in diameter. An actuator without the mechanical breakup feature will deliver around 1 g per sec, whereas the extra restriction of the breakup chamber will reduce this rate somewhat. Lower pressures can be used, and the mechanical breakup feature has made it possible to put perfumes into uncoated glass containers. The breakup chamber is also useful when the ratio of propellant to product is low. One other effect of the mechanical breakup is to reduce the chilling effect on the skin, because of the soft wide spray pattern. A vapor tap also has this effect of reducing the chilling of the skin, by diluting the issuing gas with vapor from the headspace, which is at room temperature.

Powder aerosols work best with valves having a high seating pressure, with the seat close to the terminal orifice, especially if there is more than 10 percent powder in the formulation. A fine powder with a particle size under 200 mesh is preferred, and the propellant should have a specific gravity close to that of the powder, to minimize settling out. High pressure and a low percentage of solids give the least trouble with clogged valves.

Sealed valves are available with metal seals to prevent the product from coming in contact with the valve gaskets until the seal is broken on the initial actuation. These valves have been used to minimize seepage losses when very long or very adverse storage conditions are encountered.

Metering valves deliver a predetermined amount and then automatically shut off. They do this with a double valve construction in which one valve closes when the other opens. The product that is in between is discharged by the pressure of the propellant mixed with the product. When the valve is released, the chamber is refilled by the pressure in the container. This type of valve is suitable only for products that are miscible with the propellant, although there are some that use a steel ball in the dip tube as a check valve; these are partly successful with three-phase systems. Metering valves are available in different "shot" sizes such as

50 mg, 100 mg, etc., up to 1 oz, with a variation of about 10 percent, although some may be a little more accurate than this.

Pressure filling, in which the product is put into the container and the valve secured in place before the propellant is added, has required special designs in some cases. One valve, for example, has flats on the stem which provide openings for filling instead of going through the stem orifice. Low-swell rubber compounds have been developed, and noncracking dip tubing is available for minimizing malfunctioning of valves with difficult products.

Some valves are designed to work in any position. One of these has a weighted dip tube that falls to the bottom of the liquid no matter which way the container is turned. Another has a weight in the valve that drops away from a special orifice when the package is inverted. A mixing valve that permits two materials to be brought together as they are dispensed has applications for hair coloring, and for shave cream that is heated by chemical reaction. Co-dispensing valves for hot shave creams are being offered by all the major valve manufacturers. Such valves allow an oxidizing agent such as potassium sulfite to combine with a peroxide to generate the heat. A flexible plastic bag contains the peroxide and is attached to the bottom of the valve. This is surrounded by the shave cream containing the oxidizing agent, along with the propellant. The proportions of each that go through the valve are about four parts of shave cream to one part of peroxide.

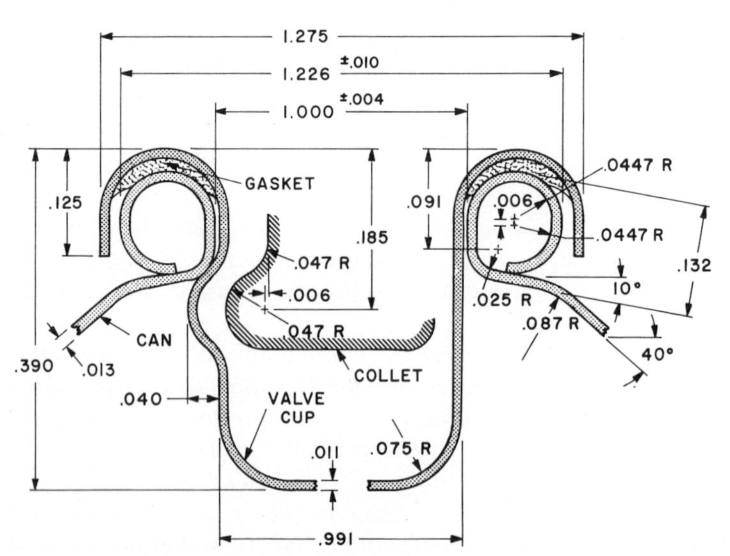

Fig. 6. Crimp dimensions. A good seal between valve and container is important to prevent leakage. The critical dimensions of the valve cup and the container opening must be carefully maintained to ensure a tight closure.

An interesting development has been the self-contained pressure unit which can be attached to the top of any type of container. Sold in this country under the trade name Preval (Precision Valve Corporation), and in France as Innovair (Geigy AGCHIM), these units have a siphon tube passing down through the center or off to the side of the propellant container. Since the product section is not under pressure, it can be made of glass, plastic, or any other material. The spray pattern is limited to a wet type of spray, since the operation depends on venturi action to suck up the liquid from the lower compartment.

Valves that are used with cans always have a 1-in.-diameter cup (see Fig. 6), but for glass containers and purse-size metal containers the valves have a metal ferrule to fit a serum vial finish, either 13-, 15-, or 20-mm size. (See Fig. 7.) The skirt length is between 0.285 and 0.325 in. in any of these sizes, to suit the particular container being used; it should be long enough to tuck well under the finish of the bottle. Top pressure during crimping should be sufficient to compress the gasket to about 40 percent of its original thickness. A good way to measure this is to grind away part of the ferrule on a sealed unit against a fine grinding wheel until the gasket is exposed.

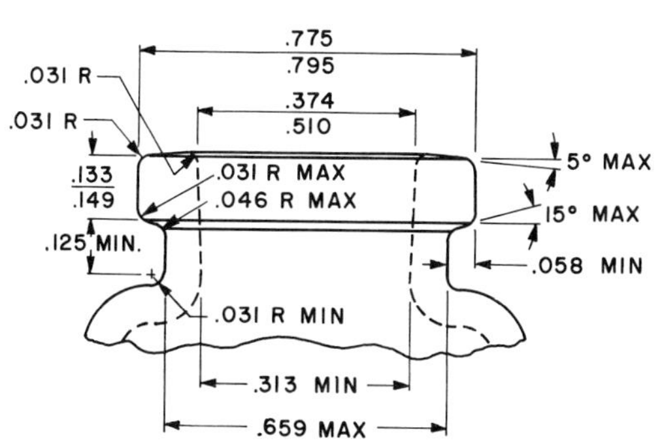

Fig. 7. Glass finish. Dimensions of a 20-mm biological finish, also known as a No. 2710 finish, are shown. Valves with metal ferrules to fit this size of opening are widely used for cosmetics and toiletries.

Dip tubes are made from polyethylene tubing and usually are cut to a precise length that reaches almost to the bottom of the container. With some products the dip tube tends to "grow," owing to absorption of solvents, and must be made shorter so that it does not seal off against the bottom of the container. Dip tubes are usually curved, simply because it is impossible to make them straight. To ensure that the maximum amount of the contents can be dispensed when the can is tilted forward, it is common practice to put a red dot on the rim of the valve cup in the direction that the dip tube curves. Then the actuator button can be oriented in the same direction as the red dot.

Valve gaskets are made of either neoprene or Buna N, depending on the formulation of the product. Neoprene is used most often with water-base systems. Actuators are made in different shapes for special purposes, such as cake decorating, vapor inhalants, and spot treatment of weeds. Other types of actuators include foam heads and spouts, upwardly directing sprays, mastitis and vaginal applicators, spreading combs, and mother-daughter transfer devices. In some cases the actuator is combined with the overcap so that the cap does not need to be removed for use, and the product is dispensed through an opening in the side of the overcap.

Costs. There is little difference in cost among the various standard valves, most of them being around 4 cents each. Metering valves and other special constructions, however, command a premium price, and in some cases they are double the cost of regular valves.

Suppliers. Aerosol Research.; Aerosol Venezola, C.A.; Avoset Co.; Bakan; CIMA; Clayton Corp: Cook Chemical Co.; Coster; Dill; Emson Research, Inc.; Fritz Albert Riegler; Genoud; Maruichi; Metal Box Co., Ltd.; Mitani; Newman-Green, Inc.; OEL, Inc.; Perfect-Safca; Precision Valve Corp.; Reddi-Wip, Inc ; Rhodia; Risdon Mfg. Co.; Samuel Taylor Ltd.; Scovill Mfg. Co.; Seaquist Mfg. Corp.; Solfrene; Sprayon Products, Inc.; Toyo Valois; Vac, Inc.; Vapsol.

CONTAINERS

There are three principal types of aerosol containers: metal, glass, and plastic. Of these metal is by far the most popular, accounting for more than 90 percent of the current production of aerosols. *Tinplate* is used for two-piece and three-piece cans, and aluminum is made into one-piece containers by impact extrusion and by spinning. (See Sec. 7, "Metal Containers.") Stainless steel, usually type 304, is used for very small containers. (See Table 4.)

For difficult products special pure tin solders are sometimes used with tinplate containers, and various interior coatings have been developed to suit specific requirements of different products. Some of the interior coatings must withstand the effect of products known as *strippers* and *perforators*: spray starches of the PVA types are perforators; window cleaners and some shave creams are strippers. Three types of coatings, usually called enamels, are used in aerosol cans. These are vinyls, epoxies, and phenolics. Sometimes combinations of enamels are used. It is a generally accepted fact that cans cannot be coated so completely that no metal is exposed. There are sure to be some voids; tests for detecting these are given on page 11-28.

TABLE 4 Relative Costs of Three-piece Cans

Nominal capacity, oz	Cost, cents
3	7
4	$7^1/_4$
6	$7^1/_2$
8	8
12	$8^1/_2$
14	$8^3/_4$
16	9

These are often critical factors in determining an adequate shelf life, as corrosion may cause leaky cans, and unfavorable reactions between container and product may result in flakes or sediment that clogs the valves. Shave cream, for example, in the presence of a solder containing lead will sometimes form a hard precipitate that gets stuck in the valve. Pure tin can be used in this case, but it does not make as strong a joint and is more costly than a 2 percent tin and 98 percent lead solder. A solder made from 98 percent tin and 2 percent antimony is coming into wide use for this purpose. Inside tab construction of the side seam is stronger than the outside tab design although cross-bead impressions help to strengthen the outside tab cans.

The nominal capacity of a can has little relationship to its real capacity, as can be seen in Table 5. The maximum recommended fill is $92^1/_2$ percent at 70°F. This is best determined by measuring the volume of a

TABLE 5 Sizes of Metal Cans

Designation*	Nominal capacity, oz	Capacity to overflow	
		fl oz	cc
202 × 214	3	5.07	150
202 × 314	6	6.76	200
Spra-tainer	6	7.27	215
202 × 406	6	7.78	230
202 × 509	9	9.98	295
Spra-tainer	12	12.68	375
211 × 413	12	14.03	415
211 × 510	14	16.40	485
211 × 604	16	18.26	540

* The first digit of each dimension designates whole inches, and the second and third digits designate sixteenths of an inch. Thus 211 means the diameter is $2^{11}/_{16}$ in.

known weight of formula with a graduated glass pressure tube at the required temperature. This can be converted to specific gravity and used to calculate the weight which would be equivalent to 92½ percent of the overflow capacity in fluid ounces. There does not seem to be any minimum fill, and there are products on the market, especially the smaller sizes, that are less than one-third full in terms of the liquid contents. As long as the label correctly states the amount in the container, it is all perfectly legal.

There are regulations covering the construction of metal cans for certain purposes. Most of the containers used for household products are "unclassified" if the pressure does not exceed 140 psig at 130°F and if the capacity is not over 19.3 fl oz. The minimum bursting strength for an unclassified container is 210 psig (1½ times the pressure at 130°F). If the pressure at 130°F is over 140 psig but less than 160 psig, a Specification 2P container must be used, according to section 73.306 of the Department of Transportation regulations, under Public Law 86-710. The 2P can must be made of steel plate not less than 0.007 in. thick and must withstand 240 psig internal pressure. The standard aerosol can made with special solder will usually meet this requirement, but it must be marked "ICC 2P." If the pressure at 130°F is over 160 psig but less than 180 psig, a 2Q container must be used. The two-piece container of the type made by Crown Cork & Seal Company is the only inexpensive can that meets this specification. An example of a product requiring a 2Q container is a refrigerator refill unit which contains straight propellant 12.

The one-piece aluminum container is more popular in Europe than in this country, because the cost is more favorable in comparison with tinplate than it is here. Although some smaller units are being used for fragrance items and pharmaceuticals, the larger units have not penetrated the American market to any appreciable extent. One-piece containers have the advantages of a clean neat appearance without seams and the ability to withstand high pressure. The normal bursting strength is above 300 psig. Larger sizes can be made than are available in tinplate containers; up to 48- and 72-oz sizes are being made in Italy. The chief disadvantage is the poor resistance to corrosion, particularly with alkalies and strong acids, and to some extent with water. Sometimes known as "monobloc" containers, these cans are impact-extruded and then necked in. A slug of metal is put into a shallow die, and a punch forces the metal to extrude back over the punch. It is then blown off the punch and sent to a trimming operation, after which it is necked in to form the seat for the valve cup.

Glass, which is resistant to attack by most chemicals, is useful as an aerosol container for corrosive products. A valve with all plastic and rubber parts is often used with a glass container, for products that would attack metal. There is a greater variety of shapes available in glass than in metal, and for this reason it is frequently chosen for cosmetics and toiletries, but so far only about 4 percent of aerosols are in glass. Some bottles are dip-coated with a plastisol to keep the glass fragments from scattering if it should be dropped, and this allows pressures up to 25 psig or more to be used, depending on the shape and size of the bottle. A 1/16-in. hole is sometimes put in the bottom of the coating to act as a vent for the escaping gas if the bottle should get broken. Uncoated glass is generally kept under 15 psig, for safety, and the lowest pressure that will give a satisfactory spray pattern is preferred. Compressed gases could be used at a higher pressure than fluorocarbons or hydrocarbons, since they have less potential energy. There are no legal limitations on the pressures that can be used in glass containers, but the aerosol industry has tried to establish safety standards, to forestall any governmental action in this area.

There are *plastic* containers for aerosols made of acetal, nylon, or polypropylene, but these have not been used to any great extent and they have accounted for less than 1/10 percent of all aerosols. Propellant 11 in concentrations of over 35 percent reduces the tensile strength of acetal containers at 120°F or above, causing distortion of the container. The thermal expansion of acetal is approximately four times that of the metal used in aerosol valves, and it may cause problems in crimping tightly enough to avoid leaks. These compatibility problems and high cost have limited their use to a few items in the toiletries field. Plastic-coated glass is about the same price as plastic and is often better suited for the purpose. The chief advantages of plastic containers are the opportunities for color and intricate shapes, plus light weight when compared with glass. However, the permeability or incompatibility of plastic containers with alcohol, essential oils, and propellants, along with the problems of crimping valve cups to plastic necks so that they do not leak, has all but eliminated this type of container from the marketplace.

Barrier containers, which consist of a bag in a can, keep the propellant separated from the product. Variations of this construction are made with pistons, accordion-pleated plastic inner containers, and thin aluminum compartments that collapse as the contents is expelled. The piston unit is very vulnerable to denting, and the plastic separators are not suitable for hydrocarbon propellants, as these small molecules can readily penetrate the plastic barrier material. These containers are particularly useful for food products such as cheese spreads. However, foods which are to be dispensed from aerosol containers should not have fibrous particles because they clog the valves and prevent them from closing properly.

Labels. The three-piece metal can is the most widely used container, but it has one serious drawback, from an aesthetic viewpoint. Because the side seam cannot easily be covered with lithography, the bare metal must be left exposed in that area. Paper labels can be used to overcome this objection; they offer a wide choice of embossing and printing techniques. (See Table 6.)

TABLE 6 Recommended Sizes of Labels for Aerosol Cans

202 × 406:
4 1/16 in. maximum height, grain parallel to this dimension
7 1/8 in. maximum length, 1/2 in. unprinted and unvarnished at one end
202 × 509:
5 1/4 in. maximum height, grain parallel to this dimension
7 1/8 in. maximum length, 1/2 in. unprinted and unvarnished at one end
211 × 604:
5 15/16 in. maximum height, grain parallel to this dimension
8 3/4 in. maximum length, 1/2 in. unprinted and unvarnished at one end

OVERCAPS

To protect the valve from accidental discharge, and to keep dust and dirt out, an overcap is usually snapped over the aerosol container. It is made to fit either the valve cup or the outside of the container. In some cases the actuator is combined with the overcap in such a way that the valve can be worked without removing the overcap. Others are equipped with pilferproof devices that prevent their being discharged before they are purchased. Still others are made into special shapes such as a fireman's hat for a fire extinguisher, a dog's head for a canine shampoo, and a simulated swirl of whipped cream for a food topping.

Cost of overcaps to fit over the rim of the can, per thousand, is about $8.50 for the 201 size, $10 for the 211 size, and $12 for the 300 size.

Suppliers of overcaps are: Aerosol Research Co.; J. L. Clark Mfg. Co.; Clayton Corp.; Eastern Cap and Closure Co.; E. J. McKernan Co.; Gilbert Plastics, Inc.; Metal Fabrications, Inc.; OEL, Inc.; Owens-Illinois; Pharmaplastics, Inc.; Reddi-Wip, Inc.; Risdon Mfg. Co.; Sterling Seal Co.; Sunbeam Plastics Corp.; VCA, Inc.; Walter Frank Organization; West Penn Mfg. and Supply Corp.; Wheaton Plasti-Cote Corp.

PROCESSES

The earliest method of filling nonaqueous products was *cold filling*. This consisted of chilling the product or concentrate and putting a measured amount in each container; then a measured amount of chilled propellant was added, and as quickly as possible the valve was put on and crimped in place. The completed package was then carried by conveyor through a hot water bath as required by DOT regulations, to detect any leakers, which would show up by bubbling under the water. Cold filling is no longer used in production to any extent because the newer method of under-the-cup filling is much more efficient.

It is not possible to chill products that contain water, as they would freeze solid. They are *pressure-filled,* by adding the concentrate at room temperature and putting the valve in place before the propellant is added. A small amount of propellant may be put in to purge the container of air before the valve is fastened in place. Fluorocarbon propellants are almost five times as heavy as air and will displace the air from the bottom up. If the trapped air were not removed, it might increase the pressure in the finished package by as much as 15 or 20 psig. The bulk of the propellant is then forced in through the valve stem. It is a slow process, and more sophisticated equipment has been developed to inject the propellant under the valve cup just before the valve is crimped in place, at speeds of more than 200 per minute. To test valves on coated glass bottles to see whether they are crimped tight enough, a torque tester can be used. With a 20-mm size it should take between 10 and 18 in.-lb to rotate the valve on the bottle. Although this is not an accurate test, strictly speaking, it is a quick, easy test that can be made at the production line.

Glass containers should be pressure-tested at 150 psig before filling. A special *puck* or cup made of plastic is used to hold each glass or aluminum container while it is being handled on the production line. Different sizes of pucks are needed for the different sizes of aerosol containers. Tinplate cans of the standard 2⅛- or 2¾-in. diameters will fit the usual feeding mechanisms and do not require pucks.

Other operations usually performed on a filling line include cleaning, code marking, label inspection, weighing, and valve testing.

SPECIFICATIONS

The valve should be specified in sufficient detail to cover all orifices, with tolerances, type of gasket material, stem and body material, dip tube dimensions and material, actuator type and orifice sizes, and valve cup coatings.

Details of the container include coatings inside and outside, type of solder if soldered construction, and kind of decoration with number of colors and coatings, including bottom and top pieces.

Specifications for the overcap should indicate the type of material; whether it should fit outside the chime or inside, or over the breast or valve cup; and any decoration that may be required.

TESTING

Before a new product is released to the trade, it should be thoroughly tested. Test methods should be as realistic as possible; that is, the container should be exhausted in short bursts, with an interval between successive bursts and not all at once. Storage conditions should be at room temperature, although quicker results will be derived at elevated temperatures. Some laboratories consider 1 week at 110°F equivalent to 1 year at room temperature. It should be emphasized that accelerated tests do not always correlate with results under normal conditions. Things to look for in stability tests are settling out of solid particles, degradation of perfume, crystal formation, lumping together of particles, corrosion, and leakage.

Control testing of component parts as well as the filled units from the production line should include visual examination of materials and coatings. Orifices can be checked with drills of the right size as go no-go gauges. Interior enamels can be tested for voids with a solution of copper sulfate, which darkens any exposed metal; bare steel will turn dark blue with concentrated hydrochloric acid followed by potassium ferrocyanide solution.

A spray pattern will show up on colored paper when the container is held a prescribed distance away; it should be tested with each new batch of containers from the production line. The particle size also should be checked by spraying on a microscope slide and using a calibrated ocular for measurement.

A flame projection test consists of placing an aerosol, conditioned at 70°F, 6 in. from a candle flame, spraying for 4 sec through the top third of the flame, and measuring the extension of the flame beyond the candle. The contents is classified as "flammable" if the flame projection exceeds 18 in. at full valve opening or if the flame extends back to the dispenser at any degree of valve opening. (See Figs. 8 and 9.)

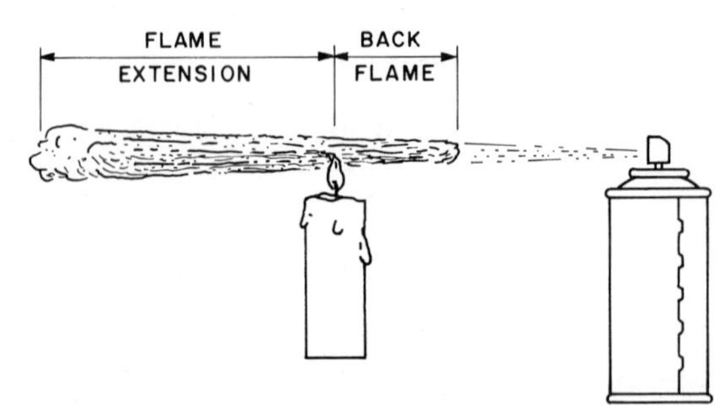

Fig. 8. Flammability test. When sprayed into a candle flame 6 in. away, the length of the flame determines whether it is "flammable" or "extremely flammable." Consult regulations for details.

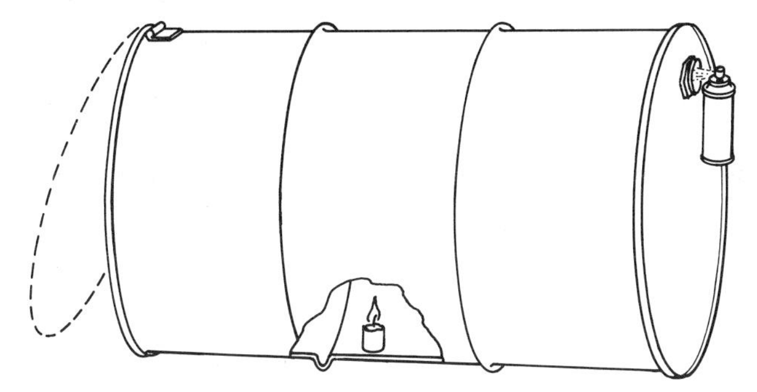

Fig. 9. Closed-drum test. A 55-gal drum with a loosely fitted hinged cover is used to determine the explosive nature of an aerosol product. A lighted candle is placed in the drum and the aerosol is sprayed through a small opening. The length of time is measured until an explosion takes place. See latest regulations for details.

A method for determining the degree of enamel coverage on the inside is to fill the container with a 5 percent solution of sodium chloride. An electrode is placed in the solution, and another against the outside of the metal can. A 5-volt dc potential is placed across the electrodes, with the positive side in the solution. If the flow of current, measured on a milliammeter, is more than 55 ma after 5 sec, the coverage is inadequate.

In order to find the location of the voids in the enamel, a concentrated solution of copper sulfate, to which has been added a few drops of sulfuric acid, will plate out the copper on contact, wherever there is bare metal. Examination under a strong glass will then clearly show where the exposed metal has been plated. The most vulnerable spots are in the fillet area of the seam, particularly at each end of the seam. For aluminum containers a 1 percent solution of copper sulfate with 0.05 percent acetic acid, and 0.005 percent sodium dioctylsulfosuccinate to remove grease, is used with a 5-volt dc current to measure the amount of discontinuity and to stain the bare areas. A reading is made after 5 sec, and the current is reversed to clear away the hydrogen bubbles, which would cause increased resistance to the flow of current.

Section 12

Fibre Tubes, Cans, and Drums

Fibre Tubes and Cans 12-1
History 12-2
Forms and Modifications 12-3
Advantages and Disadvantages ... 12-4
Processes 12-5
Materials 12-9
Costs 12-10

Fibre Drums 12-10
History 12-11
Advantages and Disadvantages ... 12-12
Forms and Modifications 12-12
Carrier Regulations for Fibre Drums 12-13
Design Criteria 12-15

FIBRE TUBES AND CANS

The basic principle in working with the composite type of structure is to use that combination of materials which is best suited for the purpose, in the minimum amounts that are necessary to accomplish the packaging objectives. The goal may be low cost, protection from environmental factors, or an attractive appearance; but whatever the purpose, the law of parsimony is the overriding principle.

Combinations of metal and paperboard or plastic and paperboard, incorporating films, foils, coatings, or adhesives where needed, are finding applications in many fields. Citrus juice cans, spice boxes, and cocoa canisters are examples among food packages, and there are similar applications in household chemicals, garden supplies, and other fields.

Expensive materials can be kept to a minimum by using them in thin layers, supported by inexpensive paperboard for strength and rigidity. Fragile components of a lamination can be buried between other more sturdy materials for protection. Functional parts can be made of the best substance for a particular need without regard for the overall needs of the package, since other elements can be chosen to balance out the structure. In this way it is possible to produce containers that will do things that no single material could accomplish.

The best-known example is the metal-end fibre-body can, for which the economy of paperboard to make up the main parts is utilized; tinplate is used for the end pieces, where rigidity is the prime consideration.

Fibre tubes are made in sizes from 0.050 in. in diameter and 0.040 in. long up to 5 ft in diameter and any length that can be handled. For packaging, however, it is usually necessary to have closures on the ends; the range of diameters for these cans is from 5/16 to 8 in.

History. Paper tubes were first used during the Civil War for ammunition. The earliest use of fibre cans for packaging is believed to have been near the end of the eighteenth century. These containers had paper bodies and ends. By the middle of the nineteenth century they were being used for certain types of munitions, and a little later for dynamite. More recently they have been used for salt, oatmeal, and similar dry products.

In the early 1950s biscuits ready for baking were introduced in a foil-lined composite can, and the market was expanded to include sweet rolls and other refrigerated dough products. The spiral-wound can is particularly well suited for this purpose, and the genius that developed this package-product concept is one of the unsung heroes of the packaging profession. The combination of high-test paper, to withstand the pressure of the rising dough, and foil, to retain the shortening and water without staining the label, was very effective and met with instant success. Housewives were fascinated with the convenience of opening the can by rapping it against the edge of the table, and popping the biscuits into the oven without further preparation.

Partly as a result of the methods and equipment that were developed to produce the refrigerated dough packages, in the late 1950s composite cans for citrus concentrates and lubricating oil were introduced. About 85 percent of the citrus concentrate containers, or about 1½ billion cans, are now this type. The oil companies have not been as quick to adopt these fibre containers because of bad experiences in the past. When fibre-bodied containers were tried before, there had been problems of leakage at the point where the metal end was seamed onto the body. There was also wicking of the oil under the foil liner where the edge of paper was exposed at the overlap. It is now possible to produce a double seam at the ends, in the same manner as for a metal can, which eliminates the leakage problem. The wicking problem also has been solved by folding the foil and paper lamination back on itself at the overlap, so that no paper is exposed. With these improvements the composite can is being adopted by some of the major oil companies, and more than half of the lubricating oil is now being packaged in fibre containers.

There are today about 200 plants making nearly $300 million worth of fibre tubes and cans. These are used for salt, cocoa, oatmeal, baking powder, scouring powder, and, more recently, dry snack foods. They are beginning to be used for paint, but the biggest potential market is for lubricating oil, which currently uses more than 2 billion containers in the 1-qt size, mostly tinplate at the present time. Beer is another big market for fibre cans, but the test results so far have not been very promising, and the economics is not so favorable as for the larger cans.

Forms and Modifications. There are three major types of bodies for composite cans: convolute-wound, spiral-wound, and lap-seam. (See Fig. 1.) In all instances the bodies are wound on a mandrel and trimmed to length. In the case of *convolute* winding, the various layers are carried over glue rollers and then fed straight into the revolving mandrel. After three to five revolutions of the mandrel the stock is cut off and wiped down. Labels may be applied, after which the tube is cut to length and the ends are attached.

The second method, which is the most popular and cheapest, is known as *spiral* winding. Because it is faster and makes better use of materials than convolute winding, it is a more economical process. Spiral winding can be used only for cylindrical shapes, whereas the convolute method is used for squares, oblongs, and ovals in addition to the cylindrical type. Spiral winding does not make as strong a container as convolute winding, and for larger packages for which resistance to the hazards of

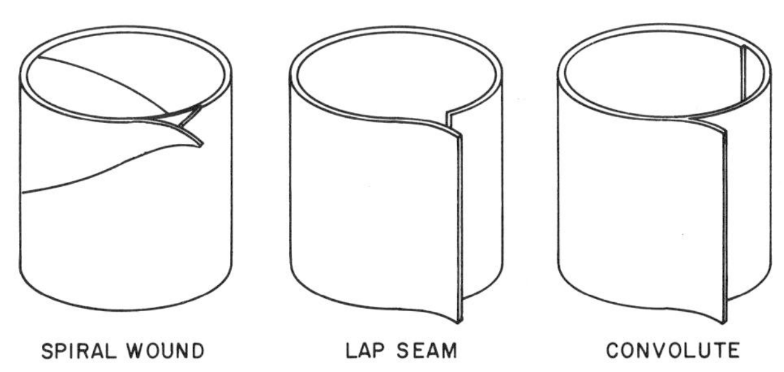

Fig. 1. Fibre tubes. Three methods of construction are used in making bodies for composite cans. Spiral winding is usually several layers of different materials with angled overlapping joints. Lap-seam bodies, as shown in the second illustration, are made from laminated material, cut into blanks and joined at the side with adhesive. Convolute winding consists of several layers coiled on top of each other, straight in from the side. Spiral-wound bodies are always round, but the other types can be oblong or oval if desired.

shipping and storage is of paramount importance, the convolute construction will outperform spiral winding.

With spiral winding, the material is carried over glue rollers, is fed at an angle to the mandrel, and is carried around the fixed mandrel by a moving belt, forming a continuous tube. The tube is cut to length as it comes off the end of the mandrel, and the tops and bottoms are added later in a separate operation, to complete the package.

Advantages and Disadvantages. Any desired amount of rigidity can be had in these composite cans by varying the thickness of the paperboard walls. In a similar manner moisture proofness can be incorporated into the structure economically by using asphalt or wax between the plies, and greaseproofness also can be added by putting glassine or foil on the inside, next to the product. Easy-opening features may be included, with tear strings or pour spouts in the body, or pull tabs and friction plugs in the end pieces. Incidentally, round metal plugs are preferable to oval plugs because they cannot fall inside, as the oval-shaped ones are too apt to do. Shoulder pieces can be formed by means of an extra tight-fitting tube placed inside the body and projecting above the top edge, so that a cover of the same diameter as the body can telescope down over the neck, making a flush joint. Sifter tops that are used with certain types of products are readily available in either plastic or metal.

There are two significant weaknesses in these composite containers, however. The joint of the end pieces with the side walls can be a source of trouble. When the metal is curled over so that it grips the body tightly around the rim, the metal bites into the paperboard, but since paperboard is soft and weak, the ends sometimes break away from the bodies in shipment. This may occur when the bottom edge of one can overrides the rim of an adjacent can and forces the metal piece downward from the body. The other hazard in shipping is a side shock, which presses the rim of one can against the body of the next can, tearing the paperboard inward from the rim.

Another point to keep in mind is the vulnerability at the overlap. It is difficult to get a secure joint, and wicking may take place where the edges are exposed to liquids or pastes. Skiving or tapering the edges will minimize this condition, but it requires close controls in the manufacturing operation.

Some can openers will not work on composite cans because of the greater thickness of the rim. A fibre body is three times as thick as a metal body, and when the ends are double-seamed, the extra thickness will not fit between the cutting wheel and the driving wheel of some types of can openers. This disadvantage has delayed the introduction of coffee and shortening in fibre packages, and although it is possible to get pull-out easy-open ends, there have been some problems with cut fingers that have made the coffee and shortening packers look for a better opening method.

In comparison with metal containers, a fibre can will provide far more thermal insulation, which may be good or bad, depending on the type of product it contains. If quick freezing is part of the process, the fibreboard will interfere with the rapid cooling. On the other hand, it will protect the contents from a temperature change that might be detrimental.

Processes. Paperboard is the basic material from which fibre cans are made, and the quality ranges from a low grade of chipboard, which is made from reprocessed scrap material, to a pure virgin sulfate *stock* suitable for food packages. Factors to consider in choosing a paper stock are tensile and bursting strength, porosity, glue holdout, and the shrinkage rate.

The three methods of winding fibre tubes will now be taken individually, starting with the spiral winding process. In this system, which is the most widely used, the paperboard is slit in narrow widths and wound into rolls 3 or 4 ft in diameter. The width of the roll is related to the diameter of the finished tube, normally between 1½ and 2 times the inside diameter. The outer ply is often a label stock with the printing angled, so that when it is wrapped around the tube, the printing becomes straight and matches up where the edges butt together.

The rolls of paperboard are mounted on *unwind stands* on both sides of the tube machine. A means is provided for splicing in a new roll as the old one is running out, without interrupting the operation. The edges are sometimes *skived* or tapered just before the adhesive is applied. (For more about skiving, see page 12-9.) The paperboard is threaded down into a glue pot, where it is coated on both sides with adhesive and the excess is doctored off by bars with notches about 3/16 in. apart, so that stripes of glue remain. The spacing and depth of the notches are such that the glue will spread into a solid film when it is pressed against the next layer of paperboard. Sometimes only one side of each ply is wet with glue.

Various kinds of *adhesives* are used for different purposes. Silicate adhesives are cheap and very effective. They are also odorless and verminproof, and they add more stiffness to the tube than most other adhesives. Dextrin adhesives are quite inexpensive and are also widely used. They do not dry very quickly, but they are satisfactory for most purposes. Animal glues are used for glassine and parchment papers, but they are more expensive than dextrins. Resin emulsion glues are fast-setting, but they are higher in price. The most expensive of all are the latex adhesives. For more information on this subject, see "Adhesives" in Sec. 9, "Closures, Applicators, Fasteners, and Adhesives," page 9-27.

After being coated with adhesive, the ribbons of paperboard travel at an angle, onto the top of the mandrel and around it in an overlapping helix. A belt goes on top of the paperboard and makes one turn around the mandrel on the outside of the paperboard tube, following the corkscrew direction of the paperboard ribbon and providing the driving force to keep the tube spiraling off the end of the mandrel. (See Fig. 2.) The mandrel does not revolve, but remains fixed. The *drive belt* is supported on two vertical pulleys, mounted on each side of the mandrel, so that the belt goes around the mandrel, then makes a half twist and wraps around one pulley, goes straight across the mandrel to the opposite pulley, and after going halfway around it, finally returns to the mandrel again in an endless circuit. The pulleys and belt make up the main part of the *winding head,* and the angle that it makes with the axis of the mandrel can be adjusted to suit the width of material and the diameter of the mandrel.

The diameter of the polished steel *mandrel* determines the inside dimension of the finished tube. These mandrels are interchangeable and can be made any desired size, so that close fits of telescoping parts can be accomplished by adjusting the diameter of the mandrels until a precise fit is achieved. One end of the mandrel is fastened solidly to the machine, but the takeoff end is unsupported. Sometimes a V-shaped trough is placed under the revolving tube to lend support, up to the

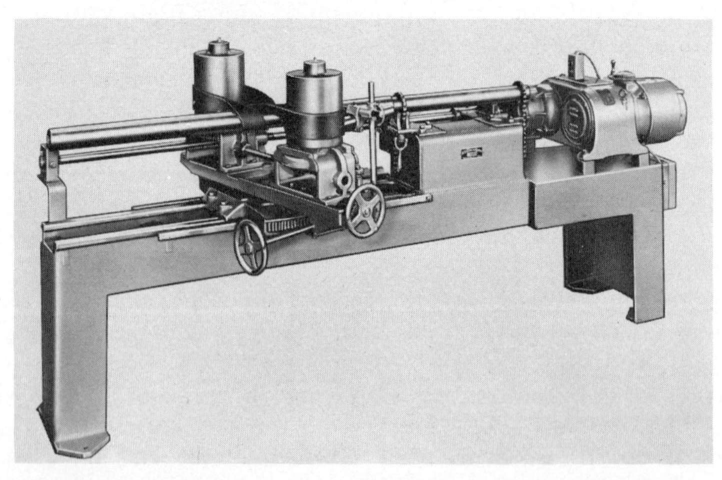

Fig. 2. Spiral winding machine. Narrow strips of paperboard are brought in from both sides of the machine at an angle and are wrapped around the mandrel by the belt which is driven by the vertical pulleys. The completed tube moves continuously off the mandrel toward the left. (*M. D. Knowlton Co.*)

cutoff section. This distance to the cutoff point is determined by several things, but mostly by the amount of time it takes for the glue to set.

A circular knife or a revolving saw is used to cut the tubing to length. The most widely used *cutoff* is a freely spinning circular knife that is pressed against the tube as it moves along the mandrel and is backed up on the opposite side of the mandrel by supporting rollers. The knife travels along with the tube until the cut is completed and the piece drops off the end of the mandrel; then the knife pulls away from the mandrel and returns to the starting point. Heavy-wall tubes are cut with a revolving saw instead of a knife. (See Fig. 3.)

In some high-production operations the pieces are fed directly into the next operation, such as recutting or flanging. Not all container bodies are *recut,* but in the case of very short bodies, or neck sections that cannot be cut quickly enough on the regular automatic cutoff section, they must be taken off in multiples and then recut into smaller units.

Sometimes the bodies are flanged out before the ends are applied, to get a more secure seal. Otherwise a burr, formed on the inside edge by the cutoff knife, might reduce the inside diameter and make it more dif-

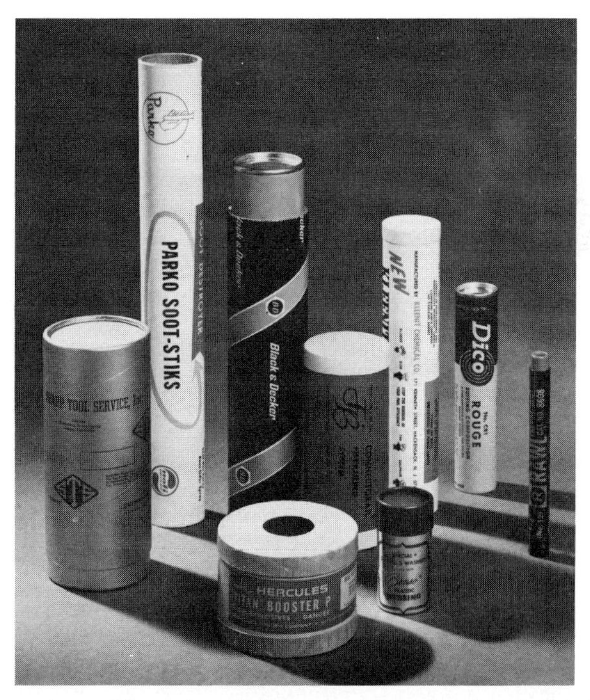

Fig. 3. Fibre containers. The variety of sizes and constructions in cylindrical fibre cans is indicated in this group picture. (*Niemand Bros., Inc.*)

TABLE 1 Standard Sizes of Fibre Cans

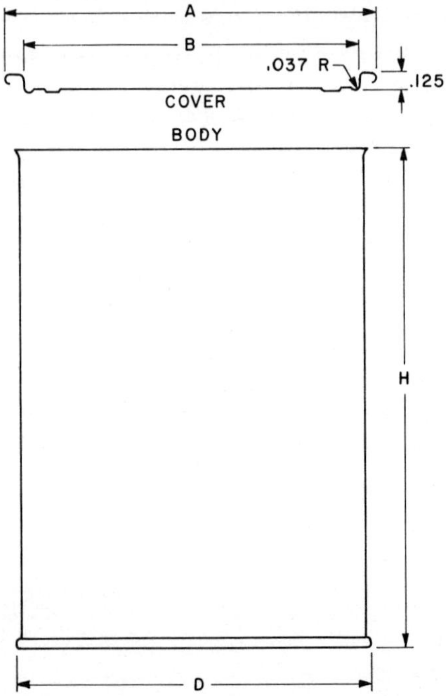

Size, in.*	Type	A Outside diameter	B Countersink diameter	D Steam diameter, max.	H Body height
202 × 314	6-oz citrus			2.220	3.855–3.890
211 × 414	12-oz citrus	2.931–2.951	2.557–2.564	2.735	4.835–4.875
401 × 509	1-qt motor oil	4.278–4.298	3.884–3.891	4.070	5.535–5.580
401 × 510	32-oz citrus	4.278–4.298	3.884–3.891	4.070	5.595–5.640
610 × 713	1-gal motor oil	6.902–6.922		6.710	7.755–7.810

* The first digit of each dimension designates whole inches, and the second and third digits designate sixteenths of an inch. Thus 211 means the diameter is 2 11/16 in.

ficult to position the end pieces. The *metal ends* are made from tinplate approximately 0.0105 in. thick or aluminum up to 0.0145 in. thick. (See Table 1.)

The number of *plies* in the body can vary from one to a dozen or more. The more plies for a given wall thickness, the greater the strength of the tube, as a general rule. However, the cost goes up as the number of plies increases, owing to the added cost of both the adhesive and the paperboard. The maximum thickness of any ply is limited by the ability of the material to bend around a small diameter without cracking, and with a minimum of springback. Citrus concentrate cans are now being made of only one ply of material by skiving or tapering the edges so that there is no extra thickness at the overlap. This provides for faster freezing of the concentrate. Resistance to side pressure is not as good with skived joints, but resistance to end pressure is better than with overlapped or staggered joints.

Materials. End pieces can be made from paperboard, plastic, or metal. Two types of *metal* are used for this purpose—steel and aluminum—and in some cases the two ends of the can may be different. For example, household can openers require the hardness and magnetic properties of steel on one end, but aluminum on the opposite end is economical at the packing plant. Aluminum has the advantages of light weight, rustproofness, and low tensile strength. The last is an advantage if a tear-open feature is needed. Tinplate as well as tin-free steel is used for the end pieces and for friction plugs, pour spouts, sifter tops, and other convenience features. More information on these materials will be found in Sec. 7, "Metal Containers."

Bodies are made from *paperboard,* or combinations of paperboard with plastic films or metal foils, combined by means of adhesives into a tube. For maximum strength, unbleached kraft paperboard is the best choice. By using this material, it is possible to get stacking and side compression strength equal to an all-metal can, for one-third the cost. Wet-strength resins and various binders can be included in the furnish used for making the paper, to provide special characteristics that may be needed in the final package. The inner and outer plies are usually selected on the basis of appearance; they are often printed or coated to improve their visual qualities. Where cost is important, chipboard is used, but more often other factors are involved and it becomes a case of balancing strength, porosity, glue holdout, and shrinkage against the base cost of the paperboard.

Aluminum *foil* offers the maximum in barrier properties, being especially effective against moisture and greases. For a more detailed description of the physical and chemical properties of metal foils, see Sec. 3, "Films and Foils." Where a specific requirement can be met with glassine paper, plastic film, or a wax coating, the economics will usually dictate the choice of these materials; further information is contained in Sec. 4, "Paper and Paperboard," and in Sec. 13, "Coatings and Laminations."

The liner and label material for a composite container may be prelaminated to the paperboard if they are too light or stretchy to be handled separately on the winding machine. The *adhesives* used for this and other operations will depend on the specific needs in each case. Silicate or dextrin adhesives are the most economical, but hot animal glues which have high tack may sometimes be required, or resin emulsions can be used for some materials that are difficult to laminate. Dry bonding and heat-activated coatings are useful where shrinkage or warping must be avoided, and many specialized adhesives have been developed to meet particular needs. If the adhesive is applied on the winding machine, it can be put on both sides or on one side only. For two-side application the paperboard is carried through an immersion tank, where it travels under one or more bars inside the tank. In the case of one-side coating, rollers are used to bring the adhesive up against the underside of the paperboard, as the ribbon of material passes over the top of the rollers. Another technique is to bring half of all the plies in from one side of the machine without adhesive, and the other plies from the other side with glue on both sides. This method uses less adhesive and minimizes shrinkage, but it is more difficult to set up.

Starch pastes, jelly gums, and other heavy bodied adhesives are not fluid enough for tube winding. Silicate adhesives have good tack, are hard and stiff when dry, and are among the lowest in cost. Animal glues are fast-setting and are often used for dense materials such as glassine paper and foil. Chemical adhesives include polyvinyl alcohol, polyvinyl acetate, casein, urea-formaldehyde, and melamine resins. All but the first are used where waterproofness is desired.

Costs. In this country we have an abundance of paperboard, films, and foils so that the savings over all-metal cans are very significant. Because most other countries do not have this advantage, the composite can is strictly an American development. Over half the cost of a composite can is in materials. Since the paperboard is used in narrow widths, it is often possible for the tube or can manufacturer to keep his costs down by buying end cuts from the paper mills. The end cuts have little value and otherwise would end up in the beater, to be remade into paper.

FIBRE DRUMS

A larger version of the fibre can is the fibre drum that is used for shipping bulk chemicals and other industrial products. Fibre drums are generally used for dry products, although with suitable plastic liners they can be used for pastes and certain types of liquids. A wide range of sizes is available from stock, with end pieces of metal, wood, or fibreboard, and body constructions that include a variety of laminations and coatings. Fibre drums are light in weight, and they have exceptional strength in proportion to their weight. Although a fibre drum is essentially a single-trip container, it is sometimes reconditioned and used for several trips. The railroads frown upon this practice, but they will allow step-down reuse; that is, liquid containers can be reused for paste, and paste containers for dry products. (See Table 2.)

History. The earliest known commercial use of fibre drums was in 1904, when a cheese manufacturer replaced his wooden drums with all-fibre containers. The industry has grown since then until there are now 50 fibre drum plants in this country producing nearly 40 million containers, worth about $90 million, each year. This is divided among the following uses: chemicals 41 percent; plastic resins 17 percent; food products 15 percent; pharmaceuticals 7 percent; soaps and detergents 6 percent; metal powder, wire, and stampings 6 percent; and rolled materials 2 percent. The balance goes into government and miscellaneous uses.

TABLE 2 Standard Sizes of Fibre Drums

Capacity, gal	Inside diam. × ht, in.	Capacity, cu in.	Capacity, cu ft	Capacity, gal	Inside diam. × ht, in.	Capacity, cu in.	Capacity, cu ft
7½	14 × 11½	1,773	1.02	24	18½ × 20¾	5,544	3.20
8½	14 × 13	1,848	1.06	26	18½ × 22½	6,006	3.67
10	14 × 15¼	2,310	1.33	28	18½ × 24¼	6,468	3.74
12	14 × 18¼	2,772	1.60	30	18½ × 26	6,930	4.01
13	14 × 19¾	3,003	1.74	32	18½ × 27¾	7,392	4.27
14	14 × 21¼	3,234	1.87	35	18½ × 30¼	8,085	4.67
15	14 × 22¾	3,456	2.01	38½	18½ × 33¼	8,894	5.14
16½	14 × 25	3,812	2.20				
17½	14 × 26½	4,043	2.33	35	20 × 26	8,085	4.67
19	14 × 28¾	4,389	2.54	38½	20 × 28½	8,894	5.14
20	14 × 30¼	4,620	2.67	41	20 × 30	9,471	5.48
				44	20 × 32½	10,164	5.88
12	15½ × 14¾	2,772	1.60	47	20 × 34½	10,857	6.28
13	15½ × 16	3,003	1.74	51	20 × 37½	11,781	6.81
14	15½ × 17¼	3,234	1.87	55	20 × 40½	12,705	7.35
15	15½ × 18½	3,465	2.01				
16½	15½ × 20½	3,812	2.20	41	21½ × 26	9,471	5.48
17½	15½ × 21½	4,043	2.33	44	21½ × 28	10,164	5.88
19	15½ × 23¼	4,389	2.54	47	21½ × 30	10,857	6.28
20	15½ × 24½	4,620	2.67	51	21½ × 32½	11,781	6.81
21½	15½ × 26¼	4,967	2.87	55	21½ × 35	12,705	7.35
23	15½ × 28	5,313	3.07	61	21½ × 39	14,091	8.15
24	15½ × 29¼	5,544	3.20	67	21½ × 42½	15,477	8.95
25	15½ × 30½	5,775	3.34				
26	15½ × 31¾	6,006	3.67	47	23 × 26½	10,857	6.28
				51	23 × 28½	11,781	6.81
24	17 × 24¾	5,544	3.20	55	23 × 30½	12,705	7.35
25	17 × 25¾	5,775	3.34	61	23 × 34	14,091	8.15
26	17 × 26¾	6,006	3.47	67	23 × 37½	15,477	8.95
28	17 × 28¾	6,468	3.74	72	23 × 40¼	16,632	9.04
30	17 × 30¾	6,930	4.01	75	23 × 42	17,325	10.08
32	17 × 32¾	7,392	4.27				

Advantages and Disadvantages. The light weight of these paperboard containers helps to reduce tare weight and consequently the shipping cost, for many commodities. Sometimes the shipping weight can be as much as 25 or 30 lb less than that of a steel drum of equivalent size. Fibre drums have a high strength-to-weight ratio, as long as they are stacked upright, but they should not be stored on their sides as their strength in that direction is rather poor. Also, even though they have a waterproof treatment on the outside, they are not designed for outside storage and should not be exposed to the weather for any length of time. These drums are easy to open and reclose, and they provide excellent protection for their contents. Disposal is no problem since they can be incinerated easily, which is an important advantage over an all-metal drum.

While they can be used interchangeably with metal drums for some products, they cannot compare in strength with a heavy-gauge steel drum. For powders and granular materials of medium value, the fibre drum is the most economical rigid shipping container. Products of low value, however, are more generally put into multiwall paper bags if a few broken packages are of no consequence. The class of products which are usually put into fibre drums is in the range of 50 cents to $2 a pound.

Liquids are now being shipped in fibre drums with special liners, and they are permitted by the railroads and trucking companies up to 55 gal, or 400 lb. Only four or five manufacturers are making this type of drum, however, and the restrictions of the Department of Transportation should be studied carefully before adopting this form of packaging. Some hazardous materials such as class B poisons also are permitted in fibre drums, but the packaging engineer should use good judgement in such cases.

Forms and Modifications. All fibre drums are made as straight cylinders, with the exception of a modified square type offered by one manufacturer. The end pieces are either 24- to 30-gauge steel, 3/11- (sic) to 7/16-in. plywood, 1/2- to 25/32-in. solid wood, or 0.090- to 0.240-in. waterproof fibreboard, according to size. The specifications are given in the Motor Freight and Rail Freight Regulations.

The bodies are convolutely wound of several plies of 0.012- to 0.016-in. fibreboard glued together, using a silicate adhesive. The outer ply may be waterproofed, usually with wax, and often a barrier material such as asphalt or polyethylene is buried between the plies. The total thickness must be 0.090 to 0.240 in. according to size, as specified in the various carrier regulations. The strength of the sidewalls is specified in the regulations in Mullen test units, which are roughly equivalent to the bursting strength in pounds per square inch. The range is from a Mullen test of 400 in the smaller sizes, up to a 1200 test sidewall for the

TABLE 3 Uniform Freight Classification Requirements of Fibre Drums for Dry Products

Maximum limit (see Note 1)		Sidewall test per sq in., lb (see Note 2)	Minimum requirements tops and bottoms (each)				
			Fiberboard outer ply waterproofed		Steel, U.S. gauge	Wood thickness	
Weight of contents, lb	Capacity, gal		Thickness, in.	Test, lb (see Note 1)		Solid, in.	3 or more ply plywood, in.
Not over 60	Not over 30	400	0.120 or 0.090	300 or 600	30	1/2	3/11
Over 60 but not over 115	Not over 45	500	0.160 or 0.120	400 or 800	28	1/2	3/10
Over 115 but not over 150	Not over 55	600	0.160 or 0.120	400 or 800	28	1/2	3/8
Over 150 but not over 225	Not over 65	700	0.180 or 0.120	500 or 1,000	26 See Note 3	1/2	3/8
Over 225 but not over 300	Not over 75	800	0.200 or 0.160	550 or 1,100	26 See Note 3	25/32 See Note 5	7/16
Over 300 but not over 400	Not over 75	900	0.240 or 0.200	600 or 1,200	24 See Note 4	25/32 See Note 5	7/16
Over 400 but not over 550	Not over 75	1,000	0.220	1,300	24 See Note 4	25/32 See Note 5	7/16

Note 1. The minimum requirements in the above table for sidewall, top, and bottom are governed by either the weight of contents in first column or by the capacity in second column, whichever requirements are higher.

Note 2. Cady or Mullen testing method for fibre components: Either of the following test methods may be used. When more than single ply, test shall be determined from the summation of the tests of individual plies; OR, when test is made on a completed drum, the punctures shall be made from the exterior to the interior surface, in which case the values for sidewall shall be not less than 80 percent of the value in the above table. There shall be a minimum of six tests and the average shall be not less than the prescribed minimum requirements.

Note 3. Bottom may be constructed of not less than 30 U.S. gauge steel when combined with paperboard having a minimum thickness of 0.110 in. and Mullen test of not less than 400 lb.

Note 4. Bottom may be constructed of not less than 30 U.S. gauge steel when combined with paperboard having a minimum thickness of 0.140 in. and Mullen test of not less than 550 lb.

Note 5. Where the fibre shell of drum is provided with a formed inside bead to support the wood head and a steel rim holds the head in place and is locked in an external groove of the fibre shell, solid wood heads may be not less than 1/2 in. thick. Such 1/2-in. heads must be reinforced by covering both sides with kraft linerboard not less than 0.012 in. thick securely glued throughout entire area of contact with a glue or adhesive which cannot be dissolved in water after the film application has dried.

Fig. 4. Fibre drum with locking band. A popular style of closure for drums is the lever-actuated band which draws the cover down tightly against the metal rim of the drum. A gasket in the cover provides a hermetic seal. The toggle action of the lever makes it easy to close, and a pilferproof wire seal can be used for security. A metal tool is available on request from the drum manufacturer to facilitate opening the latch and lifting the cover. (*Grief Bros. Corp.*)

Fig. 5. Fibre drums with slipcovers. The all-fibre drum on the right has a telescoping lid flush with the body. A piece of tape around the joint will hold the cover securely and will prevent leakage. Sizes range from ¾- to 67-gal capacity. The drum in the foreground has a metal slipcover with clip closure, and a raised ring for stacking. The lightweight body of this drum makes it a very economical container, but it should not be used for hazardous materials.

largest heavy-duty drums. It should be noted that the Mullen test of a laminated structure is somewhat higher than the test value of an individual ply multiplied by the number of plies. Code numbers for the different sizes and types are assigned by the different manufacturers, and a typical catalog designation might be set up as follows:

Type of ends	Gallons capacity	Diameter in inches	Number of plies	Liner or barrier
M ———	165 ———	7 ———	6X ———	PL
	(16½)	(17 in.)		

Steel covers are secured either by a locking ring with a toggle to draw it up tight (Fig. 4) or with several lugs around the edge. The wooden ends are held in place with nails driven through the sidewall into the wood, usually with a thin metal band for reinforcement and to protect the edge of the drum (Fig. 5). Fibreboard covers have a deep edge that telescopes down over a neck piece which is slightly smaller in diameter than the body, forming a flush joint. This is sealed with gummed or pressure-sensitive tape. Two plies of 3-in. 60-lb kraft tape are recommended.

Design Criteria. The most popular style for small-size drums is the all-fibreboard type with telescoping cover. In the larger sizes there seems to be a preference for the metal ends. A sponge rubber gasket or other similar material in the metal cover provides a hermetic seal when the locking ring is drawn tight. With the reinforcing band, the locking ring, and the cover, there are three layers of steel around the top edge to resist the hazards of shipping and warehousing. If this level of protection is not required, wooden ends or metal slipcovers are less expensive. For dust protection, a liner bag may be adequate at a lower cost than the gasketed metal cover.

The sidewalls can have any of a number of materials buried between the plies as a moisture barrier; asphalt, polyethylene, and metal foil are often specified for that purpose. If it is preferred, the barrier can be put on the inside next to the product; for oils and greases this is the proper place. Coatings of wax or plastic can also be sprayed on the inside, covering the ends as well as the sides of the drum.

With the great variety of sizes that are available, it is possible to get two or three drums of the same capacity but with different dimensions. The merit of this is that shipping and storage space can be saved by nesting the drums. For example, 55-gal fibre drums are available in 20-, 21½-, and 23-in. diameters. The savings in cost will more than offset the inconvenience of denesting and having to fill and stack drums of different heights. (See Table 4.)

TABLE 4 Typical Costs of Fibre Drums

Drum	Cost
1 gal, no barrier	$0.45
5 gal, no barrier	0.67
5 gal, with PE moisture barrier	0.92
15 gal, with PE moisture barrier	1.40
15 gal, for liquids	3.15
30 gal, no barrier	2.15
30 gal, with PE moisture barrier	2.40
55 gal, with PE moisture barrier	3.05
55 gal, for liquids	5.90

A special type of drum for hot-poured materials consists of a lap-seam body with a "free stripping" coating on the inside. This is most often used in the 13- and 55-gal sizes for roofing compound, waxes, rosins, or similar materials that are solids at ordinary temperatures. They are easier to open, with less risk of personal injury, than metal containers, and they are disposed of easily by burning.

Section 13

Coatings and Laminations

History 13-2
Coatings.................................... 13-3
  Wax Coating 13-3
  Varnish Coating 13-6
  PVDC (Saran) Coating 13-6
  Heat-seal Coating..................... 13-7
  Extrusion Coating 13-8
Laminations 13-9
  Design of Laminations 13-10
FDA Regulations 13-17
Costs 13-17

Flexible materials are used extensively for packaging a wide variety of products. The choice of papers, films, foils, and fabrics is quite broad, and selecting the best material for a particular purpose requires all the knowledge and skill of a packaging professional. The choice is further complicated by the many combinations and modifications that can be made from these basic materials—the subject of this section.

Cellophane could never have reached the prominent place it now occupies in this field if it were not for the variety of lacquers that have been developed to increase its barrier properties and make it sealable. The heat-seal coatings that have been applied to paper and foil have opened up whole new fields of pouch packaging. Coupled with these new surface treatments is the rapidly growing technology of laminating different materials to utilize the best features of each, and in this way achieve results that would not be possible with any one material alone.

HISTORY

The earliest packages of Kellogg's Corn Flakes appeared on the market in 1906, packed in a folding carton with a bag liner of plain white paper. The instructions on the package were: "To restore crispness, heat in a pan in a moderate oven." Some experimental work was done with wax as a coating material, and in 1912 a waxed outer covering known as the Waxtite wrap was added. This helped to preserve the crispness of the product and provide a real advantage over the competitive brands of cereals that were beginning to flood the market at that time.

In 1939 waxed glassine was introduced, made with higher melting point waxes that provided better moisture protection and reduced the blocking problems in hot weather. Although wax has excellent barrier properties against moisture, it is a rather poor grease and odor barrier, and it makes a very weak seal. With the introduction of polyethylene in the late 1930s and the development of microcrystalline waxes by the oil refineries shortly thereafter, the blending of coating materials helped to extend their usefulness.

Varnishes had been used to some extent as greaseproof coatings on paper, but the introduction of saran emulsion for coating purposes in 1946 was a major breakthrough. Extrusion coating, discovered in 1948, opened up still another new area for the converter. In 1953 extrusion-coated cellophane was introduced, and the following year aluminum foil; in 1955 polyester became available with polyethylene coatings. Other plastics have been added to the list, and the number is constantly increasing. However, there are only five really important types of coatings in use today: waxes, nitrocellulose, saran (PVDC), polyethylene (PE), and polypropylene (PP).

The total value of coating materials for flexible packaging in this country is about $150 million per year. This breaks down to 210 million lb of wax, 68 million lb of polyethylene, 47 million lb of saran, 29 million lb of nitrocellulose, and less than 1 million lb of polypropylene.

Laminations have been in commercial use for more than 25 years, but it is only since the early 1950s that the great multiplicity of combinations has become available. With about 20 different films to work with, and a dozen or more kinds of paper, plus metal foils and a few woven and nonwoven fabrics, the number of possible combinations becomes astronomical. From a practical standpoint the number is not quite so overwhelming, but it is still a rather formidable quantity.

COATINGS

The easiest method of improving the characteristics of paper or film is the addition of a coating. It is less costly than a lamination since a coating is generally thinner than the lightest film that could be used for the same purpose, and less material usually means lower cost.

The method of application will depend somewhat on the viscosity of the material that is used for the coating. For example, an emulsion is too thin to be extrusion-coated, but it works very well with an air knife. Polyethylene, on the contrary, works best as an extrusion coating, and an air knife will not handle such a viscous material.

Wax Coating. The oldest and still one of the most widely used coating materials is wax. Although it is somewhat brittle and makes poor heat seals, it is an excellent moisture barrier and is very economical. It has been replaced to some extent by the olefins, and by blends of wax and polyethylene if greater flexibility or stronger seals are required.

There are two methods of applying wax to paper: *dry waxing* and *wet waxing*. In the dry waxing process, the paper travels over a hot roll after being coated, so that the wax soaks into the paper and does not stay on the surface. In wet waxing the wax is chilled quickly by being run through a water bath, so that it does not penetrate but stays on the surface, giving it a glossy appearance. Paraffin is used for dry waxing, but a blend is used for wet waxing, such as 60 percent paraffin wax, 35 percent microcrystalline wax, and 5 percent polyethylene, to increase gloss and flexibility, and to lower the tendency toward blocking. (See Fig. 1.)

The earliest waxes used in packaging had a melting point around 128°F and had a great tendency to block in hot weather. The paraffin

Fig. 1. Wax coating. The two methods of putting wax on paper are illustrated schematically above. In dry waxing the paper is heated after it is coated, and the wax soaks into the paper. In wet waxing the wax is chilled before it has a chance to penetrate into the paper.

wax used today is greatly improved and has a melting point of about 135°F. Breakfast cereals require a 28-lb *glassine* paper with 8 lb of wax on each side, although some types of cereal that are more sensitive to moisture use two 20-lb sheets laminated with 5 lb of microcrystalline wax between, and with 8 lb of paraffin wax on both outside surfaces. Where the ultimate in protection is needed, aluminum foil is sometimes used in conjunction with an 8½-lb *strike-through* tissue, bonded with up to 20 lb of microcrystalline wax. The heat that is developed in heat sealing will drive the wax through the paper and provide a hermetic seal, filling up the channels formed by the wrinkles and folds in the paper.

Some of the disadvantages of wax are low seal strength, variable composition, variable barrier properties, and poor odor and grease barrier properties. Since paraffin wax is made from crude petroleum, it may contain various organic materials, depending on the source of the petroleum, which cause the composition to be variable. Even if melting points are identical, the barrier properties can vary by as much as 30 percent.

A smooth sheet such as glassine will carry a more uniform coating of wax and provide a better barrier than a rough paper, which takes a more uneven coating. The rate of cooling during the coating process also affects the barrier properties, and slower cooling yields larger crystals, which give better barrier properties. The odor and grease resistance of wax paper will depend on the type of paper used, and some types of glassine are especially recommended for this purpose.

Petroleum waxes are essentially pure hydrocarbons with molecular weights in the range of 250 to 980, and with 18 to 70 carbon atoms per molecule. These are largely straight-chain hydrocarbons, although there can be fairly large percentages of isoparaffins and cycloparaffins. These generally have the rings or branches attached near the end of the molecular chain. The amount of branching has a marked effect on the melting point and other properties of paraffin wax.

There is no clear definition of *microcrystalline* waxes, and there are sometimes even references to "semi-microcrystalline wax." One method of classification is to identify all waxes with a viscosity above 10 centistokes at 210°F as microcrystalline waxes.

One of the problems with wax coatings is a tendency to *block* or stick together under heat or pressure. This is caused by the lower melting components of the wax. With an oil content of 0.5 percent the blocking point of paraffin wax is about 105°F. With *solvent deoiling* the oil content can be reduced to 0.1 percent and the blocking point raised to around 120°F.

Variations in the *friction* of wax coatings will affect the operation of packaging machinery. Large amounts of branched and cyclic molecules cause an increase in friction, but long straight-chain molecules tend to reduce friction. Therefore, melting point data may not bear a direct relationship to the coefficient of friction, as is sometimes believed.

Another factor to be considered is *scuff resistance* of a wax coating. This is related to the composition of the wax; a high percentage of branched and cyclic molecules will produce a tough, ductile wax which has good abrasion resistance, whereas a low percentage will give a harder wax that is more scratch-resistant. These two characteristics are in conflict with each other, and it depends on how scuff resistance is defined: whether hardness is preferred to toughness.

The properties that affect *seal strength* are related to the ductility of the wax. Paraffin wax has low ductility and poor seal strength, whereas microcrystalline wax has good ductility and high bonding strength. The ductility spreads the stresses over a larger area and thereby helps to sustain the higher stresses. Quick chilling of the wax after sealing improves the strength, especially of paraffin waxes. The rapid cooling produces smaller crystals, which are not so likely to pull the wax away from the surface of the paper as they are formed. Well-sized paper makes stronger bonds than poorly sized paper, but wet-strength resin will interfere with a good seal.

The composition of the wax will affect the *gloss* of the coated sheet. A high paraffin content gives a high degree of crystallinity, which roughens the surface and reduces the gloss. A high percentage of isoparaffins and cycloparaffins will inhibit crystallization and provide a higher gloss.

The properties of wax as a *moisture barrier* are related to the crystal formation. Moisture vapor does not travel through the crystals, but around and between them. Therefore, the larger the crystals, the fewer the spaces, and the better the barrier. Paraffin wax is better than microcrystalline wax for protection against moisture vapor, and slow cooling of a *narrow-cut* paraffin wax will give the best results. It has poor flexibility, however, and if wrinkling and creasing cannot be avoided, it might be well to compromise with some microcrystalline wax in the composition to provide greater flexibility.

Wax coatings are not particularly good as barriers for *grease* and *oil.* They tend to dissolve in the oil or grease, but improvement to some degree is possible by using the microcrystalline types, by using a high melting point wax, or by slow cooling to reduce the amount of crystal surface exposed to attack.

The use of *plastic blends* is becoming more popular; the use of polyethylene for this purpose dates back to 1948. Other materials, particularly ethyl-vinyl acetate and ethylene acrylate, also are being used to improve toughness, flexibility, and adhesion. Some attention must be paid to the composition of the base waxes, however, as the gloss may be seriously affected by the additives, and adequate testing over several days should be made on any new compositions.

If waxes are overheated in the coating operation, there may be oxidation products that give a bad odor to the material. Some of the newer compounds are coated at temperatures up to 300°F, which can encourage the formation of odorous peroxides, especially if there is any copper or aluminum in the equipment. This can be controlled to some degree with antioxidants such as di-tertiary butylparacresol (BHT).

Varnish Coating. Numerous resins are used singly or in combination as coatings for paper. Some of these are thinned with alcohol, or if they contain wax, they may be thinned with naphtha; they are called spirit varnishes. Others are used without solvents and are known as *press varnishes,* which dry by oxidation rather than by evaporation of solvents. Both types are frequently used on printing presses to add a protective coating that will prevent smearing of the ink, and also to provide a glossy finish. They are the least expensive materials for the purpose, and they are easy to apply. Their effectiveness is dependent to some degree on the amount that is used. A coating of ½ lb of solids per 1,000 sq ft will not do as good a job as 2 lb per 1,000 sq ft, obviously. If it is intended only for appearance, then the lighter coat may be sufficient, but if protection from scuffing is required, the printer should be informed. If varnishes are printed only in certain areas, they are known as *spot varnishes.* They may be necessary to keep white areas from turning yellow with time, or to facilitate gluing, or to permit stamping of price marks or code numbers.

A better grade of coating can usually be obtained with *lacquers,* but the cost is likely to be a bit higher. These are made from vinyl-type resins and more volatile solvents than varnishes. Even more sophisticated are the epoxy-type coatings. They are made from thermosetting resins and require special ovens for curing, but they provide a hard, lustrous finish that cannot be obtained in any other way.

PVDC (Saran) Coating. A coating material which is rapidly growing in popularity is variously known as PVDC, polyvinylidene chloride, saran, or, when used on cellophane, polymer coating. It is really a copolymer of vinyl chloride and vinylidene chloride. First used in emulsion form for coating paper in Germany in 1956, it has only recently come into general use in this country. PVDC has a rare combination of barrier properties in that it provides good protection from grease and oil, water vapor, odors, and gases. PVDC coatings are superior to wax or polyethylene for products with fugitive flavors and aromas, such as coffee, dehydrated soups, spices, butter, margarine, and other oily and fatty foods.

The cost of the PVDC base resin is higher than that of polyethylene or most of the other commonly used coating materials, but the finished cost may be less, depending on the amount of material used and the method of application. A comparison of cost should be made on the basis of equal performance, and the higher barrier properties of PVDC will tend to offset its higher price.

There are several *processes* for applying PVDC emulsion to paper: (1) air knife coater, (2) blade coater, (3) metering rod coater, (4) gravure coater, (5) curtain coater, and (6) reverse roll coater. The first three apply an excess of material to the paper and then scrape off the surplus. This tends to put more in the valleys and less on the high spots, although the air knife has largely overcome this problem. The other methods apply a measured amount to the web; they are sometimes called contour coaters because the same amount of material is applied to the high spots as to the low spots in the paper. The curtain coater and the reverse roll coater, however, can be used only for heavy coatings and will not apply the light coatings most often used in packaging. The gravure coater tends to produce a pattern in the coating from the texture of the engraving on the cylinder. The method that seems to be growing most rapidly in popularity is the air knife technique.

More than one coat of PVDC is generally used, and there is some equipment that can apply up to six coats in one pass with curing in between successive coats. Heat is applied immediately after each coating, to evaporate the water and then to fuse the resin. The melting point varies, but the resin usually fuses at about 450 to 500°F. This is followed by cooling with a chill roll.

The paper to be coated should be smooth, with a minimum of groundwood, since groundwood contains shives that tend to pierce the coating. There should be a minimum of highly lignified fibers from hard cooks that would cause wicking. The paper should also be well sized with an internal sizing rather than a surface sizing. It may be necessary to apply a primer of a latex emulsion, if the sizing is not adequate, and to improve flexibility, a base coat of acrylic latex may also be used. It is sometimes also necessary to adjust the pH of the paper close to neutral when working with PVDC.

Other coatings that are applied from water suspensions or solutions are casein, starch, and polyvinyl alcohol. Sometimes borax or aldehydes are used to fix these coatings so that they will not go into solution in the presence of water.

Heat-seal Coating. If thin coatings of heat-seal material are used, the type of sealing equipment will be limited to flat bar machines. Rotary seal units, which are used with laminated materials, do not make strong seals with this type of packaging material, because pressure is such a critical factor and it is difficult to control with rotary dies. Even under the best of conditions, regardless of the sealing method used, the seal strength with coated materials will not be as good as with laminated materials.

Among the heat-seal coatings, *rubber hydrochloride* has a broad heat-seal range and makes a relatively strong seal. The machinability of *polyethylene* is excellent, and seals are more than adequate, even for liquids. *Polypropylene* has good transparency and stiffness, but it has a narrow sealing range, which makes it unsuited to some types of machines. Although *saran* has some desirable barrier properties, it is difficult to heat-seal. Pressure and dwell time must be closely controlled, usually by means of flat bar sealers and not rotary equipment. *Vinyl* is another material that needs accurate control of temperature and pressure. *Acetate* and the other cellulosics are sensitive to high temperatures, and they should not be used in thick buildups where heat-seal conditions may cause degradation.

If a *peelable seal* is required, it is possible to use polyvinyl chloride on one side and a coating of polyvinyl acetate on the other. Individual portion packages of jellies use this system, with a cup made of thermoformed PVC and a lacquered film for the cover. Rubber hydrochloride with PVC also makes a good peelable seal, but rubber hydrochloride tends to deteriorate with age.

To test the *adhesion* of a coating on film, scratch or cut the coating; then place pressure-sensitive tape over the scratch and perpendicular to it. When the tape is removed rapidly, the coating should not come off with it.

Extrusion Coating. The most economical way to combine thermoplastics with other flexible materials, in large quantities, is by extrusion coating. The great majority of extrusion coating today is being done with polyethylene. The extrusion process consists of forcing molten plastic through a long slit in a die to produce a ribbon of material that is laid on a web of paper or other material as it passes under the die. The temperature is kept fairly high (520 to 540°F) to secure good adhesion. There is rapid oxidation of the surface at this temperature, which is necessary to obtain a good bond, but unfortunately oxidation leads to the production of odor. If the material is to be used for food products, it will be necessary to control the process carefully to minimize the odor. The oxidation of the surface will also interfere with heat sealing when the lamination is converted into packages, but problems in this area should not be attributed to oxidation without checking other causes as well. It is more likely that heat-seal troubles are caused by slip agents or other additives than by the oxidized surface. Since the oxidation will vary from one batch to another, it may be necessary to adjust the sealing temperature and dwell time for each new lot of material that goes through a fabricating machine.

Extrusion-coated material is softer than a lamination of the same material, and this too can cause machining problems. If stiffness is necessary in order to push the material through a bag machine, it may be necessary to change to an adhesive-laminated combination. Extrusion-coated material has a greater tendency to curl than a lamination has, because of the high extrusion temperature. A primer coat is usually applied to the web that is being extrusion-coated, to improve adhesion, especially with foils which are impermeable and have no "tooth" for the plastic to grasp.

LAMINATIONS

There are five basic methods of laminating: (1) water adhesives, which require evaporation of the vehicle in the laminant to provide a bond between the substrates; (2) solvent adhesives, which also require evaporation for their effectiveness; (3) thermoplastic coatings, which take heat and pressure for laminating; (4) extrusion coating, in which a molten layer of plastic flowing from a narrow slot in a die is laid on a moving web of paper or film; and (5) hotmelt coating, which differs from extrusion coating only in being done at a lower temperature with mixtures containing wax and other low-melting materials.

Compared with extrusion coating, *adhesive* laminations, whether with water-base or solvent-type laminants, can be made more easily and the equipment is less complicated. There is a wide choice of materials that can be used, and the amount of scrap with an adhesive system is usually very little. However, delamination and discoloration are more likely to occur if the proper adhesive is not used, or the curing cycle is not adequate, or solvents are not completely removed. (See Fig. 2.)

Extrusion equipment is more complex than the cold coating machines and is a good deal more expensive. Temperature and pressure must be controlled very carefully in each zone of the extruder to get a uniform flow and avoid degradation of the material due to overheating. The extrusion die must also be well designed, carefully maintained, and accurately adjusted throughout its length to get consistent results. Once it is set up properly, however, it is the most economical method of laminating. Thin coatings of ½ mil or less can be made, whereas with laminations that are bonded with adhesives the film must be at least 1 mil thick to be machinable. The types of plastics that can be extruded in this manner are somewhat limited, and about 90 percent of the work is done with low-density polyethylene, or compounds in which polyethylene is the principal ingredient. Other materials that are being used for

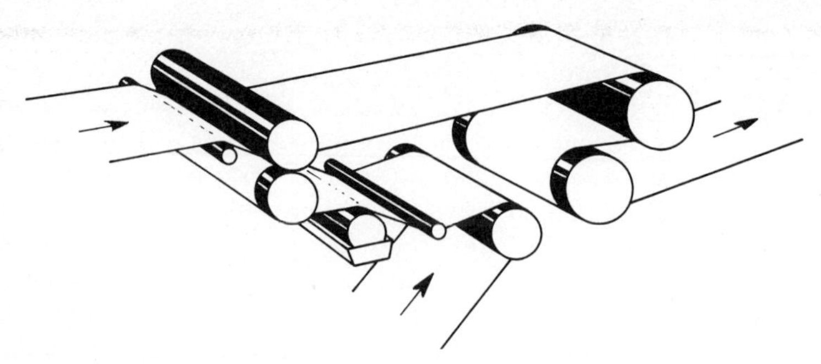

Fig. 2. Laminating. The basic process consists of combining two or more webs with adhesive. In the diagram one layer of paper is coming up from the bottom and is carried over an adhesive roller to the left. A second web coming in at the top left meets the adhesive-coated paper in the nip of the two rolls, which are one above the other on the left. The combined layers pass around the snub rolls to the right and are carried to the next operation.

this purpose include high-density polyethylene, polypropylene, and nylon.

High temperatures may cause odors owing to degradation of the coating material, but they are sometimes necessary for good adhesion. Thus the converter must work just below the point at which odors might become noticeable. The amount of scrap that is produced at the start of an extrusion run is considerable, and a run must be long to be economical. Therefore it is not very practical to run small orders by extrusion coating.

Heat lamination is used to combine coated materials by simply running them together between rollers, one of which is heated. The equipment is very simple and relatively inexpensive. Two webs of cellophane are often combined in this way to get a stiffer sheet. Saran can be combined with itself without heat or adhesive because of the characteristic way that it clings to itself. If cellophane is printed, as in the case of potato chip bags, the printing can be trapped between the two sheets for increased gloss and scuff resistance.

Design of Laminations. Before starting to choose the components of a lamination, make sure there is no single film that will do the job. A brief review of the properties of various flexible materials might be worthwhile before deciding to add the cost of conversion, if it is not absolutely necessary. (See Tables 1 and 2.) Next consider whether a coated film will do the job.

If a coating on film is the material of choice, then try to determine the best coating for the purpose. First consider the most important requirement, and then take each succeeding requirement in the order of relative significance. In this way they can be weighted and traded off

TABLE 1 Characteristics of Flexible Materials

Pouch paper	Low cost, rigidity, strength
Glassine paper	Greaseproofness, flavor protection
Foil	Moisture and gas protection, good appearance
Cellophane	Stiffness, machinability, transparency
Polyethylene	Low cost, heat-sealability
Polypropylene	Moisture barrier, stiffness
Polyvinyl chloride	Grease resistance, heat-sealability
Saran	Moisture and gas protection
Rubber hydrochloride	Grease resistance, heat-sealability
Polyester	Strength, high or low temperature performance
Nylon	Formability for deep draws, toughness

TABLE 2 Moisture Barrier Properties of Flexible Materials

Material	Water-vapor transmission*
Aclar (fluorohalocarbon):	
22A, 1 mil	0.055
22A, 1½ mil	0.046
22C, 1 mil	0.045
22C, 2 mil	0.028
33C ½ mil	0.040
33C, 1 mil	0.025
33C, 2 mil	0.015
Cellulose acetate, 1 mil	80.000
Cellophane:	
140K	0.400
195K	0.450
195M	0.650
Polyester, 1 mil	2.000
Polyethylene:	
Low-density, 1 mil	1.300
High-density, 1 mil	0.300
Polypropylene, 1 mil	0.700
Polyvinyl chloride, 1 mil	4.000
Rubber hydrochloride, 1.2 mil	1.000
Saran (PVDC), 1 mil	0.200
Two-ply waxed glassine paper	0.500
Waxed glassine paper	3.000
Waxed sulfite paper	4.000

* g loss/24 hr/100 sq in./mil at 95°F, 90 percent RH.

TABLE 3 Relative Costs of Commonly Used Flexible Materials

Material	Thickness	Cost per 1,000 sq in.
Pouch paper. . .	25 lb	$0.012
Glassine paper. .	25 lb	0.016
Polyethylene. . .	0.001 in.	0.016
Waxed paper . .	29 lb	0.020
Aluminum foil .	0.00035 in.	0.022
Cellophane . . .	195 gauge	0.033
Polyester.	0.0005 in.	0.059
Saran	0.001 in.	0.066
Nylon	0.001 in.	0.090

against the cost factors that are involved with each particular material and method of application. Such things as appearance, machinability, strength of seals, barrier properties, printability, and coating speeds must be considered in the light of the base cost of the resin as well as the carrier. (See Table 3.)

If it is decided that a lamination is necessary, every effort should be made to keep the number of plies to a minimum. It is also desirable to use each component in the very lightest gauge that will serve the purpose, and where bulk or stiffness is required, try to work with the lowest-cost materials that will give these properties. As a general rule the most protective ply should be nearest to the product.

If one of the plies has a pressure-dependent WVTR, however, the most protective film should be exposed to the high humidity. In the case of cellophane coated on one side, for example, the coated side should be toward the moist atmosphere. This improves the gas transmission rate, because cellulosic materials are better gas and water-vapor barriers when they are dry than when they are moist. The protective value of a sheet is more than doubled when the coating is on the proper side.

The combination of materials in a lamination is designed for a specific set of conditions, and each ply should have a particular purpose. If we are to choose the components of a lamination properly, we should know a few facts about the product to be packaged: the moisture content of the product when packaged, as well as the critical level of moisture content, maximum or minimum as the case might be; the desired shelf life and the conditions of storage that can be expected; the mass and texture of the product, and the quantities in each unit package as well as the shipping case quantity, for determining the mechanical strength

TABLE 4 Some Common Uses for Laminations

Lamination	Use
Oriented PP/K cello	Snack foods
K cello/K cello	Snack foods
Polyester/saran/PE	Meat, cheese (thermoformable)
Nylon/saran/PE	Meat, cheese (thermoformable)
Oriented PP/PE/K cello/PE	Cheese
M cello/PE	Candy
PP/K cello	Candy
Polyester/PE	Corrosive chemicals
Aclar/PE	Moisture-sensitive products

requirements. (See Tables 4 and 5.) Printing and decoration as well as any special heat-seal conditions should also be determined before setting up any specifications. (See Figs. 3 and 4.)

Factors to be considered in selecting the materials for a lamination are permeation of moisture and gases, extraction of plasticizers and stabilizers from the film by the product, absorption of ingredients of the product by the film, modification of the package by the product, and photochemical change of the product from exposure to light. (See Table 6.)

Stiffness can be supplied best by paper, usually a pouch paper or glassine. Paper is the most economical material, and if transparency is not essential, it should be the first choice.

Printability may be an important consideration, and cellophane or acetate will serve very well in this case. The best method of printing on film is the gravure process, but when it is used for printing on paper, it does not give as good quality as offset or letterpress. Printability can be evaluated by two tests that are in general use: the Geiger tone step

TABLE 5 Typical Pharmaceutical Applications of Laminations

Lamination	Application
Paper/PE	Analgesic tablets
Cello/PE	Antacid tablets, vitamin tablets
Foil/PE	Effervescent analgesics
Foil/PVC	Ointments, cough syrup
Acetate/Pliofilm	Antacid tablets, cold capsules
Acetate/foil/lacquer	Antibiotic tablets, vitamin tablets
Acetate/Aclar/PVC	Cough syrup
Acetate/foil/saran	Antibiotic capsules, ointments
Acetate/metallized Mylar/PE	Cough syrup
Cello/PE/saran	Antibiotic capsules, vitamin tablets
Cello/PE/foil/lacquer	Antacid tablets
Cello/PE/foil/PE	Analgesic tablets, ointments
Paper/PE/foil/PE	Vitamin tablets

Fig. 3. Printing press. Different methods of printing paper and film include letterpress, gravure, offset, and flexographic processes. The press shown is a flexographic printer. (*Paper Converting Machine Co., Inc.*)

test, which shows tone reproduction and graininess, and the Diamond-Gardiner dot dropout test to measure definition. The best test, however, is an actual print by the method that is intended to be used.

Thermoforming is used for packaging bulky products, and such items as frankfurters are packed in laminations that can be drawn into a female mold. Typical combinations for this purpose are 50-gauge M-27

TABLE 6 Barrier Properties of Laminations

Material	Oxygen transmission*	Water-vapor transmission†
0.002 saran/0.006 PVC	0.6	0.092
0.0015 Aclar/0.002 PE/0.0075 PVC	1.0	0.034
0.0015 Aclar/0.0075 PVC	1.1	0.035
0.002 PE/0.0075 PVC	1.3	0.170
0.0075 PVC	1.9	0.330
0.002 PE/0.005 PVC	2.6	0.200
0.005 PVC	2.7	0.520
0.001 nylon	25.0	19.000

* cc/24 hr/100 sq in. at 77°F, 50 percent RH.
† g/24 hr/100 sq in. at 95°F, 90 percent RH.

Fig. 4. Laminator. Various types of flexible material can be laminated on this machine, using solvent or aqueous adhesives, waxes, or hotmelts. (*Inta-Roto*)

Mylar/saran coating/2-mil polyethylene and 75-gauge nylon/2-mil polyethylene. Better results, in terms of leaky packages, have been obtained when the materials were laminated by means of thermosetting adhesives than when the polyethylene was extrusion-coated.

Moistureproofness is one of the chief attributes of polyethylene. Wax is economical and is widely used for moistureproofness. The permeation of a film is inversely proportional to the thickness, but it is not a straight-line function; the barrier properties do not increase quite so fast as the greater thickness might indicate. (See Table 7 in this section and Fig. 5 in Sec. 3, "Films and Foils," page 3-9.)

Gas transmission will vary from one film to another. The list of properties in Table 4 in Sec. 3, "Films and Foils," page 3-7, may help in choosing a material if gas transmission is critical. Saran is particularly useful in this regard, and metal foil is a perfect barrier, but at a slightly higher cost.

Cellophane is a good barrier against oxygen, but as it becomes moist, it loses some of its barrier properties. A coating of polyethylene will not add much in the way of a gas barrier of itself, but by keeping moisture away from the cellophane, it helps greatly in preventing the transmission of oxygen.

It should also be noted that the values of the different components of a

TABLE 7 Cost Comparison in Relation to Moisture Protection

Lamination	Cost per 1,000 sq in., cents	Water-vapor transmission*
Cello/PE Mylar/PE PP/PE	8–10	0.50–1.00
PVC/saran/PE PP/PE/cello/PE	25–50	0.25–0.35
Cello/foil/PE	20	0–0.04
Cello/Aclar/PE	90	0–0.04

* g/24 hr/100 sq in. at 95°F, 90 percent RH.

lamination are not directly additive, and the only sure way to determine the net effect is by testing the actual combination. An approximation of the permeability of a lamination can be calculated from the permeability of the component films as follows:

$$\frac{1}{\text{Total permeability}} = \frac{\text{thickness of } A}{\frac{\text{total}}{\text{thickness}} \times \text{permeability of } A} + \frac{\text{thickness of } B}{\frac{\text{total}}{\text{thickness}} \times \text{permeability of } B}$$

The directional effect of a lamination also should be taken into account; for example, cellophane that is coated on only one side will have a higher moisture transmission rate when the uncoated side is exposed to high humidity than when the coated side is in that direction. Also, with a combination of films, if the side exposed to a gas under pressure has a film that changes its transmission rate with changes in pressure, the result will be a higher total rate than if it were on the opposite side. It should be noted also that the internal tear strength, that is, the resistance to the propagation of a tear, is lower for a laminate than for the separate films from which it is made.

The workhorse of the converter is a combination of paper, foil, and polyethylene. Written in the usual shorthand of the industry it is pouch paper/PE/foil/PE. The outer ply is always given first, and the side toward the product is shown last. The paper in this case provides the tensile strength and the printing surface; the foil gives the barrier properties against moisture and gases; and the polyethylene, which is in between, joins them together and it is used again on the surface to provide a heat-seal coating. (See Fig. 5.)

A certain amount of *stiffness* is necessary in a lamination to prevent wrinkling, especially with foils which have very little springback.

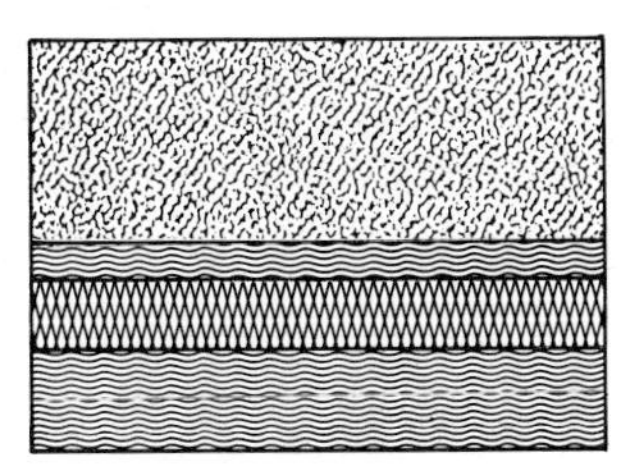

Fig. 5. Laminated construction. A cross section of a typical lamination is shown greatly enlarged. As indicated, each layer has a specific purpose. The paper side is on the outside of the package in this case, and can be printed. The inside surfaces are intended to be sealed to each other with heat.

Appearance and other requirements of the finished package also may dictate a stiff material. Paper is the cheapest and best material to add rigidity in a lamination. There may also need to be some *flexural strength* to avoid fractures of some of the layers when the lamination is draped over bulky or irregular objects. This sometimes requires a heavier foil, for example, but a word of caution here: the increased thickness may work against you by aggravating the problems of fatigue failure. There is an optimum thickness for most materials, below which the tensile strength suffers and above which the stiffness causes stress cracking.

FDA REGULATIONS

The materials used for packaging foods, drugs, and cosmetics must meet certain standards as required by law, and by the regulations issued by the Food and Drug Administration to supplement and define the intent of the law. Not only must the films, foils, and papers conform to the rules, but any adhesives, primers, inks, or release coats also must comply with the regulations. Some materials are "generally recognized as safe," referred to as GRAS materials. Others are acceptable under certain conditions; Sec. 3, "Films and Foils," gives some information for each of the individual materials. Other data are given in Sec. 18, "Test Methods," and Sec. 16, "Laws and Regulations."

COSTS

It is difficult and complicated to determine the exact cost of a lamination, but some general rules can be laid down, and a few examples will give some indication of the order of magnitude of these costs. In addition to the base cost of the materials that are used, there is the cost of processing. This varies with the type of equipment and the size of the order.

TABLE 8 Typical Costs of Laminations

Lamination	Applications	Cost per 1,000 sq in.
140 cello/0.003 PE	Citrus juice, cottage cheese	$0.13
25-lb pouch paper/0.00035 foil/0.0015 PE	Cake mix, drink mix, photo chemicals	0.15
140 cello/0.0007 PE/0.00035 foil/0.0015 PE	Photographic film, chipped beef, coconut	0.18
0.00075 saran/50-lb white sulfite paper	Cap liners	0.19
0.0005 Mylar/0.003 PE	Boil-in-bag foods, lunch meat pouches	0.19
0.00088 acetate/0.001 foil/0.0008 Pliofilm	Dehydrated foods	0.30
50-lb kraft/0.001 PE/0.0005 foil/0.0025 PE	Mil-60 government specification for case liners, small parts	0.31
0.0005 Mylar/0.001 foil/0.003 PE	Boil-in-bag foods	0.31
44 × 40 scrim/0.001 PE/0.0005 foil/0.0025 PE	Mil-72 government specification for machines, guns, etc.	0.45

There is the cost of *setting up* to run a particular combination, as well as the cost of the scrap that is wasted while bringing the equipment into adjustment. These are one-time costs, and in a long run they may be insignificant. In short runs, however, they may become a big factor in the total costs. The *running costs* are a constant figure that applies to every ream of material produced, regardless of whether the run is long or short. A typical figure would be about $250 per hour. Each of these elements will be determined by the complexity and sophistication of the machinery, and the capital cost which must be amortized over every yard of material that is produced. (See Table 8.)

For rough calculations the cost of the *materials* used can be doubled to yield a reasonably close estimate of the cost of the finished lamination. If very light gauges of film or paper are used, 2½ times would be a little closer to the real figure. If the number of operations required is known, a figure of 2 cents per thousand square inches for each pass through the machine, plus 2½ cents for profit, plus the cost of the materials will be more accurate. The thickness of polyethylene in a lamination is sometimes expressed in pounds per ream, rather than mils of thickness. For converting from one to the other, 15 lb is about equal to 1 mil of thickness. This discussion of costs may help in comparing one lamination with another for design purposes; an accurate cost, however, can be determined only by a supplier because of the many variables in the costs of adhesives, primers, and other materials used.

Section 14

Labels and Labeling

History 14-2
Types of Labels 14-3
Paper Labels 14-4
Foil Labels 14-5
Transfer Labels 14-6
Offset Printing 14-6
Silk-screen Decoration 14-7
Pressure-sensitive Labels 14-7
Printing Processes 14-8
Design of Labels 14-9
Adhesives 14-11
Costs 14-13

There are a number of reasons for using labels on packages, some of which are made necessary by law or by logic, and others of which are optional with the packager. In the first category there are laws which demand that the quantity and the manufacturer's name be clearly displayed, as well as the active ingredients in some cases and special warnings if the contents may be hazardous to the user. (See Fig. 1.) Among the optional uses of labeling there are slogans and claims that help to promote the sales of the products, trade names and descriptive phrases, instructions for use, cross advertising of other products, and many other applications of design and text to enhance the value of the package as a selling tool.

The choice of color and typography in labeling can often play an important role in the acceptance, the use, and the response of the customer, as well as the cooperation of the dealer. It might even be said

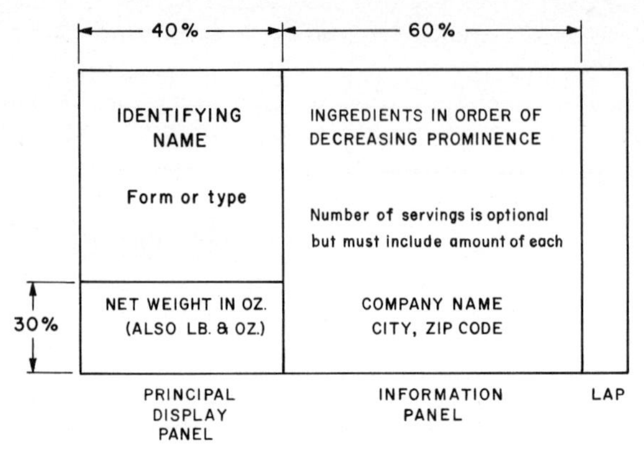

Fig. 1. Model label. Some of the requirements of the Fair Packaging and Labeling Act (21 CFR 1) are shown. The principal display panel of a cylindrical package is 40 percent of the circumference (1.7). A statement of identity in bold type and the form, if there is more than one, must be on the principal display panel (1.8). The quantity statement must be in the lower 30 percent of the main panel, and a dual statement is required as shown (1.8b). Ingredients must be listed all in one place and in descending order (1.10). The place of business must be conspicuously specified by the actual corporate name, city, state, and zip code, not necessarily on the main panel (1.8a).

that the success or failure of a packaged product, in some cases, could be attributed to the manner in which it was labeled, so significant is this aspect of packaging. The least costly item in a package may well be the most important factor at the point of sale.

HISTORY

The labeling and marking of goods has a long history, and the use of marks to indicate the source of a particular item can be traced back many centuries before Christ. Roman apothecaries are known to have dispensed an herb called *Lycium* in small jars bearing the name of the drug and the seller's name. Wines were usually sold in marked jars and bottles, until clear glass replaced the dark-colored bottles that had been used up to the end of the seventeenth century. It then became customary to have the labels hanging loose from the bottle, held by a fine chain around the neck. These markers were sometimes made of silver or ivory, and they have become collector's items; reproductions of these are sometimes seen on modern whiskey decanters.

The descriptive type of labeling that is so generally used today, on the other hand, has had a relatively short history. There are several reasons for this: for one thing the manufacturers and dealers of old were so familiar with the limited kinds of goods then available that they did not require tags to differentiate one from another. Besides, literacy was restricted to the wealthy, and all communication was on a person-to-person basis.

As the variety of goods increased and education became available to the working classes, it became expedient to identify things with printed legends. A group of druggists in London in 1819 adopted a set of regulations for marking poisons, one of which read:

> That on every wrapper or vessel containing any drug or preparation likely to produce serious mischief, if improperly used, the name of the article be affixed in a legible form, and as many persons can read print who cannot read writing, they would recommend that printed labels be used where possible, in preference to written ones.

In 1888, however, an English novelist wrote, "Poison that is bought at a drug store *usually* has a label on the bottle"; this would indicate that the regulation was not rigidly adhered to.

The promotional value of a label was not recognized or utilized to any degree until the beginning of the nineteenth century. It was about 1793 that Guinness ale and stout started using the Irish harp symbol to help the sales of their brewery in Dublin. At about the same time the French wineries began to print labels for their cordials and brandies with elaborate scenes. Similar designs were used also on matchboxes, bolts of cloth, food products, and medicines, and even today we have such art on many of the cigar boxes that are being used in retail trade.

The word "label" itself has been used for many things down through the ages. In heraldry it was a band across the upper part of a shield to indicate the eldest son of the family, and in clothing it usually referred to a ribbon or streamer attached to a head covering. Legal papers had a strip of fabric attached to carry a seal, which was called a label.

The growth of the label industry has been slow but steady, held back to some extent by the rapid adoption of direct printing methods for some types of containers. About 1½ billion paper labels worth $450 million are used in this country at the time of this writing. Pressure-sensitive labels are the fastest-growing segment of this industry.

TYPES OF LABELS

Various materials can be used for labeling, such as paper, foil, and fabric. It is also possible to print directly on a bottle or other container by means of the silk-screen, offset, or hot transfer processes. The choice will depend on the needs as well as the economics. The lowest in cost is usually paper labels, and excellent printing is possible on a good paper stock. For luxury items, embossed foils or silk-screen printing will provide a more elegant package, although at a higher price. Typical paper labels will cost less than ½ cent each, while direct printing is likely to be three or four times as much.

If the design calls for various degrees of halftones and shading, it will not be possible to use some of the direct printing methods, such as silk-screen printing, as they lend themselves only to solids and line work. Paper labels, however, can be printed with the most sophisticated designs and will give a faithful reproduction of the original artwork. The disadvantage of paper labels is the appearance of being "tacked on" and not being an integral part of the package. They are also vulnerable to scuffing, wrinkling, blistering, and lifted or curled edges. Silk-screen printing, by contrast, gives good coverage and a rich embossed appearance that is well suited to prestige products. In between there are different methods of offset printing that have somewhat thin ink lays and fuzzy edges; hot stamping, which is available in a wide variety of colors including bright gold, but does not always give clean sharp edges; and hot transfer from gravure printed strips that always has an unattractive shiny halo of carrier varnish. The choice must be determined by the effect that is desired, balanced against setup costs and unit prices.

Paper Labels. Most labels are printed on paper, since this is the most economical method, whether the quantities are large or small. They can be produced by any of the graphic processes such as letterpress, offset, gravure, or flexography. There is no limit to the colors and techniques that can be used, and any of the methods known to the printing trades can be applied to paper labels. See Table 6 in Sec. 2, "Folding Cartons and Setup Boxes," page 2-19, for a comparison of printing processes.

Paper labels can be die-cut or guillotine-cut. The first method provides accurate dimensions and freedom of design, such as rounded corners. It is higher in cost, however, since the labels must be cut one stack or lift at a time, and registration of printing with cutting cannot be held to very close tolerances. In *die cutting,* a cast steel cutting die, of the proper shape and about 2 in. high, is placed in position on a stack of printed sheets. A die press forces the die down through the stack. The die is lifted out by hand and the labels are pushed through the die and accumulated on a nearby table. It is then placed by hand on the next position to be cut, and the press is actuated to force the die down once again. With foil labels the die "cups" the edge and work-hardens the foil at that point, adding stiffness to the labels.

The majority of labels are *guillotine-cut;* that is, a stack of printed sheets is placed against an adjustable gauge plate and clamped in position. A shear knife is brought down through the sheets, cutting off strips of labels a stack at a time. After they are all cut in one direction, the gauge plate is moved the proper distance from the knife and the stacks are placed in the opposite direction against the plate, so that single labels can then be cut in groups from the stack of strips.

Some variations in size will occur with guillotine cutting, depending upon the skill of the operator, but they should not be more than 1/64 in. from the specified size if the work is carefully done and the equipment is in good shape. The design of the labels is limited to straight parallel edges with this method, and the corners are always square and sharp, although it is possible to do some corner cutting in a subsequent operation if desired.

The direction of the grain of paper is usually very critical to the operation of the labeler, and most often it is required that it be in the horizontal direction. Some machines, however, work best with the grain running vertically, especially in the case of pressure-sensitive or heat-sensitive adhesives. The direction of the grain can be determined by wetting one side and noting the direction of the curl. The grain direction is always parallel with the axis of the curl. (See Fig. 2.)

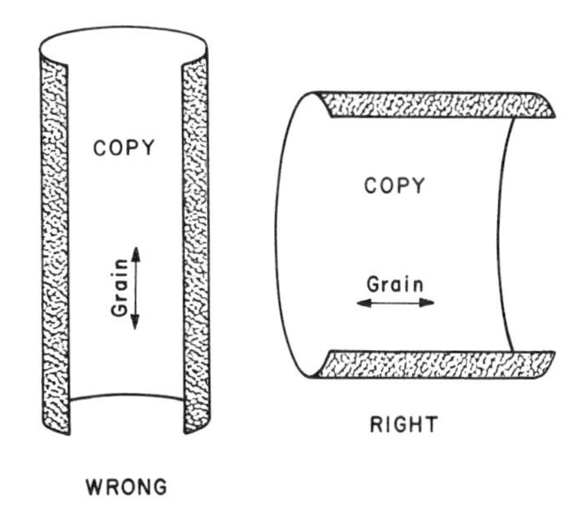

Fig. 2. Direction of grain. The grain of a label can be determined by wetting one side, so that it curls toward the dry side. The direction of the grain will be the axis of the curl as shown.

Foil Labels. It is nearly always necessary to laminate foil with paper so that the labels will work properly in the labeling machines. Even so, there will be less springback than with plain paper, because of the dead-fold characteristics of the foil, which may affect machine operation. The foil and paper together should measure 0.0025 to 0.003 in. for the best results. Laminations of the paper and foil are specified by the basis weight of the paper and the thickness of the foil in inches; for example, 40 lb (25 × 38—500) /0.0007. Adhesives that contain solvents cannot dry as quickly through the foil as through a paper label, and so heat-sensitive coatings are often used for this reason. The appearance of a well-printed foil label is superb, and the bright reflective surface adds glamour to almost any package.

Special inks are required for printing on foil, and it may be necessary to use a primer as well to get good adhesion. (See Sec. 3, "Films and Foils.") Tinted lacquers can also be used to simulate gold or copper. It is difficult to get a good white color on foil, which usually takes on a gray appearance. For this reason it may be wise to avoid large white areas. Embossing is possible, but it may interfere with stacking and feeding of the labels in the labeling machine, and it can cause blistering problems. Pebble embossing, on the other hand, is sometimes used to improve the adhesion of the label to the bottle. Pebble embossing is an overall treatment, whereas registered embossing is a spot treatment.

Transfer Labels. There are several processes for transferring heat-sensitive inks from a preprinted strip to the container that is to be decorated. These are known by the trade names of Therimage (Dennison Mfg. Co.); Di-Na-Cal (Diamond National Corp.); Howmet, Transdec (Nashua Corp.); Electrocal (Noble & Westbrook Co.). The Therimage process uses gravure printing, which requires a high initial investment in printing plates but is capable of the finest reproduction of photographic or mechanical artwork. There are also some rather high costs for holding fixtures, which must be especially designed to hold each particular style of container during the labeling operation. See Table 1 for typical start-up costs.

TABLE 1 Transfer Label Initial Costs

Machine turret	$2,000.00
Printing cylinders, each color	220.00
Lacquer cylinder	120.00
Minimum order of labels*	1,000.00

* Based on 100,000 labels for a 4-oz cylinder bottle in three colors.

The Electrocal system is more economical, but it is limited to lines and solids and does not have the capabilities for halftones and process color reproduction of the Therimage process. Tool costs are nominal and the equipment for transferring the label to the container is relatively inexpensive. Because minimum order quantities are much lower than for Therimage, it is more practical for short runs. The operating temperature at the die is around 350°F with pressures between 100 and 200 psi, depending on the size of the label, and with a dwell time of about ½ sec.

Offset Printing. The coverage is limited by the amount of ink that can be laid down with the offset process, with the result that the colors tend to be somewhat thin and lines are often broken. Offset printing is economical for long runs. Many variety store items are decorated in this manner, but offset is not often used for toiletries and other luxury items because of the poor quality of the printing.

Silk-screen Decoration. This is a slow process in which ink is forced through portions of a fine-mesh silk cloth onto the surface of a container. The initial costs are quite low, as a silk-screen costs only about $25, and the process is well suited to short-run items. The ink coverage is excellent, and the printing is slightly raised above the surface of the container, giving it a special elegance. Silk-screen printing is limited to solid colors and line work, however, and halftones cannot be produced with any degree of success. It is used quite extensively for cosmetic items. Typical cost is about 1 cent per color per side. Cylindrical containers can be printed all the way around, except for a space of about 1/4 in., with one pass on the printing machine. If more than one color is to be used, it is necessary to have an indexing socket on the bottom, or a break in the threads of the bottle, which can be picked up by the printing machine so that one color can be registered with another. (See Figs. 2, 3, 4, and 5 in Sec. 8, "Plastics," page 8-7.)

Pressure-sensitive Labels. The rubber-base adhesives that are used for pressure-sensitive labels are high in cost, and consequently these labels are rather expensive. The backing paper that protects the adhesive mass also contributes to the high cost. The labels are easy to use since no glue pots are necessary, and they lie flat without wrinkles. Various types of adhesives can be used, depending on whether the labels are to be removed easily, resist solvents or high temperatures, or are intended to remain firmly attached under all conditions. Formulations are available that are (1) permanent, (2) completely removable without residue, or (3) resealable, or combinations of these.

To prevent reapplication of a pressure-sensitive label—for example, price stickers that might be switched by unscrupulous shoppers—the edge of the labels can be nicked in several places by the label manufacturer. Then when an attempt is made to remove the label, it will tear at the notches, making it practically impossible to remove it in one piece. Special adhesives have been developed that will form a very tight bond with the surface of the container. It is characteristic of all pressure-sensitive adhesives that the bond strength increases with age, and this is particularly true of "permanent" labels.

Pressure-sensitive labels permit quick price changes to meet competitive activities, or special offers that can be added to existing packages to facilitate marketing maneuvers. Plastic containers often pose difficult problems in getting plain paper labels to adhere, but the self-stick adhesives will nearly always work; sometimes they are the only type that can be made to adhere to these tricky surfaces. (See Fig. 3.)

Fig. 3. Label press. Pressure-sensitive labels can be printed by letterpress in three colors, embossed, die-cut, stripped, slit, and rewound all in one operation, on the machine shown. (*C. A. Nielsen & Petersen*)

PRINTING PROCESSES

The most widely used method of printing labels is *letterpress.* It permits good color control, with clear sharp detail. Printing plates are moderate in cost, and makeready is fairly easy. A variation of this, in which the printing is not direct from plate to paper, but is transferred from the plate to a rubber roller and then to the paper, is called *letterset.*

Almost equal in volume to letterpress printing is the process of *lithography,* which is particularly well suited to printing on foil. Although the amount of ink that can be transferred to the stock may not always give adequate coverage, and halftones tend to be fuzzy, it is satisfactory for most purposes. The minimum quantity is about 25,000 labels.

The fastest-growing technique is flexographic printing. Press speeds are not up to some of the other processes and quality may not be the best, but it requires the least expensive equipment. For this reason it is popular in the smaller shops and is used for printing pressure-sensitive labels.

For long runs the *gravure* process is the most dependable. Color lay-down is excellent, and particularly with fluorescent inks it gives better coverage than any other process. Makeready time is practically nil and results are very consistent. Plate costs are quite high, with the result that minimum quantities are upward of a million labels.

Of limited use, *screen* printing is applicable to very short runs. A very heavy lay-down of color is possible, but this method cannot handle fine detail beyond an 85 screen.

DESIGN OF LABELS

There are four factors that determine the efficiency of applying labels to containers: the label, the container, the adhesives, and the machine. For best results standards should be established for each of these elements, and strict adherence to the specifications should be required. The dimensions, stock, and grain direction of the label need to be specified as well as the type of adhesive, viscosity, and conditions and time for storage. Surface treatment and temperature of the container are important; sometimes, if containers are brought in from a cold warehouse and sent directly to the filling line, condensation of moisture from the air causes trouble. Machine settings and general maintenance, particularly cleanup of glue pots, dilution of glue, and protection from contamination and drying out, should all be spelled out in the process specifications, and followed, for consistent results on the production line.

The choice of paper *stock* will depend on the appearance desired, method of printing, resistance to moisture or other conditions in retail stores and at the point of use, as well as resistance to the product, scuff resistance in shipment, and the requirements of the labeling equipment. The cost of the paper stock also may be a factor, but this is not usually a very significant part of the total cost. Uncoated papers are sometimes used, but most often the stock is coated on one side. Uncoated paper is less likely to curl than coated stock, but it does not present as fine an appearance. Two-side-coated papers are usually more difficult to adhere, because the glue does not penetrate readily into the paper. Back treatment is sometimes used by the label manufacturer before the application of remoistening or heat-sensitive adhesives, to prevent the solvents from penetrating the paper.

Uncoated papers come in various surface *finishes,* which in the order of decreasing smoothness are machine glaze, machine finish, English finish, vellum, eggshell, antique, and wet end embossed. There are also special off-machine finishes. These are super calendered, duplex super label, and embossed. Coated papers in the order of decreasing smoothness and gloss are cast-coated, friction, flint, and brush-finished.

For greatest accuracy labels should be die-cut, although guillotine-cut labels can be held to very close tolerances. (See Fig. 4.) Each time a die

Fig. 4. Label cutter. Stacks of large printed sheets are cut to size in this guillotine-type paper cutter. The vertical bars of the back gauge can be seen in the center. The distance from the knife to the back gauge is adjusted with the micrometer handwheel in the front. (*Chandler & Price*)

is sharpened it will "grow" slightly, so that over a period of time the labels will become larger. If a multiple die is used, it may also be necessary to have the supplier keep the labels from each die separate to ensure uniform results on the labeling machine. Die cutting also work-hardens the edges of foil so that there is less tendency for these labels to curl. A label will adhere best to a flat surface, and the greater the curvature, the more difficult it will be to keep the edges from lifting up. For this reason a label should not extend too close to a corner or the shoulder of the bottle. (See Figs. 5 and 6.) Use the lightest weight of paper *stock* that will work on the machine; usually a 50-lb (0.003-in.

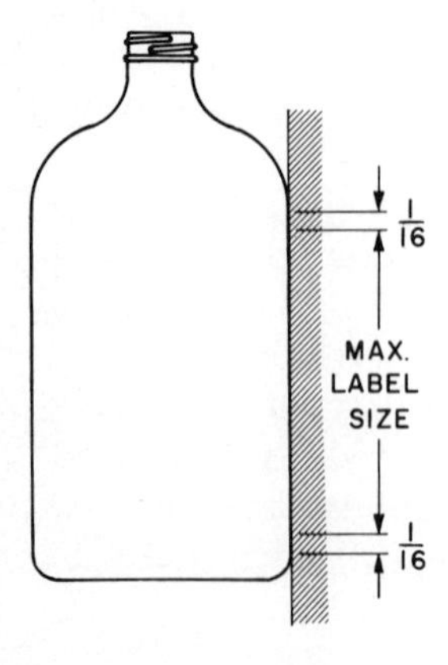

Fig. 5. Label size. Holding a flat edge against a bottle will show where the straight portion begins and ends. The label size should be less than this amount to allow for variations in placement.

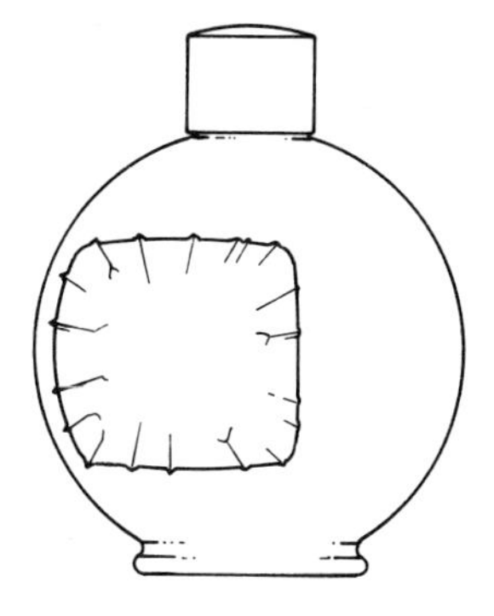

Fig. 6. Poor design. A label will not lie flat when it is on a compound curve, that is, one that curves in two directions at once, or when it extends up onto the curve of the shoulder in Fig. 5.

thick) lithograph stock, coated on one side, will give the best results, but 60-lb stock is more commonly used. The basis weight refers to the weight of 500 sheets 25 by 38 in. A tolerance of 5 percent in the thickness of the paper is generally considered an acceptable commercial tolerance. *Grain* direction is important, and this is usually specified in the horizontal direction, parallel to the printing, but there are some machines that work better if the grain runs vertically, particularly with pressure-sensitive or heat-sensitive adhesives.

The conditions of *storage* can seriously affect the efficiency of a labeling operation. Labels should be wrapped in wax paper, not tied or banded. If chipboard is cut with the labels, top and bottom, it will help to keep them flat. The amount of moisture in the air should be the normal amount under good working conditions; it is more important to avoid radical changes in humidity, such as in going from an unheated warehouse to the production floor. Most problems occur under humid conditions rather than when air is too dry, and during the summer months it is especially important to take the labels out of the machine at the end of a day's run and wrap them in wax paper.

Problems with *curling* are generally caused by uneven absorption of moisture from the glue into the paper. This can be due to excessive calendering, which affects the uncoated side as well as the coated side. It may also be the result of heavy ink coverage or penetration of the varnish coating. The cause of the difficulty can sometimes be traced back to the furnish that was used to make the paper, to which too much sizing was added. One trick that is used to help this situation is to moisten the stack of labels with a wet sponge so that the edges of the labels absorb a small amount of moisture.

ADHESIVES

The lowest-cost glues are the starch or dextrin types. They machine well and are fairly fast-drying, but they are not waterproof or iceproof when immersed. Dextrins are usually brown in color, acid in reaction, and fairly fluid in consistency. Jelly gums are widely used in the pharmaceutical industry, and they work well under adverse conditions of hot, oily, or wet containers. They are white to reddish brown in color, generally alkaline although some are neutral or slightly acid, and rather stiff and rubbery in consistency. Jelly gums have limited water resistance when immersed, but are satisfactory for high-humidity conditions. For more information on adhesives, see Sec. 9, "Closures, Applicators, Fasteners, and Adhesives."

Animal glues are used to some extent for beverage bottles because they have good resistance to immersion in ice water. They are medium brown in color, are slightly on the acid side, and have a fairly fluid body. Animal glues have good tack and machine well, with a relatively good drying speed. Casein adhesives have the best water resistance and are widely used in breweries and soft drink plants as an iceproof glue. Sometimes they are pigmented to give them a white color, but casein adhesives are actually amber to dark brown in their natural state. They are fairly fluid and slightly alkaline in reaction, have poor tack, and are slow-drying. Drying time can be speeded up by using a very thin film, and waterproofness is helped by using alcohol as a diluent.

Coverage or yield is dependent on the type of glue used and the adjustments on the machine, but some general rules can be given which will serve as a guide. Under ideal conditions a 55-gal drum (500 lb) of casein glue will label about 1½ million beer or soft drink bottles. Jelly gum will do about a million bottles, and dextrin adhesives about ¾ million under the same circumstances.

Problems are more often caused by using too thick a film of glue than for almost any other reason. A too thin film hardly ever causes trouble. Staining or discoloration of the label is often caused by an alkaline adhesive, but it can usually be corrected by changing to a neutral or acid-reacting type of glue.

Preglued labels can be the remoistening type, the heat-activating type, or the pressure-sensitive type. The remoistening kind of label is generally used for short runs and relabeling if there has been a price change or other temporary need for supplementary labeling. Heat-activated adhesives are coming into wider use on high-speed equipment to eliminate most of the problems of curling, blistering, wrinkling, and staining that sometimes occur with cold glues. They may, however, introduce other problems in the case of hot or cold bottles or if water, grease, or silicones are present. On small containers whose curved surface causes the edges of the label to lift up, it may be necessary to use very thin paper, 50-lb basis or less. Special formulations of adhesive are available which have quick stick characteristics for problem situations, or water resistance where chilled bottles may have moisture on the surface.

COSTS

It is difficult to determine the cost of producing labels, since they come in such a variety of types and sizes. However, for purposes of general information it can be said that a typical artist's sketch for a new label may cost about $75. To convert this to a black-and-white mechanical drawing, suitable for reproduction, will cost another $50. Printing plates will vary according to the process used, costing anywhere from $25 for offset plates to $1,500 for a gravure cylinder. Running cost for a small label will be about $4 per thousand, for two colors printed offset.

Section **15**

Wood Containers

History and Statistics 15-2
Characteristics 15-2
Advantages and Disadvantages 15-2
Selection Criteria 15-3
Nailed Boxes 15-4
Wirebound Boxes 15-7
  Advantages and Disadvantages 15-7
  Selection Criteria 15-8
  Assembly and Closure 15-9
Crates 15-9
  Design 15-9
Baskets 15-11
Barrels 15-12
  Selection Criteria 15-13
  Handling and Storage 15-13

One of the earliest packaging materials, and still very useful, is wood in its various forms. Although its use has been lessened by the substitution of other more sophisticated materials, it still has an important place in industrial packaging for heavy or fragile items that require rigidity and strength. For example, most bathtubs and similar enamelware are being shipped in wood, and the standard packing for auto windshields continues to be wooden containers.

The different types of packaging made from wood include baskets and hampers, tight and slack barrels, nailed wood boxes and crates, wirebound boxes, pallets and skids, and containerization units. They are made from either lumber, veneer, or plywood. Veneer is defined as wood that is less than 3/8 in. thick, regardless of whether it is sawed, sliced, or rotary-cut. The types of fasteners that are used include wire, nails, screws, staples, and bands.

HISTORY AND STATISTICS

Lumber has been used for various purposes since before the Christian era. The earliest sawmill in this country was set up in Jamestown, Virginia, in 1625, followed by one in Berwick, Maine, in 1631. By the early part of the nineteenth century the center of the lumber industry had shifted to the Midwest, followed by the development of the softwood industry in the southeastern states and the opening of timberland in the Northwest.

The U.S. Forest Service started a program of tests for the improvement of wooden shipping containers at Purdue University in 1905, and the Forest Products Laboratory was established at Madison, Wisconsin, in 1910. Since that time various branches of the government, notably the military departments, have accumulated a large amount of data on the design of wood boxes and crates.

Most of the lumber used throughout the world is produced in the United States, U.S.S.R., Canada, and Japan in that order. Packaging consumes 15 to 20 percent of the timber cut in this country in the form of lumber and paper.

CHARACTERISTICS

Wood is a structural material developed by nature to support the foliage and fruit of the tree, and it is remarkably strong for its weight. Being a natural material, it is not very uniform in its physical characteristics, however, and it becomes necessary to select and treat it in a manner that will make it useful as a packaging material. Some types of wood are better than others, and certain parts of a tree are more useful for packaging than other parts. Even the growing conditions of a particular tree will have an effect on its strength and other qualities.

Fortunately, by selecting the proper variety, sorting it for knots and other defects, drying it carefully, and sometimes laminating it to make it into plywood, we are able to get a fairly uniform material for our purpose. It is about 15 times as strong with the grain as it is across the grain, and it has a tendency to split when we fasten it together to make a box or crate, and to shrink and warp on standing, but it is nevertheless a valuable component for large packages.

ADVANTAGES AND DISADVANTAGES

With a good strength-to-weight ratio, wood is an economical structural material. It does not require very sophisticated equipment to make a box or crate, and for very rigid structures in small quantities it is the material of choice. For small packages or for large quantities, however, wood does not lend itself to high-speed operations or automatic assembly; it therefore has a high labor factor in relation to material costs. It is also bulky and often presents a problem of storage space and shipping cubage.

If rigidity, stacking strength, protection from the hazards of shipping, and light weight are essential, it is difficult to find a better material than wood. But if protection from moisture, rapid assembly, low cost, ready availability, or attractive appearance is more important, then wooden containers may not be the best answer.

SELECTION CRITERIA

There are about 1,000 species of trees in the United States, of which about 100 are commercially useful, but only about 10 are really important. Wood varies in density from 0.32 to 1.15; the heavier woods are stronger and have greater nail-holding power, but they are harder to work and have a greater tendency to split and shrink. As a result of work done by the Forest Products Laboratory, wood for box construction is divided into four groups, according to strength and nail-holding power. Group I includes the softwoods and light hardwoods such as fir, pine, spruce, cedar, chestnut, willow, bass, and poplar. Group II includes Douglas fir, hemlock, yellow pine, and tamarack, which have greater nail-holding power but more tendency to split than group I, and therefore require smaller nails spaced closer together. Group III contains the intermediate hardwoods such as black ash, gum, sycamore, and elm, which are about equal in strength and nail-holding power to the

TABLE 1 Grouping of Commercial Box Woods

Group I		Group II	Group III	Group IV
Alpine fir	Magnolia	Douglas fir	Black ash	Beech
Aspen	Noble fir	Hemlock	Black gum	Birch
Balsam fir	Norway pine	Larch	Maple (soft or silver)	Hackberry
Basswood	Redwood	North Carolina pine	Pumpkin ash	Hickory
Buckeye	Spruce	Southern yellow pine	Red gum	Maple (hard)
Butternut	Sugar pine	Tamarack	Sap gum	Oak
Cedar	Western yellow pine		Sycamore	Rock elm
Chestnut	White fir		Tupelo	White ash
Cottonwood	White pine		White elm	
Cucumber	Willow			
Cypress	Yellow poplar			
Jack pine				
Lodgepole pine				

group II woods, but are less inclined to split. They are the best for box ends and cleats, and they furnish most of the lumber for wirebound and plywood boxes. Group IV includes the hardest woods, such as oak, maple, hickory, and white ash, which are difficult to work, but make the strongest wire-bound and plywood boxes. (See Table 1.) Lumber should have about 15 percent moisture for best results; it should not have knots larger than one-third the width of the board, and should not have any knots at all in the nailing area. Other defects which should not be excessive are checks, splits, cross grain, decay, and insect damage.

The thickness of lumber to be used will depend upon the group of wood from which it is made, as described above, the type of load (whether easy, average, or difficult), and the weight of the load. The type of load is classified as type 1 if it is not very heavy, is not easily damaged, and completely fills and supports the box. Type 2 or average loads support the box at several points and may be held in place with interior packing. Type 3 loads are those which tend to shift and require a high degree of protection. (See Table 2.)

TABLE 2 Typical Box Lumber Sizes for Group II Wood and Type 2 Load*

Weight, lb	Sides, top, and bottom, in.	Ends, in.	Cleats, in.
50	1/4	1/2	1/2 × 1 1/2
85	5/16	1/2	1/2 × 1 1/2
125	3/8	5/8	5/8 × 1 1/2
225	3/8	5/8	5/8 × 1 3/4
325	1/2	3/4	3/4 × 1 3/4
425	5/8	3/4	3/4 × 2 5/8
600	3/4	3/4	3/4 × 2 5/8

* Weights are maximum and dimensions are minimum recommended.

NAILED BOXES

A modified beam formula was developed by C. A. Plaskett of the Forest Products Laboratory to determine the minimum thickness of the lumber for the top, bottom, and sides of a nailed wood box:

$$\text{Thickness, in.} = \frac{1}{8}\sqrt{\frac{\text{gross weight of box and contents, lb}}{\text{width of top or sides across grain, in.}}}$$

There are various methods of constructing a nailed wood box, depending upon the type of service required. A style 1 box is the simplest, consisting of sides, top, bottom, and ends without cleats. Style 4 is similar, but has two vertical cleats on each end. Style 5 is the same,

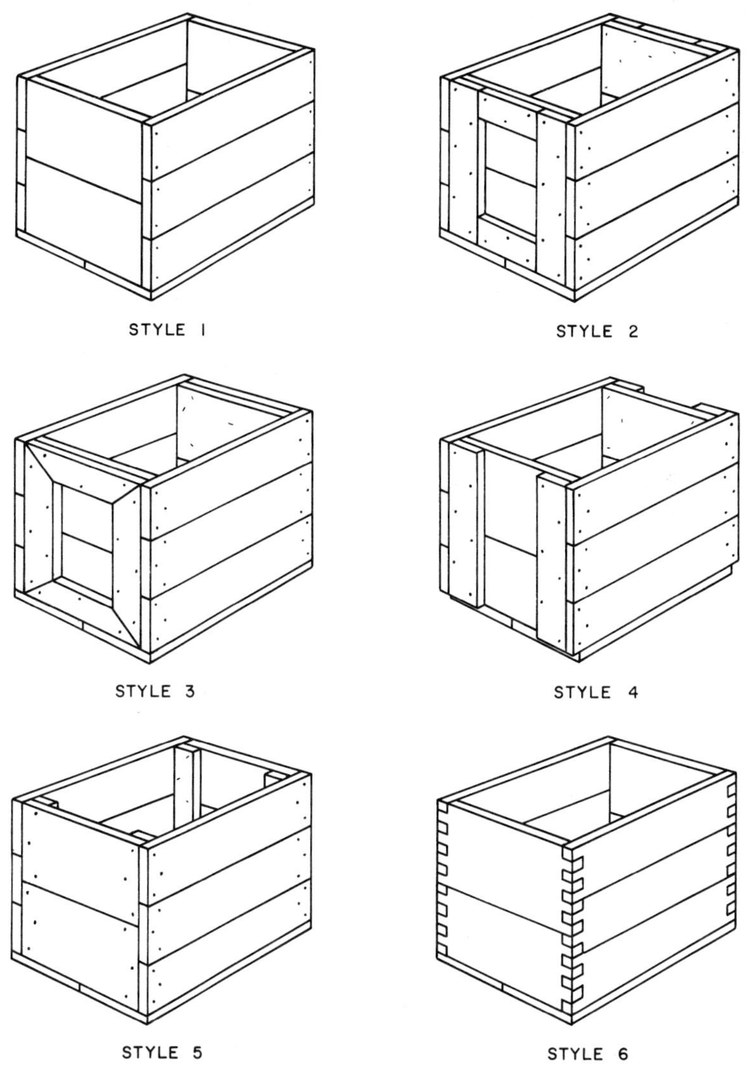

Fig. 1. Styles of nailed wood boxes. Style 1 is used for type 1 and 2 loads (see "Selection Criteria," page 15-4) up to 60 lb. Style 2 can be used for loads up to 600 lb. Style 2½ (not shown) is similar except that the vertical cleats are notched to receive slightly longer horizontal cleats. The load limit is the same. Style 3 also can be used for contents weighing up to 600 lb. Style 4 has only two cleats on each end and has a weight limit of 200 lb. Style 4½ (not shown) has horizontal cleats instead of vertical cleats, and the weight limit is 200 lb. Style 5 has interior vertical cleats, either rectangular or triangular, and can hold up to 200 lb. Style 6 has a weight limit of 100 lb. End pieces and cleats are usually 1½ to 2 times the thickness of the sides, top, and bottom, except that style 6 is always the same size throughout. There is also a Style 7 (not shown), similar to style 5 inverted. The contents is attached to the loose bottom piece, and the sides and top form a hood. Skids are fastened to the bottom. This style of box is used for loads up to 1,000 lb.

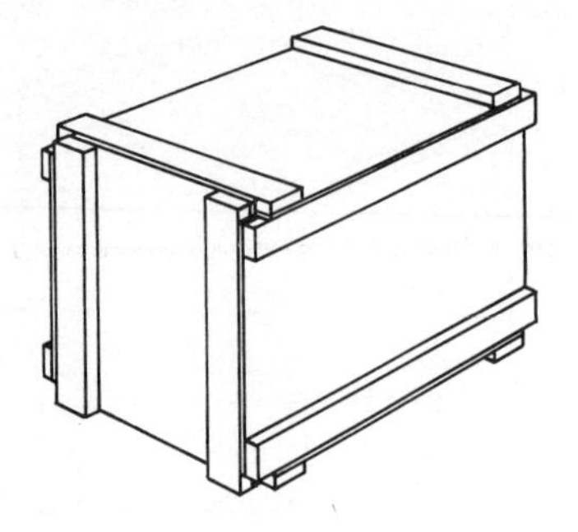

Fig. 2. Cleated panel box. A solid panel of fibreboard or plywood is reinforced with wood cleats. Other available styles vary in the number and placement of the cleats. Loads can be 150 lb or more, depending on the construction.

except that the cleats are put on the inside. Styles 2, 2½, and 3 have four cleats on each end, placed along the four edges. The only difference among them is the way the ends of the cleats are notched or mitered together. Style 6 is made without cleats, but with the vertical edges tenoned and glued. The style 1 box is adequate for loads up to about 60 lb, styles 4 and 5 will take up to about 200 lb, and styles 2, 2½, and 3 are designed for up to 600 lb or more. (See Figs. 1 and 2.)

The size and spacing of the nails is one of the most important factors in the ultimate strength of the box. The size of the nails will depend on

TABLE 3 Size of Nails for Wood Boxes

Type of wood	Thickness of wood holding points of nails, in.										
	3/8	7/16	1/2	9/16	5/8	3/4	7/8	1	1¼	1½	1¾
Group I	4d	5d	5d	6d	7d	8d	9d	10d	12d	16d	20d
Group II	4d	4d	5d	5d	6d	7d	8d	9d	10d	12d	16d
Group III	3d	4d	4d	5d	5d	6d	7d	8d	9d	12d	12d
Group IV	3d	3d	4d	4d	4d	5d	6d	7d	9d	10d	12d

TABLE 4 Spacing of Nails for Wood Boxes

Size of nails	Distance between nails	
	Side grain, in.	End grain, in.
3d to 6d	2	1¾
7d	2¼	2
8d	2½	2¼
9d	2¾	2½
10d	3	2¾
12d	3½	3
16d	4	3½

the species of wood and the thickness. The spacing will be determined by the size of the nails and whether driven into the side grain or end grain of the wood. (See Tables 3 and 4.)

Coated nails have 40 percent more holding power than plain nails, and side-grain nailing is nearly twice as strong as end-grain nailing. Where it is possible to clinch the nails, as in attaching cleats, the withdrawal resistance is increased 50 to 150 percent. Cleats should not go quite to the edge of the box, as shrinkage may leave them protruding and they may be knocked off. Steel bands will discourage pilferage and permit a one-third reduction in the thickness of lumber used for top, bottom, and sides. The bands should be placed one-sixth the length of the box from the end, with bands in between for very long boxes. (See Fig. 3.)

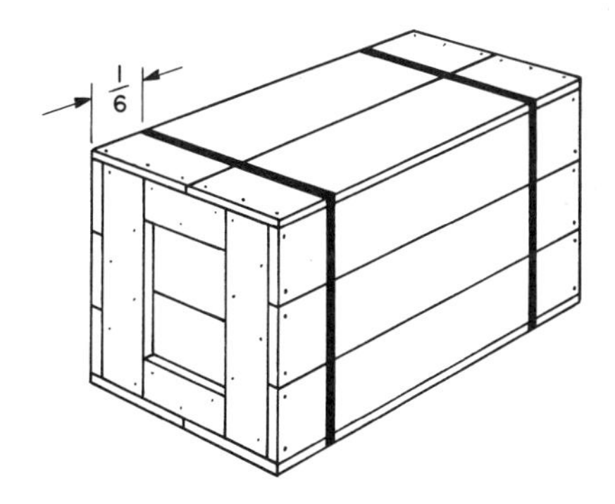

Fig. 3. Strapping. The stress on the nails is relieved and bulging is minimized by having straps in the correct position. These should be one-sixth the length of the box in from each end.

WIREBOUND BOXES

Very thin lumber is used to make wirebound boxes, and wires around the girth of the container are stapled to the wood at frequent intervals. Wood cleats are placed at the ends, and sometimes in between. These cleats are mitered and may also be tenoned, so that when assembled they will lock together. (See Fig. 4.)

Wirebound boxes are shipped and stored in the flat to conserve space, and in this form they are called shooks. The ends are usually shipped separately. Hardwoods from groups III and IV are used, and knots are avoided as much as possible, as well as divergence of grain, which should not be more than 1 in. in 8 in. of length. The wood should be used dry, with moisture not over 15 percent and preferably around 10 percent.

Advantages and Disadvantages. Wirebound boxes are lighter in weight than nailed boxes and have great tensile strength as a result of the reinforcing wires that go completely around the outside. Only

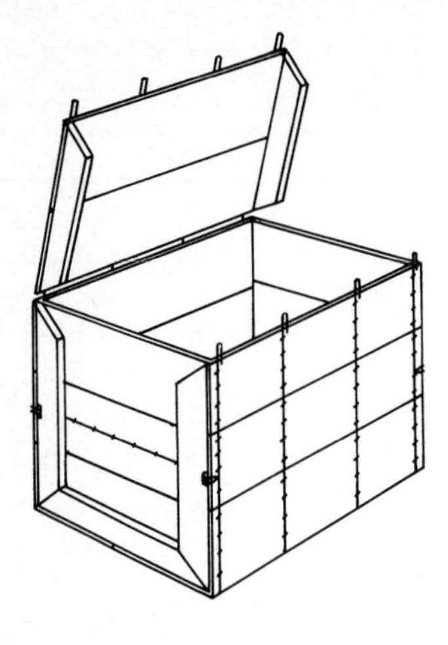

Fig. 4. Wirebound box. Thin face boards are reinforced by binding wires that are stapled in place at frequent intervals, and by wood cleats on the ends. The cover is closed by putting the wire loops through mating loops and bending them over. There are various styles, depending on the size and type of load. With proper-size lumber and wires, loads up to 500 lb may be put into these boxes.

about one-half the amount of lumber is required to provide the same service as an equivalent nailed wood box. Wirebound boxes do not have as good resistance to puncture, however, and they do not stack as well, but they are more economical. Whereas nailed wood boxes can be made in any carpenter shop, and are therefore more readily available, wirebound boxes require very sophisticated equipment for attaching wires and cleats. There are fewer plants capable of making wirebounds, and so greater shipping distances add to their cost.

A large quantity of wire-bound boxes must be purchased at one time to make it worthwhile to set up the equipment, in contrast to nailed boxes, which are economical in small quantities. If the quantities are sufficient, however, wire-bound boxes are more economical to purchase because they can be assembled more quickly and with less skill than other types of rigid shipping containers. Also, if storage space is limited, they can be stored in the flat until ready to be used.

Selection Criteria. The thin lumber that is used for wirebound boxes will spring under impact and absorb shocks that would otherwise be transmitted to the contents. It is the best type of box for lightweight fragile items, but is not recommended for goods weighing over 400 lb. The most economical shape for a given amount of cubic contents is with the length 2 to 2½ times the width.

There are four styles of wire closures to hold the cover in place for shipment: style 1 has straight wires that are twisted together; it cannot be opened and reclosed as readily as the other styles. Style 2 has the wire turned back to form loops that are hooked into each other and bent back. Style 2A also has loops, but in addition to being turned back, the end of the wire is twisted on itself for greater security. Style 3 has a looped wire closure, the same as style 2, but in addition has wires across

TABLE 5 Wire Size for Wirebound Boxes

Gross weight, box and contents, lb	Steel wire gauge
50	13
100	12
150	11
200	10
300	9
400	8

the ends in place of battens; it is not recommended for severe conditions of shifting loads, frequent handling, or overseas shipments. (See Table 5.)

Assembly and Closure. The blank or shook is delivered flat to the point of assembly. The sides should be lifted up from the bottom slightly before folding up. The sides are then folded at right angles to the bottom. The ends are next nailed to the side cleats with nails that go three-quarters of the way through the cleats, spaced about 2 in. apart. The side cleats should be nailed to the adjacent battens with sevenpenny nails, about 4 in. apart. After filling, the cover is brought down and the wires or loops are brought together and tightened. One sevenpenny nail is driven through the cleat into the end of each batten, top and bottom.

CRATES

The essential difference between a crate and a box is in the emphasis on the cleats and battens to take the stresses and support the contents. In a crate the sheathing is of minor importance, and it may be entirely omitted in some instances. A crate is most often used for items that are too large for a plain wood box. It may be nailed or screwed together, and will often incorporate runners on the bottom to allow a fork truck to get under and pick it up. Steel bands can be used to reinforce the corners, or they may be put around the girth of the crate. Metal strap-hangers, timber connectors, and other hardware from the building trades will be found useful for very large jobs.

Design. The diagonals, struts, and long members of a crate should be selected in much the same way that an engineer designs a Howe truss for a bridge. The crate designer has the advantage of being able to use sheathing to strengthen his structure. The same choice of lumber and fastenings as described under "Nailed Boxes," page 15-4, applies to crates as well. (See Table 6.)

TABLE 6 Size of Lumber for Crates

Load, lb	Spacing of cleats and battens, in.	Size of cleats and battens, in.	Size of main structural members, in.		
			2-ft spacing	4-ft spacing	6-ft spacing
100	54	$3/4 \times 2 1/4$	$3/4 \times 2 1/4$	$3/4 \times 2 1/4$	$3/4 \times 2 1/4$
200	48	1×2	1×3	1×4	1×5
500	42	1×3	1×4	1×5	1×6
1,000	36	1×4	1×5	1×6	2×4
2,000	36	1×4	1×6	1×6	2×4
5,000	24	1×4	1×6	2×4	2×4
10,000	24	1×6	2×4	2×4	2×4

The importance of having diagonal members in every panel cannot be overemphasized. A framework with diagonal members will be 10 times as strong, by actual test, as a framework with the same pieces of lumber placed parallel or perpendicular to each other. A long narrow crate should be divided up into approximately square sections by struts, and each of these sections should have a diagonal from corner to corner. The diagonals in adjacent panels should go in opposite directions, so that there are as many one way as the other throughout the crate. Diagonals must resist about the same stresses in tension and compression as the vertical and horizontal pieces, and should therefore be made of the same size of lumber. (See Figs. 5 and 6.)

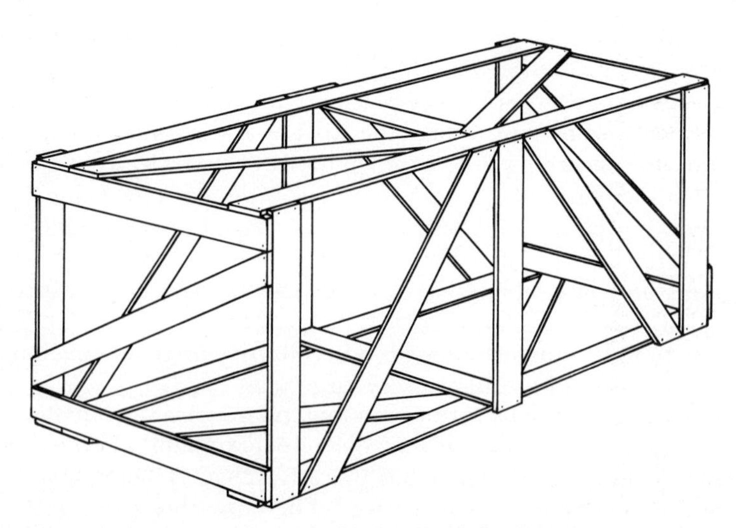

Fig. 5. Crate. A simple type of crate is shown, suitable for light duty. Very heavy loads would require a much stronger construction, with screws or bolts instead of nails.

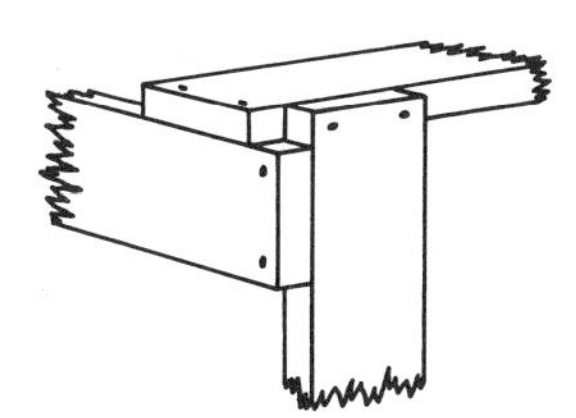

Fig. 6. Three-way corner. The strongest corner for a crate is the three-way corner shown. Every nail goes into side grain, and each member is locked in by the other two members so that the nails are not likely to work loose.

The crate should be large enough to completely enclose the contents with no parts protruding, and articles must be fastened so that they cannot move within the crate. Polished surfaces should be at least an inch away from any part of the crate. Detachable parts should be removed and packed in a cloth or heavy plastic bag or corrugated box, which should be secured inside the crate. Movable parts must be braced or blocked, and one should not depend on latches or locks to hold doors and drawers in place. Twine, wire, or a good grade of pressure-sensitive tape should be used to keep grilles and covers in place.

Items having legs should be suspended so that the legs are at least an inch away from any part of the crate. Place supports under the strongest part of the article, and put blocks or bracing on the sides and top, as near the center of gravity as possible.

BASKETS

Fresh fruits and vegetables are often packed in baskets made from thin wood veneer. The baskets come in several different styles, and in sizes that were established by the Standard Container Acts of 1916 and

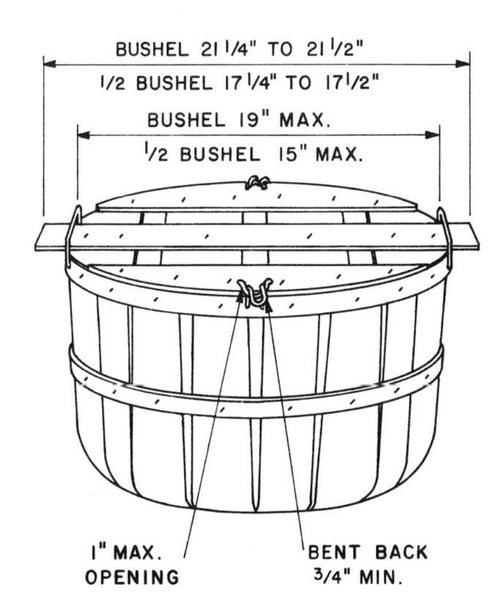

Fig. 7. Basket. Continuous stave basket with round bottom is used for citrus fruits and other fresh produce. Some baskets have a flat bottom with continuous staves, or with a solid or built-up bottom. Cover is held in place with wire bails and loops or with cross wires or sheet metal fasteners. Dimensions shown are recommended for safe handling.

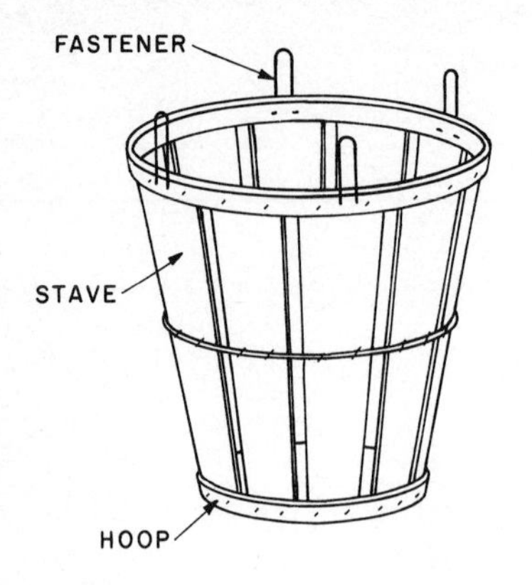

Fig. 8. Hamper. Made from veneer that has been rotary-cut against a stationary knife, these fully set up containers can be nested to save space. Various sizes from 1/8 to 1½ bu are made to dimensions prescribed in the U.S. Standard Container Acts of 1916 and 1928.

1928. Berry baskets, known in the trade as "small goods," are made in pint and quart sizes as well as tills (larger oblongs). The climax style is oval with a board bottom, and comes in 2-, 4-, and 12-qt sizes. Round baskets are made with a rounded bottom, also known as continuous stave baskets, or with a board bottom. They usually have a handle on each side and can be supplied with stave covers. Hampers come in the same sizes as the round-bottom baskets, but are taller and narrow. (See Figs. 7 and 8.)

BARRELS

Originated more than 2,000 years ago, barrels embody several sound engineering principles. The staves are arched in two directions, forming the ideal shape for maximum strength. The bulge makes it easy to roll and to upend the barrel conveniently. The flat ends provide a stable bottom for storage, and the hoops may be worked down toward the bulge to draw the staves together and make the barrel leakproof, at the same time locking the head into the grooves in the staves. A barrel is like a container on wheels. It can be rolled easily by one man, and because of its bilge it pivots readily and can be guided in any direction. Upending is accomplished by rocking back and forth on the bilge, and the chime provides a convenient handle for giving the final tug to put the barrel on its end. (See Fig. 9.)

A barrel differs from a drum in being bulged. Other terms are *cask,* which is a large tight barrel; *keg,* a small tight barrel of 10 gal or less; *tierce,* a barrel of 42-gal capacity; *firkin,* a tub that holds 56 lb of butter, but, strictly speaking, a *tub* is smaller than a firkin and holds only about 4 gal. A *pail* is larger at the top than the bottom and, by definition, has a handle or bail for carrying; a *kit* is an upside-down pail that is smaller at the top than the bottom.

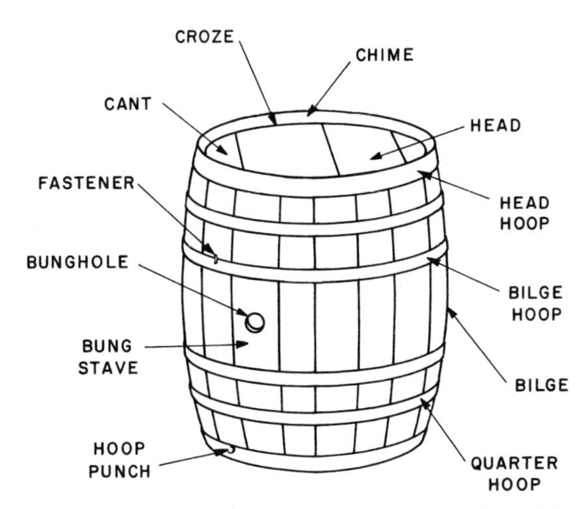

Fig. 9. Barrel. Made in various sizes from 1- to 60-gal capacity, from different woods and many types of construction. Hoops can be steel, wire, or wood, and the number of hoops will vary according to size and type of barrel. Specifications are given in Department of Transportation regulations, sections 178.155 through 178.161.

Barrels are generally divided into two classes, tight barrels to hold liquids and slack barrels for dry products. Tight barrels are usually coated on the inside, depending on the type of product to be packed. Wax is often used as a lining material for aqueous products, silicate of soda for oily materials, and glue for alcohol-based liquids other than food products.

Selection Criteria. The staves are made of many varieties of wood, and the choice is largely a matter of the most economical type for the job. One word of caution, however, in the case of food products: certain kinds of wood, particularly in the pine family, may impart an undesirable flavor or odor to the contents. Hoops are made of steel, wire, wood, or rope, and sometimes combinations of these. (See Figs. 10 and 11.)

Some standard designs are identified by their most frequent use, such as the *meat and poultry barrel* used for meat cuts and dressed birds, usually without a cover. The *glass and pottery barrel* is similar, but with closed head. Others are known as nail kegs, apple barrels, and so on.

Handling and Storage. A barrel should not be moved by rolling it on its edge or chime; it should be rolled only on its bilge or side. A loaded barrel should not be dropped, for although it is quite strong, it is not indestructible. Slack barrels should be stored on end, and tight barrels on the bilge, with the bung up. Do not remove the head with a hammer, but loosen the top hoop and lift the head out.

To close a barrel, drive the head hoop up about ½ in. all around. The head is started in the croze or groove at one point, and it is gradu-

Fig. 10. Raising a keg. The jack ring holds the sawed pine staves in proper position. A windlass near the top draws in the head of the keg for putting on the heading hoop. An experienced cooper can assemble about 100 kegs per hour. (*Grief Bros. Corp.*)

Fig. 11. Cutting the croze. After raising the barrel and heating it to put a permanent bend in the staves, a crozer puts a groove in the howel or channel near the end on the inside of the staves, to hold the barrelhead in place. (*Grief Bros. Corp.*)

ally worked down into place by tapping with a hammer all around from that point. If the head is in several pieces, start with the middle piece and work toward the sides. The head hoop is then driven down to draw the tops of the staves in tight, and threepenny nails are spaced 6 in. apart, through the hoop and into the head.

To remove a bung, tap the bung stave sharply close to the bung until it pops out. If it does not come out easily, use a bung chisel or bung puller. If vent holes are necessary, use a tenpenny nail rather than a drill, to avoid getting chips into the interior. When replacing a bung, have the grain running the same way as the stave or head board. To take the head out of a barrel, see whether one chime is deeper than the other and then take the head out of the end with the thinner chime. First remove the nails from the top hoop and drive the hoop up 1/2 in. or so. Then tap the head lightly until it falls into the barrel.

Section 16

Laws and Regulations

Basic Law 16-2
Packaging Law 16-2
History 16-3
Package Structure 16-5
Federal Trade Commission 16-5
Interstate Commerce Commission 16-5
Food and Drug Administration 16-7
Department of Commerce 16-7
Package Printing 16-7
Model Laws 16-11
Food and Drug Administration 16-11
Federal Trade Commission 16-13
Department of Transportation 16-14
Industry Standards 16-15
State Laws 16-15
Post Office 16-15
Carrier Rules 16-17
Air Cargo 16-18
Express Shipments 16-19
Trademarks, Patents, and Copyrights 16-19
Trademarks 16-19
Patents 16-20
Copyrights 16-23

Packaging law and its attendant regulations make up a body of information that is stupendous and overwhelming. The threads of authority run from Capitol Hill to the local fire department, and from the State Weights and Measures Bureau to the Bridge and Tunnel Authority, in a web that seems intended to confuse and trap the unwary package designer. However, it is not quite the jungle that it might appear to the novice, and we will attempt to lay down some guidelines to avoid the pitfalls in following the rules and regulations on packaging.

As it would not be practical to reproduce here the full text of the laws that are applicable to packaging, we can cite only brief passages to illustrate the discussion. When sections of the law are lifted out of context in this way, they may be misleading; so, as any good lawyer might counsel, the reader should consult the very latest and most complete text on the subject before making any major decision. Furthermore, the information given here is intended only for instructive purposes and is not to be taken as authoritative.

BASIC LAW

It is the function of our legislatures to determine the need for laws and to draft and enact the statutes that govern our society. Of necessity the law can provide only the broad outlines of policy that have been agreed upon by the legislature; it cannot cover every situation in detail or anticipate all contingencies. The responsibility for working out the fine details is usually delegated to an agency of the government that has been set up for the purpose. This form of legislation comes under the general heading of Administrative Law in the textbooks, as opposed to Criminal Law or Corporate Law.

A particular government agency is usually designated in the law to prepare the supplemental regulations to be followed by industry in order to comply with the law. A proposed regulation is published in the *Federal Register,* and after a specified amount of time during which industry can submit arguments and proposals, it may be revised and published as official regulations. It then has the force of law, by virtue of the authority delegated in the original law, and to the degree that it has interpreted the intentions of the legislature, it is binding upon all concerned.

An administrative agency of the government is not empowered to make laws, but only to interpret them. When it promulgates rules and regulations, it is sometimes found to be in error, and the published documents are declared invalid. Nor does the agency have any punitive powers to enforce the rules that it issues. Only the courts can decide guilt or innocence and impose penalties. But this is small consolation to the person that is affected, when it gets right down to actual cases. These rule-making bodies are nearly always right, whether we like it or not. They also have recourse to the courts in order to make their actions effective. So for all practical purposes a regulation actually has the force of law, and we must be guided accordingly.

PACKAGING LAW

Several aspects of packaging are covered by strict rules, and the designer should become familiar with the general nature of regulations in order to avoid costly changes that may come up if the packages are found to violate the law. These regulations fall into three broad categories: (1) weights and measures, (2) adulteration, and (3) public safety. In the general area of weights and measures the laws are designed to ensure that the purchaser knows the terms of his purchase and that he receives full value for his money. The customer must not be deceived or misled by the shape or size of a package, or by the printed information that is part of the package.

The subject of adulteration is covered by laws that are primarily concerned with the wholesomeness of the product, and only incidentally with the part played by the container. It is, nevertheless, quite essential that the materials and construction of the package do not seriously affect the product by adding any foreign material or by causing an undesirable change in the product. This is very important for food products and even more critical for drugs that have been precisely defined in an NDA (new drug application) for government approval before marketing.

In the interest of public safety there are regulations for the transportation of hazardous materials which are directed toward the carriers rather than the shippers. However, the carriers have their own rules, which put the responsibility for properly identifying and packaging these materials on the manufacturer.

HISTORY

In the eighth century the Arabian merchants used measuring cups and bowls certified by a bureau of standards (Dar al-'Ayar) and marked with their official seal in the form of small gobs of glass indicating the measure. In medieval England there were laws regulating the size of loaves of bread, and the Statutum de Pistoribus (1266) and the Usages of Winchester (1275) further required each baker to place his own mark on the bread he baked, to help in enforcing these laws. An ordinance of the Bottle Makers of London, 1373, prescribed that "every bottle maker shall place his mark on bottles and other vessels made of leather" to ensure full measure.

Government inspection was applied to grocers (spice dealers) early in the sixteenth century, but it was poorly administered and corrupt in its execution. In 1592 the Grocers' Company of London published "A Profitable and Necessarie Discourse for the meeting with the bad Garbelling (inspection) of Spices, used in these daies," in which they urged that the names of the inspectors be put on the seals for certifying the grades.

England adopted the Grocers and Apothecaries Act of 1557 to govern the sale of dangerous items, but it was not enforced and was soon forgotten. In an attempt at self-regulation, a group of chemists and druggists met in London in 1819 and proposed a set of regulations that would require warning labels on "any drug or preparation likely to produce serious mischief, if improperly used." These regulations also provided that such items be sold by persons of "sufficient age and experience to judge of the importance of the great caution necessary."

In 1790 Thomas Jefferson, then Secretary of State, recommended the development of a system of trademark registration, and government protection of the marks so registered, as a device to improve standards of manufacture in this country. His plea received little attention, however, and the United States did not pass its first registration act until 1870. This was declared unconstitutional, but another was passed in 1874, receiving Supreme Court approval in 1876.

The Meat Inspection Act of 1906 required the marking of grades of meat on the package, and in 1916 the Standard Container Act was passed covering poultry and poultry products. The Virus-Serum-Toxin Act of 1913 regulated the packaging of certain veterinary products, and in 1914 the Federal Trade Commission was established to control textiles and their packaging and, upon request, to investigate deceptive and unfair trade practices in any type of packaging and advertising.

The Federal Seed Act of 1939 specifies the information that must appear on agricultural seed packages. Similar controls are exercised over pesticides by the Federal Insecticide, Fungicide, and Rodenticide Act of 1947; over wool products through the Wool Products Labeling Act of 1941; and over fur garments through the Fur Products Labeling Act of 1952. The Caustic Poison Act of 1921 covered the labeling of some hazardous materials; it was followed by other laws which broadened this coverage, namely, the Insecticide, Fungicide, and Rodenticide Act; Dangerous Cargo Act; Hazardous Substances Labeling Act; and the Child Protection Act.

The Federal Trade Commission, pursuant to its authority to prevent unfair methods of competition, controls the labeling of all goods in interstate commerce. The Alcohol Tax Unit of the Treasury Department has certain powers over the labeling of beer and liquor, and the Animal Industry Quarantine Division of the Department of the Interior controls the labeling of veterinary medicines.

The Food, Drug, and Cosmetic Act of 1938, with the Food Additives Amendment of 1958, dealt with label copy and the placement of information in considerable detail, and the Fair Packaging and Labeling Act of 1966 went even further in defining the package requirements. Other pertinent laws are contained in the Internal Revenue Code, which pertains to alcoholic products and narcotics, and the Tariff Act, which regulates imports. In addition many states have passed laws for local control of such things as fertilizers, paints, bedding, and many other commodities and their packages, and even cities may have ordinances to regulate the sale and distribution of packaged goods.

An attempt to bring some uniformity into these local statutes was made by the National Conference on Weights and Measures, representing enforcement officials from all the states, and sponsored by the U.S. Department of Commerce's National Bureau of Standards. A Model State Weights and Measures Law was drafted by the Conference in 1965. This model law, which has been adopted by a majority of states without change, is an excellent guide to follow in designing labels. It should be noted, however, that section 12 of the Fair Packaging and Labeling Act of 1966 makes invalid all state laws "which are less stringent than or require information different from" the FPLA.

The individual states were given the right to control the sale and distribution of goods within their borders by an act of Congress in 1836. Since that time there have been a vast number of laws and regulations which are different from the federal laws, and must be considered if a package is to be distributed nationwide. A "doctrine of preemption," fostered by many state officials, maintains that *fulfilling federal law does not exempt from compliance with state laws.* This has never been tested in the Supreme Court and probably would not be upheld, but it cannot be ignored by the package designer if he wants to keep out of trouble.

PACKAGE STRUCTURE

Federal Trade Commission. The scope of activities of the Federal Trade Commission which affect packaging is generally related to deceptive packaging or to unfair competition. This agency does not seek out infractions of its regulations, with the exception of descriptive labeling of textile products. In most other areas it acts only upon the request of individuals or companies who feel that deceptive or unfair methods are being used.

A slack-filled container is considered *deceptive* if it has a false bottom or an excessive amount of headspace or outage. No maximum amount has been established as excessive, but 15 percent is often taken as an upper limit for headspace with liquid products. A package that is designed to look like a foreign import, although it is actually made domestically, is a form of deception. A foreign package which has the place of origin hidden is also considered misleading. In finding a manufacturer guilty of violating its regulations, the FTC does not consider it necessary to prove *intent* to deceive but only that in fact he does deceive.

Interstate Commerce Commission. The original purpose in creating the ICC was to control economic matters relating to transportation, such as mergers, operating authorities, and rates. They have become involved, however, in regulating the movement of dangerous articles as a result of the Crimes and Criminal Procedures Act of 1948. This necessitated the specification of containers and labeling for dangerous articles. The container construction regulations apply only to products that can destroy life and property.

A Department of Transportation (DOT) was set up in 1967 to take over responsibility from the ICC for the regulation of safety and the regulation of standard time zones. It is now their function to set the standards for the containers to be used for transportation of *explosives, corrosives, flammables,* and *poisons.* Etiological materials (microorganisms) are not covered, and it is up to the shipper to use his own judgment as to proper packing for viruses and bacteria for shipment.

The best source of information regarding a container for a particular product is found in the tariffs that are issued by the various carriers. The tariffs for rail and water transportation of hazardous materials are published by R. Graziano, 1920 L Street, N.W., Washington, D.C. 20036; for truck transportation the tariffs are published by William Herbold, Issuing Officer, National Motor Freight Classification Board, 1616 P St., N.W., Washington, D.C. 20036; and for air transportation domestically the publisher is Airline Tariff Publishers, Inc., 1825 K St., N.W., Washington, D.C. 20006, and for overseas it is International Air Transport Association (IATA), 1155 Mansfield St., Montreal 2, Quebec, Canada. There are other regulations for postal and express shipments, as well as state and local requirements, but the DOT regulations are a dependable guide in most cases.

The dangerous articles tariffs provide a list of the most common materials that require special packaging. They define the materials and structure of the container, maximum size allowable for each container, and the maximum number of containers in any one shipment. There are differences, for example, between the size and number of packages permitted on a passenger plane and on a cargo plane. There are also general categories of dangerous materials listed, usually marked "n.o.s." (not otherwise specified). There are two important principles to keep in mind: First, the method of packing and labeling must be on the basis of the hazard involved, and not on the end use; thus a paint thinner which is classified by chemical name as well as by paint product, requires the safer method. Second, a product which is listed by name as well as by group requires the method that will assure the safest shipment, usually the classification by name. Further, it must be understood that hazardous materials should be packed to arrive safely, and not merely to comply with rules. The regulations are only a means to achieve this end. If you are uncertain about the type or degree of hazard involved in a shipment, consult the useful reference book *Dangerous Properties of Industrial Materials,* 3d ed., by N. Irving Sax (Van Nostrand Reinhold Company, New York, 1968).

Food and Drug Adminstration. The Food Additives Amendment of 1958 defines an *adulterant* as any substance that may become a component of food, or that may affect the characteristics of food, which includes materials used in the manufacture of containers and packages. But affecting the characteristics of food does not include physical effects such as preserving the shape or preventing moisture loss. Previous to this amendment it was fairly easy to get a letter from the FDA endorsing a material or ingredient of a package for foods. Under the new law it became the responsibility of the manufacturer to prove the safety of a packaging material and to get approval before using it with foods. It should be noted that the FDA does not approve containers as such, but only the materials used in the containers.

A list of substances, sometimes called a "white list," containing those materials *generally recognized as safe* (GRAS) has been published by the FDA. In the opinion of qualified experts they are safe under the conditions of use, assuming they are of good commercial grade. The list, part of which is given in Table 1, is constantly being expanded by the FDA. There are prior sanction lists which contain materials that were previously approved by the FDA. These are often specific formulas for particular uses and may be applicable beyond the stated limits. In case of doubt, the FDA should be consulted for a ruling.

A material that is not included under GRAS or prior sanction, and is intended to be used with food, must be tested by the manufacturer, and the data must be submitted to the FDA. The method of testing is not usually defined by the FDA but left up to the petitioner. In general the test methods should simulate the actual conditions of storage and use as nearly as possible. For many indirect food additives a maximum of 1 ppm has been established, but for cancer-producing substances there is no allowance.

Department of Commerce. Under the Fair Packaging and Labeling Act of 1966 the Department of Commerce was given the authority to determine whether there is undue *proliferation* of the weights, measures, or quantities of commodities in packages. The Department of Commerce does not have the power to enforce this provision of the law at the time of this writing, but must report its findings to Congress, which may then provide the necessary legislative enforcement powers. In planning the quantity of a product to be put into a package, it might be well to conform to sizes of packages already on the market, in order to avoid a problem later on.

PACKAGE PRINTING

The requirements for labeling and marking packages for foods, drugs, cosmetics, and devices are covered by a great number of federal,

TABLE 1 Partial "White List" of Food Additives Permitted in Packaging Materials by FDA

SUBSTANCE	LIMITATION	SUBSTANCE	LIMITATION
Acetyl tributyl citrate	Plasticizer	Iron caprylate	Drier
Acetyl triethyl citrate	Plasticizer	Iron linoleate	Drier
Aliphatic polyoxyethylene ethers	Paper (see Note 1)	Iron naphthenate	Drier
1-Alkyl (C_6–C_{18}) amino-3-propane monoacetate	Paper (see Note 1)	Iron oxide	Paper
Alum, ammonium, potassium, or sodium	Paper	Iron, reduced	Paper
Aluminum hydroxide	Paper	Iron tallate	Drier
Aluminum oleate	Paper	Itaconic acid, polymerized	Paper
Aluminum palmitate	Paper	Japan wax	Cotton
Aluminum mono-, di-, and tristearate	Stabilizer	Lecithin (vegetable)	Cotton
Ammonium chloride	Paper	Linoleamide (linoleic acid amide)	Release agent
Ammonium citrate	Stabilizer	Linseed oil	Drying oil in finished resins
Ammonium potassium hydrogen phosphate	Stabilizer	Locust bean gum (carob bean gum)	Paper
Beef tallow	Cotton	Magnesium carbonate	Paper
Borax	Paper adhesives, sizes, coatings (see Note 1)	Magnesium chloride	Paper
Boric acid	Ditto	Magnesium hydrogen phosphate	Stabilizer
Butadiene-styrene copolymer	Paper	Magnesium hydroxide	Paper
Butylated hydroxanisole	Antioxidant, 0.005%	Magnesium glycerophosphate	Stabilizer
Butylated hydroxytoluene	Antioxidant, 0.005%	Magnesium phosphate	Stabilizer
p-tert. butylphenyl salicylate	Plasticizer	Magnesium stearate	Stabilizer
Butylphthalyl butyl glycolate	Plasticizer	Magnesium sulfate	Paper
Butyl stearate	Plasticizer	Manganese caprylate	Drier
Calcium acetate	Stabilizer	Manganese linoleate	Drier
Calcium carbonate	Stabilizer	Manganese naphthenate	Drier
Calcium chloride	Paper, cotton	Manganese tallate	Drier
Calcium glycerophosphate	Stabilizer	Melamine formaldehyde polymer	Paper
Calcium hydrogen phosphate	Stabilizer	Methyl and ethyl acrylate	Paper
Calcium hydroxide	Paper	Methyl acrylate (polymerized)	Paper
Calcium oleate	Stabilizer	Methyl ethers of mono-, di-, and tripropylene glycol	Paper (see Note 1)
Calcium propionate	Antimycotic	Methylparaben (methyl p-hydroxybenzoate)	Antimycotic
Calcium ricinoleate	Stabilizer	Myristo chromic chloride complex	Paper
Calcium stearate	Stabilizer	Mono-isopropyl citrate	Plasticizer
Calcium sulfate	Paper	Nitrocellulose	Paper
Carboxymethyl cellulose	Cotton	Nordihydroguaiaretic acid	Antioxidant, 0.005%
Casein	Paper	Oleamide (oleic acid amide)	Release agent
Castor oil, dehydrated	Drying oil in finished resins	Oleic acid	Paper, cotton
Cellulose acetate	Paper	Palmitamide (palmitic acid amide)	Release agent
Chinawood oil (tung oil)	Drying oil in finished resins	Polyethylene glycol 400	Release agent and paper
Chromium complex of perfluoro octane sulfonyl glycine	Paper, waxed (see Note 1)	Polyethylene glycol 1500	Release agent
		Polyethylene glycol 4000	Release agent
		Polyvinyl acetate	Paper

Substance	Use
Clay (kaolin)	Paper
Cobalt caprylate	Drier
Cobalt linoleate	Drier
Cobalt naphthenate	Drier
Cobalt tallate	Drier
Copper sulfate	Paper
Dextrin	Paper
Diatomaceous earth filler	Paper
Dibutyl sebecate	Plasticizer
Di-2-ethylhexyl phthalate	Plasticizer (for food of high water content only)
Diethyl phthalate	Plasticizer
Diisobutyl adipate	Plasticizer
Diisooctyl phthalate	Plasticizer
Dilauryl thiodipropionate	Antioxidant, 0.005%
Dimethylpolysiloxane	Release agent (see Note 2)
Diphenyl 2-ethylhexyl phosphate	Plasticizer
Disodium cyanodithioimidocarbamate with ethylene diamine and potassium N-dithiocarbamate and/or sodium 2-mercaptobenzothiazole	Paper slimicide (see Note 1)
Disodium hydrogen phosphate	Stabilizer
Distearyl thiodipropionate	Antioxidant, 0.005%
Epoxidized soybean oil	Plasticizer (see Note 3)
Ethyl acrylate and methyl methacrylate copolymers of itaconic or methacrylic acid	Paper, waxed (see Note 1)
Ethyl cellulose	Paper
Ethylphthalyl ethyl glycolate	Plasticizer
Ethyl vanillin	Paper
Ferric sulfate	Paper
Ferrous sulfate	Paper
Fish oil (hydrogenated)	Cotton
Formic acid (or sodium salt)	Paper
Gelatin	Cotton
Glycerides (mono- and di-) from glycerolysis of edible fats or oils	Paper
Glycerine	Paper
Glyceryl monooleate	Plasticizer
Guar gum	Paper, cotton
Gum guaiac	Antioxidant, 0.005%
Hexamethylene tetramine	Paper, setting agent for protein, including casein (see Note 1)
Hydrogen peroxide	Cotton
1-(2-hydroxyethyl)-1(4-chlorobutyl)-2-alkyl (C_6–C_{17}) imidazolinium chloride	Paper (see Note 1)
Polyvinyl alcohol (high or low viscosity) resins	Paper (for fatty foods only)
Potassium citrate, mono-, di-, and tribasic	Stabilizer
Potassium oleate	Stabilizer
Potassium pentachlorophenate	Paper, slime control agent (see Note 1)
Potassium sorbate	Paper
Potassium stearate	Stabilizer
Potassium trichlorophenate	Paper, slime control agent (see Note 1)
Propionic acid	Paper
Propylene glycol	Paper
Propyl gallate	Antioxidant, 0.005%
Propyl paraben (propyl p-hydroxy benzoate)	Antimycotic
α-Protein	Paper
Pyrethrins with piperonyl butoxide	Paper, in outside plies of multiwall paper bags (see Note 1)
Pulps	Paper, from wood, straw, bagasse, and other natural sources
Rubber hydrochloride	Paper
Silicon dioxides	Paper
Soap (sodium oleate, sodium palmitate)	Paper
Sodium acetate	Cotton
Sodium aluminate	Paper
Sodium benzoate	Antimycotic
Sodium bicarbonate	Paper
Sodium carbonate	Paper, cotton
Sodium chloride	Stabilizer
Sodium citrate, mono-, di-, and tribasic	Paper, cotton
Sodium hexametaphosphate	Paper
Sodium hydrosulfite	Paper
Sodium hydroxide	Paper, cotton
Sodium pentachlorophenate	Paper, slime control agent (see Note 1)
Sodium phosphoaluminate	Paper
Sodium propionate	Antimycotic
Sodium pyrophosphate	Stabilizer
Sodium silicate	Paper, cotton
Sodium sorbate	Paper
Sodium stearate	Stabilizer
Sodium sulfate	Paper, cotton
Sodium tetrapyrophosphate	Stabilizer and cotton
Sodium thiosulfate	Paper (0.1% in salt)
Sodium trichlorophenate	Paper, slime control agent (see Note 1)

TABLE 1 Partial "White List" of Food Additives Permitted in Packaging Materials by FDA (continued)

SUBSTANCE	LIMITATION	SUBSTANCE	LIMITATION
Sodium tripolyphosphate	Paper, cotton	Tin stearate	Stabilizer, not to exceed 50 ppm tin as a migrant in finished food
Sorbic acid	Antimycotic	Titanium dioxide	Paper (see Note 1)
Sorbitol	Paper	Triacetin (glycerol triacetate)	Plasticizer
Sorbose	Cotton	Triethyl citrate	Plasticizer
Starch, acid modified	Paper	2.4.5-Trihydroxy butyrophenone	Antioxidant, 0.005%
Starch pregelatinized	Paper	Urea	Paper, cotton
Starch, unmodified	Paper	Urea-formaldehyde polymer	Paper
Stearamide (stearic acid amide)	Release agent	Vanillin	Paper
Stearato-chromic chloride complex	Paper	Vinylidine chlorides, polymerized	Paper
Stearic acid	Cotton	3-(2-Xenoyl)-1.2-epoxypropane	Plasticizer
Stearyl citrate, mono-, di-, and tri-	Plasticizer	Zinc chloride	Cotton
Sulfamic acid	Paper	Zinc hydrosulfite	Paper
Talc	Paper, cotton	Zinc orthophosphate	Stabilizer, not to exceed 50 ppm zinc as a migrant in finished food
Tall oil	Drying oil in finished resins, and cotton	Zinc resinate	Stabilizer, not to exceed 50 ppm zinc as a migrant in finished food
Tallow, hydrogenated	Cotton	Zinc sulfate	Paper
Tallow flakes	Cotton		
Tartaric acid	Cotton		
Thiodipropionic acid	Antioxidant, 0.005%		

Note 1. Under conditions of normal use, these substances would not be expected to migrate to food, based on available scientific information and data.

Note 2. Substantially free from hydrolyzable chloride and alkoxy groups, no more than 18 percent loss in weight after heating 4 hr at 200°C; viscosity, 300 to 600 centistokes at 25°C; specific gravity, 0.96–0.97 at 25°C; refractive index 1.400 to 1.404 at 25°C.

Note 3. Iodine number, maximum 6; and oxirane oxygen, minimum, 6.0 percent.

state, and local laws and regulations. They apply generally to retail or consumer types of products only, and there are usually exemptions for professional or institutional packaging, such as prescription and veterinary drugs, diagnostic reagents, and similar items that are not usually sold "over the counter." The Fair Packaging and Labeling Act, for example, applies only to packages "for consumption within the household." In a few cases it is necessary to obtain prior approval of printed packaging materials before they can be offered for sale. This is true of meat and poultry products, alcoholic beverages, pesticides, and new drugs.

Model Laws. A good starting point in writing copy and making a layout for printed materials would be the Model State Weights and Measures Law, and the Uniform State Food and Drug Law. These have been adopted by a majority of the states without change and have formed the basis for the more recent federal laws. The Model State Weights and Measures Law was adopted in 1965 by the National Conference on Weights and Measures, representing enforcement officials from all the states. Copies are available from the Executive Secretary, National Conference on Weights and Measures, National Bureau of Standards, Washington, D.C. 20204. For information on the Uniform State Food and Drug Law, write to the Association of Food and Drug Officials of the United States, P.O. Box 1494, Topeka, Kansas 66603.

Food and Drug Administration. A number of requirements for the print copy come under the jurisdiction of the FDA. The *identity* of the commodity must appear prominently on the main panel. The name must be clear and understandable, and not misleading in any way. A descriptive phrase also may be necessary to identify the particular form of the product within the package. If a proprietary name is used for a drug, it is necessary also to use the generic name at least once in association with the proprietary name. (See Fig. 1 in Sec. 14, "Labels and Labeling," page 14-2.)

The weight, measure, or count also must appear on the main panel, and this *quantity* statement should be in the lower 30 percent of the display space (except where this panel is less than 5 sq in.), in lines parallel to the base. There must be clear space above and below equal to the type height, and on either side, equal to twice the width of the letter N. A double statement is often necessary to comply with this portion of the regulation, which requires that the total number of ounces be followed by the largest whole unit (pint or pound), with the remainder in ounces. A fractional or decimal figure is permitted, such as "1.5 lb" or "1½." *Av.* or *fl.* must be used to differentiate dry weight from liquid measure, and the words *net weight* also must be used. Several exceptions to the above have been granted by FDA, such as carbonated beverages, coffee, butter, and eggs. The quantity statement on an aerosol must be the amount delivered including the propellant, when the instructions for use are followed.

The name of the *manufacturer* and the place of business, including zip code, must be on the package, but not necessarily on the main panel. The street address is not required if the company is listed in the telephone directory.

When a list of *ingredients* is required, as when the name of the product itself is not explanatory, they must be listed in order of decreasing prominence. For drug products the requirements are given in section 502(e) of the Federal Food, Drug, and Cosmetic Act. A hazardous product must list each component that contributes substantially to its hazard. If two ingredients are similar, and one is less than 20 percent of the mixture, a quantitative statement must be given. The ingredients may be printed in any location on the package, provided they are all on the same panel. In the case of drug products, the intended action of the active ingredient also must be given, such as "to reduce fever."

There are several laws that require *warnings* to be printed on the package. Such statements as "Warning—May be habit-forming" are designated in the Federal Food, Drug, and Cosmetic Act, and "Danger—Poison" is required for toxic materials by the Federal Hazardous Substances Act. (See Table 2.)

Drugs and devices must have adequate *directions* for use, so that a layman can use them safely. These should include the disease for which

TABLE 2 Excerpts from Labeling Regulations for Hazardous Substances (21 CFR 191)

Item	Type size	Notes
SIGNAL WORD	18 point, all capitals	1, 2
Word "POISON" (when required)	18 point, all capitals	1, 2
STATEMENT OF HAZARD(S)	12 point, all capitals	1, 2
All other items of label information required	10 point, case optional	1, 3

Note 1. ". . . The size of the . . . (Item) . . . shall be of a size bearing a reasonable relationship to the other type on the . . . panel involved . . . ; but shall not be less than . . . (above) . . . point type, ..."

Note 2. ". . . , unless the label space on the container is too small to accommodate such type size. When the size of the label space requires a reduction in type size, the reduction shall be made to a size no smaller than is necessary and in no event to a size smaller than 6 point type."

Note 3. ". . . , unless the available label space requires reductions, in which event it shall be reduced no smaller than 6 point type, unless because of small label space an exemption has been granted under section 3(c) of the Act and Paragraph 191.63."

TABLE 3 Type Size for Quantity Statement on Package

Panel size, sq in.	Type height, in.
0–5	1/16
5–25	1/8
25–100	3/16
100–400	1/4
400 and up	1/2

the product is intended, the usual quantity per dose, method of administration, and similar information. If the quantity statement is blown or molded into a bottle or jar, it should be 1/16 in. larger than the minimum sizes shown in Table 3.

Federal Trade Commission. The responsibility for preventing *unfair competition* in packaging and advertising falls upon the Federal Trade Commission. Prior to 1966 the only definite regulations of the FTC on packaging pertained to textile products. Now some of the provisions of the Fair Packaging and Labeling Act of 1966 come under their jurisdiction, but not if they are controlled by another administrative agency, as is the case of foods, drugs, devices, cosmetics, meat, poultry, tobacco, seeds, veterinary biologicals, pesticides, and alcoholic beverages. This leaves such things as soaps, detergents, cleaners, stationery, and sundry other household items, whether packaged or merely labeled, to be controlled by the FTC.

Transparent wrappers need not be printed, as long as the required information shows through. The requirements for a net quantity statement are identical with FDA provisions mentioned above. The word "new" in promotional copy on a package can be used only for 6 months after a product enters into the general retail market. *Deceptive practices* include misleading statements, such as price marking with an inflated figure to suggest a larger markdown than is actually being given. The test is whether the price is that for which the product is substantially sold in representative markets. If a domestic package is made to look foreign, or if a foreign package has its place of origin hidden, it violates FTC regulations.

Price marking of packages by the manufacturer comes close to violating the federal antitrust and trade regulation laws, which forbid a manufacturer or distributor from setting retail prices. Although "fair trade" laws permitted manufacturers to set retail prices within those states that had such laws, the practice of manufacturing "preticketed" products has been attacked as illegal by the Federal Trade Commission.

Department of Transportation. When material is to be transported between states, it falls under the jurisdiction of the Department of Transportation. This adminstrative agency is especially concerned with the proper labeling of *hazardous* materials that are carried over state lines, and it has published certain guidelines to help the shipper to meet his obligation. Compliance with these regulations does not automatically protect the shipper from damage claims, since they are intended only as minimum standards. The shipper is also expected to use good judgment in packing and marking his goods. If more than one label is indicated, only the label for the greatest hazard should be used, except in the case of a tear gas.

Flammable liquids require a red label of a particular size and shape on the shipping container as specified in the regulations. The definition of a flammable liquid in the Federal Hazardous Substances Act (1960) amended by the Child Protection Act (1966) is any substance with a flash point of 80°F or below, by the Tagliabue (Tag) open-cup test. This test consists of gradually raising the temperature at a specified rate until a lighted match will ignite the vapors in the cup.

Flammable solids and oxidizing materials require a yellow label; they should not be stored near white-label products, which is the designation for acids and alkalies. A green label is used for nonflammable compressed gases. Radioactive materials are labeled according to the amount of radiation at the surface of the package in rems.

Poisons are classified according to type and severity. The great majority of such shipments are class B poisons, requiring a white label with skull and crossbones in red. A class B poison is one which has an LD_{50} of 50 mg or less, that is, a *l*ethal *d*ose for 50 percent of the test animals when fed at the rate of 50 mg or less per kilogram of body weight. Class A poisons are poison gases, which take a label similar in color to the class B label, but with the words "Poison Gas" in large letters. A class C poison is a tear gas and has that inscription in the same color combination as the class A and class B labels.

Explosives do not require warning labels under the present law, except for express or air shipments, unless they are for laboratory examination. Etiological materials (disease-producing organisms) are not covered by regulations; only the air transport tariffs specify the packing and labeling to be used for air shipments. This does not absolve the shipper of any responsibility in these cases, but rather puts the burden on him to design suitable markings to suit the hazards involved.

The Department of Transportation is permitted to use "the services of carrier and shipper associations, including the Bureau for Safe Transportation of Explosives and Other Dangerous Articles," in accordance with section 834, Public Law 86-710. The Bureau of Explosives is under the direction of Mr. R. Graziano and handles nearly all of the changes to the DOT regulations. Therefore the Rail Carrier Tariff and the Water Carrier Tariff issued by Robert Graziano, 1920 L Street, N.W., Washington, D.C. 20036, have become the standard reference works for packaging of hazardous materials.

Industry Standards. Certain basic principles have been developed in the chemical industry for labeling hazardous materials. These are generally covered in the standard warning labels, but careful consideration should be given to the following points:

1. The name of the product must appear on every package. The chemical name is more important than the trade name in this case.
2. Warnings can be made part of the product label, or they can be used separately. In either case there are several areas to be covered, as follows:
   a. Use the word "Danger" for the most serious hazards, "Warning" for less critical conditions, and "Caution" for the least hazardous materials.
   b. Give a definition of the danger, such as "Causes Burns." Note that hazards include toxicity, corrosiveness, oxidation, skin irritation, explosiveness, flammability, and radiation.
   c. Include precautions to avoid injury, for example, "Keep Away from Open Flame." This can be omitted if the previous statement is self-explanatory. Note, however, that injury may result from inhaling dust or vapors, absorption through the skin, getting into food products, or combining with other materials in a shipment to create a hazardous condition.
   d. Give instructions in case of exposure, such as "Rinse with Water."
   e. In special cases it may be necessary to include instructions for disposing of the empty container.

The warning labels should be fastened securely to the primary container, and not to a wrapper that may be discarded.

STATE LAWS

Nearly all the states have regulations for the transportation of explosives, but only ten have any regulations for other dangerous materials.

POST OFFICE

There are four classes of mail, and the rates vary according to the type of service offered. *First-class mail* consists of letters, cards, and packages containing personal messages, and packages sealed against inspection.

The minimum size for letters and cards is 3 by 4½ in. There is no maximum size. Shapes other than rectangular are nonmailable. The rates are 6 cents per ounce regardless of distance. Single cards are 5 cents, double cards are 10 cents, and airmail is 10 cents.

Second-class mail consists of newspapers and magazines. The rates are by the pound and vary according to distance. Costs range from 1¼ cents per pound to 14 cents per pound, but not less than ½ cent per copy. Special rates are allowed for religious, educational, and other nonprofit organizations.

Third-class mail is composed of bulk mail which is over 50 lb total, or over 200 identical pieces. Minimum size is 3 by 4½ in. for cards and envelopes, but there is no minimum for packages. Maximum weight up to but not including 1 lb is allowed for each piece. Shapes other than rectangular are nonmailable. An indicia or imprint must be printed in the upper right corner of the main panel with the name of the sender, the post office where mailed, and the word paid. Single pieces up to but not including 16 oz also may be sent third class, at a lower rate than fourth class.

Fourth class (parcel post) is used for small individual packages over 1 lb. For less than 1 lb, third class is more economical. Parcel post is one of the cheapest forms of transportation for miscellaneous shipments. Rates are based on weight and distance, with special rates for heavy books, catalogs, manuscripts, films, and records.

Maximum weight is 40 lb, and maximum size is 84 in. (maximum length and girth at the thickest part combined), if mailed between first-class post offices. When packages are mailed to or from smaller post offices, the maximum weight is 70 lb, and the maximum size is 100 in.

Special services that are provided by the Post Office include airmail; registration, which includes insurance; certification, which does not include insurance but provides a record of delivery; special handling, which in effect puts a package in the first-class category; and special delivery, which provides for immediate delivery out of the receiving post office.

Since 1954 the Post Office Department has placed the responsibility on the sender to determine the mailability and proper packaging of his products. Nonmailable materials include poisons, explosives, and flammable or other dangerous articles. Exceptions are granted, upon request, for hazardous articles that are not outwardly or of their own force dangerous or injurious to life, health, or property. The test of adequate preparation and packaging is whether the contents are safely preserved under the ordinary hazards of mail handling and transportation. The act of mailing is considered the shipper's assurance that his product is not dangerous to life, health, or property.

CARRIER RULES

The control and enforcement of certain federal laws are delegated to the Department of Transportation. Among these duties is the approval of the tariffs issued by the various carriers. When these tariffs are filed with the DOT by the carriers, they achieve the effect of a statute, under the provisions of the Interstate Commerce Act. The Uniform Classification Committee, 202 Union Station, Chicago, Illinois 60606, was set up by the railroads to protect rail carriers from gross damage claims by requiring high standards of construction for shipping containers. This committee issues the tariffs for the railroads, including package specifications, as its Uniform Freight Classification.

The National Motor Freight Traffic Association, Inc., has a National Classification Board equivalent to the railroads' Classification Committee. They also prepare the tariffs for filing with the DOT for freight handled "over the road." Copies of these tariffs can be obtained from the American Trucking Associations, 1616 P Street, N.W., Washington, D.C. 20036. Although there are other tariffs for air freight, waterway freight, parcel post, and express, the rail and truck specifications are the most significant for the packaging engineer. It should be noted that these tariffs are not the same as the dangerous articles tariffs mentioned previously under Department of Transportation regulations, and both sets are needed if hazardous materials are being shipped. Also, fresh fruits, berries, melons, and vegetables are not covered by Rule 41, but come under special tariffs which can be obtained from W. T. Jamison, 516 West Jackson Boulevard, Chicago, Illinois 60606.

The most significant rule for the packaging engineer is Rule 41 of the Uniform Freight Classification, or its equivalent Rule 222 of the National Motor Freight Classification. This rule defines the minimum standards for corrugated boxes. Emphasis should be on the word "minimum," as this is often taken as a standard, whereas it may be quite inadequate for some situations; a good designer will specify his packing to suit a particular set of conditions, and not merely to comply with the regulations. When articles are found in transportation which do not comply with Rule 41, an increase in freight charges of 10 percent on carload shipments and 20 percent on less-than-carload shipments may be imposed, in accordance with Rule 5.

Nearly all corrugated boxes are made to conform with the specifications in Rule 41. Thus a "200 test" box which is permitted to carry up to 65 lb is made from two 42-lb paperboard facings and a corrugated medium of about 26-lb basis (not specified), as defined in the rules. See Sec. 5, "Corrugated Fibreboard," for more information on corrugated boxes. It has been argued that the Uniform Classification Committee is arbitrary and unscientific in its specifications, but even if this is true, the rules have nevertheless helped to bring order out of chaos. There can be no doubt that damage claims have been substantially reduced as a result of these specifications, and they have therefore achieved the purpose for which they were intended. The loss and damage rate is currently around $3 billion annually, with claims against the carriers running about $1/2 billion. Without Rule 41 it would be considerably higher.

Other rules of the Uniform Freight Classification which may be of special interest to the packaging engineer are as follows:

Rule 6 states simply that packages containing fragile articles or articles in glass or earthenware must be marked fragile—handle with care or with similar precautionary marks.

Rule 21 defines articles which may be shipped "nested."

Rule 22 defines various types of finishes on wooden articles such as furniture.

Rule 40 covers containers other than corrugated or solid fibreboard. This includes metal drums and pails, wood barrels and boxes, and multiwall paper bags.

Rule 49 provides for test shipments of experimental containers, with the approval and under the control of the Uniform Classification Committee, which is in Chicago, Illinois. Each package must be marked "Test Container," and a permit signed by the chairman of the Uniform Classification Committee must be given to the carrier's agent at the point of origin. This permit may have a 6-month time limit or other restrictions which the committee may deem necessary. The burden of accumulating performance data is on the shipper, not the committee. Tests are usually specified to be performed in standard railcars without cushioned underframes.

The test data are submitted to the Uniform Classification Committee, which approves or disapproves the application for "number packages." These are commodity-by-commodity exceptions to the constructions given in Rule 41. For example, shipping containers sealed with hotmelt adhesives covering less than the required 50 percent of the inner flaps are allowed for a limited number of commodities under Package No. 1445.

Air Cargo. Air cargo is similar in its requirements to Railway Express if it is carried on cargo planes. On passenger planes, however, there are more restrictions than on cargo planes as to size and type of materials that can be accepted. There are no specifications for package construction, and the airlines will accept any type of container which is adequate for the purpose. Nearly all explosives are prohibited from air transport, and all extremely dangerous poison gases, as well as liquids that emit poisonous vapors, are forbidden. Flammable compressed gases may be carried in cargo planes, in limited quantities, but not in passenger planes. The Official Air Transport Restricted Articles Tariff published by Airline Tariff Publishers, Inc., 1825 K Street, N.W., Washington, D.C. 20006, is endorsed by nearly all domestic airlines. A similar tariff for export shipments is published by International Air Transport Association (IATA), 1155 Mansfield Street, Montreal 2, Quebec, Canada.

Sizes of packages for cargo aircraft can be up to 60 by 80 by 90 in. and can weigh 1,000 lb or more, but for passenger planes they are usually restricted to about 20 by 24 by 44 in. and weights under 200 lb.

Express Shipments. The requirements for packaging of articles to be shipped by express are essentially the same as for rail freight. Packages under 35 lb, however, need not comply with construction requirements, but boxes for fragile items are subject to a drop test procedure in Rule 18. Detailed specifications are given also for packing such things as furniture, lamps, shades, and citrus fruit. The Official Express Classification and the Air Express Tariff can be obtained from REA Express, Inc., 219 East 42nd Street, New York, New York 10017.

TRADEMARKS, PATENTS, AND COPYRIGHTS

Trademarks. The principal function of a *trademark* is to indicate the origin of goods. It says, in effect, that the goods are "made by" a particular manufacturer. Secondary purposes are to guarantee the quality of the goods and to create and maintain a demand for the product. Unlike a patent, the act of registering a trademark does not establish any rights to the mark. It does, however, help to strengthen the rights which are established through commercial use of the trademark. The only way that ownership of a particular mark can be established is through wide use on a continuing basis. The more it is used in public, the stronger it becomes. If the use is discontinued, the rights to the mark may be lost.

The fee is $35 and the registration will remain in effect for 20 years, with unlimited renewals permitted for additional periods of 20 years, The mark is published in the *Official Gazette* of the Patent Office, at the time of application or renewal. The *Official Gazette* is published weekly; an annual index to it is also published and obtainable from the superintendent of documents.

The Trademark Act of 1905 authorized the registration of trademarks used in interstate commerce. Previous laws applied to trademarks in commerce with foreign nations and therefore were of little value. The Lanham Act of 1947 made the following changes: provisions for registering service marks and certification marks; incontestability of trademark registrations under certain conditions; cancellation of registrations after the sixth year if an affidavit of use is not filed during the 6 years; and others.

Any word, name, symbol, device, or combination thereof that is used by a manufacturer or merchant to distinguish his goods from those manufactured or sold by others can be considered a trademark. A name or mark which is descriptive of a product is not nearly so strong as a coined name that is meaningless. Thus Kodak is a far better name for a camera than Handy would be. A person's name does not make a good trademark, since any persons with the same name would have some rights to that name as a trademark, and it would be difficult to stop them from using it. A geographical or other location name would also be difficult to protect.

It is helpful to use the word "trademark" or the initials T.M. in conjunction with the name or symbol, to serve notice of the claims for exclusive use. If it is used broadly enough, without protest from other manufacturers, it strengthens the rights of the user. It is not necessary to use T.M. with the mark every time it is used; once is enough on each package component, if it is in a prominent place.

There is a risk of having the trademark become a generic term for a group or class of products. This has happened with such well-known words as thermos, mimeograph, escalator, cellophane, saran, and aspirin (in the United States). To prevent this, it is well to use the name only as an adjective, and not as a noun, and always to capitalize it; do not use two trademarks together. Printing the word "brand" after the trademark also helps.

Patents. On April 10, 1790, President George Washington signed the Patents Act. For the first time in history an idea became a form of property, and an inventor had the right to derive whatever benefit he could, for a fixed period of time. Patents had been granted previously by acts of the state legislature, but there were no general laws, and patents were difficult to obtain. The practice of numbering patents was started in 1836. Design patents were first granted in 1842, and plant patents in 1930.

Nearly 3½ million patents have been granted in addition to about ¼ million design patents and over 2,000 plant patents in the United States.

There is an important difference between a *patent* and a trademark. The filing of a patent gives a high degree of protection immediately, without prior use or publicity. A trademark, on the other hand, must be in commercial use when it is registered, and the act of registering adds little to the value of the mark.

Any person who has invented a process, machine, composition of matter, ornamental design, or new variety of plant that meets certain requirements may obtain a patent. The filing fee is $65 plus $10 for each claim, and the total Patent Office fees will probably be over $200. A patent is granted for 17 years and cannot be extended. What is granted is not the right to make, use, or sell, but only the right to exclude others from making, using, or selling the invention; ordinarily the inventor already has the right to make, use, or sell anything he wishes, if it does not violate any law. It is no longer required by law that an article be marked "patented," but before an infringement can be proved, it is necessary to serve notice of the existence of the patent; marking the patent number on the item serves as a form of notice to the public at large. The use of "patent applied for" has little value in a legal sense, but it may be a deterrent to anyone who might want to copy an idea.

There are several reasons for obtaining a patent, such as financial reward, prestige, or advantage over a competitor. Too often, however, patents are obtained on trivial improvements that are not worth the time, effort, and money that are necessary to secure them. Try to determine whether it is going to be worthwhile before starting an application for a patent.

It is more difficult to define invention than to list some of the things that are not considered invention, although even here there have been exceptions: small advances that would be obvious to a person having ordinary skill in the art; a mere exercise of skill; substitution of materials; making things portable or adjustable; change of degree or form; superior workmanship; unification or multiplication of components; new uses for something existing; an aggregation of existing things; or substituting mechanical action for manual. However, greater economy or efficiency and new combinations of old elements may be allowed, particularly if they solve a long-felt want as demonstrated by prompt and general adoption. It must be a new and useful process or device, and not merely an idea for such a process or device; it must be "new," it must be "useful," and it must be fully described in the application.

The first step, after deciding to file for a patent, is to establish the time of conception. In this country at the present time, the date of conception is more important than the date of filing. A written description, preferably with sketches, should be signed and dated in the presence of a witness. The witness should write some phrase such as "Read and understood by me on the date opposite my name." Choose a witness who is not likely to be a co-inventor, but one who can corroborate your own testimony as to conception, diligence, and successful testing.

A search of the patents already issued, called a *novelty* search, should be made to see whether your invention has already been patented. This is usually done by an expert in the field, and he should be supplied with an accurate description of the invention and given a limit on the time or money to be expended. There is no guarantee that every pertinent patent will be found, but a group of patent papers will be furnished that are close to the subject. This will help to determine whether it is worthwhile to file an application for a patent, and it may also be of benefit by showing how the claims can be made more definitive and perhaps stronger.

It is sometimes desirable to have a search made for other reasons. When no invention is involved, but the solution to a problem is being sought, it is often helpful to have a *state of the art* search made. This will reveal the many approaches to the problem that have been made by inventors, and it may suggest an answer that had not been previously considered. Another reason for making a patent search is to determine the *validity* of an existing patent. If a manufacturer has infringed, or intends to infringe, on an issued patent, he may want to determine the likelihood of its being upheld in a court of law. This kind of search can be very expensive, since it attempts to find prior conception through publication in foreign patents or even in technical journals. Any mention of the idea which predates the filing of the patent in question may be sufficient to invalidate it.

A patent consists of two main parts—the specification and the claims. The *specification* is for the purpose of teaching the reader how to make and use the invention, and it often has drawings to clarify the description. This makes the *claims,* which are the heart of the patent, more understandable. Some of the claims should be as broad as possible, and others will need to be more specific.

The contents of the application, including the serial number and the filing date, are kept confidential by the Patent Office until the patent is granted, unless there is *interference* due to two applications having similar claims being filed at the same time.

The Patent Office examiner will read the application when it reaches the top of the pile—perhaps several months after filing. He will search the patent files and the literature to see whether the features claimed in the application have been published before. The examiner may then reject certain of the claims and cite the references which support his opinion, in a letter called an *office action.* Nearly all applications are rejected after the initial filing. The office action should be answered within 6 months by a letter called an *amendment,* asking the examiner to cancel or reword the claims in question, or to divide out portions of the application as a separate filing. The amendment must be a bona fide attempt to advance the case, and not for the purpose of postponing the granting of the patent. This exchange of actions and amendments may be repeated several times until the examiner states that the rejection is final. An appeal can be made to the Board of Appeals if desired. The whole process takes an average of 2½ years.

A patent application must be filed within 1 year after the invention is made public, either by being described in a printed publication, being used in public, or being offered for sale. Otherwise the inventor loses his rights to a patent. Patents are granted for about two out of every three applications that are filed.

The *Official Gazette,* published weekly by the Patent Office, contains a claim and a figure from the drawings of each patent granted during that week, with other news and information pertaining to patents. The annual index gives the names of patentees in alphabetical order and a list of patents by subject.

Copyrights. To protect the "right to copy" for the author of literary, musical, artistic, or dramatic works, it is necessary to publish the work with the notice of copyright required by law, and then promptly send the necessary copies to the Copyright Office in the Library of Congress. If a work is published without the required notice, copyright is lost forever.

Information cannot be copyrighted; only the form in which it is presented can be. Facts that are expressed in different words, or figures and charts constructed from published material but in a different way would not be considered plagiarism. The term of a copyright is 28 years, and it can be renewed for an additional 28 years.

Lectures and other material intended for oral presentation may be copyrighted before publication, but if they are published at a later date, a second registration then becomes necessary. Labels can be copyrighted if they contain an appreciable amount of original text or pictorial matter. This does not mean mere listings of ingredients or contents, simple variations of typography or coloring; the text must be more like an extensive prose composition. Brand names, trade names, or slogans cannot be copyrighted, no matter how artistically they are done. To be copyrighted, labels must contain the word "Copyright" or the symbol © adjacent to the owner's name. The year date is not required on a label, as it is on literary works.

Unpublished manuscripts may be copyrighted, but they are protected by *Common Law Literary Property,* which is a matter of state law and remains in force as long as the work is unpublished. The penalties for plagiarism are severe, but the criteria for determining infringement are rather indefinite. A quotation of 50 to 200 words, if it were not the most essential part of the work, would not be considered plagiarism in most cases. However, a court could rule that the copyright holder had suffered, and award the statutory penalty of $1 per infringing copy.

Section 17

Cushioning

Selecting the best type of cushioning and the right amount for a particular packaging application depends, in most instances, only on three or four bits of information. There are even some rule-of-thumb figures that can be substituted for some of the variables, so that the process of designing a package for a fragile item need not be very complicated. This is not to deny that there are some complex forces involved; for a deeper study of the subject we recommend the classic report of Dr. R. D. Mindlin of Columbia University, "Dynamics of Package Cushioning," published by Bell Telephone Laboratories, Inc., in 1945. In most cases, however, a simplified approach to the problem will provide a completely adequate solution.

First we must know something about the fragility of the item to be packaged; then we must consider the hazards it will be exposed to; and finally we must understand the characteristics of the cushioning materials that are available, so that a choice can be made of the most economical material that will do the job. Starting with the item itself, it is essential to know how much abuse it can take without being seriously damaged. If we can translate this into a *G factor,* we will have one of the elements we need for solving our problem. The G factor is the minimum force required for damage, given as the number of *g* (acceleration of gravity, or 32.19 ft per sec per sec). This is another way of expressing the minimum drop height that would break the item without any cushioning, in terms of the number of seconds it takes to reach the ground; or to put it another way, the number of times its own weight that it takes to crush an item is its G factor. This last, however, does not take into account any unsupported parts that would be damaged by whip action. (See Tables 1–3.)

TABLE 1 Typical Forces on Packages in Transit

Freight car in motion	$1\frac{1}{4}$G
Freight car coupled at 9 mph	18G

TABLE 2 Normal Vibrations in Various Methods of Transportation

Freight car in motion	$2\frac{1}{2}$–$6\frac{1}{2}$ cps
Ships	10–100 cps
Aircraft, propeller-driven	20–60 cps
Truck on rough road	20–70 cps
Truck on smooth road	70–200 cps

TABLE 3 Examples of Fragility Factors

Precision instruments	15G
Electronic equipment	25G
Typewriters and cash registers	50G
Television receivers	75G
Refrigerators and washing machines	100G
Machinery	125G

We also need to know the *weight* of the object being packaged and the *area* to be supported by the cushioning. If the area is not the same on all sides, the safest thing is to take the smallest area. Dividing the weight by the area W/A gives the *static stress* in pounds per square inch.

Let us now take up the matter of the potential *hazards* that our package is likely to meet in shipment. A package under 10 lb can be tossed like a basketball, but a heavier unit can be raised only waist-high and then dropped. Since most of the charts of cushioning materials are based on a 30-in. drop, we might as well take this as our standard, realizing that for a light package we can use 25 or 33 percent more cushioning, and for a very heavy item that much less than the charts indicate. (See also Table 6 in Sec. 8, "Plastics," page 8-31, for polyurethane foam.)

The next piece in our puzzle is the thickness of a particular *cushioning material* that will be necessary to satisfy the requirements given above. This information is given in the G vs. W/A curves which are supplied by the manufacturers of the various materials. A few of these curves are given for reference in Fig. 1. Draw a horizontal line from the G factor (fragility) of the item to be packaged, and a vertical line up from the W/A figure that we derived from the weight and area, and note the point where they intersect. Any curve which passes below this point indicates that the material can be used in that thickness.

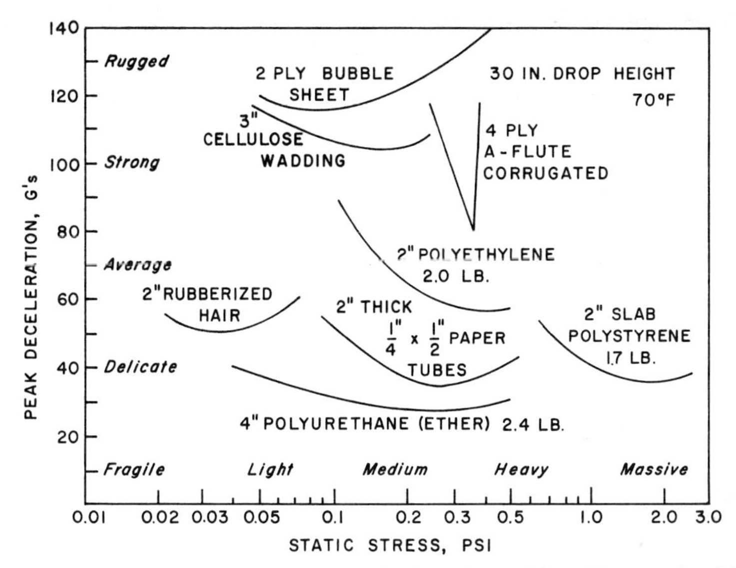

Fig. 1. Cushioning effectiveness. A random selection of materials to illustrate the different behavior of various materials. Other thicknesses, drop heights, and temperatures will move the curves up or down, but will not greatly affect their positions from left to right.

A sample problem may help to illustrate this:

weight = 15 lb
bearing area = 10 × 10 in.
fragility = 50G
potential hazard = 30-in. drop

$$\frac{W}{A} = \frac{15}{10 \times 10} = 0.15 \text{ psi}$$

Referring to Fig. 2 for polyurethane foam, 2 lb per cu ft density, we find where 0.15 intersects the line from 50G. We see that the curve for 2-in. thickness dips below this point, indicating that this will satisfy the requirements. A few minor considerations ought to be taken up at this point. One is the matter of *creep* or cold flow under a load over a long period of time. Whatever this amounts to should be added to the thickness of the cushioning material. Thus if a 2-in. thickness of urethane foam would be satisfactory for immediate shipment, but in long-term storage would settle down 33 percent, then we should use a 3-in. thickness so that there will still be the required 2-in. thickness after it has settled.

Other points to be considered are: (1) the effects of abrasion of the cushioning material on the item; (2) the settling of the cushioning material from repeated impacts; (3) resonance or periodic vibrations along

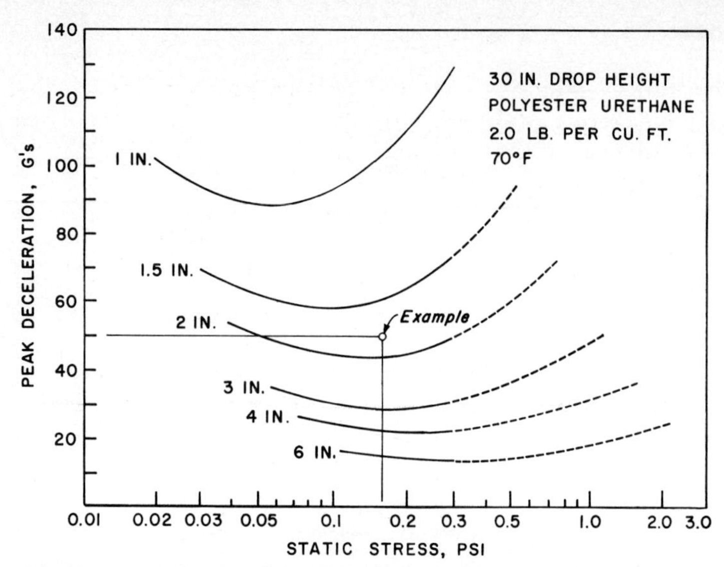

Fig. 2. Polyurethane curves. Dynamic cushioning of polyurethane in various thicknesses. If the fragility of a particular item is such that it can withstand 50 G's and the weight per unit area of bearing surface is 0.15 psi, the intersection will be above the curve for 2-in. thickness, as shown. Any curve below this point may be used. Creep, which becomes a significant factor in the dotted portion of the curve, may be as much as 75 percent loss of thickness.

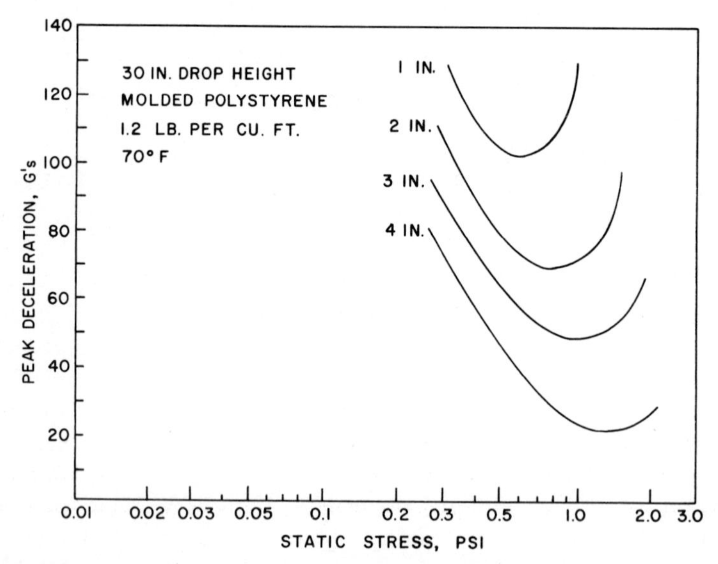

Fig. 3. Polystyrene curves. Dynamic cushioning of molded expanded polystyrene in various thicknesses. Find the point where the fragility of the item intersects the dead weight. The curve that comes closest, but dips below this point, indicates the most efficient thickness of 1.2-lb density material for a 30-in. drop height. If the height is reduced by half, static stress can be doubled. If the density is increased, it improves the load-bearing capacity only slightly, but it also reduces the shock absorption by a small amount.

with harmonics, which are rarely a serious problem, but may help to explain some of the unpredicted failures that occur every once in a while; (4) the mass of the cushioning itself; this is a negligible factor, but for precise calculations one-third of the weight of the cushioning should be added to the weight of the article; (5) internal damping of shock vibrations; this is a plus factor and varies with different materials, but it can be ignored for all but sophisticated investigations; and (6) temperature and humidity, which may alter the protective qualities of certain materials. Plastics tend to stiffen at low temperatures and soften at elevated temperatures. Cellulosic materials are affected by moisture, and if this is expected to be a factor, the supplier of the cushioning should be consulted for data under these conditions. (See Fig. 3.)

Section 18

Test Methods

Introduction 18-1
Fundamentals 18-1
Current Practice 18-2
Design of Experiments 18-3
  Kinds of Tests 18-4
  Number of Tests 18-4
Brief Descriptions of Common Tests ... 18-4

INTRODUCTION

An essential part of any packaging program is the testing and evaluation of the complete packaged unit, as well as the various components. It is good economics to determine the optimum design in the beginning and to maintain uniform performance throughout the life of the package. A good test program will indicate the results to be expected in the field, and it will yield dividends far in excess of its original cost; and good management demands an objective evaluation of every step in the packaging operation.

FUNDAMENTALS

Testing must be planned ahead of time, to avoid unnecessary work, save time, and ensure satisfactory results. This means deciding what the problem is, what method is best for obtaining the desired information, and what materials and equipment will be needed. The method selected ought to be consistent with standard practice in other organizations, it should be reproducible by other people at other times, and it should be free of personal bias as much as possible. The work must be carefully and thoroughly done, and accurate records must be kept. Interpretation of the results is especially important, and this takes knowledge and skill that come only with considerable training and experience.

In the design of tests, one should assume that the methods and results will be challenged at some future time, and one should so plan the work that the results will be defensible under any circumstances. This takes a little forethought and meticulous attention to details. Without advance planning, all the effort and expense for the tests may be completely wasted.

A distinction should be made at this point between performance-type tests and properties tests. It is always preferable to evaluate packages in their final form and under actual conditions of manufacturing, storage, transportation, and end use. This may not always be practical, but every effort should be made to approach these ideal conditions in testing the performance of a package, and to follow up at a later date with a more realistic test if necessary. But sometimes answers are needed in the development stages, when materials and environment are not accessible, or in some cases time will not permit the kind of testing that might be desired. Then it becomes necessary to do the work piecemeal and to extrapolate the results in an attempt to predict the final performance of the package, on the basis of the various properties.

Another basic decision to be made at the start of a test program is whether to seek average typical results, or to determine the magnitude of the effects from the most extreme conditions and then try to interpolate the results in between. The resolution of this dilemma must be found in the amount of product and packaging materials that can be used for test purposes. If the quantity is small, there is a statistical risk that a serious defect may not be discovered in trying to achieve a set of average results. For example, a small number of containers for glassware may pass a simulated average shipping test with no breakage, and yet have a potential for 10 percent damage that would show up only in a large-scale test. In this case it might be wiser to simulate the worst hazards of shipment and set the level of acceptance at a lower point; that is, use a height of 24 in. instead of 18 in. for a drop test, but allow a breakage rate of, say, 8 percent to be acceptable in the test.

CURRENT PRACTICE

The ultimate objective is to relate the test results to actual conditions, but this ideal situation can never be fully realized. It is better to recognize that laboratory conditions can never simulate actual experience in every detail, and to make the necessary allowances, taking advantage of the rapidity and reproducibility of the controlled environment in the laboratory. It is preferable to relate the results to a known standard, such as a package that is performing satisfactorily in the field, putting this through the same tests and using it as a control. It may be desirable in that case to run the tests "to destruction" to get a good comparison.

Accelerated tests are very valuable for getting information ahead of normal test results. Thus, higher temperatures will usually speed up chemical reactions, so that 1 month at 37°C may correspond to a year at room temperature, and 3/4 hr of vibration on a shake tester could be equivalent to a thousand-mile truck ride. It must be emphasized, however, that the results obtained under the artificial conditions cannot safely be related directly to standard conditions, but can be taken only as an indication of what may occur in normal situations.

There seems to be no agreement among the experts as to what constitutes a good test program. Although several trade associations have instituted a series of test methods intended to be used as reference standards, most laboratories continue to set their own procedures.

Certain criteria have been established by government agencies, such as the Department of Transportation's specifications for containers to be used with hazardous materials. In order to reduce damage claims, the Uniform Freight Classification has also set minimum standards for corrugated boxes and other containers that may be shipped by rail.

DESIGN OF EXPERIMENTS

Packaging tests must stand up to the scrutiny of anyone who may wish to challenge the results. If the tests are planned with this in mind, the worker will not only earn the confidence of others, but will do better work with less effort and will have the satisfaction of knowing that his answers are right ones. The techniques available for this purpose are mainly in the area of statistical analysis, which can become highly complex, but for our purposes it need not be too difficult for the average technically minded person.

Since tests and observations of results are often influenced by the individual's own experience and opinion, it becomes necessary to eliminate these human factors as much as possible in order to get results that are purely objective. There are three important decisions to be made at the outset, if we are to have a well-designed experiment: first, the question to be answered must be objectively formulated and clearly stated; second, the choice of experimental method must be made; and third, the number of tests to be performed should be predetermined. All this *must* be done before the test is started, as any change that is made from the original plan during the course of the tests may throw doubt on the validity of the final results.

Kinds of Tests. There are many kinds of tests for many different purposes. Our concern here is for tests to be used in the development of package structures, and the following list will be devoted mainly to this type of evaluation. A brief mention of other tests would include *subjective* testing that takes in color perception, trademark recognition, psychological impact, and similar consumer reaction tests. *Quality control* is a highly specialized area which makes use of probabilities and statistics to apply limited information to large amounts of material. *Process* tests are used to judge the merits of materials or structures in the course of their fabrication, so that adjustments can be made toward a more suitable end product. All these are important in their places, but a list of such tests would be endless. We will therefore confine ourselves to those tests likely to be of interest to the packaging practitioner.

The following list contains a brief description of some of the methods used for evaluation of packages and packaging materials. For more complete descriptions we refer the reader to the publications of TAPPI, ASTM, PI, and government agencies. The ASTM numbers given with each of the following descriptions may not always refer to similar tests; they are included to help the reader find other sources.

Number of Tests. Sufficient tests should be performed to develop a pattern of results. When the answers are clearly repeating the results of previous tests, it can be said that a pattern has been formed, and nothing will be accomplished by continuing with more of the same tests. A statistical examination of the results, as described in Sec. 19, "Quality Control," can be used to confirm that sufficient data have been collected to support the conclusions.

BRIEF DESCRIPTIONS OF COMMON TESTS*

Abrasion. See *Rub Resistance.*

Antistatic. See *Static.*

Beach Puncture. The corner of a cube, mounted on a pendulum, is swung through the material. The upward swing compared with the downward swing gives the energy absorbed. ASTM D781.

Bending. See *Score Bend.*

Blocking. The lowest temperature at which a material adheres to itself under pressure. ASTM D918.

Boil-in. A test for boil-in-bag structures. One-inch strips are sealed

* ASTM numbers are for reference only and may not be equivalent.

across, and a weight puts peel stress on the bond in wet steam at 250°F for 30 min.

Bond. Scotch tape is rubbed down on, and perpendicular to, a knife cut, and quickly pulled away at right angles to the surface. There should be no separation of layers where they are joined.

Brightness. Reflectance of a surface compared with magnesium oxide and reported as a percentage. Thus 90 brightness is 90 percent of the reflectance of magnesium oxide. ASTM D985.

Burst. See *Mullen Test.*

Charpy Impact. A notched bar of plastic is supported at both ends with the notched side down. A falling weight strikes the bar over the notch. Results are reported as energy absorbed. ASTM D256.

Clarity. Ratio of light transmitted through specimen to light transmitted through air, times 100. Grandiner-U.S.I. meter sometimes used. ASTM D1746.

Cohesion. See *Tensile.*

Cold Flow. See *Creep.*

Color. Value, hue, and chroma are compared with a standard in a north light. Gloss, transparency, and opacity also are noted. Tristimulus values are measured by instruments with filters that pass only X (red), Y (green), and Z (blue) elements. Metamerism is a match under one light and a mismatch under another light.

Compression. A fibreboard box is sealed as though for shipment, usually without contents. A uniformly distributed load is gradually increased until sudden buckling occurs. Total load at failure is reported.

Creep. Also called cold flow or deformation under load. Plastic is subjected to steady tension or compression at a constant temperature. Movement is plotted against time, and the slope of the curve is reported at the beginning, in the middle, and near failure.

Crushing. See *Compression.*

Delamination. A knife cut is made through one layer of substrate. Pressure-sensitive tape is rubbed down across the cut. The tape is quickly pulled perpendicular to the surface. Any separation is reported as failure.

Density. A powder is carefully sifted until it half fills a 100-ml graduated cylinder. The cylinder is dropped 1 in. onto a wood surface three times. Final volume divided by weight = density in milliliters per gram.

Drop. A fibreboard box, prepared as for shipment, is dropped on one bottom corner at the joint, on the three edges radiating from this corner, and on the three flat faces adjacent to this corner. The distance for all drops is 30 in. for packages weighing up to 10 lb, 24 in. up to 20 lb, 18 in.

up to 50 lb, and 12 in. for packages weighing up to 100 lb. ASTM D775.

Bags are tested with a flat drop, alternating front and back, starting at 24 in. and increasing 6 in. with each drop. The seventeenth drop is at 10 ft. Highest drop without failure is reported. ASTM D959.

Durometer. A special point, which extends below a pressure foot, is pressed against rubber, or other resilient material to be tested, which is at least 1/4 in. thick. Penetration of the spring-loaded indenter is read on a dial with a scale from 0 to 100. ASTM D1706.

Elongation. The maximum stretch in a film that will still return to its original size, as a percentage of original length.

Extraction. One hundred square inches of plastic is exposed to 200 ml of *n*-heptane at processing temperature. If more than 5 mg is extracted, it is reported in milligrams per square inch. ASTM F34.

Fade. Color stability is checked by exposure to ultraviolet light. An open-flame sunshine carbon arc is about 3 times as strong as sunlight. An enclosed UV carbon arc is 7½ times as strong as sunlight. Forty hours of sunlight with one half covered with black paper should not cause a marked difference between the two halves.

Fatigue. Number of cycles of fluctuating stress, below the elastic limit of the material, that will produce failure.

Fingernail Test. See *Hardness.*

Flame Treatment. Plastic bottle is dipped in distilled water. It should not shed water in less than 15 sec.

Flash Point. The temperature at which a vapor will ignite from an open flame.

Flat Crush. Ten square inches of corrugated fibreboard is loaded at a rate of 1 in. per min until the flutes are crushed. The maximum force is reported in pounds per square inch.

Flavor. Two identical standards and one test specimen are examined by a test panel. If the test specimen is chosen as the odd sample significantly more often than one-third of the time, it is not a good match. In preference tests two samples are used. A preference by 75 percent of the panel is significant, 80 percent is very significant, and 85 percent is exceptional.

Gas Transmission. Volume in milliliters, per 100 sq in. per 24 hr at 1 atm pressure and 75°F, passing through a film 1 mil (0.001 in.) thick. ASTM E96.

Gloss. Percentage of light reflected at a 70° angle. ASTM D1223.

Grease Resistance. Time required to penetrate folded or unfolded material by an oily or greasy product with dye added for test purposes. Turpentine is used as a reference standard. ASTM D722.

Hardness. Well-sharpened pencils of various grades are used to scratch the surface. The softest pencil that leaves a permanent indentation is reported, such as 3H hardness. See also *Rockwell Hardness.*

Haze. The percentage of scattered light that deviates more than 2½° from a straight path while passing through a transparent film. ASTM D1003.

Heat Distortion. Temperature at which a standard-size bar deflects 0.010 in. under a stress of 66 psi.

Hoop Stress. Circumferential stress in a cylinder due to internal pressure. Can be calculated by Barlow's formula:

$$\text{Hoop stress in psi} = \frac{\text{pressure in psi} \times \text{outside diameter in inches}}{2 \times \text{wall thickness in inches}}$$

Identification. See "Liners" in Sec. 9, "Closures, Applicators, Fasteners, and Adhesives," page 9-16, and Tables 15 and 17 in Sec. 3, "Films and Foils," pages 3-51 and 3-53.

Incline Impact. A dolly is pulled up a 10° incline until it is 3 ft above the floor. A package is placed on the dolly and allowed to ride down until it strikes a solid wall at floor level. ASTM D880.

Insect Infestation. Packages are exposed to 500 insects per cubic foot, for 6 months, with adequate ventilation and culture media to sustain them. Saw-toothed grain beetles, confused flour beetles, and flat grain beetles are generally used. Penetration by five or more is considered failure.

Izod Impact. A bar of plastic 2½ in. long, with a sharp notch near the middle that reduces the thickness to 0.400 in., is held rigidly by one end up to the notch in a vertical position. A pendulum strikes above the notch on the notched side. The upward swing compared with the downward swing gives the energy absorbed, reported in foot-pounds per inch of notch width. ASTM D256.

Label Adhesion. Labeled containers are aged for 1 week and then put into a desiccator for 24 hr. Removal of label should cause some fiber tear.

Mar Resistance. See *Rub Resistance.*

Melt Index. The amount of plastic in grams that can be forced through an 0.0825-in. orifice in 10 min, when heated and under pressure of 2,160 g.

Modulus of Elasticity. Ratio of stress to strain. Rate of deflection as a function of the applied force.

Moisture-Vapor Transmission. See *Gas Transmission.*

Moldproofness. Samples in a petri dish floating in water at 80°F in the vicinity of moldy bread should not show growth within 30 days.

Mullen Test. Pressure in pounds per square inch required to force a rubber diaphragm through a round hole against a specimen of paper firmly clamped around the edges of the hole. Results are reported as Mullen units. ASTM D774.

Notch Sensitivity. A measure of reduction in load-carrying ability caused by scratching or notching.

Odor. Shredded material with a small amount of water is warmed in an oven, in a quart jar. Panel members smell the material while it is warm and judge concentration of odor.

Oil Penetration. See *Grease Resistance.*

Organoleptic. See *Flavor* and *Odor.*

Parcel Post. A 4 ft drop test on corners and flat surfaces is recommended by the Post Office.

Peel Bond. Separation of plies at the adhesive interface without affecting the ply. Test is performed at an angle of 180° and at a rate of 6 in. per min.

Permeability. See *Gas Transmission* and *Porosity.*

pH. Acidity or alkalinity on a scale of 1 to 14 in which 7 is neutral, below 7 is increasingly acid, and above 7 is increasingly alkaline. The pH is the logarithm of the reciprocal of hydrogen-ion concentration in gram atoms per liter.

Pinholes. Number of holes per square foot visible to the unaided eye when material is held in front of a strong light.

Porosity. Time in seconds for a column of air to be forced through 1 sq in. of paper under a constant weight. Usually performed on a Gurley densometer.

Puncture. See *Beach Puncture.*

Rigidity. See *Stiffness.*

Rockwell Hardness. Penetration of a steel ball into a material that is at least 1/4 in. thick, measured on a dial indicator 15 sec after the load is removed. Size of ball and load varies, and is reported accordingly as B, C, or R scale. ASTM D785. See also *Hardness.*

Rub Resistance. Number of strokes for obvious smearing or removal of ink or coating when material is rubbed against itself, or against kraft paper, under a 4-lb weight. A Sutherland ink rub tester is sometimes used for this test. Suggested minimums are 50 strokes for grocery items, 100 strokes for pharmaceuticals, and 200 strokes for cosmetics.

Score Bend. Resistance to bending, or springback, of a folding carton score. Performed on a PCA (Ohio) score bend and physical force tester.

Scotch Tape. See *Delamination.*

See-through. See *Clarity.*

Slip. Tangent of the angle at which one layer of film under a weight begins to slip on itself as it is slowly tilted. ASTM D1894.

Smoothness. Measure of the escape of air between a paper surface and a metal plate. A Gurley instrument is sometimes used.

Softening. See *Vicat Softening.*

Specific Gravity. Density (mass per unit of volume) of a material, divided by that of water. Grams per cubic centimeter is the same as density, for all practical purposes.

Static. A plastic container is rubbed three times with cleansing tissue. The static charge is measured frequently with a Kiethly voltmeter to determine at what point the charge decays to less than 2,000 volts. The static dissipation time is reported.

A soot chamber uses a fan to circulate smoke from burning filter paper soaked with toluene. Plastic containers are rubbed with tissue and placed in the chamber for 24 hr at 15 percent relative humidity before introducing the soot. Attraction for the soot is rated from none to heavy.

Stiffness. Force required to bend sample at a specific angle. A Taber stiffness tester is sometimes used. ASTM D747.

Stress Cracking. A narrow strip of sheet plastic is kept under bending stress at 120°F while immersed in a liquid (Igepal CO 630 is standard). Any visible crack within 1 week is reported as a failure.

Sutherland Ink Rub. See *Rub Resistance.*

Taste. See *Flavor.*

Tear. Force required to continue a split in paper or film. Usually performed two ways, in the machine direction and in the cross direction. ASTM D689.

Tensile. Force required to stretch material at a constant rate to the breaking point. Heavy sections are reported in pounds per square inch of cross section; thin webs in pounds per inch of width. An Instron tester is sometimes used for this test. ASTM D638.

Torque. Screw caps require a rotating force to effect a seal on a container. The removal torque is measured in inch-pounds, 5 min after application. For metal caps on glass containers this should be at least one-fourth the cap size. Example: 6 in.-lb for 24-mm cap.

Vibration. A filled package is vibrated at 250 cycles per min (or 1*g*) for 45 min, on its normal bottom. Condition of contents is reported as salable or unsalable.

Vicat Softening. The temperature at which a rod of 1 sq mm cross section can be forced 1 mm into plastic. ASTM D1525.

Water-Vapor Transmission. See *Gas Transmission.*

Wax Pick. Resistance to picking of paper fibers when tested with a graded set of waxes. The higher the wax number, the greater the resistance.

Wet Strength. Strength of a material or an adhesive bond after soaking in water 24 hr.

Section 19

Quality Control

Introduction	19-1
Frequency Distribution	19-3
Process Control	19-4
AQL	19-5
Significance of Results	19-5
Sampling	19-7
Size of Sample	19-9
Specifications	19-10
Standard Samples	19-10
Inspection	19-10
Disposition of Rejected Material	19-11

INTRODUCTION

In the process of designing a new package, seeing it through the various stages of development, and ultimately following it to its final destination, the packaging engineer may have to deal at some point with the problems of quality control. In this section we will discuss a few of the basic principles in order to acquaint the reader with the terminology and techniques that are used for this work. It is not possible to cover such a broad subject fully in these few pages but it is felt that a brief treatment may help to familiarize the reader with the broad aspects of inspection and control. For a more comprehensive coverage, we suggest the *Industrial Engineering Handbook* edited by H. B. Maynard (2nd ed., 1963), which has a section devoted to this subject, or *Statistical Quality Control* by E. L. Grant (3rd ed., 1964) both of which are published by McGraw-Hill Book Company.

Quality control is a system of *specifications, inspections, analysis,* and *recommendations.* In the first phase it involves the establishment of criteria for judging the attributes of a product or a package, such as function, appearance, safety, or economy, which make it suitable for a particular purpose. The specifications set up for this purpose are based on knowledge and experience, plus some judgment as to the relative importance of each attribute. It is a serious mistake to set rigid limitations where they are not necessary, just as it is wrong to provide more latitude than the situation warrants. Tight restrictions will not only increase costs and reduce productive output, but also shift the emphasis away from some of the more critical points, which may result in a general lowering of overall quality. The fewer the number of different things that need to be measured, the greater will be the attention that can be given to each of them individually. When more information becomes available from additional experience, there should be as much effort toward relieving some of the specifications which may be too tight as there is in holding close tolerances and in adding new specifications.

After specifications have been established, it is necessary to inspect the products and packages against these specifications to see whether they are within the intended limits. It is not usually necessary to test every piece; for example, in a match factory it would not be desirable to strike every match to ensure that it is functional. It would be far better to try one out of every thousand, and assume that the rest are of equal quality. The extent to which inspection should be carried and the degree of confidence that can be derived from the results require the application of statistical principles.

When the inspection of a group of items is completed, the data must be organized and examined to determine what action should be taken. The dictionary defines control as exercising restraint or direction, or keeping within limits. Quality control implies that quality will be kept within the limits set forth in the specifications. This is done by applying the necessary action to ensure that each unit produced will be satisfactory. Hence products and packages coming from an operation that is conducted in accordance with good quality control techniques can be accepted without further inspection. It is assumed that the instruments are accurate, that the work is carefully done, and that good records are kept. In addition to the instruments' being accurate, they must have the proper degree of sensitivity. There is a *rule of ten* which states that an instrument should have divisions which are 10 percent of the unit of measure given in the specification. If the specified limits are given in thousandths of an inch, the finest division on the measuring instrument should be one ten-thousandth.

It is not enough to inspect and analyze; there must also be a recommendation for action. Either the pieces will be accepted and used, or they will be rejected and returned. There are also some intermediate courses that could be taken in certain cases. The pieces could be accepted with a warning that corrective steps be taken in the future, or they could be rejected with the stipulation that sorting or rework is required to make them acceptable.

Throughout the whole process of quality control there is an underlying discipline that is based upon the mathematical principles of probability. The true value of quality control is inherent in the ability to apply small bits of information to a large system; to show trends and to predict consequences from a limited amount of information; to determine the condition of large quantities of material from small samples, and to know with a precise degree of certainty just how much this information can be trusted. To understand the methods that are used to obtain this knowledge, we will now look at some of the techniques used in statistics.

FREQUENCY DISTRIBUTION

If a group of pieces are accurately measured, there will be found to be small differences among them. In spite of our best efforts to make pieces exactly alike, there will always exist some small variations that cannot be eliminated. When these differences have been reduced to the very smallest variations that are possible, they are called the "inherent variability." In order to know whether this minimum has been reached, the frequency of the variations can be plotted as a distribution curve. A normal curve has only one peak and is symmetrical about the center; that is, the greatest number of measurements should fall in the center, and fewer and fewer points should be found as the distance from the center increases. The curve that is formed in this way is sometimes called a tin hat curve, because of its shape. (See Fig. 1.)

A useful way to express such a distribution pattern is by means of the "standard deviation" which is the square root of the mean of the squares of the individual deviations, generally expressed by the Greek letter σ (sigma). The standard deviation measures the expected spread of

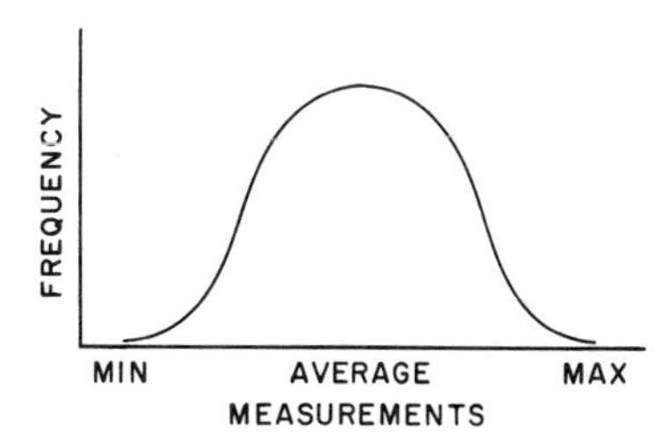

Fig. 1. Frequency curve. Normally any value such as length, weight, or other measurement will be found most frequently close to the desired figure. The number of pieces that are above or below this value will be fewer and fewer, as the deviation becomes larger. When these numbers are plotted on a graph, we get a histogram that follows the general shape shown here.

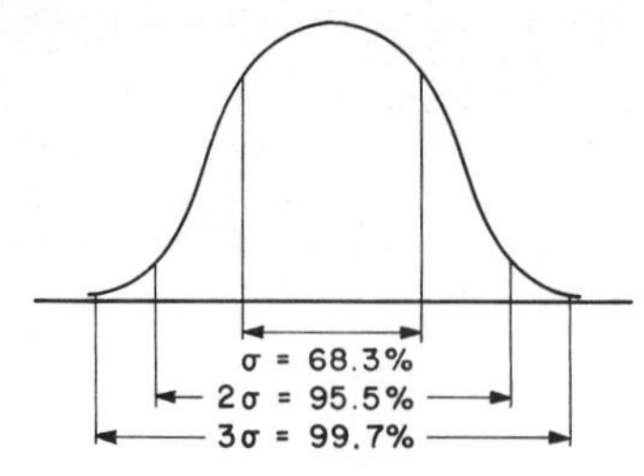

Fig. 2. Standard deviation. The root-mean-square deviation of a set of measurements provides the expected distribution of those measurements, represented by σ. Any measurements outside the three-sigma limits indicate an "unnatural" condition that is capable of being controlled.

measured values as shown in Fig. 2, and it is used to set the limits for control. If the standard deviation is applied to a process and more than a small percentage of products are found to be outside these limits, it can be assumed that something is happening beyond the inherent variability previously mentioned and that it can be controlled. The "three-sigma" limits, which can be expected to include 99.73 percent of all measurements, are called the "natural" tolerance.

Properties which can be measured accurately are called *variables.* These include such things as dimension, strength, chemical properties, or electrical characteristics. Sometimes there are qualities which cannot be measured, but are simply judged as good or bad, acceptable or defective, go or no-go. These are called *attributes.* Measuring quality on a continuous scale, by variables, gives a better indication of the true quality of the item. It tells you whether the quality is at the midpoint or on the low side and just barely passable. To get the same degree of control on a go or no-go basis would require at least 10 times as many pieces. Although it is possible to plot a curve for attributes similar to the one for variables by means of the binomial distribution technique, or the Poisson distribution method, it is generally considered more practical to control by variables, where it is necessary to watch trends.

PROCESS CONTROL

If we turn the distribution curve on its side, and project the lines from the middle and from the three-sigma points, we have a control chart. (See Fig. 3.) The value of a control chart is that it monitors a manufacturing process to ensure that the quality of the product is maintained at a constant level. By plotting the measurements of each of the samples as points on a chart, moving from left to right with time, we can see not only whether they are within limits, but also whether there is a trend in one direction or another. Such a trend might indicate potential trouble in the making, and an investigation should be made to see whether the trouble can be headed off. This is one of the values of maintaining such a chart: that problems can often be detected before they become serious.

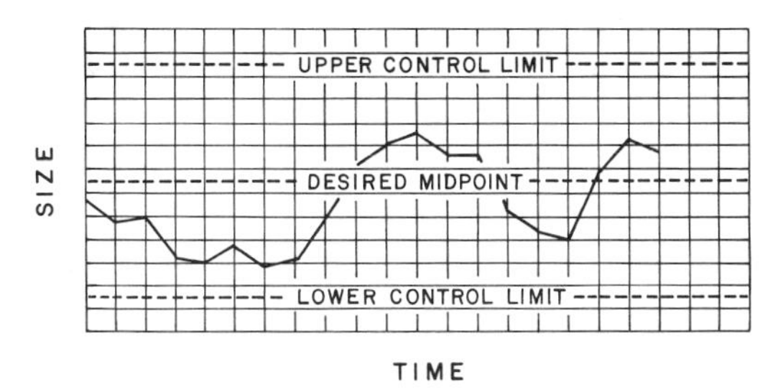

Fig. 3. Process control. The average size variations of a container can be plotted on a continuing basis to detect any trend which might foretell a problem in the making.

If any points are plotted which are beyond the three-sigma limit, we can say with 99.73 percent certainty of being correct that the cause of the variation was something other than normal; this is called an "assignable cause." Also it is a general rule that if seven consecutive points are all on one side of the midpoint, an assignable cause is at work.

AQL

Purchased material is inspected basically to determine whether it is acceptable or rejectable. The criteria that are used for measuring this material are the ranges that are set up in the specifications. However, it is not enough to establish the limits within which the measurements must fall; we must also predetermine what the inspector shall do if he finds some samples that are outside the specifications.

Whether attributes or variables are being checked, there are sure to be cases in which an occasional piece is outside the limits. The inspector will need to know at what point to consider the lot unacceptable. For this he requires an acceptance quality level (AQL), which is a percentage figure for the allowable number of defects. This will vary for different types of defects, depending upon their seriousness. In Sec. 6, "Glassware," for example, the types of defects that are found in glass bottles are classified as "critical," "major," and "minor." In such a case we might choose 0.25 percent for critical defects, 1.5 percent for major defects, and 4 percent for minor defects.

SIGNIFICANCE OF RESULTS

In any type of testing it may be necessary to determine whether enough tests were made to support the conclusions that were drawn. This can be calculated by a form of mathematics that is used mostly by statisticians. If the variations between individual results are so large as to throw doubt on the final average, the t test can be applied to see whether the variations could be considered normal.

A set of measurements can be called significant if it has a low probability of occurring by chance or, to put it another way, if it has a high probability of being caused by some outside force. The probability of a coin falling heads is ½, and of a die rolling one particular number is ⅙. The total of all the probabilities is always 1; that is, the probability of a die rolling any of the six numbers is the sum of each of the probabilities of rolling a particular number. If we wish to test a set of figures for significance, we can use the t test. t is the ratio of a variable quantity to the standard deviation of that quantity, usually written:

$$t = \frac{X}{S(X)}$$

Note that $S(X)$ does not mean S times X but rather the standard deviation of X. The arithmetical average of a set of data is $\bar{X}$, called "X bar." The average is simply the sum of all the measurements divided by the number of measurements. Then the pertinent t is:

$$t = \frac{\bar{X}}{S(\bar{X})}$$

The average $\bar{X}$ is easily calculated, but the standard deviation is a little more difficult.

Using the Greek letter Σ (sigma) to denote "sum of," and n to denote "the number of" measurements, we can find the standard deviation:

$$S(\bar{X}) = \sqrt{\frac{n\Sigma X^2 - (\Sigma X)^2}{n^2(n-1)}}$$

The measurements must be reduced to the simplest form, so that the differences show up as simple digits. For example, 2.498, 2.500, and 2.501 become −0.002, 0.000, and 0.001, which can be still further simplified to −2, 0, and 1. If we add these together, we have:

$$\Sigma X = -2 + 0 + 1 = -1$$
$$n = 3$$
$$\bar{X} = -1/3$$
$$\Sigma X^2 = 4 + 0 + 1 = 5$$
$$S(\bar{X}) = \sqrt{\frac{3 \times 5 - 1}{9 \times 2}} = \frac{3.74}{4.24} = 0.882$$
$$t = \frac{\bar{X}}{S(\bar{X})} = -\frac{0.33}{0.882} = -0.377$$

In Table 1 find the "degree of freedom" in the left-hand column, which is equal to $n - 1$ or, in this case, 2. Reading across from this

TABLE 1 *t* Test of Significance*

Degree of freedom†	Degree of confidence‡											
	0.90	0.80	0.70	0.60	0.50	0.40	0.30	0.20	0.10	0.05	0.02	0.01
1	0.158	0.325	0.510	0.727	1.000	1.376	1.963	3.078	6.314	12.706	31.821	63.657
2	0.142	0.289	0.445	0.617	0.816	1.061	1.386	1.886	2.920	4.303	6.965	9.925
5	0.132	0.267	0.408	0.559	0.727	0.920	1.156	1.476	2.015	2.571	3.365	4.032
10	0.129	0.260	0.397	0.542	0.700	0.879	1.093	1.372	1.812	2.228	2.764	3.169
20	0.127	0.257	0.391	0.533	0.687	0.860	1.064	1.325	1.725	2.086	2.528	2.845
40	0.127	0.255	0.388	0.529	0.681	0.851	1.048	1.303	1.684	2.021	2.423	2.704
120	0.126	0.254	0.386	0.526	0.677	0.845	1.040	1.289	1.658	1.980	2.358	2.617

* Lack of significance indicates more work to be done to find causes or to increase the degree of confidence. More complete tables for the value of *t* will be found in textbooks on statistics.

† Degree of freedom is one less than the number of measurements $(n - 1)$.

‡ Results should fall in the far left column to be of any value.

figure, find *t*, which in our case is 0.377; we find that this has a degree of confidence between 0.70 and 0.80, or about 0.75. Therefore a value of *t* as large as 0.377 could occur from chance causes nearly 75 percent of the time, and so it is not very significant. We want a value of less than 10 percent before we can read any meaning into our figures.

Evidence of a difference when none really exists is called a *type I* error. If the evidence indicates there is no difference when one actually does exist, it is known as a *type II* error.

SAMPLING

To determine whether a shipment of material is acceptable, it is necessary to know what percentage of defects would be found in the entire shipment. This can be calculated very accurately by examining a small amount and extrapolating the results, provided the methods that are used follow certain basic principles. The manner in which samples are taken is an important part of this process. Any shipping cases which show obvious shipping damage should be excluded from sampling. Except for these, all items in a shipment must have an equal chance of being selected for testing. This is important, and the validity of an inspection report is dependent upon the manner in which the samples were chosen. Any method which provides a truly random sampling is satisfactory, such as rolling dice, flipping a coin, spinning a wheel, or drawing numbers from a box.

A widely used technique for selecting samples is the table of random numbers. (See Table 2.) To use this table, start at any place and move in a straight horizontal, vertical, or diagonal line as far as possible, and then turn in either direction. This will tell you which pieces in a lot to take as samples. Each item in the lot, or population as it is called by the statisticians, is considered to have a number in sequence, that is, 1, 2,

TABLE 2 Random Numbers

65 48 11 76 74	17 46 85 09 50	58 04 77 69 74	73 03 95 71 86	40 21 81 65 44
80 12 43 56 35	17 72 70 80 15	45 31 82 23 74	21 11 57 82 53	14 38 55 37 63
74 35 09 98 17	77 40 27 72 14	43 23 60 02 10	45 52 16 42 37	96 28 60 26 55
69 91 62 68 03	66 25 22 91 48	36 93 68 72 03	76 62 11 39 90	94 40 05 64 18
09 89 32 05 05	14 22 56 85 14	46 42 75 67 88	96 29 77 88 22	54 38 21 45 98
91 49 91 45 23	68 47 92 76 86	46 16 28 35 54	94 75 08 99 23	37 08 92 00 48
80 33 69 45 98	26 94 03 68 58	70 29 73 41 35	53 14 03 33 40	42 05 08 23 41
44 10 48 19 49	85 15 74 79 54	32 97 92 65 75	57 60 04 08 81	22 22 20 64 13
12 55 07 37 42	11 10 00 20 40	12 86 07 46 97	96 64 48 94 39	28 70 72 58 15
63 60 64 93 29	16 50 53 44 84	40 21 95 25 63	43 65 17 70 82	07 20 73 17 90
61 19 69 04 46	26 45 74 77 74	51 92 43 37 29	65 39 45 95 93	42 58 26 05 27
15 47 44 52 66	95 27 07 99 53	59 36 78 38 48	82 39 61 01 18	33 21 15 94 66
94 55 72 85 73	67 89 75 43 87	54 62 24 44 31	91 19 04 25 92	92 92 74 59 73
42 48 11 62 13	97 34 40 87 21	16 86 84 87 67	03 07 11 20 59	25 70 14 66 70
23 52 37 83 17	73 20 88 98 37	68 93 59 14 16	26 25 22 96 63	05 52 28 25 62
04 49 35 24 94	75 24 63 38 24	45 86 25 10 25	61 96 27 93 35	65 33 71 24 72
00 54 99 76 54	64 05 18 81 59	96 11 96 38 96	54 69 28 23 91	23 28 72 95 29
35 96 31 53 07	26 89 80 93 54	33 35 13 54 62	77 97 45 00 24	90 10 33 93 33
59 80 80 83 91	45 42 72 68 42	83 60 94 97 00	13 02 12 48 92	78 56 52 01 06
46 05 88 52 36	01 39 09 22 86	77 28 14 40 77	93 91 08 36 47	70 61 74 29 41
32 17 90 05 97	87 37 92 52 41	05 56 70 70 07	86 74 31 71 57	85 39 41 18 38
69 23 46 14 06	20 11 74 52 04	15 95 66 00 00	18 74 39 24 23	97 11 89 63 38
19 56 54 14 30	01 75 87 53 79	40 41 92 15 85	66 67 43 68 06	84 96 28 52 07
45 15 51 49 38	19 47 60 72 46	43 66 79 45 43	59 04 79 00 33	20 82 66 95 41
94 86 43 19 94	36 16 81 08 51	34 88 88 15 53	01 54 03 54 56	05 01 45 11 76

3. . . . Numbers in the table that are not applicable or have been previously used are ignored, and the next suitable number is used Three- or four-digit numbers can be found by combining columns, as long as a consistent pattern is followed.

A "lot" for inspection purposes should be considered as the quantity delivered by one vehicle; it is usually recorded on a single lot ticket. The number of cases of small items to be opened for sampling should equal the square root of the total cases plus 1. If there are fewer than 10 cases, they should all be opened and sampled. If there are several die positions, or different mold cavities, they should all be represented in nearly equal quantities in the sample. If it is found that this is not the case, then additional samples should be taken until there is an adequate number for each position in the tool.

SIZE OF SAMPLE

How large a sample should be taken will depend upon the need for accuracy and reliability. One of the simplest methods for determining sample size is to take the square root of the total quantity in the lot. This tends to give sample quantities that are too small for small shipments, and too large in the case of large lots. At least 100 samples are needed to get statistical results that are meaningful. Even then, a lot with 2½ percent defective pieces will check out as though it were all perfect 4 percent of the time; that is, only 96 percent of the time will at least one defective piece appear in the sample group. Using 200 or more samples will greatly increase the chances of picking up the all-important critical defects. Many companies have adopted the government standards for sample size which are given in MIL-STD-105D. This is available from the Superintendent of Documents, Washington, D.C. 20402, for 40 cents. A simplified version of one of the plans is given in Table 3.

Conventionally a 95 percent probability of acceptance (5 percent probability of rejection) is considered the "producer's risk," while a 10 percent probability of acceptance is usually taken as the standard "consumer's risk."

TABLE 3 Single Sampling Plan for Normal Inspection*

Lot size†	Sample size‡	Acceptable quality level§								
		0.15	0.25	0.40	0.65	1.0	1.5	2.5	4.0	6.5
Over 2	2	1	1	1	1	1	1	1	1	1
8	3	1	1	1	1	1	1	1	1	1
15	5	1	1	1	1	1	1	1	1	1
25	8	1	1	1	1	1	1	1	1	1
50	13	1	1	1	1	1	1	1	1	2
90	20	1	1	1	1	1	1	1	2	3
150	32	1	1	1	1	1	1	2	3	5
280	50	1	1	1	1	1	2	3	5	7
500	80	1	1	1	1	2	3	5	7	10
1,200	125	1	1	1	2	3	5	7	10	14
3,200	200	1	1	2	3	5	7	10	14	21
10,000	315	1	2	3	5	7	10	14	21	21
35,000	500	2	3	5	7	10	14	21	21	21
150,000	800	3	5	7	10	14	21	21	21	21
500,000	1250	5	7	10	14	21	21	21	21	21

* Inspection by attributes (go or no-go) at level II (normal) for acceptance or rejection on the basis of a single sampling.

† Number of units in a batch or shipment.

‡ Number of units to be inspected that are selected at random from each lot.

§ Acceptable if the number of defective units in the sample are this amount or less, rejectable if above.

SPECIFICATIONS

The heart of a quality control system is the specification. This is what determines the degree of precision that must be used in the manufacturing department. Since it affects costs, production rates, sales volume, and many other aspects of the business, a great deal of thought should be given to these important documents. The specification is a communication from the designer to the purchasing department, and ultimately to the supplier. It also tells the inspector what to look for, which methods to use to check critical points, and how to deal with variations that may show up in the final product.

The supplier should be furnished an up-to-date copy of the specification, so that he can make a thorough inspection in his own plant before the parts are shipped. The customer's inspection department should really be considered as an extension of the vendor's quality control department. In some cases the certification of quality by the vendor is actually accepted at face value, and no further inspection is made by the purchaser. This has the obvious advantage of reducing the cost of inspection, but it can be done with confidence only if the supplier is known to be reputable and if there is a history of consistently good material.

STANDARD SAMPLES

The packaging engineer should furnish a sample of all new items to the inspection department. It may be a handmade sample or a first-piece sample from production tools. In subsequent runs the inspector will replace this with an up-to-date sample that can be used for reference.

It is not possible to cover every contingency in the written specification, but a standard sample will often help to answer any questions that may arise. The engineer should assume that the inspector has never seen the item before and therefore cannot be expected to know its function or how it relates to other components. The engineer should also realize that an inspector does not have the background information to determine the relative importance of the various characteristics of a part, and it is up to the designer to educate the inspector so that he can do an intelligent job of examining the items for defects.

INSPECTION

Each item in the sample should be examined visually for any obvious defects. One or more should be handled in a way to simulate its end use; for example, it should be fitted into its mating part, a cap should be tried on a bottle, and a box should be filled with its contents. Some defects do not show up when measured by gauges, but become apparent only when the parts are put to their intended purpose. Too often such faults are overlooked by the inspector because he is intent on following the letter of the specification, and he neglects to verify that the piece is functional.

The methods and tools to be used will depend on the item being tested. A few of the more common tests are listed in Sec. 18, "Test Methods." For repetitive tests an investment in limit gauges will be repaid many times over. There is also a great deal less fatigue in working with a good set of tools than there is in reading the fine graduations on an all-purpose tool for long periods of time. A standard sample should also be kept for reference. This can be a piece from a previous lot that has been processed and found to be satisfactory, or it may be an approved sample supplied by the engineer or designer.

Accurate records of quantities, dates, lot numbers, and other pertinent information should be kept. If necessary, a number should be assigned to the group of material that was tested, and samples should be filed under this number for future reference. The number of defects should be recorded so that trends can be noted and different sources can be compared. When defects are tabulated for determining acceptance or rejection, and there is more than one defect in a single piece, only the most serious defect should be counted and the others should be ignored. However, all defects should be entered in the records for analytical purposes.

DISPOSITION OF REJECTED MATERIAL

When material is found unacceptable, it should be ticketed with a distinctive marker so that it does not get out of the quarantine area. The purchasing department should be notified so that negotiations can be started with the vendor, and disposition arranged. The accounting department also should be informed so that payment can be held up pending settlement of the negotiations. Since the planning department may want to change their schedule or reorder the material, they also should receive notice.

Section 20

Machinery and Equipment

General Considerations 20-1
The Systems Approach 20-2
The Product 20-3
The Market.............................. 20-4
The Plant 20-4
Personnel 20-5
The Package 20-5
Layout and Selection of Equipment 20-6
The Machine Manufacturer 20-10
Types of Machines..................... 20-11
Operation Controls 20-16
Logic Systems 20-17
Standard or Custom Design......... 20-18
Speed or Flexibility 20-19
Foreign Sources 20-19
Ordering and Scheduling.............. 20-20
Financial Analysis..................... 20-20
Buy or Lease 20-25
Contracts and Specifications 20-26
Scheduling 20-30

GENERAL CONSIDERATIONS

In the production of packaged goods varying degrees of mechanization are required, depending on the type of manufacturing and the volume and diversity of the product line. Many businesses start out in a small way, with hand assembly of the product into its container, and as the volume increases, they find it necessary to add mechanical equipment. This may take the form of a conveyor belt or holding fixtures, or it may be a sophisticated assembling machine, according to the needs of the situation. In any case some decisions have to be made, and engineering work will need to be done. The following pages are intended to give a few pointers for the planning and evaluation of this type of mechanical equipment for the packaging department.

The packaging machinery business is a thriving industry at the present time. The total sales are approaching $500 million per year, and this could easily double in the next decade, at the present rate of growth. The largest volume in terms of number of units is in sealing machines such as tapers and staplers. When measured in terms of dollars, the largest category is liquid filling machines, followed by case handling equipment and labelers, in that order. There are about 100 machine builders who specialize in packaging machinery, along with countless other smaller shops that build custom machines on special order. The biggest purchasers of machinery at the present time are the beverage bottlers. The baking industry also is a large user of automatic equipment, mostly wrappers and baggers. (See Table 1.)

TABLE 1 Packaging Machinery Annual Production

Type of machine	Machines built	Average cost*
Gluers, tapers, staplers	13,500	$ 175
Labelers	9,700	2,600
Code markers	7,000	1,250
Shrink wrappers	5,300	2,250
Heat sealers	3,400	1,250
Bag fillers and closers	2,900	3,750
Liquid fillers	2,600	18,000
Cartoners	2,000	8,800
Wrappers and bundlers	1,600	10,500
Case sealers and loaders	1,400	12,500
Vacuum formers	1,300	6,400
Form-fill-seal	1,100	25,000
Dry fillers	1,000	8,250
Count fillers	900	3,600
Viscous fillers	800	10,000
Cappers	700	7,650
Checkweighers	600	6,500

* 1967 sales figures divided by shipments for the same period.

As machinery becomes more sophisticated, with better controls and higher speeds, the cost must necessarily go up accordingly. Not only does the initial cost go up as these machines become more complicated, but the maintenance costs are higher and the mechanics who do the setting up and adjusting must be better trained. It also follows that downtime is a much more significant factor with an expensive piece of equipment, and the planning and scheduling functions take on added importance. The trend is definitely in this direction, and the demand for higher speeds, tighter quality specifications, and faster changeovers makes it increasingly necessary to do a thorough analysis before committing funds for this purpose.

The Systems Approach. As machine speeds increase, and light weight containers replace older, heavier packaging components, the need for an overall coordinated approach to production operations becomes important. The incompatibilities of the different units in an assembly line may not be apparent at slower speeds, but when we get to higher rates of production, particularly with unstable plastic containers or where glass breakage is a factor, a smooth uninterrupted flow becomes very essential to an efficient operation.

Not only must the various machines be coordinated, but there should also be a good marriage of packaging materials with the equipment for combining them into the finished units. This may require some redesign of the package configuration, and the selection of materials might have to be made on the basis of machinability rather than the aesthetic qualities. (See examples in Figs. 10 and 11 in Sec. 6, "Glassware," page 6-18.)

There are three different ways of achieving this type of coordinated operation: (1) by balancing the various pieces of equipment so that they are synchronized in a close-coupled, smooth-flowing production line, (2) by developing special machines that can do the complete packaging operation from start to finish all on one machine, and (3) through changes in the design of the package to simplify the packaging process and eliminate production bottlenecks. In the first case it is necessary to eliminate or at least modify the infeed and outfeed sections of each unit, and to interconnect the drives so that the starting, stopping, and running speeds are synchronized. In the second case there are many types of form-fill-seal machines that take bulk material and form it into packages, fill and seal each unit, and deliver them in groups ready to be put into the final shipping container. An example of the third type of redesign is the incorporation of the glue with the label, as in the case of heat-activated label stock. This eliminates glue pots and simplifies the placing and wiping operations.

The Product. One of the most important factors in the packaging system is the nature of the product that must be handled: What are its physical and chemical characteristics, its value in relation to other economic factors, and its susceptibility to degradation? It is just as important to have a complete set of specifications for the product as it is to have them for the packaging materials. This may seem elementary, but it is surprising how often this simple fact is overlooked.

In the case of dry products, does the density remain constant, or will the equipment have to be adjusted frequently? The condition might warrant a weigh filler instead of a volumetric filler. Does the product "bridge" or stick to the walls so that a vibrator is needed? Is it free-flowing, or must it be forced into a package? Is it dusty, or does it give off vapors which would require a collection system? Must the product be protected from temperature or humidity conditions, or from contamination? Will it be damaged by augers or pusher bars?

The value of the material may determine whether it would be cheaper to overfill than to pay for a more accurate but slower type of filling machine. Food products and pharmaceuticals have special problems of sanitation that must also be considered in the design of packaging machines. The purpose of all this is to point up the importance of having an accurate description of the product, with the standards and tolerances that have been set by the quality control department.

You may also want to look into the possibility of a change in the product to suit the needs of a higher-speed operation, or tighter limitations on the physical characteristics of the product for more trouble-free packaging. This is the time to optimize all of the elements of product, package, and equipment. Some companies find that they can increase efficiency by grading their product, that is, separating the large pieces from the small pieces and running them separately. Keeping the granules separate from the fines will often give a better flow pattern for a powder product and provide a more uniform fill.

The Market. It is essential to know how a machine is to be used at the beginning, and what changes may occur in the method of operation in the future. Is the equipment intended for a test market operation, or is it to be used for an established product? If it is to be used for a limited introduction, can the equipment be upgraded in size and speed or integrated into a full-scale production line at a later date? There should be some estimate of the initial volume of production as well as the expected demands for the next several years.

It is necessary to know also the anticipated pack size or product variations, and to plan for any changes that are likely to occur in the requirements of the marketplace. If there is a chance that a large economy size may be added to the line, or that a sampling program might require a miniature version of the standard package, it would be well to include this in the original specifications. It may not be possible to predict these changes with any certainty, but the history of other similar products should serve as an indication of what might occur.

The Plant. The conditions at the site will often have a bearing on the particular machine that is chosen for a packaging operation. Some of the things that can influence this decision are space limitations, available utilities, safety requirements, sanitation problems, and dust or fumes. A thorough study of the proposed location should be made to see that the floor will not be overstressed, and that there is room to bring the machine into the building. If it is necessary to have a supply of compressed air, or water, or power of the right phase and voltage, make sure that these can be provided at the proper time. If dust or fumes will be a problem, it may be necessary to plan for ventilating to the outside. If explosive vapors are involved, there must be special windows and doors that will blow out of their frames, along with other safety precautions required by ordinances and regulations.

Personnel. The degree of sophistication that should be sought in a machine may be determined by the level of skill of the operators who are available to run the equipment, and by the knowledge and experience of the maintenance crew. It may not be wise, for instance, to purchase a packaging machine with an elaborate electronic control system if your staff does not include a good electronics man. It might be better in that case to choose a mechanical system that could be adjusted by the regular maintenance personnel. If the operation of the equipment takes a high level of intellegence or exceptional dexterity, be sure that your labor force can supply the caliber of people you will need.

Some machine companies will arrange to train your people in the operation and maintenance of their equipment. This can be very worthwhile, and if such service is available, it should be included in the plans. At the very least, the head of the maintenance crew should visit the supplier's plant during the course of construction and learn all he can about the machine from the people who are building it.

The Package. The packaging materials must be accurately specified before the equipment for handling them can be precisely defined. The size and type of each component should be carefully detailed in writing, with supplemental drawings and models where necessary. Examples of the kind of information that is required are given in Fig. 10 in Sec. 2, "Folding Cartons and Setup Boxes," page 2-13, and Table 8 in Sec. 6, "Glassware," page 6-22. It would also be wise to try to anticipate future changes when writing these specifications. There will probably be attempts to economize at a later date by using lighter-weight materials and eliminating certain elements of the package. On the other hand, it may become desirable at some time in the future to add extra labels, flaps, inserts, seals, and easy-opening or reclosing features. All these things should be considered, and while it may not be possible to predict every change that could occur, at least some thought should be given to such contingencies.

If it is possible to make any changes in the package that would improve the efficiency of the machine, they should be given serious consideration at this time. The scores of a folding carton might be changed from creased scores to perforated or cut scores, to reduce the force required for setting them up, and thereby minimize the chance of jamming. An overwrap may need to be stiffer or have a calendered surface to suit a high-speed operation. Inks and coatings that might accumulate on the working surfaces can be changed or eliminated. The small neck of a bottle, that would have a poor filling rate and might require special nozzles, could be enlarged. A shape that is unstable and would need pucks to keep the containers upright will increase the handling problems. And be sure that the machine builder warrants his equipment to operate with the materials as they are defined in the specifications.

Storage of packaging materials may be more critical with a faster or more sophisticated machine. Check to see that the shipping cases are strong enough to protect the contents from damage. Labels should be wrapped in wax paper to protect them from changes in humidity and sealed with tape, not simply fastened with string or rubber bands. Metal containers must have strong separators to prevent them from being dented, and folding cartons should be packed in trays that will keep them from getting warped or losing their prebreak.

LAYOUT AND SELECTION OF EQUIPMENT

The planning of a packaging line must start with the materials involved, the quantities that are to be handled, and the people who will make it work. These three elements are of equal importance, and none of them can be neglected in designing a machine layout. The materials are important because the end product consists of the product and the package; it is the whole reason for being; without it you have nothing. The ultimate objective should be a production line that controls the container as it moves along, without marking the printed surface, without chipping or scratching the container, restraining each unit at the transfer points, without denting or creasing any part of the package. These ideals will pay off in improved morale, pride in line performance,

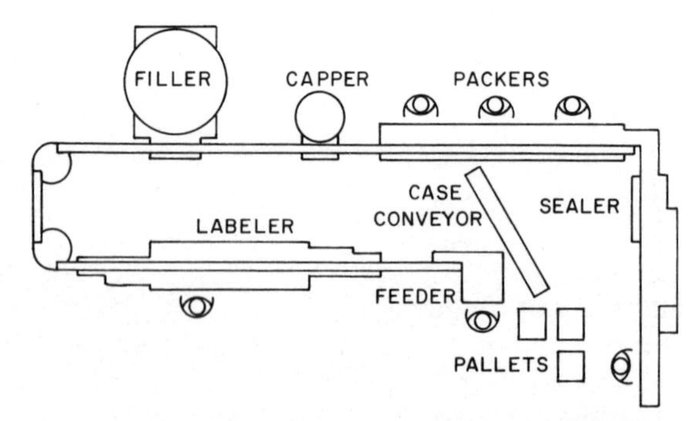

Fig. 1. Equipment layout. Bottles in reshipper cases are brought to the line on pallets. The operator dumps the bottles onto the feeder and puts the empty case on the case conveyor. The bottles are labeled, filled, and capped automatically. Packers take the empty cases from the case conveyor and pack the filled bottles. Filled cases are then sealed automatically, and the operator stacks them on a pallet. In this layout the distance the empty cases travel is minimum, and the pallets are unloaded and reloaded in the same area, for maximum efficiency.

good housekeeping, and adherence to quality standards, all of which add up to a high degree of efficiency. An example of a packaging line layout is shown in Fig. 1.

The quantities of packages to be produced in a given period of time will be a major influence in the selection of equipment. Not only the immediate needs of the production operation, but the projected requirements for at least the next 5 years should be taken into account. This will largely determine the degree of sophistication that will be required, but it will still leave some room for alternatives. For example, the economics of a fully automated system can be measured against different levels of semiautomatic production to see which yields the lowest unit cost. (See Figs. 2 and 3.) When capital costs are added to operating costs, a compromise is suggested. Perhaps two slow-speed lines would be more economical than one high-speed line, or a three-shift operation might reduce the capital investment that would be required for a one-shift operation. Usually, however, a faster machine on a one-shift basis turns out to be the most practical solution.

A straight production line is always preferable to one that turns or doubles back. (See Fig. 4.) If space limitations or the flow of materials makes it necessary to deviate from this, try to avoid dead plates, which increase the stresses on the package units. Instead use rotary plates or twist plate conveyors at the turns, and live conveyors between machine stations. (See Fig. 5.) The spacing mechanisms such as worms, helixes, and star wheels should be made of plastic rather than metal, to minimize the scratching and scuffing of the containers passing through them.

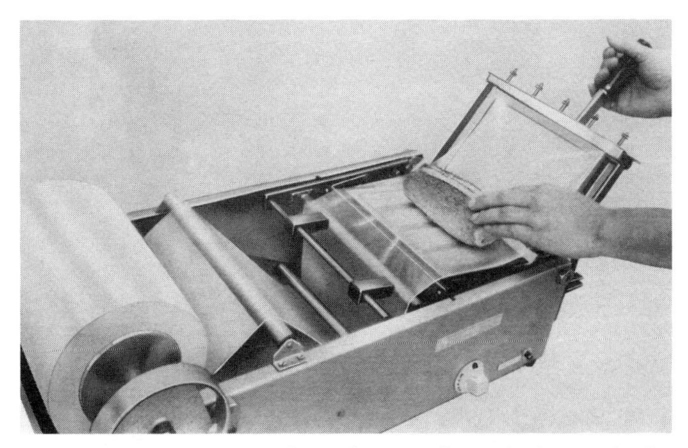

Fig. 2. Semiautomatic wrapper. For low volume or frequent changes in size, a hand-actuated machine is inexpensive and well suited for a small operation. But output is low and labor costs are high in comparison with the machine in Fig. 3. (*Packaging Aids Corp.*)

Fig. 3. Automatic wrapper. A high-speed machine is used where the volume justifies the high capital investment. This unit operates without attention, except for replacement of rolls of film. Compare with Fig. 2. (*Hayssen Mfg. Co.*)

Guide rails also should be covered with plastic, preferably Teflon sheet stock and not a thin tape or film.

When layout prints are obtained from the machinery manufacturer, they should be checked against the floor plan. Column locations and ceiling heights should be accurately determined. If Board of Health rules or Good Manufacturing Practices regulations are applicable, they should be taken into account. Allow sufficient space with accumulation

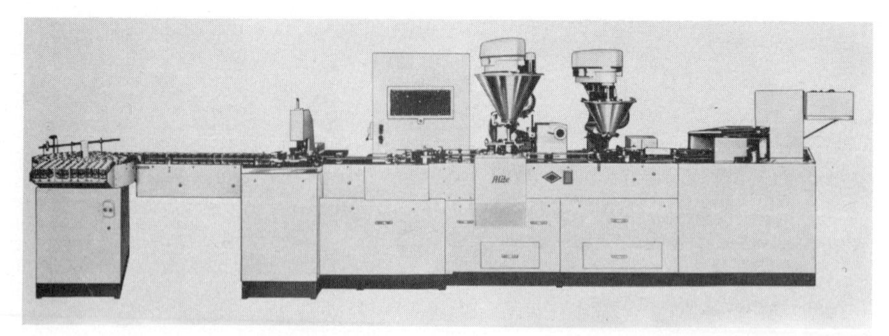

Fig. 4. Packaging line. Machines are combined to provide a coordinated production operation. In this picture cans for baking powder are dumped on the unscrambler on the left; they move by means of a conveyor to the code marker, then to the tare weigher, and to the bulk filler, which can be identified by the cone-shaped hopper; then to the dribble filler, which has the smaller cone-shaped hopper, on to the weighing station, and finally to the lidding machine. (*Arenco-Alite Ltd.*)

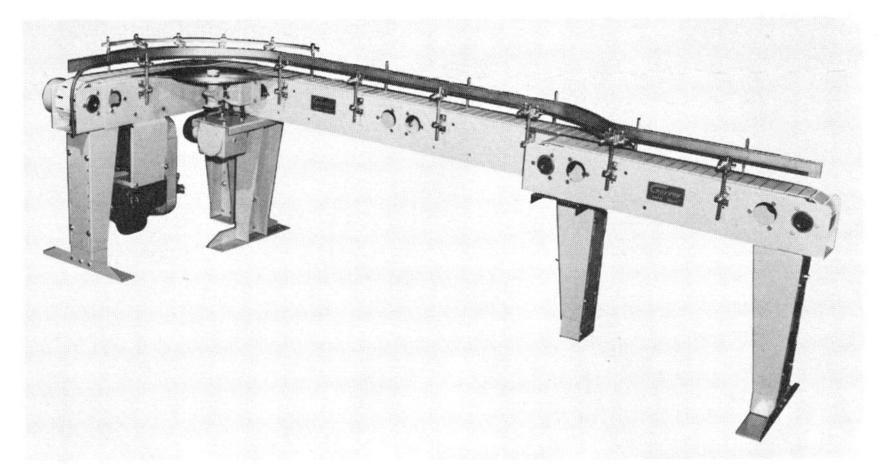

Fig. 5. Conveyor. Machines are connected by conveyors, which bring a steady supply of containers and take away the completed units. This permits the accumulation of a backlog, which helps to even out line surges, and the conveyor can also be used to bypass a machine that is temporarily out of production. (*Garvey Products Corp.*)

stations between connected machines for intermachine surges or bypassing in the event of a shut down.

Ease of cleaning and provisions for sanitation and prevention of product mixing are becoming more essential with the tightening of controls by the Food and Drug Administration. The problems of salmonella in food products and the danger of contamination or mixing of medicines are causing great concern among regulatory officials. Machines must be built with sufficient access panels easily removable with a few thumbscrews, and whole sections of the machine that can be taken out for easy cleaning. More fabricated frames should be used instead of castings, to avoid projections and rough surfaces that collect material.

Success or failure of a new installation will depend to a great extent on the people who are responsible for the day-to-day operating of the equipment. It would be false economy to skimp on anything that contributes to the comfort and welfare of the line personnel, supervisors, or mechanics. There should be provision for adequate light (75 ft-c in general, 100 for inspection areas), ventilation, safety, low noise level (90 db is maximum), good housekeeping, easy maintenance, convenient access to trouble spots for cleaning jams and making adjustments, and guards or special painting to dress up the machine, with color coding in key areas. See that there is adequate aisle space, no obstructions on floors, no slippery floors, no sharp or projecting objects or points on machines, sufficient overhead clearance, and enough space to turn trucks. Any features such as these and others that will build better personnel relationships will be a good investment and should be given a prominent place in the overall plan.

It is good practice to keep key people, such as line supervisors, union representatives, and maintenance mechanics, informed while the project is still in the planning stages. A series of progress meetings to pave the way toward a smooth transition, will avoid misunderstandings that could create personnel problems. For additional information on plant layout and equipment, see H. B. Maynard (editor), *Industrial Engineering Handbook*, sec. 8 (2nd ed., McGraw-Hill Book Company, New York, 1963).

The Machine Manufacturer. The choice of a machine supplier ought to be based on his past performance in the packaging field, particularly his experience with similar types of products; and although the fine features of a control system or a clever feed mechanism are significant, it is more important to look beyond these mechanical niceties and to examine the reputation of the manufacturer in the packaging field. Follow up on the references that are furnished by the machine builder, and if possible, go and see the machines operating in a production situation. (See Fig. 6.)

You should also look to see how well the company is organized, from the president on down. Inquire about their development programs;

Fig. 6. Cartoner. A versatile machine which takes flat cartons from the magazine at the far right, erects them and closes the bottoms, conveys them past the operators who pack various items into the cartons, and closes the tops. (*R. A. Jones & Co., Inc.*)

these will provide clues as to their progressiveness and their interest in being competitive. Check into their building expansion programs; these will indicate how well their business is doing and whether they will be around when you need them.

The packaging material supplier can often be helpful in selecting the equipment to handle his particular material. Most of the larger companies have one or more equipment specialists on their staffs to advise their customers in such matters. Some of them have even gone into the equipment business themselves in order to improve their service. Of course it is to their advantage to furnish this machinery, if in doing so they can be assured they will get the major share of the packaging material business. Since the profit potential in materials is far greater than the return on the equipment, the supplier may be willing to make some concessions on the cost of his equipment, and this aspect should be explored by the engineer.

Types of Machines. There is a great variety of equipment available to do just about any kind of packaging job that is required, and the task of choosing the best machine for a particular set of conditions can be a very

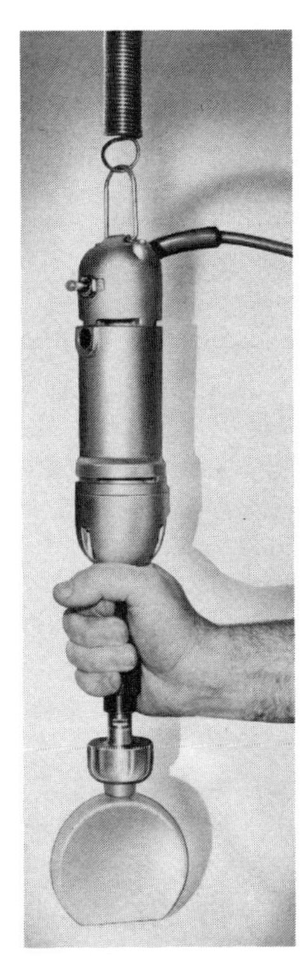

Fig. 7. Cap tightener. The simplest method of applying screw closures is placing them on the bottles by hand and tightening them with the motor-driven chuck shown. A slip clutch can be adjusted to provide just the right amount of torque. Compare with Fig. 8. (*National Instrument Co., Inc.*)

complex undertaking. Some machines are of the simplest types, and for small quantities they are well suited for the purpose. Liquid fillers, heat-sealers, and cappers, to name a few, can be completely manual, so that the cost is kept low and the operation is quite flexible. At the other end of the scale are the highly sophisticated machines used in the beverage and food industries which can operate at high rates of speed and with great accuracy. In between are all degrees of automation, flexibility, and speed. (See Figs. 7 and 8.) Auxiliary equipment such as conveyors, accumulators, elevators, and inspection devices are also being offered in great profusion. It would not be possible to discuss all these, or even to enumerate them in a book of this kind, but a few examples will be given to illustrate some of the different methods that can be used to solve a particular problem. From this it will be seen that there can be many approaches to accomplish the same purpose, and it will require some discretion on the part of the engineer to choose wisely from among them.

Fillers. There are five basic types of filling machines: (1) liquid level fillers, (2) liquid volumetric fillers, (3) dry volumetric fillers, (4) dry weighers, and (5) counters. Counters are of several types: (1) hand

Fig. 8. Capper. This fully automatic machine places the caps on the bottles and tightens them to the desired torque at a high rate of speed. No operator is needed, other than to keep the hopper filled with caps. Contrast this with Fig. 7 to see the broad range of mechanization that can be used for different situations. (*Consolidated Packaging Machinery Corp.*)

paddle, (2) draw board, (3) drum, (4) disk, and (5) slot, all of which are cavity fillers, and (6) column, (7) unit counters, and (8) direct-from-tablet compression machines. Constant level liquid fillers make a better-appearing product, provide drip control, are lower in cost per station, and are more likely to have a no-container-no-fill feature than the volumetric-type fillers.

The major problem in filling *dry products* is the variation in density. It may be necessary to resort to weigh filling in extreme cases, but this is considerably slower than bulk filling. The scales in a weighing machine cannot make more than 20 measurements a minute, but volumetric fillers can operate at up to 400 a minute. Several weighing stations can be used to increase the output, but it is still a very slow operation by comparison. A combination of the two is sometimes used, in which most of the product is filled volumetrically and the remainder is dribbled in. There is even a sensing device that notes when the density changes, by checking the bulk fill and automatically making adjustments so that the slower dribble fill does not take too long to complete. There are also machines that fill in several stages to allow time for the air to escape, assisted by vibrators.

An auger filler will normally give an accuracy of ±2 percent, a net weigher is accurate to 1/10 percent or less, and a gross weigher will hold to ±2/10 percent. Volumetric filling of dry products by means of an auger, or with a rotating plate that has openings of the proper size, requires the least expensive equipment and is the least complicated. (See Fig. 9.) Accuracy is not very good in these systems, although vacuum filling can help with light fluffy powders, and especially with hazardous products. Vacuum filling is slower than other volumetric methods, and this increases costs. Net weight filling, in which the product is weighed in a bucket and then dumped into the package, is the most accurate, but the slowest. With gross weight filling the product goes directly into the package which is resting on the scale. The scale is adjusted to allow for the average tare weight of the container. It is the best method for sticky materials that might cling to the weigh bucket, or for products that must settle or be compressed, but it is not quite as accurate as net weighing because of variations in the weight of the package.

The two main types of *liquid fillers* are volumetric and constant level. For viscous liquids, pastes, and certain food products, the volumetric principle is the only practical method. In volumetric fillers there is a choice among the piston type, gear pump, worm screw pump, and weighing station. For thin liquids (6,000 centipoises or less) of which the value of the product is not too high and a full appearance is important, a constant level filler will provide a faster, and more economical system.

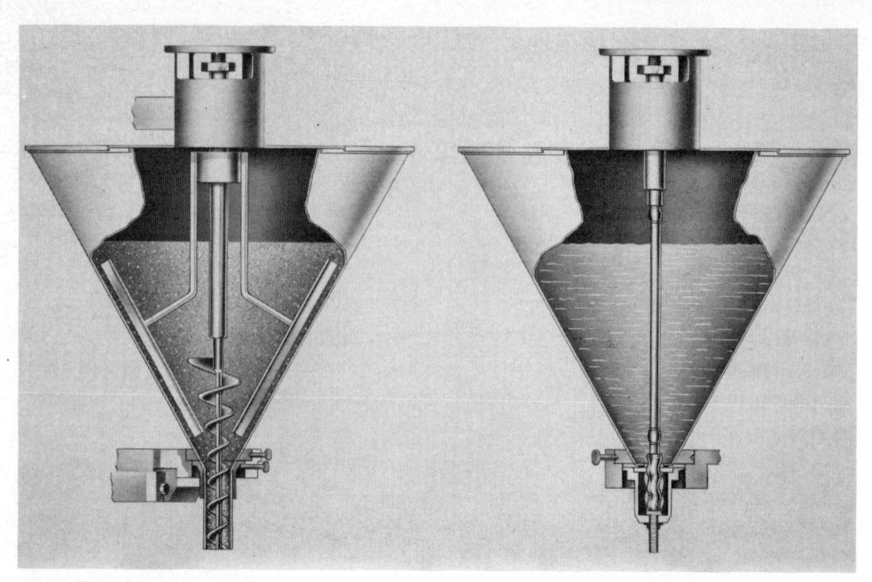

Fig. 9. Auger fillers. Each pitch of an auger has a specific volume, and by adjusting the number of turns it makes, the precise amount desired will be delivered. The auger filler is particularly suited to difficult products like cake mixes and finely ground coffee. A stirrer scrapes the sides to prevent cavitation, as shown on the left. Viscous fluids also can be filled by auger, as shown on the right. (*M. G. Diehl Mateer*)

The simplest liquid filler is the gravity type, in which the supply tank is above the container and a vent tube is put into the container at the proper height to carry off the excess when the liquid level reaches that point. Constant level fillers may also be of the vacuum type, in which the filling head seals against the mouth of the container and draws a vacuum. This provides the force to bring the liquid from the storage tank, and when the level in the container reaches the vacuum tube, the excess is siphoned off and an automatic valve shuts the flow. For faster fills, pressure in the filler bowl may be combined with the vacuum in the container, but this requires more complicated controls. Bottom-up filling starts with the nozzle at the bottom of the container, and the nozzle is moved up as the container becomes full. When the nozzle is kept below the liquid level by this method, there is less turbulence, which is important when filling foamy liquids. Flexible containers such as plastic bottles and oblong metal cans cannot be filled under vacuum, because they would be distorted by the outside pressure; a level-sensing filler must therefore be used. In a machine of this type a stream of low-pressure air passes down the center of the filling nozzle, and when the liquid level reaches the tip of the nozzle, the liquid blocks the air flow, causing pressure to build up and actuate a switch to shut off the liquid.

Flexible Packaging. The numerous machines to handle flexible packaging materials fall into several general classifications: (1) vertical form-fill seal, (2) horizontal form-fill-seal, (3) vacuum form-fill-seal, (4) controlled atmosphere packaging, (5) preformed bags and pouches, (6) intimate wraps, and (7) overwraps. The choice of a machine will depend on the type and materials of the package, as well as the characteristics of the product.

If the product is free-flowing, either dry or liquid, the vertical form-fill-seal machine is usually the best choice. In this type the packaging material feeds down over a former and is made into a partially sealed envelope. The product is directed down into this package through a filling tube, by gravity or force feed. The seal is then completed and the finished unit is cut off, falling into a tote tray or onto a conveyor belt, to be taken to the next packaging station.

Either a single or a double web of paper or film can be made into a package by forming it into a tube and sealing the edges together as it moves down through the machine. If a single web is used, a thin flexible material is necessary to withstand the sharp bend over the forming mandrel. For dry products on which a perfect seal every time is not essential, the single-web package is more economical of material, since there is only one longitudinal seam. A double web is usually preferred for liquids, however, because all seals are made through only two thicknesses of material, and there is less chance of leaks. With the single web there are extra thicknesses where the back seam and cross seam come together, requiring more heat to penetrate and effect a seal and increasing the chance of open channels where the thickness changes.

For products that are in chunks or slices, the horizontal style of equipment is usually preferred. Horizontal form-fill-seal machines require a certain amount of stiffness in the packaging material, so that the partly sealed pouch will open up easily to receive the product. It cannot be too stiff, however, or it may not provide the rounded opening necessary for filling. Either a single web, folded at the bottom, or two webs of packaging material can be used, just as in the vertical machine.

Vacuum packaging with flexible material involves removing the air from around the product. This can be accomplished in either of two ways: by pumping the air out of the package, or by compressing the walls of the package to force the air out. The purpose is usually to reduce the amount of oxygen below the 2 percent level.

Gas-flush packaging utilizes evacuation of the air and replacement with an inert or harmless gas. Nitrogen is inert and is the best for most purposes; it is also the most expensive. Carbon dioxide can be substituted for nitrogen in some cases, or a mixture of nitrogen and carbon dioxide can be used as a compromise. If desired, the gas can be heated and a partial vacuum will be produced when the package cools.

Preformed bags or pouches are higher in cost, but in some circumstances purchasing these ready-made packages is the most practical method. If the sales volume or the speed of the operation does not warrant automatic equipment, or if there are changes in the product or the nature of the product does not lend itself to sophisticated machinery, it might be better to consider a semiautomatic operation. Fragile items, textile products, and odd-shaped units are not generally suited to high-speed operations, but there are various inexpensive filling and sealing devices for handling such products.

Intimate wrapping includes such things as twist wrapping of candies, chub wrapping of soft cheese, and wrapping quarter-pound sticks and pound prints of butter. This type of packaging takes highly specialized machinery that is designed just for the particular purpose. It does not allow a very wide choice of equipment, and the variety of packaging materials also is quite limited. As a rule these packages are made with light weight, highly flexible films and papers.

Overwrapping machines are used for items that are enclosed in a carton, tray, or U-board. These may be loose-wrapped or tight-wrapped; that is, they may be spot-sealed or the entire wrapper can be adhered to the carton. In the latter case it is usually for the purpose of ensuring that the product identification will not be lost by discarding the wrapper. Cold glue can be used for this, or a heat-seal coating may be applied to the wrapper stock. Cold glue is lower in cost, but it is a slower operation, it is messy, and it requires extra labor to supply the glue and for cleaning up.

Operation Controls. A decision as to the type of controls to be used on a machine can be a serious one requiring some study and evaluation. The advantages of electronic or fluidic controls may be worth the added problems of educating the setup and maintenance personnel in their operation. At the same time it could require some extra people, since a good electronics man may not be able to make mechanical adjustments as well as a trained mechanic, and the decision to use one of the more sophisticated systems will be dictated by the caliber of the men in your maintenance department. Do not be dazzled by the glamour of these sophisticated devices; in general it is better to avoid electronic or fluidic controls, and to use mechanical controls that can be adjusted by a mechanic, if it is at all possible.

But where speed, reliability, precision, and resistance to moisture, vibration, heat, or dust are required, there is no denying the advantages of a good electronic system. The cost may be two or three times as much as the conventional systems that have been used in the past, but with line speeds getting up to over 1,000 per minute, and the increasing demand by the government to avoid errors in the labeling of drug products, there may be no other choice but to use the more sophisticated method.

Electronic units can replace conventional relays, timers, and stepping switches to control drives, brakes, valves, starters, and counters. They can perform a decision-making function with signals from push buttons, limit switches, pressure switches, photocells, or pulse generators. There are many and varied applications for these devices; the following examples will help to illustrate some of the ways they can be used: (1) labeling operations require that the correct label be used, that if a bottle is missing, a label will not be presented, that there should not be a misprinted label, and that the label should be applied straight; (2) temperature control is important in food wrapping, and wide swings in sealing temperatures can be minimized with the faster response of electronic controls; (3) case packing and palletizing machines must employ proximity switches and fill detectors that are not affected by dust, vibration, or moisture; and (4) pieces of glassware on high-speed lines must be kept from touching each other if breakage is to be avoided, and electronic controls can pre-position the bottles and synchronize the machines so that they will be evenly spaced throughout the entire operation.

Logic Systems. One of the advantages of electronic and fluidic circuits is the ability to solve problems and make decisions. Unlike their human counterparts, they are never late or absent, do not get tired, will work overtime and not file a grievance, and never take sick leave or go on a vacation. The method by which they operate is based on three elements: *and, or,* and *not.* These three elements will open or close when they receive a particular set of signals or inputs. The *and* element will produce an output only when every one of several inputs is present. If we want to be sure, for example, that the product, leaflet, and carton are all in position, we use the *and* circuit. An *or* element is made so that any one of several inputs will trigger an output. Thus a jam at any one of several points will actuate the *or* circuit and signal the machine to stop. The *not* element is so arranged that it will give an output only when an input is absent, as would be the case when a stop button is not depressed. Two or more of these functions can be built into one unit for controlling more complicated operations.

Electronic Controls. The decision-making function in this case is performed on the basis of signals or inputs from push buttons, limit switches, photocells, or other sensing devices. A system of relays, tubes, amplifiers, resistors, and condensers is arranged to analyze the input and to control the operation. For instance, an empty box on a scale can start the filling cycle, and when it reaches a certain point, it can shut off the bulk feed and start the dribble feed. When the proper weight is reached, the dribble feed stops on signal and a pusher moves the box out of the way and brings in another empty one.

Fluidic Controls. In a hazardous environment where electrical controls would not be permitted, or where steam and caustic solutions are used for cleaning, as in a food plant, fluidics offer a good alternative. They operate in milliseconds, at a rate not so fast as the nanosecond response of electronics, but much faster than most other methods. Machine operations that cycle over 10 times a minute are suited to fluidic control. At less than that, power losses become a significant factor, since the air flow is constant whether they are working or not. Using shop air at 3 lb pressure will cost about ½ cent per shift per element. (See Fig. 10.)

Fluidic elements are compact; they can be made as small as 1/100 cu in. The average mechanic with experience in pneumatic devices is usually able to do troubleshooting and repairing of fluidic systems. Cutting into a live fluidic line is not disastrous, nor does a wrong connection send up a curl of blue smoke. The test equipment is also simple in comparison with electronic gear.

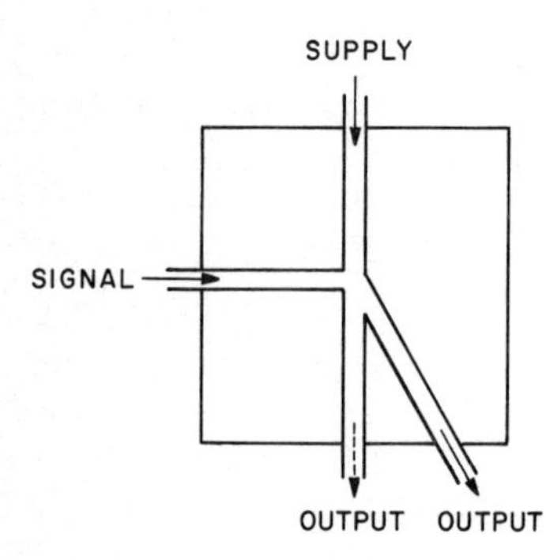

Fig. 10. Fluidic control element. Low-pressure air enters from the top and passes down through the straight leg until an input pulse from the left diverts it to the other leg. When the signal stops, the flow reverts to the original leg. In other types of elements the flow continues in the new channel, even after the signal has stopped, and a signal from the opposite side is required to change its course.

Typical applications for fluidic controls are counting, fill indicating, closure verifying, safety interlocking, and detecting web breaks. They can be installed easily on existing equipment as well as designed into new machinery. The same air supply can be used to operate a cylinder, blow a whistle, or actuate a switch, and hybrid systems which combine fluidics with electromechanical devices can also be used.

Standard or Custom Design. When a stock machine is available that will meet the requirements, there is no good reason for designing and building a special machine. On the other hand, if there are some unusual operations to be performed that would require major changes to a standard model, it might be better to start from scratch. The trend is toward more custom-built machines, and 85 percent of the food and drug manufacturers are using some especially built equipment.

There are several reasons for designing and building a special machine in preference to buying an existing one: (1) available machines are not flexible enough, (2) they are not fast enough, (3) the package is outside the range of standard models, (4) the package cannot be modified to suit regular machines, (5) limited space requires a special design, and (6) standard machines are not consistent and dependable in operation.

It goes without saying that a special machine will have the added cost of engineering and design, while a piece of commercial equipment has already been completely engineered and tested and its cost has been spread over a number of machines. This is partly offset by profit margins, shipping costs, and contingency fees that are part of the price of a stock machine.

The other big disadvantage of custom-made equipment is the longer debugging period. In any machine there are some minor adjustments to be made after it is set up and ready to operate. With a special design that is completely new and untried, there will be considerably more secondary development work to be done, and it may even require the rebuilding of whole sections of the machine. This is not unusual, and it should be anticipated.

Speed or Flexibility. The engineer often has to make a choice between a fast single-purpose machine and a slower machine that can be adjusted to make several different-size packages. Speed is more essential on high-volume production, whereas adjustability is the important factor with low-volume items; it is usually necessary to sacrifice one to get the other. High-speed machines take up less floor space than the equivalent number of low-speed machines. But when a high-speed line is down, production is seriously crippled, whereas one slow unit of a group can be down without affecting the output of the entire department quite as much.

In trying to get the maximum flexibility, however, it may not be wise to try to cover every possible contingency. Sometimes in trying to encompass too wide a range of sizes, it is necessary to sacrifice the efficiency on the popular sizes, and this may not be practical. It is usually better in that case to go for additional slow-speed lines, so that some of the lines can remain set up for the large-volume items.

Foreign Sources. An increasing number of packaging machines are being brought in from other countries. There are several reasons for this: costs are often lower, because labor rates in most foreign countries are only about one-third of those in this country; some machines are more sophisticated, either as a result of better engineering or because their technology is farther advanced, as in some areas of plastics fabrication; and in some cases it is felt the machines are more solidly built and will last longer. (See Fig. 11.) It is probably true that better craftsmanship is being put into some of the equipment built in foreign countries, and although it is difficult to generalize, it is widely accepted that foreign machines usually run slower than those made in this country;

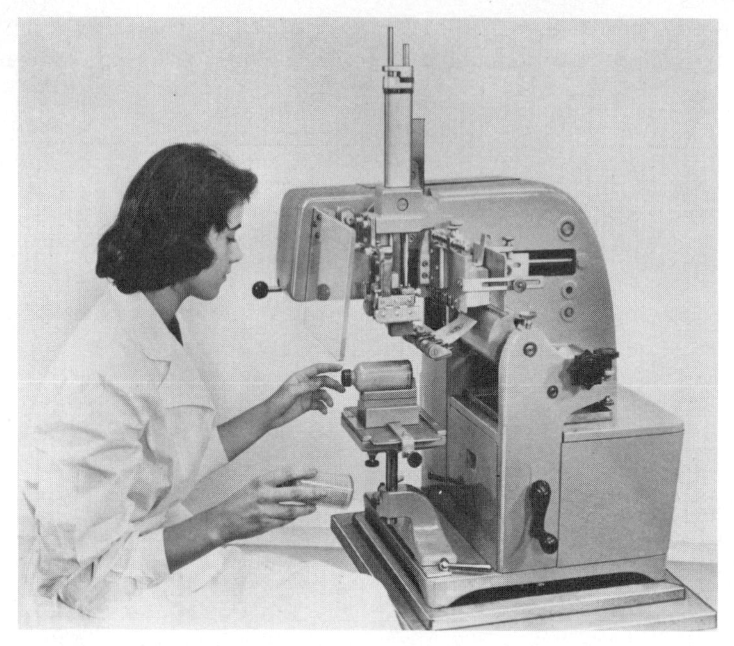

Fig. 11. Labeler. A semiautomatic machine for labeling small bottles. A good operator can do 60 bottles per minute, and changeover to a different size can be made in a few minutes. This is an example of the fine craftsmanship in some of the foreign-made machines. (*Jagenberg-Werke AG*)

that they require more skill to set up and adjust; and that they tend to be heavier and more rugged.

If the purchase of a foreign machine is being considered, there are several points to be kept in mind. Delivery may take considerably longer, since the machinists in other countries work much slower than they do in this country. The metric system of measurement may pose some problems when it comes to fabricating replacement parts in case of a breakdown. Although spare parts may be carried by a local agent, there is some risk that it will be necessary to go back to the factory, and this usually means a long delay. Special care will be necessary in writing the purchase contract to make certain of such things as motors suited to the type of current at the site and avoiding patent infringements and licensing violations.

ORDERING AND SCHEDULING

Financial Analysis. Before purchasing any equipment for the packaging line, it is generally necessary to make an economic study to see whether the expenditure will be justified. This entails an analysis of all the expenses involved, balanced against the savings that will accrue by the elimination of hand labor. Some of the items to be counted as cost factors are the following:

1. Base cost of the machine
2. Interest, or loss of earnings on invested capital
3. Increase in taxes
4. Insurance
5. Depreciation
6. Floor space
7. Extra attachments
8. Change parts
9. Replacement parts
10. Engineering
11. Travel and telephone
12. Machine design changes
13. Package redesign
14. Tighter material specifications
15. Delivery
16. Installation
17. Debugging
18. Materials for testing
19. Maintenance
20. Supervision
21. Training of mechanics and operators
22. Operating personnel
23. Premium pay for operators
24. Imbalance of labor
25. Changeover downtime
26. Downtime for malfunction and shortages
27. Scrap losses
28. Added inspection

When all these items have been considered and evaluated, and the total cost is spread over the required number of years, it must be compared with the real savings that will accrue. For each operator that is eliminated as a result of mechanization, there will be a saving in salary plus some overhead costs. Note that not all overhead costs can be dropped in this case. A certain amount of the fringe benefits and supervision will be eliminated, but heat, light, and other building costs will go on just the same. Among the items that can properly be included in savings are the following:

1. Salaries
2. Fringe benefits
3. Supervision
4. Washroom space
5. Parking space
6. Bookkeeping
7. Administration
8. Resale value
9. Tax savings
10. Quality (more accurate count, etc.)

An important decision that will have to be made is the length of time to be allowed for amortizing the cost of the equipment. This varies with different companies and different situations; the average length of time for most companies is about 3 years. This means that if a machine will not pay for itself within 3 years, it is not considered a good investment.

If the product is a new one, and the chance of survival in the marketplace is questionable, it might be safer to figure on a 1-year payout, in order to justify properly the expense of a new piece of equipment. On the other hand, if the product is well established and there is little likelihood of a change that would make the equipment obsolete, it might be reasonable to amortize the costs over a 5- or even a 10-year period. This is rather rare in today's fast-changing market, and while there may be a strong temptation to stretch the payout period to justify a new machine, this kind of thinking should not be allowed to influence the final decision.

Since capital costs can be taken from taxable income, it is sometimes said that machinery can be bought with 48-cent dollars. While this may not be entirely true, since depreciation is a diminishing factor, there is some justification for shading the costs to reflect a fair amount of the tax benefits. The tax rate for most corporations is around 52 percent of profit. Depreciation of a machine can be charged to operating expense, so that for each dollar deducted from taxable income, there is a saving of 52 cents in taxes. Both the tax credit and the fast write-off provisions of the tax law also should be taken into account in calculating these savings.

There may be some other extenuating circumstances which would weigh in favor of buying a piece of equipment, even when the dollar figures do not appear to justify the added expense. Such things as better housekeeping, improved quality, a different type of package, or a shortage of labor may be quite valid reasons for investing in a new machine. There may even be such reasons as a desire for a showcase operation, a shortage of washrooms, or a lack of parking space that would swing the balance toward mechanization of an operation. These are all perfectly good arguments, and they are not to be taken lightly.

In calculating the output of a machine, do not neglect the factor of downtime. When a piece of equipment is shut down for adjustment or lack of supplies, the operating personnel are still being paid, in most cases. If the reliability of a machine is calculated at 0.98 and the line consists of four machines, the reliability of the line as a whole is the fourth power of 0.98 or about 0.92. To this should be added the risk factor of having a breakdown at a critical time. If the lack of a replacement part will bring on a back order situation, or if it will result in missing a seasonal or promotional market peak, it can cause a serious economic loss.

Total Cost of Equipment. The real cost of owning equipment is more than just the purchase price plus the operating costs. The fact that capital is being tied up instead of earning interest in other investments must be taken into account. Also the variable character of the different elements that make up the total figure should be considered. They are not usually straight-line functions, but have a changing rate that makes it difficult to break them up into nice neat pieces. Thus it would be incorrect from an accounting viewpoint to amortize a new machine at a steady rate over a specified number of years. Since capital costs decrease with time as a result of diminishing interest charges, and operating costs go up as maintenance and replacement costs increase, there is a variation in costs with time. A recommendation to purchase a new piece of equipment might be perfectly justifiable from an engineering viewpoint, but unless it is presented to management in the right terms, it will run the risk of being rejected as a poor investment. While it is not necessary that an engineer understand all the intricacies of accounting procedures, it helps to be able to use the methods and terminology of the accountants so that he can more easily relate a proposal to the current investment plans.

There are several ways to evaluate proposals for new equipment. *Payout* is a method for measuring the time required to recover the original investment. It is a quick, easy gauge for measuring the risk involved, but it has certain disadvantages. It does not allow for earnings beyond the recovery period, it is not related to any minimum acceptable rate of return, it does not take into account the time value of money, and it cannot be related to other investment opportunities.

Another method for evaluating requests for capital funds is the *return on investment,* either the return on gross investment or the return on average investment. The first uses gross investment as a divisor, and the second takes half the depreciable assets as the divisior. In either case the annual earnings are related to the investment over the expected life of the equipment. The "expected life" is where the calculations can lead you astray, because the value of earnings that can be reinvested is ignored, as is the compounding of interest. Return on investment must be calculated from increased productivity and reduced cost of product, package, handling, transportation, or marketing.

There is also a *discounted cash flow* method of calculating the costs of equipment. On this basis an investment must generate enough additional profit or savings to recover the original capital plus a certain minimum return in earnings. This becomes difficult to calculate when the cash flow in each direction occurs at different times, although compound interest tables can be used for present values, and a reciprocal table for future values.

Whichever method is used for figuring the cost of equipment, it is necessary to combine the capital cost and the operating cost to get the full cost of ownership. Figure 12 shows how the declining capital cost is added to the increasing operating cost to get a combined figure that reaches its lowest point in this example at about the sixth year. This may not be the end of its economic life, since replacement cost will be somewhat higher than the lowest point on the curve, and this extends the useful life of the machine another 3 years in this case. It is not essential to work out this cost curve with any degree of accuracy; it is necessary only to know that it exists, so as not to put too much emphasis on a straight-line payout plan.

A piece of equipment loses its value with age, so that all or part of the original investment is lost. What remains of the original value at any

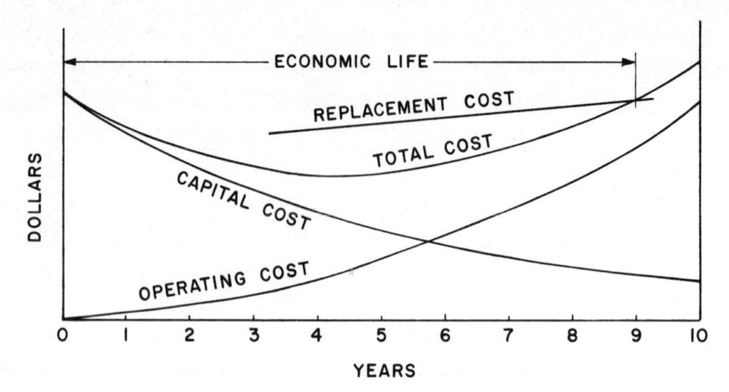

Fig. 12. Total cost. As capital cost is depreciated, operating costs tend to go up, although not usually in the smooth curve shown. The full cost of ownership is the sum of the two, which is shown as a total cost curve. When this becomes greater than the replacement cost, it is time to get a new machine.

point is called the *salvage value*—the amount that could be realized by sale or by trade-in. A method of prepaying this loss out of the return on investment by *depreciating* the original value according to some standard formula yields a *book value,* which is a declining value with time. There are several ways of calculating depreciation: (1) the straight-line method, in which the equipment is half depreciated in half its useful life; (2) the sinking-fund method, which takes into account the salvage value of the equipment and gives a half-life depreciation of about 40 percent; (3) the double declining balance, which depreciates the value by two-thirds at the halfway point, and is the most popular method; and (4) the sum-of-the-integers method, in which the equipment is 75 percent depreciated in half its expected life. Note that interest should be charged also on the undepreciated balance, since this money could be earning a return if it were invested elsewhere.

The effect of taxes on capital investment must be taken into account. In a growth company where the payment of dividends is not in proportion to its profits, a fast write-off is desirable. If the funds that are made available in this way are reinvested in other improvements, taxes can be deferred in proportion to their continued rate of reinvestment. On the other hand, if there is anticipation of increased profits in the future, it might be better to defer the depreciation by using a straight-line method of accounting. The tax rate for a typical corporation is about 52 percent of profit. Depreciation can be charged to expenses rather than being capitalized, so that 52 percent of the deductions for depreciation that would normally be paid out in taxes can be retained. Net interest costs can be deducted in their entirety. These are not so significant as the savings that are made in write-offs, and since interest is charged against the book value of the equipment, it is a diminishing credit.

One of the side effects of tax write-offs is to shorten the economic life of a piece of equipment. Not only does it reduce the time it takes to reach the low point in the annual total cost and the annual average cost, but it reduces the replacement cost and moves the crossover point in Fig. 12 more toward the left.

Buy or Lease. There are many companies, both large and small, that prefer to lease packaging machinery rather than buy it. If the product is standard and is expected to have a good future, outright purchase is undoubtedly the best choice. The tax benefits and the opportunity to make alterations to the equipment weigh heavily in favor of outright purchase. In some special cases, however, it is more prudent to rent a piece of machinery, or even a whole production line. The conditions which favor leasing are (1) products of uncertain future, (2) new and unproven machines, (3) one-time promotions or other short-run situations, (4) supplemental equipment for seasonal or peak demands, (5) government contracts which allow full write-off of the rental cost, and (6) shortage of capital or lack of approval by top management.

Some companies have a rule of thumb that if the purchase price is less than the cost of a 5-year lease, they will buy it outright. Otherwise they favor a rental arrangement. This is particularly true in industries where rapid changes in packaging could make a machine obsolete in a short time.

Lease-purchase agreements are the most common type of arrangement. Under this plan the user pays a fixed rate per month or per year, and at the end of a specified period either owns the machine outright or can buy it for a nominal sum. The tax benefits under such a plan are limited to the amount paid at the end, plus the interest portion of the rental charges. In a direct purchase the entire amount can be written off.

Other leasing plans are (1) quick-termination contracts that can be paid up at any time, (2) decreasing payments and a nominal rental after a certain number of years, (3) the "evergreen" type of decreasing payments with no definite termination, and (4) "humpback" contracts that increase the charges to coincide with anticipated sales curves. Some machinery manufacturers prefer to turn over their leasing contracts to a finance company. There are a number of leasing corporations that will handle these financial arrangements. The charge for this service is about 5 percent added to the prime interest rate.

There is an important difference between a rental arrangement and a leasing agreement that should be understood. A rental is usually for a short period and can be terminated at any time, normally without penalty. A lease is for a longer period, sufficient to pay the total cost of the machine. If it is terminated before that time, the lessee is obligated to find a buyer and make up any difference between the sale price and the balance of the lease agreement.

Contracts and Specifications. A good workable set of specifications is the purchaser's best assurance that the machinery he buys will fit specific production line requirements and deliver trouble-free performance. These specifications must be complete and accurate, and they should include the following points:

1. Define clearly the purchaser's basic requirements and any special needs that may be occasioned by such things as safety, sanitation, dust or explosion hazards.
2. Determine what is "standard" with the vendor, and set forth what is acceptable as substitutions or deviations.
3. Provide the manufacturer with mechanical and electrical information to eliminate any need for guessing or assumptions about conditions at the site.
4. Establish a plan for testing, acceptance, and subsequent handling of complaints for both the vendor and the purchaser.
5. Provide legal protection for both buyer and seller, and define ownership of new designs and patentable ideas.
6. Inform other departments concerned, such as maintenance, production, purchasing, and plant engineering.
7. Provide a permanent record for purposes of reordering.

It is imperative to put into writing all the details that have been agreed upon, and the understanding of exactly what is to be furnished, and when, and how. It is particularly necessary to know what the manufacturer furnishes as standard on his equipment, and to spell out those additional or special features that the packer requires to meet his particular needs. It is in this area that most of the misunderstandings between buyer and seller occur.

The cost of special features often adds considerably to the purchase price, simply because it takes a great deal more engineering and fabrication time for such extras, The design cost of the basic machine can be spread over many units, but a custom-made attachment must absorb all the one-time costs. (See Fig. 13.)

One area that should be carefully detailed in the purchase specification is the electrical wiring and controls. This can include such things as types of enclosures, code markings, cables and harnesses, and location of components. Try to avoid brand names in writing specifications, as this can increase costs and cause delays. It is much better to describe the operational characteristics that are required, and let the machine builder choose the source of the items. Types of belts, chains, sheaves, and sprockets can be covered in a general way, but it is best not to be too specific. Those areas which will require steam cleaning should be made of

Name: Capitol B-144 Liquid Filler **Date:** June 11, 1971

Location: Building 6 **Revised:** July 7, 1971

Supplier: Capitol Tool Company

SPECIFICATION

General

Furnish one Model B-144 rotary pressure-vacuum automatic liquid filling machine which will transfer bottles from a chain conveyor onto a filling mechanism, fill the bottles, and discharge them back onto a chain conveyor. The filler is to have side opening filling stems.

The machine is to be capable of handling and filling the following glass bottles:

Nominal size	Fill rate	Fill point	Drawing number
12 oz	180 bpm	5¼" up	B-1015
20 oz	140 bpm	6¾" up	B-1016
32 oz	100 bpm	7¼" up	B-1017

Change parts for each of the above bottles are to be furnished with the machine. Six extra filling stems for each size must be provided by the machine supplier.

The liquid to be filled has a specific gravity of 0.994, a viscosity of 3,000 cps, and it foams when agitated or aerated. The pH is 6.5 to 7.0, and it is a water solution, light blue in color.

Electrical specification E-224, furnished by the purchaser, is hereby made a part of this specification. Paint specification P-117 is also a part of this specification.

Conveyors

All infeed and discharge guide rails, infeed worms, and starwheels to be of plastic or plastic-clad metal which will not scratch or mar the bottles. Edges of Formica elements to be coated with epoxy resin. Infeed worm to take choke feed bottles on a conveyor and space them to enter the starwheel without breaking or jamming.

Drive shafts, chains, sprockets, belts, and pulleys to be guarded. Access panels are to be fitted with wing screws for

Fig. 13. Typical specification. After consultation with the manufacturer, the details of construction and operation of the machine are written into a specification. Illustrated is the first of several pages that would make up such a document.

noncorrosive materials such as stainless steel, aluminum, copper, or brass, or they should be heavily plated. Guards should be required in those areas where there are heaters, moving parts, or electrical connections.

Changeover methods can be spelled out. Some machines are

adjusted by moving guide rails, turning handwheels, and sliding collars into position. This is quick and easy to do, and sometimes the machine operator can be trained to make the adjustments. However, such machines are not usually so solidly built as the ones that require change parts. It may take a little longer to replace these components than it would to simply reposition them, but it usually yields more consistent results and is worth the extra effort.

Drives will require some explanation. First decide whether they must be synchronized with other machines, and then determine whether this should be done mechanically or electrically. Some allowance should be made for declutching in case it becomes necessary to hand-feed while the machine is being worked on. It is a good idea to specify variable-speed drives, so that the operators can be paced to maximum efficiency, and it is also possible that later on they might be run at a higher speed. See Table 2 for a checklist of other items.

TABLE 2 Checklist of Items to Be Included in a Purchase Contract

A. Basic machine
  1. Name and model
  2. Materials of construction (corrosion-resistant, etc.)
  3. Right-hand or left-hand operation
  4. Utilities required
    a. Electrical current (phase)
    b. Air (pressure)
    c. Water (temperature)
  5. Drives and controls
    a. Motors
    b. Starters
    c. Safety switches
    d. Controls
  6. Lubrication system
    a. Automatic
    b. Centralized
  7. Appearance
    a. Guards
    b. Paint
    c. Plating
  8. Auxiliary equipment
    a. Counters
    b. Heaters
    c. Indicators
    d. Clutches

B. Change parts
  1. Quantity
  2. Interchangeability
  3. Marking

C. Terms of contract
  1. Price
  2. Rate of payment
  3. Delivery date
  4. Method of shipment
  5. Demonstration on line
  6. Test period
  7. Defective workmanship
  8. Defective parts
  9. Service

D. Operational guarantee
  1. Speed
  2. Packaging materials
  3. Tolerances
  4. Operators required
    a. Number
    b. Skill

E. Acceptance criteria
  1. Length of runs
  2. Number of changeovers

F. Data for installation
  1. Weight
  2. Electrical load
  3. Certified drawings
  4. Wiring diagrams
  5. Operating manual
  6. Lubrication and maintenance data

G. Spare parts
  1. Maintained by user
  2. Maintained by manufacturer

Lubrication systems should be covered; a central system is usually preferred, but any method that is convenient for the operator to manipulate on a routine basis should be satisfactory. If it is too difficult to operate, the machine may suffer from lack of attention. Safety controls, such as interlocks or overload and jam switches which will automatically stop the machine in the event of a malfunction, will prevent damage to the machine and minimize downtime. Fill control will have an important effect on the cost of product for the life of the machine. It is essential that the allowable tolerances be spelled out, and that the amount of scrap be not above a certain minimum figure. Label placement and cap tightening torque are other details for which tolerances should be fixed when drawing up a contract.

Testing in the supplier's plant is usually expected, but it should be mentioned in the contract. The purchaser is required to furnish all packaging materials at no cost, sufficient for an hour's run, or whatever is agreed upon. The package components that are furnished should be carefully sorted to be sure they are as nearly perfect as possible. It is unfair and misleading to expect a machine to function with defective materials. If there is more than one supplier of packaging materials, representative lots from each should be tested, but they should not be intermixed. There must be enough for the infeed and takeoff mechanisms to be properly evaluated, at full operating speeds. The duration of the test should be predetermined in terms of time or quantity of pieces, and the machines should be expected to operate without a single miss for that period. If there is a jam or skip during the test, it should be repeated until there is faultless operation for the prescribed time. The duration of a dry run without product or packages, which can be for several hours, also should be defined.

Repair service should be looked into. This comes high, but it might be worthwhile to have a factory-trained mechanic to make major repairs in an emergency. The availability and terms for this service can be made a part of the purchase contract. Training of personnel also should be part of the agreement, as well as the initiation of a maintenance program. A complete set of blueprints, a parts list, and an operating manual ought to be included, and the buyer should insist upon it.

Some arrangement should be made for spare parts. Either the buyer keeps a supply of critical parts such as shear pins, belts, and chains, or the manufacturer agrees to maintain a supply which can be shipped on short notice. It is often a very serious matter when a packaging line is shut down while waiting for a replacement part. This is particularly true in the case of foreign machines unless a complete stock is carried at a distribution point in this country. Spare parts that are likely to be needed for normal replacement should be ordered at the same time as the machine. There is a better chance that they will fit properly if they are machined at the same time, and the cost will be less if they are made with the same shop setup. It is not practical to try to cover every contingency, and therefore it is essential that the machine builder maintain the necessary patterns and jigs to supply other parts promptly in case of a breakdown.

Security clauses will bind the seller not to disclose to outside sources any information regarding product, package, or machine, and should be made a part of the contract. Also the exclusive use of special designs and the period of exclusivity can be included where applicable.

Scheduling. The target date for getting the new equipment into production may be determined by a need to meet competition, or the introduction of a new product or variation of an existing product, or by a demand for quantities that are beyond the capacity of the existing facilities. It is very possible that the start-up date will be the most important factor in the selection of packaging equipment. If one type of machine is available sooner than another, this could easily become the deciding influence in the final choice.

The promise date of the manufacturer is the starting point for setting up a schedule. If a firm delivery date has not been determined, it will be necessary to use an average figure. For standard machines the average delivery time is about 6 months, and for special or customized designs it is more likely to be 14 to 16 months. To this must be added the time required for uncrating and moving the machine into place. There is usually some wiring and piping to be done, and perhaps some carpentry or masonry work. Some of this can be done prior to delivery, if there is sufficient information, but it would be safer to figure on at least a week for installation, after the delivery date.

Some alterations may have to be made to the equipment after it arrives—either planned changes to suit the conditions on the site or corrections that become necessary in matching up a machine to other components in a production line. An extra week is not too much to allow for such contingencies.

Then there is the debugging period. There is scarcely ever a piece of equipment that functions properly the first time it is tried. If it is a standard machine which is identical to a number of other machines, the chances are good that it will take only a couple of weeks to get it operating at peak efficiency. If it is a one-of-a-kind, custom-built machine for a special purpose, it may take many months before it reaches its rated speed and is functioning as intended.

A critical path or PERT chart will be helpful in keeping track of the different events, so that pressure can be applied in the right place, and if a delay is inevitable, the affected people can be notified at the earliest possible moment.

If all the preliminary work is carefully done, and a reliable manufacturer is chosen, the chances are good that the final results will be satisfactory. As line speeds increase and machines become more complicated, the need for careful planning becomes more acute. As long as an orderly consideration of all the factors involved is started early enough, and if the contract and specifications are written to cover all points, and a complete understanding is established between the various parties concerned, both the buyer and the seller of the equipment can be assured of a profitable and successful venture.

Index

Abrasion of cushioning, **17**-3
Abrasion testing, **18**-4
Accelerated testing, **18**-3
Acceptance quality level (AQL), **19**-5
Acro-art synthetic paper, **4**-21
Additives (*see* Plastics, additives for)
Adhesives, **9**-25
for composite containers, **12**-5, **12**-9
for corrugated, **5**-5, **9**-30, **9**-31
dextrin, test for, **9**-29
for foil, **3**-58
for glass, **6**-13, **6**-15
heat-activated, **14**-12
labeling, **9**-30
yield for, **14**-12
for nylon, **3**-20
for polycarbonate, **3**-22
pressure-sensitive, **14**-7
principles of adhesion, **9**-25, **9**-26
properties of, table, **9**-28
remoistening type of, **14**-12
types of, **9**-27
Adulteration of product, regulations for, **16**-2, **16**-3, **16**-7
Advertising media, table, **1**-19
Aerosols, **11**-1
actuators, **11**-18, **11**-19
allowable loss for, **11**-6
barrier containers, **11**-25
cans, **7**-16
interior coatings for, **7**-18
manufacturing of, **7**-10
pressure distortion, **7**-16
Aerosols, cans (*Cont.*):
standard dimensions for, tables, **7**-17, **11**-23
crimp dimensions, illustrated, **11**-20
diluents, **11**-15
dip tubes, **11**-21
filling methods, **11**-7
food aerosols, **11**-3, **11**-14, **11**-25
gas laws, **11**-15
glass containers, **11**-25
glass finish, illustrated, **11**-21
headspace, **11**-14
inhalation therapy, **11**-3
labels, **11**-26
leakage, **11**-6
legal limitations, **11**-25
new developments, **11**-8
orifice sizes of, table, **11**-18
overcaps, **11**-26
plastic containers, **11**-25
pressure drop-off, illustrated, **11**-9
principles of operation, **11**-4
processes, **11**-26
propellants, **11**-5, **11**-8, **11**-11, **11**-12
blends, **11**-11
compressed gases, **11**-14
flammability, **11**-13
fluorocarbons, **11**-10, **11**-12
hydrocarbons, **11**-13
systems for, **11**-15
toxicity, **11**-10
regulations, **11**-14, **11**-24, **11**-28, **16**-12
safety aspects, **11**-3

Aerosols (*Cont.*):
selection criteria, **11**-7
specifications for, **11**-27
temperature, effects of, illustrated, **11**-9
tests, **11**-6, **11**-8, **11**-14, **11**-27 to **11**-29
valves, **11**-16
Aesthetics, **1**-21
Air cargo regulations, **16**-18
Air express regulations, **16**-19
American Trucking Associations, **16**-17
Amortizing the cost of machinery, **20**-21
Anchorage of cellophane coating, **3**-12
Annealing:
aluminum tubes, **7**-26
glassware, **6**-8, **6**-12
Antistatic agents in plastics, **8**-44
Antistatic testing, **18**-4
Applicators for tubes, **7**-29
Assignable cause in quality control, **19**-5
Attributes in quality control, **19**-4
Avogadro's law, **11**-15

Bags, **10**-1
breakpoint between drums and, **10**-9
burlap sacks, **10**-11
design considerations, **10**-5
dimensions, illustrated, **10**-3
export multiwall sacks, **10**-7
extensible paper, **10**-8
folding a sleeve, illustrated, **10**-4
greaseproofness, **10**-8
length and width ratio, **10**-7
mailing, **10**-10
mesh, **10**-11
moistureproof paper, table, **10**-7
multiwall sacks, table, **10**-9
pasted sacks, **10**-6
plastic film, **10**-5, **10**-10
plies, number of, **10**-7, **10**-8
polyethylene sacks and, **10**-9, **10**-10
polypropylene, **10**-10
printing, **10**-11
produced per year, **10**-1
reinforced materials, **10**-2
sacks, definition of, **10**-2
size limitations, table, **10**-9
stepped-end construction, **10**-7
stitched, **10**-6
styles of, **10**-3, **10**-4
testing, **10**-11
valve type, **10**-6
weight of contents, maximum, **10**-8
wet-strength, **10**-8
Bails for pails, **7**-23
Baler bag, **10**-3
Barrels (*see* Wood containers)
Barrier properties of laminations, table, **13**-14
Base weights, tinplate, **7**-5
Basebox, definition of, **7**-4
Basis weight of paper, **14**-5, **14**-14
Baskets (*see* Wood containers)
Beach puncture test, **18**-4
Beating, effect of, in papermaking, **4**-8
Bender in paperboard, **4**-17, **4**-18
Bleached and semibleached kraft paper, **4**-14
Bleached sulfate boxboard, density and stiffness of, **4**-10
Bleaching of paper, **4**-7
Bliss box, corrugated, illustrated, **5**-2
Blisters, carding methods, illustrated, **8**-82
Blocking test, **18**-4
Blooming of glassware, **6**-5
Blow molding (*see* Plastics, blow molded)
Bogus paper, **4**-17, **5**-4
Bond test for coatings and laminations, **18**-5
Book value of machinery, **20**-24
Book-wrap style, corrugated, illustrated, **5**-2
Borosilicate glass, **6**-5, **6**-13
Bottle caps (*see* Caps)
Bottles:
glass (*see* Glassware)
plastic (*see* Plastics, blow molded)
Bottom mold marks:
in glassware, illustrated, **6**-17
in plastics, illustrated, **8**-63
Boxboard (*see* Paperboard)
Boxes:
corrugated (*see* Corrugated boxes)
folding, **2**-1
(*See also* Folding cartons)
setup, **2**-21
Boyle's law, **11**-15
Brightness of paperboard, **4**-15
Brightness test, **18**-5
Bristol-Lund formula, **2**-10
Buckets (*see* Pails)
Bulge in paperboard, **4**-19
Bulge factors in folding boxes, illustrated, **2**-8
Bundling with cellophane, **3**-14
Burlap sacks, **10**-11
Burst test, **18**-5

Calendering of paper, **4**-12
Cans, **7**-10
aerosol, **7**-16
interior coatings for, **7**-16, **7**-18
pressure distortion of, **7**-16

Cans, aerosol (*Cont.*):
standard dimensions for, table, **7**-17
aluminum, **7**-14, **7**-15
annual production of, **7**-2
cemented side seams, **7**-12
corrosion inhibitors, **7**-9
cost of, table, **7**-12
decoration for, **7**-9
depalletizing, **7**-13
drawing and ironing, illustrated, **7**-15
embossing, **7**-9
fabrication of, **7**-9, **7**-11
filling equipment, **7**-14
friction-plug, **7**-18, **7**-19
joints and seams in, **7**-12
linings, **7**-8
manufacture of, **7**-2
pull tabs, **7**-14
reshippers, **7**-13
sanitary-style, **7**-2
screw-top, **7**-18, **7**-20
sealing compound, **7**-10
seamless, **7**-20
shaped, **7**-13
shipping weights, table, **7**-15
sizes of, table, **7**-13
slipcover, **7**-20
terneplate, **7**-5
tin coating, **7**-2
tin-free steel, **7**-6
tinplate, **7**-2
base weight, tables, **7**-4, **7**-5
composition, illustrated, **7**-3
cost of coating weights, table, **7**-6
tempers of, table, **7**-4
types of containers, illustrated, **7**-19
Cap tightener, illustrated, **20**-11, **20**-12
Capital costs of equipment, **20**-22
Caps, **9**-2
collapsible tube, **7**-29, **9**-5
costs of, table, **8**-55
crown, **9**-9
liners for, **9**-12 to **9**-19
mold for, illustrated, **9**-8
quality control, **9**-11
quantities used, by type, table, **9**-3
specialty, **9**-9
standard dimensions for, table, **9**-4
tightness of, table, **9**-6
Card stock, **4**-2
Carding methods for blisters, illustrated, **8**-82
Cargo plane regulations, **16**-6
Carrier rules, **16**-17
Cartoner, illustrated, **20**-10
Cartons, folding (*see* Folding cartons)
Casks, definition of, **15**-12
Cellophane (*see* Films, cellophane)
Cellulose acetate (*see* Coatings; Films; Plastics)
Center special slotted container (CSSC), illustrated, **5**-2
Charles' law, **11**-15
Charpy impact test, **18**-5
Chemical pulp, **4**-6
Chemical reaction of glassware, **6**-5
Chemical resistance of plastics, **8**-39
Chipboard, **4**-17
Clarity, test of, **18**-5
Closures, **9**-1
(*See also* Caps)
Coarse paper, **4**-13
Coatings:
acetate, **13**-8
adhesion, test for, **13**-8
cellophane, **3**-12
dry waxing, **13**-3
extrusion, **13**-8
glassware, **6**-13
table, **6**-7
heat-seal, **13**-7, **13**-8
peelable seal, **13**-8
polyethylene, **13**-8
polypropylene, **13**-8
processes, **13**-7
PVDC (saran), **13**-6, **13**-8
rubber hydrochloride, **13**-8
saran (PVDC), **13**-6, **13**-8
types of, **13**-2
varnish, **13**-6
vinyl, **13**-8
wax, **13**-3 to **13**-5
wet waxing, **13**-3
Cockling of paper, **4**-5
Collapsible tubes, **7**-24, **8**-74
applicators, **7**-29
carton, illustrated, **2**-10
closures, **7**-29, **9**-5
decoration, **7**-27
filling methods, **7**-26
as food containers, **7**-25
metal, **7**-29
costs of, **7**-29
laminated, **7**-26
linings for, **7**-29
manufacturing of, **7**-26
materials, **7**-25
packaging, **7**-26
peel coat, **7**-29
plastic necks, **7**-30
standard dimensions of, **7**-28
styles of tips, illustrated, **7**-30
plastic, **8**-74

Collapsible tubes, plastic (*Cont.*):
standard dimensions of, illustrated, **8**-75
Color testing, **18**-5
Colors:
of glass, **6**-5
for protection from light, **6**-6
Combination run of folding cartons, **2**-14
Composite containers (*see* Fibre cans)
Composite films, **3**-2
Compression test, **18**-5
Compressive resistance of corrugated fibreboard, **5**-5, **5**-7
Consultants, **1**-9
Consumer packaging, **1**-7
Containerboard, **4**-16
Containerization, **1**-4
Contract packager, **1**-10
Control testing, **18**-3
Controls, machinery, **20**-16
Conveyors, **20**-9, **20**-27
Copy (text on packages), **1**-22
Copyrighted labels, **16**-23
Copyrights, **16**-23
Cords in glassware, **6**-4, **6**-8
Cork properties, table, **9**-13
Cork stoppers, **9**-19
Corrosion inhibitors in cans, **7**-9
Corrosion resistance of aluminum, **7**-32
Corrugated boxes, **5**-1
adhesives, **5**-5, **9**-30, **9**-31
board construction, **5**-2
bursting resistance of, table, **5**-3
center special, illustrated, **5**-2
closing and sealing, **5**-16, **5**-17, **9**-21
coating, **5**-14
combining, **5**-5
compressive resistance, **5**-5, **5**-7
corrugated medium, **4**-16, **5**-4
corrugations, **5**-3, **5**-4
cost calculations, **5**-15
crush resistance, **5**-6, **5**-7
curtain coating, **5**-14
cushioning, **5**-6, **17**-3
cylinder kraft, **4**-16, **5**-5
design considerations, **5**-11
direction of corrugations, **5**-4
double-wall, **5**-2, **5**-7
drop test, **5**-18, **18**-5
fatigue factor, **5**-6, **5**-8
fitting to folding cartons, **5**-10
flat crush values, table, **5**-6
flutes: height of, table, **5**-3
selection of, **5**-5
fourdrinier kraft, **5**-5
full overlap, illustrated, **5**-2
glued joints, **5**-10
Corrugated boxes (*Cont.*):
hot-melt coating, illustrated, **5**-14
humidity factors, table, **5**-8
incline-impact test, **5**-18
interior packing parts, illustrated, **5**-12
interlocking in a stack, **5**-11
jute, **5**-5
linerboard, **4**-16, **5**-3
long-term storage, **5**-6
manufacturer's joint, **5**-4, **5**-10
medium, **5**-2
pallet patterns, **5**-11
printer-slotter, **5**-11
printing, **5**-11 to **5**-13
proportions, most economical, **5**-11
regular slotted container, **5**-2
reinforced tape, **5**-17
revolving drum test, **5**-18
scoring allowances, **5**-8
scoring wheels, **5**-9
sealing costs for, table, **5**-17
setup charges for, table, **5**-16
single-faced, **5**-2
size designation, illustrated, **5**-8
very small boxes, **5**-6
stacking strength, **5**-4, **5**-6 to **5**-8, **5**-10, **5**-18
stiffness ratio, **5**-5
stitched joint, **5**-4, **5**-10
strength at score line, **5**-6
styles of, **5**-2
taped joints, **5**-10
testing, **5**-18
tolerance, dimensional, **5**-9
triple slide box, illustrated, **5**-2
triple-wall, **5**-2
unbalanced sheet, **5**-3
vibration test, **5**-18
water resistance, **5**-14
Cost of equipment, **20**-21, **20**-23
Cost estimates, **1**-25, **1**-26
Crates (*see* Wood containers)
Creep of cushioning, **17**-3
Creped paper, **4**-14
Crown caps, **9**-9
Crush resistance of corrugated fibreboard, **5**-6, **5**-7, **18**-5
Curtain coating of corrugated fibreboard, **5**-14
Cushioning, **17**-1
abrasion of, **17**-3
corrugated fibreboard, **5**-6, **17**-3
creep of, **17**-3
forces on packages in transit, **17**-2
fragility factors, **17**-2
harmonics, **17**-5
materials, selection of, **17**-3

Cushioning (*Cont.*):
periodic vibrations, **17**-3
polystyrene foam, **17**-4
polyurethane foam, **17**-4
resonance, **17**-3
settling, **17**-3
shock damping, **17**-5
vibrations, **17**-2, **17**-5
Custom design machinery, **20**-18
Cylinder board, **4**-10
Cylinder kraft, **5**-5
Cylinder paper machine, **4**-8

Deception, **1**-20, **16**-5
Defects of glassware, **6**-18, **6**-22, **6**-23
Delamination test, **18**-5
Densities:
of films, table, **3**-52
of folding paperboard, table, **4**-17
of liquids, **3**-52
of powder product, **18**-5
Department of Commerce, U.S., **16**-7
Department of Transportation (DOT), U.S., **16**-6, **16**-14
Depreciation of machinery, **20**-24
Design of experiments, **18**-3
Dextrin adhesives **9**-29
Di-Na-Cal decoration, **14**-6
Display, **1**-19
Disposal of fibre drums, **12**-12
Dissolvo soluble paper, **4**-21
Doctrine of preemption, **16**-5
Double-seam fabrication, illustrated, **7**-11
Double-wall corrugated, **5**-2, **5**-7
Drop test, **18**-5
Drums, steel, **7**-21
cost of, table, **7**-24
dimensions of, illustrated, **7**-23
linings for, **7**-21
locking ring, **7**-22
manufacturing of, **7**-21
open-head, **7**-22
regulations for, **7**-21
returnable, **7**-21
single-trip, **7**-21
sizes of, **7**-22
threaded openings, **7**-22
tight-head, **7**-22
types of, **7**-22
weights of, **7**-22
Dry wax paper, **4**-16
Dry waxing, **13**-3
Dust attraction for plastics, **8**-38

Ecology and pollution, **1**-6
Economic life of machinery, **20**-24
Educational programs in packaging, **1**-8
Electrocal decoration, **14**-6
Electrolytic tinplate, **7**-4
Electronic controls, **20**-16, **20**-17
Elongation test, **18**-6
Enamels for cans, **7**-8
Envelopes:
definition of, **10**-2
sizes and styles of, illustrated, **10**-3
Environmental pollution, **1**-6
Equipment (*see* Machinery)
Estimating costs, **1**-26
Ethics, **1**-11
Etiological materials, regulations for, **16**-15
Expenditures for packaging, **1**-1
table of, **1**-2
Explosives, regulations for, **16**-14
Export multiwall sacks, **10**-7
Export problems, **1**-16, **1**-17
Express regulations, **16**-19
Extensible paper, **4**-14, **10**-8
Extraction test, **18**-6
Extrusion of plastics, **8**-4
Extrusion coatings, **13**-8

Fade test, **18**-6
Failure analysis of glass, illustrated, **6**-10
Fair Packaging and Labeling Act, **16**-11
Fatigue factors in corrugated fibreboard, table, **5**-8
Fatigue test, **18**-6
FDA (*see* Food and Drug Administration)
Federal Trade Commission, **16**-5, **16**-13
Fiber-free surface of paper, **4**-15
Fibre cans, **12**-2
convolute-wound, **12**-3
lap-seam, **12**-3
skiving, **12**-4
spiral-wound, **12**-3
standard sizes, table, **12**-8
Fibre drums, **12**-10
code numbers, **12**-15
costs of, table, **12**-16
disposal, **12**-12
freight classification of, **12**-13
hazardous materials, **12**-12
liquids, **12**-12
moisture barrier, **12**-15
nesting, **12**-15
pressure-sensitive tapes, **9**-22
standard sizes of, table, **12**-11
step-down reuse, **12**-11
style of closure, illustrated, **12**-14
suitable products, **12**-12
waterproofed, **12**-12
Fibre tubes, **12**-2, **12**-3

Fibreboard, corrugated (*see* Corrugated boxes)
Fill points, glassware, **6**-16
Filling lines:
  cans, table, **7**-14
  glassware, **6**-14
Filling machinery, **20**-13
Filling speeds, glassware, **6**-18
Films, **3**-2, **8**-4
  Aclar, **3**-47
  Alathon polyethylene, **3**-34
  bake-in foods, **3**-23
  Bexphane polypropylene, **3**-37
  Biax polystyrene, **3**-46
  Bicor polypropylene, **3**-37
  Bi-Poly-S polystyrene, **3**-46
  boil-in-bag, **3**-19 to **3**-21, **3**-23
  Capran nylon, **3**-21
  Celanar polyester, **3**-25
  cellophane: anchored coating for, **3**-12
    for baked goods, **3**-14
    bundling, **3**-14
    for candy, **3**-14
    for coatings, **3**-12
    code designations for, **3**-13
    dry conditions, **3**-12
    excess drag on machine, **3**-16
    for fresh meat, **3**-13, **3**-14
    for fresh produce, **3**-13
    for frozen foods, **3**-14
    for greasy products, **3**-14
    humectants in, **3**-12
    identification, **3**-51
    looseness in wrapping, **3**-15
    nitrocellulose coated, **3**-7
    polymer coated, **3**-7
    puckered, **3**-16
    for retarding rancidity, **3**-13
    sealing, **3**-14, **3**-15
    shrinkage, **3**-13
    sizing, **3**-12
    softeners, **3**-12
    solubilities, table, **3**-53
    static, **3**-16
    storage conditions, **3**-15
    uses for, typical, table, **3**-14
    yield of, **3**-16
  cellulose acetate, **3**-7, **3**-16, **3**-51, **3**-53
  cellulose acetate-butyrate, **3**-47, **3**-51, **3**-53
  cellulose nitrate, **3**-47, **3**-51, **3**-53
  cellulose propionate, **3**-7, **3**-53
  cellulose triacetate, **3**-47
  chlorotrifluoroethylene, **3**-47
  Clopane PVC, **3**-40
  Cobex PVC, **3**-40
  cold forming, **3**-22
Films (*Cont.*):
  composite, **3**-2
  Conolene polyethylene, **3**-34
  cut edges, sticking caused by, **3**-36
  Daran polyvinylidene chloride, **3**-44
  decoration for, **8**-6
  deep-drawn, **3**-20
  densities of, table, **3**-52
  dielectric sealing, **3**-7
  Durethene polyethylene, **3**-34
  dust attraction of, table, **3**-7
  Dynafilm polypropylene, **3**-37
  Ediflex starch film, **3**-47
  Estane polyurethane, **3**-49
  ethyl cellulose, table, **3**-7, **3**-17
  EVA copolymer, **3**-47
  extrusion, **3**-29, **8**-4
  Fandflex polyvinyl chloride, **3**-40
  fluorohalocarbon, **3**-47
  Fortiflex polyethylene, **3**-34
  for fresh produce, **3**-17
  gauge numbers of, **3**-2
  Ger-Pak polyethylene, **3**-34
  gloss, table, **3**-7
  H-film, **3**-48
  haze, table, **3**-7
  heat-and-serve, **3**-24
  heat-seal temperatures of, table, **3**-7
  impact strength of, **3**-7
  ink adhesion, test for, **3**-37
  irradiation, table, **3**-3
  Jodapak, **3**-34
  Kardel polystyrene, **3**-46
  Katheron polyethylene, **3**-34
  Kel-F, **3**-47
  Kodar polyester, **3**-25
  Korad methyl methacrylate, **3**-48
  Koroseal polyvinyl chloride, **3**-40
  Kypex polyvinyl chloride, **3**-40
  Methocel soluble, **3**-19
  methyl methacrylate, **3**-48
  methyl cellulose, **3**-7, **3**-19
  Noryl, **3**-48
  nylon, **3**-7
  Olefane polypropylene, **3**-37
  orientation of, **3**-4, **3**-6
  Oriex polyvinyl chloride, **3**-40
  Panta-Pak polyvinyl chloride, **3**-40
  Parylene poly-para-xylylene, **3**-48
  for pastry wrapping, **3**-17
  permeability of, **3**-7, **3**-9
  polycarbonate, **3**-7, **3**-22
  polyester, **3**-7
  polyethylene, **3**-25
    blown tubular, illustrated, **3**-29
    colored, **3**-32
    density of, effect of, table, **3**-32

Films, polyethylene, density of (*Cont.*):
high-density, table, **3**-7
low-density, table, **3**-7
heat-sealing, **3**-32, **3**-33
optical properties of, **3**-28
printing, **3**-32
roll print designation, illustrated, **3**-33
shrink, **3**-32
slip characteristics of, table, **3**-32
stiffness of, table, **3**-28
stress cracking, **3**-27
stretch films, **3**-32
treatment of surface, **3**-30
uses of, **3**-26
Polyox, **3**-49
polypropylene, **3**-34
cold-weather difficulties, **3**-35
heat-sealing difficulties, **3**-35
properties of, **3**-7
shelf life of, **3**-35
polystyrene (*see* styrene *below*)
PPO, **3**-49
properties of, table, **3**-7
PVC (polyvinyl chloride), **3**-7, **3**-37
degradation of, **3**-39
heat-sealing, **3**-39
odors, **3**-39
optical properties of, **3**-40
plasticizers, **3**-39
rigid and flexible, **3**-39
radiation, table, **3**-4
Relpro polypropylene, **3**-37
Relthene polyethylene, **3**-34
Resinite polyvinyl chloride, **3**-40
Reynolon polyvinyl chloride, **3**-40
in rolls, illustrated, **3**-30
rubber hydrochloride, **3**-7, **3**-40, **3**-51, **3**-53
Rucoam polyvinyl chloride, **3**-40
saran (PVDC), **3**-42
properties of, table, **3**-7
slip agents, **3**-43
solvents, **3**-44
Scotchpak polyester, **3**-25
sealing, **3**-5, **3**-36
shrink, **3**-4, **3**-5, **3**-8
silicone, **3**-50
sleeve wrapping, **3**-6
Snugpak rubber hydrochloride, **3**-42
specific gravity, **3**-52
sterilization, **3**-19 to **3**-21
styrene, **3**-44
heat-sealing, **3**-46
metalizing, **3**-46
properties of, table, **3**-7
static, **3**-46
Styroflex polystyrene, **3**-46
Sumilite polyvinyl chloride, **3**-40

Films (*Cont.*):
Surlyn ionomer, **3**-48
synthetic paper, **4**-21
tear strength of, table, **3**-7
Teflon, **3**-47
temperature range of, table, **3**-7
Terylene polyester, **3**-25
Teslar polyvinyl fluoride, **3**-50
test for ink adhesion, **3**-37
Thermalux polysulfone, **3**-49
transparency, table, **3**-7
treatment: aging of, **3**-37
illustrated, **3**-31
Trithene chlorotrifluoroethylene, **3**-47
Trycite polystyrene, **3**-46
Udel polypropylene, **3**-37
Velon polyvinyl chloride, **3**-40
Videne polyester, **3**-25
Visolyte polystyrene, **3**-46
Vitafilm polyvinyl chloride, **3**-40
Vypro polypropylene, **3**-37
Watahyde polyvinyl chloride, **3**-40
water-soluble, **3**-19, **3**-49
water-vapor transmission, table, **3**-7
for windows in envelopes, **3**-17
yield of, table, **3**-2
Zytel nylon, **3**-21
Fine papers, **4**-13
Finish sizes, definition of, **9**-3
Finishes:
of boxboard, table, **4**-18
glassware, **6**-2, **6**-19, **6**-20, **6**-24, **11**-21
Firkin, definition of, **15**-12
Flame treatment test, **18**-6
Flash point test, **18**-6
Flat crush test, **18**-6
Flat crush values of corrugated, table, **5**-6
Flavor retention, table, **3**-58
Flavor test, **18**-6
Flexible packaging machinery, **20**-15
Flint glass, **6**-1
Fluidic controls, **20**-17, **20**-18
Flutes per foot, corrugated, **5**-3
Foamed plastics (*see* Plastics, foamed)
Foil:
aluminum, **3**-54
adhesives for, **3**-58
alloys, **3**-54, **3**-55
coatings, **3**-58
definition of, **3**-54
degassing, **3**-57
forming, **3**-56
labels, **14**-5
moisture-vapor transmission, table, **3**-57
oxide defects of, **3**-56
pack rolling, **3**-56

Foil, aluminum (*Cont.*):
pinholes, **3**-56
primers, **3**-59
printing, **14**-6
properties of, table, **3**-58
reflectivity of, **3**-56
residues of oil, **3**-60
rolling oil, **3**-60
scalping, **3**-57
smooth sidewall trays, **3**-60
strike-through lamination, **3**-58
for surface finishes, table, **3**-57
temper of, **3**-55
tensile strength of, **3**-55
tinfoil, **3**-54
Folding boxboard (*see* Paperboard)
Folding cartons, **2**-1
artwork for, **2**-20
boxboard, **2**-2
bulge factors for, illustrated, **2**-8
collapsible tube carton, illustrated, **2**-10
combination run, **2**-14
creasing and cutting, illustrated, **2**-16
decoration for, **2**-15
die rub-off, illustrated, **2**-15
dimensions of, illustrated, **2**-13
gluing and folding, **2**-18
grain, **2**-11
humidity, effect of, illustrated, **2**-11
laminations, **2**-5
made ready, printing press, illustrated, **2**-15
materials for, **2**-1
nomenclature of, illustrated, **2**-13
plastic, **2**-5
printing processes, table, **2**-19
properties of, table, **2**-10
reverse-tuck, **2**-5
rigidity of, illustrated, **2**-8
samples of, **2**-11
score lines: illustrated, **2**-12
prebroken, table, **2**-17
seal-end type, **2**-5
shipping and storage of, **2**-18
specifications for, **2**-20
stacking strength of, illustrated, **2**-9
stiffness of, **2**-5
styles of, illustrated, **2**-5 to **2**-7
testing of, **2**-12, **2**-20
thickness of board, table, **2**-8
Food and Drug Administration (FDA), **8**-5, **16**-7, **16**-11, **16**-12
Forces on packages in transit, **17**-2
Foreign machines, **20**-19, **20**-29
Forest Products Laboratory, **15**-2, **15**-3
Fourdrinier kraft, **5**-5
Fourdrinier machine, **4**-8
Fragility factors, **17**-2
Frequency distribution, **19**-3
Friction-plug cans, **7**-18, **7**-19
Full-overlap slotted container, **5**-2
Furnish for paper, **4**-8

G factor, **17**-1
Gas-flush packaging, **20**-15
Gas permeability of plastics, **8**-38
Gas transmission test, **18**-6
Gauge numbers of films, **3**-2
Glass Container Manufacturers Institute (GCMI), finishes on glassware, **6**-19, **6**-24
Glassine paper, **4**-15
Glassware, **6**-1
adhesives for labeling, **6**-13, **6**-15
aerosols, **11**-21, **11**-25
alkali in glass, **6**-5
allowance in mold design, **6**-9
analysis of fractures, **6**-9, **6**-10
annealing, **6**-8, **6**-12
blank mold, **6**-12
blooming, **6**-5
blowing operations, illustrated, **6**-12
borosilicate, **6**-5, **6**-13
bottom marks, illustrated, **6**-17
chemical reaction, **6**-5
chemistry, **6**-3
coatings for, table, **6**-7, **6**-13
colors, **6**-5, **6**-6
composition of, table, **6**-3
cords, **6**-4, **6**-8
defects in, **6**-18, **6**-22, **6**-23
design considerations for, **6**-8
design faults in, illustrated, **6**-18
double-gobbing, **6**-2
fill points, **6**-16
filling lines, **6**-14
finishes, **6**-2, **6**-19, **6**-20, **6**-24, **11**-21
flint glass, **6**-1
GCMI (*see* Glass Container Manufacturers Institute)
glass plant, illustrated, **6**-11
glass tank, illustrated, **6**-4
headspace, **6**-15
impact energy, **6**-18
IS (individual section) machine, **6**-2
illustrated, **6**-11
labeling, **6**-13
leaching, **6**-5
lehr, illustrated, **6**-11, **6**-12
light transmission, **6**-6
line breakage, **6**-14
locating bar on, **6**-13
lubricity of coatings for, **6**-8

Glassware (*Cont.*):
luster decorating, **6**-15
Lynch machine, **6**-2
manufacturing, **6**-10
microcracks, **6**-7
Miller press-and-blow machine, **6**-2
molds, **6**-15, **6**-16
notch sensitivity, **6**-7
Owens machine, **6**-2
parison, **6**-9, **6**-12
pharmaceutical, **6**-5
pressing, **6**-19
quality control, **6**-27
resistance to chemical action, **6**-4
sample size, **6**-27
seeds and blisters in, **6**-4, **6**-23, **6**-26
shock bands, illustrated, **6**-9
soft grade, **6**-4
sorting, **6**-12
stippling, **6**-8, **6**-16
strength of, **6**-6, **6**-7
stronger glass, **6**-3
sulfur treatment, **6**-5
surface treatment, **6**-14
tempered, **6**-6
thermal shock, **6**-4
tolerances, **6**-19, **6**-22
trademarks, illustrated, **6**-17
tubing products, **6**-28
Glazed papers, **4**-14
Gloss in wax coatings, **13**-5
Gloss test, **18**-6
Glued joints in corrugated, **5**-10
Gluing corrugated boxes, **5**-16
GRAS regulations, **16**-7, **16**-8
Grease resistance test, **18**-6
Greaseproof papers, **4**-15
Greaseproofness of bags, **10**-8
Greasy products in cellophane, **3**-14
Groundwood pulp, **4**-5
Gummed tape, **9**-20

Hardness test, **18**-6
Hardwood, definition of, **4**-2
Harmonics in cushioning, **17**-5
Hazardous materials:
in fibre drums, **12**-12
regulations for, **16**-3, **16**-6, **16**-12
Haze test, **18**-7
Headspace, **6**-15, **16**-5
Heat-and-serve films, **3**-24
Heat-distortion test, **18**-7
Heat-seal coating, **13**-7
Heat-sealing of films, **3**-32, **3**-39, **3**-46, **13**-8
Heat-transfer printing, illustrated, **8**-7
Hoop stress, **18**-7
Hot-dip coatings, **3**-17
Hot-dipped tinplate, **7**-4
Hot-melt on corrugated, illustrated, **5**-14
Hot-stamp printing, illustrated, **8**-7
Howmet labeling, **14**-6
Humectants in cellophane, **3**-12
Humidity, effect of:
on boxboard, **2**-11
on cellophane seals, **3**-15
on corrugated, **5**-8

IATA (International Air Transport Association) regulations, **16**-6, **16**-19
Impact energy of glassware, **6**-18
Impact extrusion, illustrated, **7**-27
Impact strength:
of films, **3**-27
of plastics, table, **8**-39
Incline-impact test, **18**-7
Industrial packaging, **1**-7
Infrared spectroscopy, **3**-52
Injection molding of plastics, **8**-4, **8**-19, **8**-45
Ink adhesion test, **3**-37
Insect infestation test, **18**-7
Inspection, quality control, **19**-10
Institutional packaging, **1**-7
Interior corrugated parts, illustrated, **5**-12
Interstate Commerce Commission, **16**-5
Invention, **16**-21
Irradiation of films, table, **3**-3
IS (individual section) machine for glassware, **6**-2, **6**-11
Izod impact test, **18**-7

Joints and seams in cans, **7**-12
Jute fibreboard, **4**-16, **5**-5

Keg, definition of, **15**-12
Kit, definition of, **15**-12
Kraft, **4**-6
corrugated medium, **5**-4
paper, **4**-13
process, **4**-6

Labeling, **14**-1
adhesion test, **18**-7
adhesives, **9**-30, **14**-11
aerosols, **11**-26
copyrighted, **16**-23
costs of, **14**-13
designs, **14**-9

Labeling (*Cont.*):
  die cutting, **14**-4
  foil labels, **14**-5
  glassware, **6**-13
  grain direction, **14**-5, **14**-11
  label cutter, **14**-4
    illustrated, **14**-10
  label size, illustrated, **14**-10
  machinery, illustrated, **20**-20
  offset printing, **14**-6
  paper stock, **14**-10
  pressure-sensitive, **14**-7
  printing, **14**-4, **14**-8
  promotional value, **14**-3
  regulations for, **14**-2, **16**-12, **16**-13
  silk screen printing, **14**-7
  storage conditions for, **14**-11
Lacquers, **13**-6
Laminations, **3**-59
  applications for, table, **13**-13
  barrier properties of, table, **13**-14
  characteristics of materials, **13**-11, **13**-12
  costs of, **13**-12, **13**-16, **13**-18
  design of, **13**-10
  directional effects of, **13**-16
  laminating machine, illustrated, **13**-15
  methods for, **13**-9
  permeability calculations of, **13**-16
  printability, **13**-13
  process, basic, illustrated, **13**-10
  stiffness, **13**-13
  thermoforming, **13**-14
  uses for, table, **13**-13
Laying out copy for corrugated, **5**-13
Layout of machinery, **20**-6
LD_{50}, **16**-15
Leaching of glassware, **6**-5
Leasing machinery, **20**-25
Light transmission of glassware, illustrated, **6**-6
Liner facings for caps, **9**-16
Linerboard, **4**-16, **5**-3
Linerless caps, **8**-27
Linings for cans, **7**-8
Lint-free paperboard, **4**-20
Liquids of different densities, **3**-52
Locating bar for glassware, **6**-19
Logic systems, **20**-17
Low-pressure polyethylene, **8**-24
Lubricants in plastics, **8**-43, **8**-68
Lubricating oil, **12**-3
Lumber, **15**-2
Luster decorating of glassware, **6**-15
Lynch machine for glassware, **6**-2

Machinery, **20**-1
  alterations, **20**-30
Machinery (*Cont.*):
  amortizing the cost of, **20**-21
  annual production of, table, **20**-2
  canning, table, **7**-14
  contracts, **20**-26, **20**-28
  controls, **20**-16, **20**-18
  conveyors, **20**-7, **20**-9
  costs of, **20**-2, **20**-21 to **20**-24
  custom design of, **20**-18
  debugging, **20**-30
  depreciation of, **20**-22, **20**-24
  economic life of, **20**-24
  economic study of, **20**-20
  economics of, **20**-7
  electronics, **20**-16, **20**-17
  fillers, **20**-12 to **20**-14
  financial analysis of, **20**-20
  flexibility of, **20**-19
  foreign machines, **20**-19, **20**-29
  installation of **20**-30
  layout of, illustrated, **20**-6
  leasing of, **20**-25
  line operation of, illustrated, **20**-8
  lubrication of, **20**-29
  maintenance of, **20**-5, **20**-29
  overwrapping, **20**-15, **20**-16
  papermaking, **4**-8, **4**-14
  rental of, **20**-25
  safety controls for, **20**-29
  sanitation, **20**-9
  scheduling delivery, **20**-30
  single purpose machine, **20**-19
  site for, **20**-4
  spare parts for, **20**-29
  specifications for, **20**-27
  supplier selection, **20**-10
  synchronizing, **20**-3
  systems approach, **20**-2
  testing, **20**-29
  types of, **20**-11
  warranty, **20**-6
Mailing bags, **10**-10
Management of packaging, **1**-8
Manila-lined board, **4**-6
Manufacturer's joints in corrugated, **5**-4
Mar resistance of plastics, **8**-39
Mar resistance test, **18**-7
Medium for corrugated, **4**-16, **5**-2
Melt index test, **18**-7
Mesh bags, **10**-11
Metal:
  aerosol cans (*see* Cans, aerosol)
  alloys: aluminum, **7**-31, **7**-32
    tin and lead, **7**-26
  aluminum, **3**-53, **7**-31
    cans, **7**-15
    corrosion resistance of, **7**-32
    foil, **3**-54, **3**-55

Metal, aluminum (*Cont.*):
orientation or grain of, **7**-31
temper of, **7**-31
trays, **7**-32
tubes, **7**-26
work-hardened, **7**-31
aluminum-coated steel, **7**-6, **7**-7
aluminum foil (*see* Foil, aluminum)
aluminum-killed steel, **7**-4
annealing aluminum tubes, **7**-26
base cost of tube metals, **7**-25
cans (*see* Cans)
cemented side seams, **7**-8
chrome-coated steel, **7**-6
CMQ tinplate, **7**-5
enamels, **7**-8
foils (*see* Foils)
passivated steel, **7**-6, **7**-7
plug closures, **12**-4
steel for containers, table, **7**-7
Straits tin, **7**-6
tin-free steel, **7**-7
tinplate, **7**-7
tubes: bottom closure, illustrated, **7**-30
capacity, table, **7**-28
laminated tin and lead, **7**-26
neck size of, table, **7**-28
production of, annual, table, **7**-25
sealing, **7**-30
standard dimensions of, table, **7**-28
vapor coating, **7**-8
welded side seams, **7**-8
Metalizing films, **3**-40
Microcrystalline waxes, **13**-4
Military packaging, **1**-7
Miller machine for glassware, **6**-2
Model law, **16**-5, **16**-11
Modulus of elasticity test, **18**-7
Moisture-vapor transmission (MVT), **13**-11, **18**-7
in plastics, **8**-37
Moistureproof paper, **10**-7
Mold inhibitors in paper, **4**-15
Molder's marks in plastics, illustrated, **8**-63
Molding machine for plastics, **8**-46, **8**-51 to **8**-53
Moldproofness test, **18**-7
Molds, **8**-49
blow molding, **8**-60
caps, **9**-8
glass, **6**-15
plastics, **8**-49, **8**-51
thermoforming, **8**-77
Mullen test, **18**-7
Multiwall sacks, **9**-22, **10**-3, **10**-9
MUP (molded urea plastic) colors, **8**-35
Mylar polyester film, **3**-25

Nailed wood boxes (*see* Wood containers)
National Motor Freight Classification Board, **16**-6
Rule 222, **16**-17
Natural tolerance in quality control, **19**-4
Newsback paperboard, **4**-17
Newsprint paper, **4**-6
Nonbending chipboard, regular number for, **4**-19
Nonpolar plastics, **8**-24
Notch sensitivity in glassware, **6**-7
Notch sensitivity test, **18**-8
Number packages, **16**-18
Numbers, regular, for paperboard, **4**-18

Odor and flavor retention, **8**-6
Odor test, **18**-8
Odors in film, **3**-39
Offset printing, illustrated, **8**-7
Oil penetration test, **18**-8
Open-head drums, **7**-22
Optical properties of films, **3**-28, **3**-40
Organoleptic tests, **18**-8
Orientation of films, **3**-4, **3**-6, **3**-38
Overwrapping machinery, **20**-16
Owens machine for glassware, **6**-2

Pails:
bails for, **7**-23
definition of, **15**-12
metal, **7**-23, **7**-24
wood, **15**-12
Paper, **4**-5
back liner, **4**-17
basis weight of, **14**-5, **14**-14
beating curves, illustrated, **4**-8
bender, **4**-17
bleached kraft, **4**-14
bogus, **4**-17
calendered, **4**-12
card stock, **4**-2
coarse papers, **4**-13
consumption, per capita, **4**-2
corrugated medium, **4**-16
costs of, **4**-21, **5**-15
creped paper, **4**-14
density of, **4**-12
extensible paper, **4**-14
fiber-free surface, **4**-15
fine papers, **4**-13
finishes, **14**-9
fourdrinier machine, **4**-9
furnish, **4**-8
glassine, **4**-15
greaseproof, **4**-15
jute, **4**-16

Paper (*Cont.*):
kraft, **4**-13
kraft process, **4**-6
labeling, **14**-5, **14**-10
machine, **4**-9
machine-glazed, **4**-14
mold inhibitors, **4**-15
newsprint, **4**-6
parchment, **4**-15
porosity test, **18**-8
pulp, **4**-5, **4**-6
rancidity prevention, **4**-15
release coatings, **4**-15
shrinkage and expansion, **4**-4
soda process, **4**-6
stiffness, **4**-10, **4**-12
strength of, when wet, **4**-15
substitutes for, **8**-86
sulfate process, **4**-6
sulfite process, **4**-6
surface, illustrated, **4**-16
synthetic papers, **4**-21
types of, **4**-13
Tyvek, **4**-20
waxed papers, **4**-15, **4**-16
wet-strength, **4**-14
Paperboard, **4**-17
bending board grades, **4**-18
boxboard, table, **4**-18, **4**-19
brightness, table, **4**-14, **4**-19
bulge, **4**-19
chipboard, **4**-17
containerboard, **4**-16
costs of, table, **4**-20
cylinder board, table, **4**-10
cylinder machine, **4**-8
folding boxboard, **2**-2, **2**-8, **2**-10, **4**-17, **4**-18
gauges, table, **4**-18
linerboard, **4**-16
lint-free, **4**-20
manila-lined, **4**-6
newsback, **4**-17
nonbending, **4**-18, **4**-19
numbers for, regular, **4**-18
papermaking, **4**-4, **4**-11
setup boxes, **2**-21
solid bleached sulfate, **4**-10, **4**-18, **4**-19
standard finishes for, table, **4**-18
sterile packages, **4**-20
top liner, **4**-17
ultra low density, **4**-19
wet machine board, **4**-18
Parcel post test, **18**-8
Parchment paper, **4**-15
Parison:
glass, **6**-12
Parison (*Cont.*):
plastic bottle, **8**-57
programmed, illustrated, **8**-59
Passenger plane regulations, **16**-6
Passivated steel, **7**-6, **7**-7
Pasted sack, **10**-6
Patents, **16**-20
Payout of machinery, **20**-23
Peel bond test, **18**-8
Peel coat for tubes, **7**-29
Peelable seal coatings, **13**-8
Permeability, **3**-9, **8**-23, **8**-38, **13**-16, **18**-8
Pilferage, **1**-20
Pinhole test, **18**-8
Plastic films (*see* Films)
Plastic folding cartons, **2**-5
Plastic papers, **4**-21
Plastics, **8**-2
ABS, **8**-7, **8**-41
Abson ABS, **8**-9
acetal, **8**-9
acrylic, **3**-53, **8**-10
additives for, **8**-4, **8**-68
antistatic agents, **8**-44
fillers, **8**-18
lubricants, **8**-43, **8**-68
plasticizers, **3**-39, **8**-41, **8**-68
stabilizers, **8**-43, **8**-68
toxicity, **8**-43
aerosol containers, **11**-25
Alathon polyethylene, **8**-24, **8**-26
Alkathene polyethylene, **8**-26
amorphous, **8**-3
Ampacet polyethylene, **8**-26
Anar polyethylene, **3**-34
Arodure urea, **8**-36
Arothane polyurethane, **8**-32
bags, **10**-5, **10**-10
Bakelite phenolic, **8**-19, **9**-2
Beetle urea, **8**-36
Bexoid cellulose acetate, **8**-13
Bexphane polypropylene, **8**-28
Bextrene polystyrene, **8**-30
Blapol polyethylene, **8**-26
blisters, carding methods, **8**-82
blow molded, **8**-4, **8**-56
blow ratio, **8**-64
bottle collapse, **8**-25, **8**-59
coatings, **8**-71
costs of, **8**-65, **8**-66
design of, **8**-58
extruder screw, illustrated, **8**-63
finish dimensions, **8**-61
machines, **8**-58, **8**-62, **8**-64
molds, **8**-60
processes, **8**-57, **8**-65
production per year, **8**-56

Plastics, blow molded (*Cont.*):
programmed parison, **8**-59
stress cracking, **8**-25
testing, **8**-71
brittleness, table, **8**-39
C-11 copolymer, **8**-41
calendered film, **8**-4
Carag polyethylene, **8**-26
Carina polyvinyl chloride, **8**-35
Carinex polystyrene, **8**-30
casting, **8**-4
Celcon acetal, **8**-10
cellulose acetate, **8**-11, **8**-12
cellulose propionate, **8**-13
chemical properties, definition of, **8**-36
chemical resistance, table, **8**-39
Chem-o-thane polyurethane, **8**-32
Chempol polyurethane, **8**-32
clarity, table, **8**-38
cold brittleness, **3**-35
cold forming, **3**-32, **3**-36, **8**-20
collapse of bottles, **8**-25, **8**-59
collapsible tubes, **8**-74, **8**-75
compression molding, **8**-4, **8**-19, **8**-53, **8**-54
copolymers, **8**-4, **8**-7, **8**-8, **8**-40
cost of, **8**-38, **8**-49
creep test, **18**-5
cross-linked, **8**-3, **8**-22
crystalline, **8**-3
Cycolac ABS, **8**-9
decoration of, **8**-6
Delrin acetal, **8**-10
Durez phenolic, **8**-19
dust attraction, table, **8**-38
Dylan polyethylene, **8**-24
Dylene polystyrene, **8**-30
Dylite expanded polystyrene, **8**-85
EEP foam, **8**-86
Elvax polyvinyl chloride, **8**-35
Escon polypropylene, **8**-28
ethyl acrylate, **8**-10
ethyl cellulose, **8**-14
ethylene butene, **3**-47
EVA, **8**-41
expanded (*see* foamed *below*)
Expandofoam polyurethane, **8**-32
extrusion, **8**-4
E-Z Flow polystyrene, **8**-30
fillers, **8**-18, **8**-44
film (*see* Films)
foamed, **8**-4, **8**-84
mold, illustrated, **8**-87
in place, **8**-32
polystyrene, **8**-85
polyurethane, **8**-31, **8**-32, **8**-86
Fortiflex polyethylene, **8**-24, **8**-26

Plastics (*Cont.*):
Fosterene polystyrene, **8**-30
Geon polyvinyl chloride, **8**-35
Hi-Fax polyethylene, **8**-26
high-pressure polyethylene, **8**-22
hinge, **8**-27
Hostalen polyethylene, **8**-26
Hostalit polyvinyl chloride, **8**-35
impact strength, table, **8**-39
injection mold, illustrated, **8**-46, **8**-50
injection molding, **8**-4, **8**-19, **8**-45, **8**-46, **8**-49, **8**-50
injection molding design, **8**-46
ionomers, **3**-48
Irvinil polyvinyl chloride, **8**-35
Isofoam polyurethane, **8**-32
isotactic form, **8**-26
Kardel polystyrene, **8**-30
Kenron polyvinyl chloride, **8**-35
Lexan polycarbonate, **8**-21
Lorkalene polystyrene, **8**-30
Lucite acrylic, **8**-11
Lumarith cellulose acetate, **8**-13
Luparen polypropylene, **8**-28
Lupolen polyethylene, **8**-26
Luran polystyrene, **8**-30
Lustrex polystyrene, **8**-30
Lux-Foam polyurethane, **8**-32
mar resistance, table, **8**-39
Marvinol polyvinyl chloride, **8**-35
materials, selection of, **8**-5
MEP foam plastic, **8**-86
Merlon polycarbonate, **8**-21
Mesa phenolic, **8**-19
methyl methacrylate, **8**-10
mold costs for, **8**-51
molder's marks, illustrated, **8**-63
molding machines, **8**-46, **8**-51 to **8**-53
molds, **8**-49, **8**-51, **8**-60, **8**-77, **9**-8
monomers, **8**-3
MUP colors, **8**-35
nonpolar, **8**-24
Nopofoam polyurethane, **8**-32
nylon, **8**-15, **8**-16
Nypel nylon, **8**-17
Olefane polypropylene, **8**-28
Opalon polyvinyl chloride, **8**-35
Pacrosir acrylic, **8**-11
permeability, table, **8**-38
Petrothene polyethylene, **8**-26
phenolics, **8**-17, **8**-18
Plastacele cellulose acetate, **8**-13
polyallomers, **3**-48
polyamides, **3**-19, **8**-15
Polyart synthetic paper, **4**-21
polycarbonate, **3**-21, **3**-53, **8**-19, **8**-21
polyester, **3**-23, **3**-51, **3**-53

Plastics (*Cont.*):
polyethylene, **8**-21
bags, **10**-9, **10**-10
blow-molded, **8**-65
coatings, **13**-8
crystals, illustrated, **8**-22
density of, table, **8**-66
film, **3**-25
high density, **8**-24
identification of, table, **3**-51
low density, **8**-21
melt index, table, **8**-67
permeability of, **8**-23
solubilities, table, **3**-53
weight loss in bottle, **8**-23
polyethylene oxide, **3**-48
Polyfilm polyethylene film, **3**-34
Polyflex polystyrene film, **3**-46
polymer coated cellophane, **3**-12
polymers, **8**-3
Polyox polyethylene oxide, **3**-49
poly-para-xylylene, **3**-48
polyphenylene oxide, **3**-49
polypropylene, **8**-26
bags, **10**-10
coatings, **13**-8
crystals, illustrated, **8**-28
film, **3**-34
hinges, **8**-48
identification of, **3**-51
nucleated, **8**-27
polystyrene, **8**-29, **8**-69
expanded, **8**-85, **17**-4
film, **3**-44
foam, **8**-85, **17**-4
identification of, table, **3**-51
polysulfone, **3**-49
Polython polyethylene film, **3**-34
polyurethane, **8**-30
film, **3**-49
foam, **8**-31, **8**-32, **17**-4
polyvinyl alcohol, **3**-49, **3**-53
polyvinyl chloride, **8**-33
blow-molded, **8**-67
film, **3**-37
identification of, table, **3**-51
plasticizers, **8**-42
solubilities, table, **3**-53
polyvinyl fluoride, **3**-50
polyvinylidene chloride, **3**-42
press welding, **3**-23, **8**-21
printability of, table, **8**-38
Pro-Fax polypropylene, **8**-28
Propafilm polypropylene, **3**-37
Propathene polypropylene, **8**-28
properties of, table, **8**-39
PVC (*see* polyvinyl chloride *above*)

Plastics (*Cont.*):
reinforcements, **8**-44, **8**-45
Resfurin urea, **8**-36
Resilo-Pak expanded polystyrene, **8**-85
Restirolo polystyrene, **8**-30
Riblene polyethylene, **8**-26
rotational molding, **8**-72, **8**-73
Rucoblend polyvinyl chloride, **8**-35
SAN copolymer, **8**-41
Santofome expanded polystyrene, **8**-85
selection of, **8**-37
sheet (*see* Films)
shrinkage, **8**-48
Sicaloid cellulose acetate, **8**-13
Siritle urea, **8**-36
skin packaging, **8**-82
slush casting, **8**-5, **8**-73
spin welding, **8**-21
Stafoam polyurethane, **8**-32
Stamylan polyethylene, **8**-26
Stanfoam polyurethane, **8**-32
stiffness, table, **8**-39
stress cracking, **8**-23, **8**-59
styrene (*see* polystyrene *above*)
Styrofoam polystyrene, **8**-85
Super Modulene polyethylene, **8**-26
synthetic paper, **4**-21
Synvarol urea, **8**-36
Syplast urea, **8**-36
tear strength, table, **8**-39
Tenite I cellulose acetate, **8**-13
terpolymers, **8**-7
thermoforming, **8**-76
costs of, **8**-83
design of, **8**-77
laminations for, **13**-14
methods for, **8**-78
molds, **8**-77
processes, **8**-79
Thermothane polyurethane, **8**-32
Udel polypropylene, **8**-28
Ultramid nylon, **8**-17
Unifoam polyurethane, **8**-32
Uralane polyurethane, **8**-32
urea, **8**-35
urethane, **8**-31, **8**-86, **8**-88, **17**-4
Vestron polystyrene, **8**-30
Vinoflex polyvinyl chloride, **8**-35
Vygen polyvinyl chloride, **8**-35
X-tal nylon, **8**-17
Zytel nylon, **8**-17
Pollution, air and water, **1**-6
Presentation to management, **1**-27
Press varnishes, **13**-6
Pressed glassware, **6**-19
Pressure-sensitive tapes (*see* Tapes, pressure-sensitive)

Price marking, **16**-13
Printability tests, **13**-13
Printing, **8**-6
  bags, **10**-11
  heat-transfer process, **8**-7
  hot stamping, **8**-7
  legal, **16**-7
  offset, **8**-7
  polyethylene, **3**-32
  silk-screen, **8**-7
Printing press, illustrated, **13**-14, **14**-8
Probability in quality control, **19**-3, **19**-6
Process control, quality control, **19**-4
Produce, films for fresh, **3**-13, **3**-14, **3**-17
Publications on packaging, **1**-28
Pull tabs on cans, **7**-14
Pulping, **4**-3, **4**-6, **4**-7
Puncture test, **18**-8
Purchasing, **1**-27
PVC (*see* Plastics, polyvinyl chloride)
PVDC (*see* Coatings, saran; Films, saran; Saran)

Q-Kote synthetic paper, **4**-21
Q-Per synthetic paper, **4**-21
Quality control, **19**-1
  acceptance quality level (AQL), **19**-5
  assignable cause, **19**-5
  attributes defined, **19**-4
  for caps, **9**-11
  certification, **19**-10
  consumer's risk, **19**-9
  for disposition of rejected material, **19**-11
  frequency distribution, **19**-3
  for glassware, **6**-27
  inspection, **19**-10
  natural tolerance, **19**-4
  probability, **19**-3, **19**-6
  and process control, **19**-4, **19**-5
  producer's risk, **19**-9
  rule of ten, **19**-2
  sampling, **19**-7 to **19**-9
  significance of results, **19**-5
  size of sample, **19**-9
  specifications, **19**-2, **19**-10
  standard deviation, **19**-3, **19**-4
  standard samples, **19**-10, **19**-11
  statistical principles, **19**-2, **19**-3
  t test of significance, **19**-6, **19**-7
  tools, **19**-11
  type I and *type II* error, **19**-7
  variables defined, **19**-4

Radiation of films, table, **3**-4
Rancidity prevention, **3**-13, **4**-15
Random numbers, quality control of, **19**-8
Random sampling, **19**-7
Raoult's law, **11**-16
Reference publications on packaging, **1**-28
Regular slotted container (RSC), illustrated, **5**-2
Regulations, **16**-4
  for adulterants, **16**-7
  for aerosols, **11**-14, **11**-24, **11**-28
  air cargo, **16**-6, **16**-18, **16**-19
  air express, **16**-19
  Bureau of Explosives, **16**-14
  for copyrights, **16**-23
  Department of Commerce, **16**-7
  Department of Transportation, **16**-6, **16**-14
  for etiological materials, **16**-14
  for explosives, **16**-14
  express, **16**-19
  Fair Packaging and Labeling Act, **16**-11
  Federal Trade Commission, **16**-5, **16**-13
  for fibre drums, **12**-13
  Food and Drug Administration, **16**-7, **16**-11
  for glass surface treatment, **6**-14
  GRAS, **16**-7
  for hazardous materials, **16**-6, **16**-12
  IATA, **16**-6, **16**-19
  labeling, **14**-2
  model laws, **18**-11
  National Motor Freight Classification Board, **16**-6
  package structure, **16**-5
  for passenger planes, **16**-6
  for patents, **16**-20 to **16**-22
  post office, **16**-15
  state laws, **16**-15
  for trademarks, **16**-19, **16**-20
  for warning labels, **16**-14, **16**-15
  for weights and measures, **16**-11
Reinforcements in plastics, **8**-44, **8**-45
Release coatings on paper, **4**-15
Rental of machinery, **20**-25
Research in packaging, **1**-7
Reshippers for cans, **7**-13
Resonance in cushioning, **17**-3
Reverse tuck (*see* Folding cartons)
Rigidity of folding cartons, **2**-8
Rigidity test, **18**-8
Rockwell hardness test, **18**-8
Roll forms of film, **3**-30
Roll print designation, **3**-33
Rotational molding of plastics, **8**-72, **8**-73
Rub resistance test, **18**-8
Rubber stoppers, **9**-14, **9**-20
Rule of ten, quality control, **19**-2

Rule 41, Uniform Freight Classification, **16**-17
Rule 222, National Motor Freight Classification, **16**-17
Rust protection in cans, **7**-2

Sacks (*see* Bags)
Sampling, **19**-7
SAN plastic copolymer, **8**-41
Sanitary-style cans, **7**-2
Saran (PVDC), **3**-42
  coatings, **13**-8
  identification, table, **3**-51
  solubilities, table, **3**-53
Scarab urea, **8**-36
Score bend test, **18**-8
Score line strength of corrugated, **5**-6
Scoring allowances for corrugated, **5**-8
Screw caps (*see* Caps)
Screw-top cans, **7**-18, **7**-20
Scuff resistance of wax coatings, **13**-5
Sealing:
  costs of, for corrugated, **5**-17
  of film, **3**-5, **3**-36
  of metal tubes, **7**-30
Sealing cellophane, **3**-14, **3**-15
Sealing compound for cans, **7**-10
Seals, **9**-10
Seamless cans, **7**-20
Semichemical pulp, **4**-7
Setup boxes, **2**-21 to **2**-25
Shock cushioning, **17**-5
Shockbands in glassware, illustrated, **6**-9
Shrink bands, **9**-10
Shrink films, **3**-4, **3**-5, **3**-8, **3**-32, **3**-38
Shrinkage of plastics, **8**-48
Significance of results in testing, **19**-5
Silk-screen printing, **8**-7, **14**-7
Single-faced corrugated, **5**-2
Single-trip drums, **7**-21
Single-use impact-extruded containers, **7**-30
Size designation of corrugated boxes **5**-8
Sizing on cellophane, **3**-12
Skin packaging, **8**-82
Sleeve packing with shrink film, **3**-6
Slip agents, in films, **3**-12, **3**-28, **3**-43
Slip characteristics of film, table, **3**-32
Slip test, **18**-8
Slipcover cans, **7**-20
Slush casting of plastics, **8**-5, **8**-73
Smoothness test, **18**-8
Soda process for paper, **4**-6
Soft grade of glass, **6**-4
Softeners for cellophane, **3**-12
Softening point test, **18**-8
Softwood, definition of, **4**-2
Solder, types of, **7**-12
Solid bleached sulfate, **4**-10, **4**-18 to **4**-19
Soluble films, **3**-19, **3**-53
Soluble paper, **4**-21
Specific gravity of films, **3**-52
Specific gravity test, **18**-8
Specifications, **1**-27, **2**-52
  of machinery, **20**-27
  in quality control, **19**-2, **19**-10
Spin welding plastics, **3**-23, **8**-21
Spiral winding, **12**-5
Spiral winding machine, illustrated, **12**-6
Spirit varnishes, **13**-6
Springwood, **4**-2
Squeeze bottles, **8**-56
Stabilizers in plastics, **8**-43
Stacking strength:
  corrugated boxes, **5**-4, **5**-8, **5**-10, **5**-11, **5**-18
  of folding cartons, **2**-9
Standard deviation in quality control, **19**-3
Static in films, **3**-16, **3**-46
Static test, **18**-9
Statistical quality control, **19**-1 to **19**-3
Steel sheet, thickness of, table, **7**-22
Stepped-end bags, **10**-7
Sterile paperboard packages, **4**-20
Sterilizable films, **3**-19 to **3**-21
Stiffness:
  of corrugated, **5**-5
  of folding cartons, **2**-5
  of plastics, table, **8**-39
Stiffness test, **18**-9
Stippling of glassware, **6**-16
Stitched joints in corrugated boxes, **5**-10
Stoppers, **9**-19, **9**-20
Strawboard corrugating medium, **5**-4
Strength of glassware, **6**-3, **6**-6, **6**-7
Stress cracking of plastics, **3**-27, **8**-23, **8**-25, **8**-59, **18**-9
Stretch bands, table, **9**-10
Stretch films, **3**-32
Strike-through laminations, **3**-58
Strippable coatings, **3**-17
Styling of packages, **1**-21
Styrene (*see* Films, styrene; Plastics, polystyrene)
Sulfate bleached paperboard, **4**-18
Sulfate paper **4**-6, **4**-10, **4**-19
Sulfite process for paper, **4**-6
Sulfur treatment of glassware, **6**-5
Supercalendering of paper, illustrated, **4**-13
Suppliers service, **1**-10
Surface treatment:
  of glass, **6**-14

Surface treatment (*Cont.*):
of plastics, **8**-24
Sutherland ink rub test, **18**-9
Symbols and numbers in glassware, **6**-17
Synthetic papers, **4**-21

t test of significance, **19**-7
Taped joints in corrugated, **5**-10, **5**-16, **5**-17
Tapes, **9**-20
gummed, **9**-20
pressure-sensitive, **9**-21
all-purpose, table, **9**-24
costs of, table, **9**-25
for fibre drums, **9**-22
for multiwall bags, **9**-22
properties of, table, **9**-23
testing of, **9**-22
Taste test, **18**-9
Tear strength of plastics, **8**-39
Tear test, **18**-9
Temper of aluminum, **7**-31
Tempered glass, **6**-6
Tempers of tinplate, **7**-4
Tensile test, **18**-9
Terneplate, **7**-5
Test shipments, **16**-18
Testing, **18**-1
accelerated tests, **18**-3
adhesion of coatings, **13**-8
aerosols, **11**-6, **11**-8, **11**-14, **11**-27 to **11**-29
bags, **10**-11
control, **18**-3
corrugated boxes, **5**-18
design of experiments, **18**-3
"to destruction," **18**-3
folding cartons, **2**-12, **2**-20
glass, **6**-14
identification, **3**-51, **4**-16
kinds of tests, **18**-4
machinery, **20**-29
magnitude of, **18**-2
number of tests, **18**-4
performance-type test, **18**-2
plastic bottles, **8**-71
plastics, **8**-6
pressure-sensitive tapes, **9**-22
properties, **18**-2
statistical, **18**-3
subjective, **18**-4
vibration, **18**-3
Text on packages, **1**-22
Therimage decoration, **14**-6
Thermal shock of glassware, **6**-4
Thermoforming, **8**-76
Thermoplastic plastics, **8**-3, **8**-4
Thermosetting plastics, **8**-3, **8**-4, **8**-17, **8**-19, **8**-35
Tierce, definition of, **15**-12
Tight-head drums, **7**-22
Tin-free steel, **7**-6, **7**-7
Tinfoil, **3**-54
Tinplate for cans, **7**-3 to **7**-5, **7**-7
Torque, **9**-6, **9**-7, **18**-9
Toxicity of stabilizers, **8**-43
Trademarks, **16**-19
in glass, illustrated, **6**-17
Transdec decoration, **14**-6
Transfer molding, **8**-19, **8**-55
Trays, aluminum, **7**-32
Treatment of plastic surface, **8**-24
Triple slide box, illustrated, **5**-2
Triple-wall corrugated, **5**-2
Tub, definition of, **15**-12
Tubes (*see* Collapsible tubes)
Tubing glassware, **6**-27, **6**-28
Twist wrap, **3**-43
Type I and *type II* error, **19**-7
Types of pharmaceutical glass, **6**-5
Tyvek synthetic paper, **4**-20, **4**-21

Ultraviolet (UV) stabilizers in plastics, **8**-43
Uniform Classification Committee, **16**-17
Uniform Freight Classification, Rule 41, **16**-17

Vacuum packaging machinery, **20**-15
Valve opening bags, **10**-6
Valves for aerosols, **11**-16, **11**-18
Variables in quality control, **19**-4
Varnish coatings, **13**-6
Varnishes:
press, **13**-6
spirit, **13**-6
Vending machines, **1**-17
Veneer in wood boxes, **15**-1
Vibration, **18**-9
cushioning, **17**-5
testing, **18**-3, **18**-9
in transportion, **17**-2
Vicat softening, **18**-9
Vinyl coatings, **13**-8
Vinyl plastic, **8**-33
Viscose shrink bands and caps, **9**-11

Warning labels, **16**-14, **16**-15
Warpage of plastics, table, **8**-38, **8**-40
Water-resistant corrugated, **5**-14
Water-soluble films, **3**-19, **3**-49
Water-vapor transmission, **8**-37, **8**-38, **18**-9

Water-vapor transmission rate (WVTR) for plastics, **8**-37, **13**-12
Wax coatings, **9**-14, **13**-4, **13**-5
Wax pick test, **18**-9
Waxed papers, **4**-15, **4**-16
Weights and measures, **16**-2, **16**-5, **16**-11
Welded side seam cans, **7**-8
Westite cement, **7**-31
Wet-strength, **4**-14, **4**-15, **10**-8, **18**-9
Wet wax coatings, **13**-3
White List, **16**-8 to **16**-10
Windows, envelope, films for, **3**-17
Wirebound boxes (*see* Wood containers, wirebound boxes)
Wood containers, **15**-1
  barrels, **15**-12
    handling and storage of, **15**-13
    illustrated, **15**-13
    lining materials for, **15**-13
    slack, **15**-13
    tight, **15**-13
  baskets, **15**-11
Wood containers (*Cont.*):
  box lumber size, table, **15**-4
  crates, **15**-9
    design of, **15**-9
    size of lumber, table, **15**-10
    three-way corner, illustrated, **15**-11
  hamper, illustrated, **15**-12
  nailed wood boxes, **15**-6
    cleated panel box, illustrated, **15**-6
    size of nails, table, **15**-6
    spacing of nails, table, **15**-6
    strapping, illustrated, **15**-7
    styles of, **15**-5
    thickness of lumber, **15**-4
  veneer defined, **15**-1
  wirebound boxes, **15**-7
    assembly and closure, **15**-9
    illustrated, **15**-8
    selection criteria, **15**-8
    wire size, table, **15**-9
Woods, box, grouping of, table, **15**-3
WVTR (*see* Water-vapor transmission rate)